Samuel de Champlain before 1604

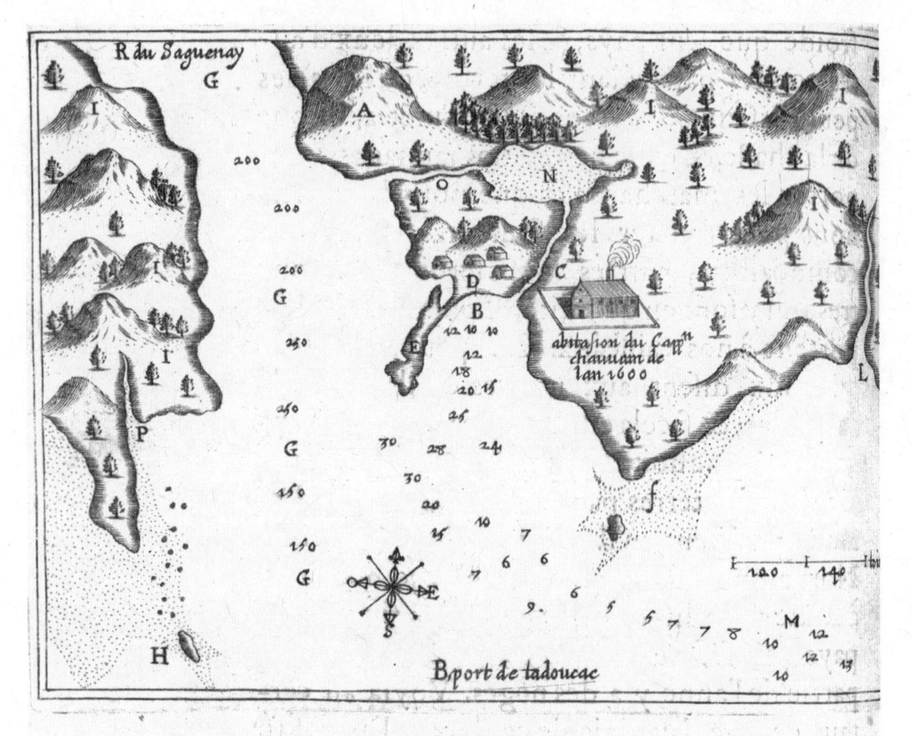

Les chifres montrent les brasses d'eau.

A Vne montaigne ronde sur le bort de la riuiere du Saguenay.
H Le port de Tadoussac.
C Petit ruisseau d'eau douce.
D Le lieu ou cabannent les sauuages quand ils viennent pour la traicte.
E Maniere d'isle qui clost vne partie du port de la riuiere du Saguenay.
F La pointe de tous les Diables
G La riuiere du Saguenay.
H La pointe aux alouettes.
I Montaignes fort mauuaises, remplies de sapins & boulleaux.
L Le moulin Bode.
M La rade ou les vaisseaux mouillent l'ancre attendant le vent & la maree.
N Petit estãg proche du port.
O Petit ruisseau sortant de l'estãg, qui descharge dans le Saguenay.
P Place sur la pointe sans arbres, où il y a quantité d'herbages.

On 26 May 1603, Champlain first stepped ashore in Canada – at Tadoussac – and stayed at the habitation built by Pierre Chauvin de Tonnetuit. At the time, this was the only European building in Canada. Built in 1600 to house sixteen men, this cabin had a central hearth and was 24 ft (7.3 m) by 18 ft (5.5 m), and 8 ft (2.5 m) high. Champlain was not impressed by its size and construction, calling it *une maison de plaisance* (a country cottage). The important *tabagie* on 27 May, presided over by Anadabijou and witnessed by Champlain and Gravé Du Pont, was held at Pointe aux Alouettes (H on the map), shown here as an island at high tide. Champlain's survey of the harbour of Tadoussac, with Chauvin's habitation on it, was probably made in 1608.

Source: *Les Voyages dv Sievr de Champlain …* (Paris: Jean Berjon, 1613), p. 172. Library of Congress, Rare Books and Special Collections Division, F1030.1.C446. Digital ID: rbfr 0012, http://hdl.loc.gov/loc.rbc/rbfr.0012.

Samuel de Champlain before 1604

Des Sauvages and Other Documents Related to the Period

EDITED BY
CONRAD E. HEIDENREICH
AND K. JANET RITCH

The Champlain Society
Toronto

McGill-Queen's University Press
Montreal & Kingston · London · Ithaca

ISBN 978-0-7735-3757-6

Legal deposit fourth quarter 2010
Bibliothèque nationale du Québec

Printed in Canada on acid-free paper that is 100% ancient forest free
(100% post-consumer recycled), processed chlorine free

This book has been published with the help of a grant from the Canadian Federation for the Humanities and Social Sciences, through the Aid to Scholarly Publications Programme, using funds provided by the Social Sciences and Humanities Research Council of Canada.

McGill-Queen's University Press acknowledges the support of the Canada Council for the Arts for our publishing program. We also acknowledge the financial support of the Government of Canada through the Canada Book Fund for our publishing activities.

LIBRARY AND ARCHIVES CANADA
CATALOGUING IN PUBLICATION

Samuel de Champlain before 1604 : Des Sauvages and other documents related to the period / edited by Conrad E. Heidenreich and K. Janet Ritch.

Co-published by: Champlain Society, Toronto, 2010.
Text in French and English translations.
Includes bibliographical references and index.
ISBN 978-0-7735-3757-6 (MQUP ed.).
ISBN 978-0-7735-3756-9 (Champlain ed.)

1. Champlain, Samuel de, 1567–1635. 2. Canada – History – To 1663 (New France). 3. Indians of North America – Canada. 4. America – Discovery and exploration – French. I. Champlain, Samuel de, 1567–1635 II. Heidenreich, Conrad E., 1936– III. Ritch, K. Janet (Katharine Janet), 1955– IV. Champlain Society

FC332.S19 2010 971.01'13 C2010-904713-3

Set in 10.2/13 Adobe Minion Pro and 10.5/13 Adobe Garamond Pro
Book design & typesetting by Garet Markvoort, zijn digital

CONTENTS

APPENDICES

MAPS AND TABLES

MAPS

TABLES

PREFACE

1 INTRODUCTION

The first edition of Champlain's Works *by The Champlain Society*

Early in October 1905, a few months after The Champlain Society was founded, its president Byron Edmund Walker (Sir Edmund after 1910) appointed a publications committee, headed by Professor George Wrong, to draw up a list of potential publications. Almost the first on the list was "Documents Relating to Champlain," to be edited by Dr. Henry Percival Biggar, chief archivist for Canada in Europe, stationed in Paris and London. Champlain's writings are of undeniable importance to Canadians. They are the first good geographical, cartographic, and ethnographic descriptions of eastern Canada by the man who laid the groundwork for its exploration and firmly established the French presence in the country. The original plan was to have a work in four volumes, including a printing for the general public, ready for the Champlain Tercentenary, which was to be held in Quebec City in 1908. Owing to lack of funds, nothing happened until Sir Edmund obtained a publishing grant of $5,000 through the good offices of the prime minister, Sir Wilfrid Laurier. The grant came to The Champlain Society through the National Battlefields Commission – of which Sir Edmund was a member – which was organizing the Champlain Tercentenary events.[1]

Soon after the translators and editors set to work, they realized that this was a far more difficult project than expected. Although the translators thought they had a workable French text in Charles-Honoré Laverdière's 1870 edition of Champlain's *Œuvres*, which was supposed to have been compiled from the various editions and printings done during Champlain's lifetime, this was far from the case. Well into the project, The Champlain Society realized that it would have to hire a specialist to collate the variations between the known copies. After several attempts to find such a person, some ten years after the project had started, the Society finally settled on Professor J. Home Cameron, a specialist in seventeenth-century French at the University of Toronto. Since there were few original copies

1 Heidenreich, *Champlain and The Champlain Society*. This book is a treatment of the founding of the Society and its involvement in editing and publishing the six Champlain volumes and the portfolio of maps.

of Champlain's books in Canada, Cameron visited archives in the United States, France, England, and Scotland, all on his own account. In retrospect, it was in large measure because of him that the project was a success. Another problem, all translators agreed, was that Champlain wrote poor French. Yet Champlain's French compares favourably with that of other explorers during this and earlier periods. Apart from J. Home Cameron, the translators exhibited a bias against the simpler, "undistinguished," "disorderly," and "formless" French of Champlain, presumably in comparison with the classic French texts of the seventeenth century which the professors would have taught. At the same time, they were relatively inexperienced with exploration literature. But even to experienced scholars such as Professor Cameron and native French speakers such as William Ganong and William LeSueur, the texts presented formidable difficulties.

As the project advanced, so did the translators' quarrels over translation, interpretation, and style, especially with the chief editor, Henry Percival Biggar. Rather than leaving Champlain's ambiguities, an effort was made to guess what he was trying to say, and rather than leaving his "laboured pedestrian style," an effort was made to make it more intelligible to a modern reader. Biggar, who had the final decision on the text, was frequently accused of making Champlain "a better writer than he actually was," and it was sarcastically said that "Champlain wrote better English than French." Another accusation was Biggar's "irritating habit" of "correcting" his translators by occasionally substituting words which they considered incorrect.

A third problem that developed was that the various translators were also expected to act as content editors. Some knew little about the history of the period, and only the botanist, Professor William F. Ganong, knew anything about the Native cultures and the natural environment that Champlain had observed. In Ganong's case, this was the east coast of Canada and its Mi'kmaq inhabitants. Especially misleading was some translators' habit of making assumptions about Native cultures that led them to wrong translations. For example, no distinction was made in English when Champlain wrote *coupperant la teste* [tête] (cutting off the head) and *coupperant la peau de la teste* (cutting off the skin of the head, i.e., the scalp); both were interpreted as taking the scalp, on the incorrect assumption that Natives only scalped. The word *robbe* (robe) was often translated as "beaver robe," on the faulty assumption that all Native robes were made of beaver skins. The Iroquoian group that Champlain called the *Antouoronons* were translated as the Onondaga, which they were not, and the French word *salubre,* used by Champlain to describe the water in the Great Lakes from Algonquin accounts, meaning "wholesome" or "fresh," was translated as "brackish" and "salty," following the translation by Richard Hakluyt, published by Samuel Purchas in 1625.

The Champlain Society edition finally appeared in six volumes between 1922 and 1936, a prodigious task that exhausted its participants and cost far more money than had been allocated. Only 550 sets were printed, half for Champlain Society members and half for subscribing libraries. The sets for sale to the general public, which had been promised and which the Battlefields Commission had subsidized, were never printed. Neither was the introduction to the six volumes that was to contain a biography of Champlain, a discussion of Champlain's original texts, his writing abilities, and many other subjects. There have been no new English editions since then.[2]

Although the original Champlain Society edition has stood the test of time remarkably well, we felt it was time to present an updated edition of his first book, *Des Sauvages*, and related documents for the 400th anniversary of the founding of Quebec by Champlain, under instructions from Pierre Du Gua de Mons, with an enabling monopoly granted by King Henri IV.

Des Sauvages *and other documents related to the period before 1604*

The purpose of this project is to publish a new edition of Champlain's first book, *Des Sauvages*, and the earliest documents pertaining to him, in celebration of the 400th anniversary of the founding of Quebec. Where possible, each document has been newly translated, incorporating the latest research on Champlain, New France, and the Native societies he observed that has been done since the original Champlain Society edition of 1922. Also included are newly discovered documents and those not previously translated. Rather than providing a review of what previous authors have written on Champlain, we base our observations and explanations almost entirely on the primary documents and what they suggest to us. Because of significant gaps in that record, it is inevitable that interpretations differ among historians.

The organization of this volume follows previous Champlain Society volumes. It is divided into two major parts. The first is an interpretive essay that deals with Champlain, his family, his acquaintances, and his life, set in the milieu in which he lived up to the end of 1603. This is followed by an analytical essay that pertains to Champlain's French, the editorial procedures followed in this volume, the problems relating to the establishment of a translatable text, and the printing

2 The original Champlain texts can be viewed at http://www.canadiana.org/eco/index.html. The Champlain Society edition is available on http://www.champlainsociety.ca/.

history of the two editions of *Des Sauvages*. Aspects of these essays are elaborated, where helpful, in a series of five appendices.

The second part is devoted to the translated and edited documents pertaining to Champlain. These are accompanied by the original French and Spanish texts, supported by notations on the meaning and use of words and phrases.

We begin with a selection of brief statements from the earliest biographies of Champlain preceded by a short introduction on the growing output of such material, which became a veritable industry by the early twentieth century. Except in a few biographies based on his actual writings, Champlain would not have recognized himself in most of these.[3] They exhibit the myth-making – or unmaking – tendencies to which Champlain's life has been continually subjected.

The army pay records are the first documents that we currently have of Champlain. In these records, kept by the paymaster of the army, Gabriel Hus, under orders of maréchal Jean d'Aumont, he is already named Sieur Champlain and Samuel de Champlain. His rank and function was that of a *fourrier* and *aide du sieur Jean Hardy maréchal des logis de l'armée du Roy*. These records have been translated for the first time.[4]

We do not include his *Brief discours* (Voyage to the West Indies)[5] here because it does not relate directly to Champlain's later career in Canada; its authenticity has been called into question, and some statements in the manuscript are contentious in terms of verifiable facts. There is no reason to doubt that he took such a voyage, but since Champlain's presence on a Spanish ship was irregular enough for a Frenchman recently at war with Spain, it seems unlikely that he would jeopardize his own security by attempting to keep a private journal with maps. In fact, some specialists on this topic are of the opinion that he was on board clandestinely, probably as a spy for Henri IV. Since there are many statements in the manuscript that can be verified, it seems reasonable to assume that the foundation of the *Brief discours* may have been laid by Champlain's oral *rapportage* to Henri IV upon his return from the West Indies and Spain, probably recorded and fleshed out by a scribe.

Upon his return from the West Indies, Champlain visited and took care of his sick *oncle provençal*, Guillermo Elena (Guillaume Allene), in Cádiz early in 1601. Here he witnessed a gift from his uncle to him. The text of the gift (*donación*), presented here, is the first complete critical edition and translation of this im-

3 Document A, "Early Biographies of Champlain"

4 Document B, "Personnel and Pay Records"

5 Biggar, *The Works*, 1:3–80. The original title of the manuscript is *Brief discours des choses plus remarquables que Sammuel Champlein de brouage a reconneues aux Indes occidentalles* … etc. Facsimile of first page in Biggar, ibid., facing p. 3.

portant document, formerly misidentified as a "will," and the subsequent power of attorney (*poder*), which has come to light for the first time.[6]

Following his voyage to the West Indies and his interlude in Spain, Champlain returned to France, where he became involved in the 1603 voyage to the St. Lawrence. We include a translation of the enabling documents[7] for this expedition, as well as later complaints against Champlain by the merchants of Saint-Malo regarding his presence on the voyage.[8]

The central part of this volume is a new collation and translation of the two editions of *Des Sauvages*, Champlain's first book.[9] *Des Sauvages* follows closely the orders that King Henri IV gave to the leader of the expedition, François Gravé Du Pont, and to Aymar de Chaste, under whose auspices the expedition took place. These orders were to conclude an agreement with the Montagnais that would open the St. Lawrence Valley to peaceful French settlement; to conduct a resource survey of the St. Lawrence Valley and comment on its suitability for settlement; and to question Natives about the possibility of finding a westward route to the Orient across the Lachine Rapids. The latter two aims were largely Champlain's responsibility.

The first translation of the abbreviated version of *Des Sauvages* by Pierre-Victor Cayet in 1605 has also been included here.[10] In order to complete the texts of *Des Sauvages* printed during Champlain's lifetime, we have included the first English translation of *Des Sauvages*, by Richard Hakluyt, published in 1625 by Samuel Purchas.[11]

In his later writings Champlain made a number of retrospective statements that shed some light on his life before 1604. These statements not only confirm his earlier writings but often furnish the reader with deeper insight into Champlain's own perspective. We have brought these statements together, newly translated and edited.[12]

In these early documents Champlain comes to us as a shadowy figure who reveals almost nothing about himself and those close to him. This lack of information has led to much guesswork and mythologizing. At this point, we feel that it is unlikely that additional documents still exist that could illuminate his early life, but we have presented some hypotheses that may explain some of it.

6 Document C, "Gift from Guillermo Elena"

7 Document E, "Decrees and Commissions Preparatory to the Voyage of 1603"

8 Document D, "Factum of the Merchants of Saint-Malo against Champlain"

9 Document G, collated and translated text of *Des Sauvages*, [1603], 1604

10 Document H, "Of the French Who Have Become Accustomed to Being in Canada," 1605

11 Document I, "The Voyage of Samvel Champlaine of Brouage, made unto Canada," 1625

12 Document F, "Excerpts from Champlain's Works Related to Events before 1604"

Champlain's later career

After 1603, Champlain's life and career firmed up into the path he would follow for the rest of his life. From 1604 until late in 1607, he worked for Pierre Du Gua de Mons,[13] *lieutenant général pour le roy* in Acadia, doing the tasks of a surveyor and geographer. These tasks, although similar to those he had performed in 1603, were done much more rigorously and described in greater detail. On the Atlantic coast, from Cape La Have to Nantucket Sound, he honed his training and talents looking for potential harbours, resources, and places to settle. The most important sites were surveyed, latitude and magnetic declination were recorded, and the entire coastline mapped.

In 1607 Du Gua de Mons lost his commission in Acadia. He was reappointed on 7 January 1608 but decided to move his operations back to the St. Lawrence. Champlain was chosen as his *lieutenance* [sic] *pour le voyage*, with the task of erecting a settlement (Quebec) and engaging in exploration. Gravé Du Pont was to further develop good Native relations and manage the fur trade. With the assassination of Henri IV in 1610, the Canadian enterprise lost its major supporter. Champlain returned to France, got married, and sought out influential members of the new regime to continue the work that Henri IV, de Chaste, de Mons, the trader Pierre Chauvin de Tonnetuit, and he had begun. On 13 November 1612 the new king, Louis XIII, appointed Henry II, prince de Condé, as his viceroy to New France, and on 22 November Condé appointed Champlain as his lieutenant. During his stay in France while these events were taking place, Champlain finished writing his second book, *Les Voyages dv Sievr de Champlain Xaintongeois, capitaine ordinaire pour le Roy, en la marine* … bearing the date 1613.

Early in 1613, Champlain returned to Canada, no longer as the lieutenant of a lieutenant general but as lieutenant to a viceroy with the de facto powers of a governor. He now assumed much broader administrative responsibilities and was also determined to push his exploration west and north. His first journey west of the Lachine Rapids is recorded in the *Quatriesme Voyage du Sr de Champlain* undertaken up the Ottawa River in 1613. The *Quatriesme Voyage* was written as a separate book, with its own pagination and index, but was appended at the end of *Les Voyages* (1613). The two books appeared in a single volume early in 1614.

In 1615 Champlain travelled to the Huron country and south across the east end of Lake Ontario in the company of a Huron-Algonquin war party to attack the Iroquois. Late in the year he returned to the Huron and stayed with them over the winter, which permitted him to make the first ethnographic observa-

13 The name of Pierre Du Gua de Mons is often spelled "Monts." We have chosen to use "Mons" because Champlain consistently spelled the name that way.

tions of this important *nation*. These events form the bulk of his book *Voyages et Descovvertvres faites en la Novvelle France, depuis l'année 1615*, published in 1619 and reprinted with some alterations in 1620 and 1627. The *Quatriesme Voyage* and *Voyages et Descovvertvres*, which constitute Champlain's forays into what is now Ontario, document his last voyages of exploration.

After 1616 Champlain's role changed. He ceased exploring, probably in large part because of his increased administrative duties and a wound suffered in the 1615 assault on the Iroquois. His role was to administer the growing colony, keep good Native relations, and promote the fur trade. In 1628 he received his last appointment. The letter from the king addressed Champlain as *nostre cher & bien amé le sieur de Champlain, commendant en la Nouuelle France, en l'absence de nostre tres-cher & bien-amé cousin le Cardinal de Richelieu.*[14] He was now *commandant* (lieutenant governor) of New France to Cardinal Richelieu, the most powerful man in France after the king.

Champlain's career and that of New France took a severe jolt when Quebec was taken by the English in 1629. Champlain described this "conquest" in some detail, but the English documents pertaining to this operation and the considerable efforts made to have Canada returned to France have not been treated adequately. While he was in forced "exile" in France, Champlain wrote his last book, *Les Voyages de la Novvelle France Occidentale, dicte Canada*, published in 1632 and reissued with alterations in 1640. In it, he reviews and summarizes his life to 1631. Included in this volume is the *Traitté de la marine*, an interesting compendium of what Champlain knew about leadership and seamanship near the end of his career. While he was in France between 1629 and 1633, Champlain liquidated his French assets, evidently with the intention of settling permanently in Canada.

In 1633 Champlain returned to Canada, a journey he documents in the *Relation du voyage du sieur de Champlain en Canada* (1633). Champlain's life from 1633 to his death on 25 December 1635, is documented through passages in the *Jesuit Relations* and *Le Testament de Champlain* (17 November 1635).

The importance of Des Sauvages

Des Sauvages, the shortest of Champlain's books, contains his observations made between his departure from Honfleur on 15 March 1603 and his return to *Le Havre de Grace* on 20 September of the same year. He began writing the text from his notes in July, after leaving the Lachine Rapids, in order to be prepared to submit the completed text to the referee in the royal chancellery upon arrival

14 The letter from Louis XIII of 27 April 1628 is reproduced in Biggar, *The Works*, 5:288.

and consequently receive the licence by 15 November. The book is the substance of observations made under the orders of King Henri IV and Aymar de Chaste, who organized the expedition. The first edition seems to have been published upon Champlain's return, after his audience with the king and the issuance of the licence. Although the title page is undated, general consensus and a brief statement by Champlain in a later work places it in the same year, 1603. The second edition followed in 1604.

After the voyages of Jacques Cartier and his predecessors, Canada had achieved a miserable reputation. The winters could kill Europeans, and the Natives were unfriendly and might kill those who survived the cold and scurvy. The Lachine Rapids made westward exploration difficult if not impossible, just as the falls at Chicoutimi made travel north up the Saguenay River impossible. The St. Lawrence River was not a passage west, but a cul-de-sac. Worse, there were no resources that lent themselves to easy exploitation. In the opinion of Europeans who had travelled to Canada, the people living in the eastern forests and along the coast were impoverished *Sauvages* who could not be exploited for their wealth because, unlike the people of Mexico and Peru, they had none. Moreover, they made poor slaves because they died from infectious European diseases. Given that all these negative impressions were believed to be true, the untested question remained whether it was possible for Europeans to build settlements and live in Canada. The task set for Champlain and the others on the 1603 voyage was therefore to determine if the St. Lawrence Valley was suitable for colonization and to see if, in spite of Cartier's failed efforts, it was possible to explore westward from the Lachine Rapids to the Orient.

Des Sauvages is one of the more important – perhaps even the most important – books in Canadian exploration literature. It is the first of a series of books written by the person who was to have a great influence on the founding and evolution of Canada, as well as on the imagination of the Canadian people. In this deceptively slim volume we are given a glimpse into the first arrangement made between a European power and a group of Canadian Natives to cooperate in mutually beneficial endeavours. This agreement set out a policy for the future evolution of New France; it advocated that it was preferable to colonize peacefully rather than by force, unlike the Spanish genocide in South America and the modus operandi on the Atlantic seaboard settled by the Dutch and English. This meant reciprocal obligations between the French and their new neighbours. These obligations, however, included unfortunate but unavoidable military alliances that drew the French into the Iroquois wars. The resource survey conducted by Champlain was essentially accurate and convinced the French court that the settlement of Canada was possible. He correctly predicted the lands along the

St. Lawrence west of Quebec that could sustain agriculture and the sites where major settlements could be located: Quebec, Trois-Rivières, and Montreal.

Champlain was one of the first Europeans to trust Natives to give him reliable verbal and cartographic accounts of the areas he wished to explore. This process of information gathering became standard for all French explorers after him. In fact, most later French explorers knew, through Native geographical accounts, roughly what they would find wherever they were going, even before they set off from their settlements or interior posts. Champlain was also the first to state unequivocally that the interior of the continent could only be explored by canoe and with Native cooperation. Although in retrospect this seems an obvious observation, it was as profound as it was original, because no European had thought of it before him; and, as it turned out, it was the only correct solution. What it meant was that the French had to develop peaceful relations with the Native groups whose lands they wanted to explore, and to place themselves into Native hands and Native technology in order to do so. In fact, to explore the interior, away from the relative comforts of life along the St. Lawrence, meant dispensing with aspects of European culture and technology, and adopting aspects of the culture of their Native neighbours. This was not easy for any Europeans to do. Through Champlain's observations in 1603 and his own explorations, conducted by joining travelling Native groups during his tenure in Canada, he established the principles by which the successful exploration of the Canadian wilderness had to be carried out. Neither the Dutch nor the English had the flexibility in their contemporary cultures to make the adjustments necessary for successful inland exploration until much later. One of the results was that by 1685, when the first English-Dutch expedition reached Lake Ontario guided by French renegades, the French had already explored and mapped the entire Great Lakes system, the Mississippi River south to the Gulf of Mexico, and two of the major river systems to James Bay. These French successes were based on Champlain's observations made in the summer of 1603, to which he added additional ones over succeeding years.

In 1622 Champlain ruminated upon his achievements and concluded: "*Tous hommes ne sont pas propres à risquer, la peine & fatigue est grande; mais l'on [n']a rien sans peine: c'est ce qu'il faut s'imaginer en ces affaires; ce sera quand il plaira à Dieu: de moy, ie prepareray tousiours le chemin à ceux qui voudront aprés moy, l'entreprendre.*"[15] ("All men are not suited to take the risk; the trouble and fatigue are great, but nothing is to be had without toil. That is what one must think in these affairs; it will be when it pleases God. As for me, I will always prepare the

15 Champlain, *Les Voyages, 1632*, pt. 2, 41; Biggar, *The Works*, 5:73

way for those who will wish, after me, to carry it on.") His achievements began with the insights he expressed in *Des Sauvages*. Over succeeding years, as he expanded his activities, he formulated additional guiding principles that led to the exploration of Canada, and laid the foundations for permanent European settlement. These were his gifts to those who continued after him.

2 ACKNOWLEDGMENTS

As in any labour of love, the gestation period is always tumultuous, with many unexpected turns, new avenues, and occasional reversals. First and foremost in this birthing travail, we express our thanks to Professor Roger Hall, who served as Champlain Society editor for this volume, and to Andreas Motsch, French professor at Victoria College in the University of Toronto. We would like to express our sincere thanks to our bilingual copyeditor Carlotta Lemieux, whose attention to detail and rigorous application of the fifteenth edition of the *Chicago Manual of Style* brought much-needed systematic order to our manuscript.

The generous contributions of many scholars and students outside The Champlain Society must likewise be acknowledged. Individuals in both categories shared not only their own research but also their time and advice upon request. For help with Spanish, we thank Professor Stephen Rupp, Nubia Soda, Juan Luis Suarez (seventeenth-century specialist in legal Spanish at the University of Western Ontario), and all the welcoming employees of the provincial archives in Cádiz, Spain, especially D. Manuel Ravina Martín, director of the Archivo Histórico Provincial de Cádiz.

For help with French publishing history, we thank Professor Grégoire Holtz, seventeenth-century specialist, University of Toronto, and Laurence Augereau, PH D; for help with Champlain's French from a linguistic point of view, Robert Taylor, professor emeritus at the University of Toronto, and Marie-Rose Simoni-Aurebou, directrice de recherche émérite, CNRS. We also acknowledge the helpful work of Professors Réal Ouellet and Alain Beaulieu on Champlain's early-seventeenth-century French, as published in their glossary to Champlain's *Des Sauvages*.[16] *Beaucoup de mercis* also to historian, politician, and publisher Denis Vaugeois for his unsolicited support and helpful comments.

A very special *expression de notre reconnaissance profonde* goes to Eric Joret, Jean-Philippe Millot, and the Archives départementales d'Ille-et-Vilaine, Archives régionales de Bretagne, Rennes, for generously reproducing their documents pertaining to Champlain's activities in Brittany. We would also like to thank Peter Ingles and Claude Beradelli of Radio-Canada/CBC for generously

16 Beaulieu and Ouellet, *Des Sauvages*, 239–42

sharing information and providing us with their contacts in the Archives départementales d'Ille-et-Vilaine in Rennes.

We have corresponded with Douglas Hunter for the last four years through an exchange of hundreds of e-mails. He has been more than generous in sharing with us his often provocative ideas and intimate knowledge of French/English/Dutch activities in early-seventeenth-century North America, especially with reference to the possible influence that the writings of Edward Hayes (1602) may have had on *Des Sauvages*.

For help with verifications and finding documents in Paris (France), we are grateful to Marie-Bethsabée Zarka; and for helping to establish or proofing the various texts in French and English, we thank the following student assistants: Margaret Schotte, Jane Bau, Alizabet Shtelman, and Grégoire Rouleau.

Byron Moldofsky and Mariange Beaudry of the University of Toronto cartography office applied their legendary skills to the final drafts of our four maps. It is always a pleasure and a learning experience to work with them. A special thanks from Janet Ritch to Victoria University (Toronto) for the use of space in the Northrop Frye Centre to pursue her research and writing.

Lastly, what would we have done without the magnificent Internet sites Gallica, mounted by the Bibliothèque nationale de France, and Early Canadiana Online (ECO), mounted by the non-profit organization Canadiana.org, formerly the Canadian Institute for Historical Microreproductions? Gallica makes available some 90,000 works and 80,000 images relating to French history and culture, while ECO has an ever-expanding collection, which at present includes some 37,600 documents totalling more than 2,800,000 pages pertaining to pre-twentieth-century Canada. Both collections contain all the original printings of Champlain's works. These huge digital libraries provide an essential service that cuts research time and the expense of travelling to distant libraries. On behalf of all researchers, we thank you.

3 NOTES ON SPELLING AND USAGE

- In addition to the conventional uses of italics, they are employed for place names and quotations from original sources in French, Spanish, and German for which the orthography has not been modernized.
- Modern French is in the same typeface as English.
- All modern Canadian geographical names are taken from Canada, Department of Energy, Mines, and Resources, *Canada Gazetteer Atlas* (Ottawa: Macmillan of Canada, 1980). In this atlas, all the geographical names have been approved by the respective provinces and territories and have not been translated from one language to another. French-language

place names were also checked on the 2008 road map issued by the province – *Québec, La Province* – published by JDM Géo, Inc., 5790 Donahue, Ville St-Laurent, Quebec.
- Square brackets inserted in quotations (e.g., [Méridien]) indicate additions to an original text. Square brackets preceded by a solidus indicate a change in foliation in early manuscripts (e.g., /[f. 34]) or pagination in printed editions (e.g., /[p. 34]). Presumed authors and dates also are given in square brackets (e.g., [1603]).
- Pagination in Champlain Society editions is indicated as follows: {34}.
- Abbreviated references are given in the footnotes. Full references are given in the References among the end matter.
- The term "Natives," meaning the original inhabitants of North America, is capitalized, like Europeans, Asians, and Africans.
- Names of Native groups given in their language are in italics (e.g., *Ouendat*) but are in roman type if the name is of European origin (e.g., Huron) or if it has been corrupted from the original Native language (e.g., Wyandot).
- The spelling of the names of Native groups follows that developed by the Smithsonian Institution in Bruce Graham Trigger's *Handbook of North American Indians: Northeast* (1978) and June Helm's *Handbook of North American Indians: Subarctic* (1981).
- The term *Sauvages* used by the early French writers for North American Natives will remain as is and not be translated into "savages" because the meaning of the word has changed. For a full discussion of this term, see part 2, the introduction to Document G, "*Des Sauvages, or, Voyage of Samuel Champlain,* [1603], 1604."

For a complete discussion of spelling and usage, see part 1, B, "Textual Introduction to *Des Sauvages,*" no. 4, "Editorial Principles and Procedures."

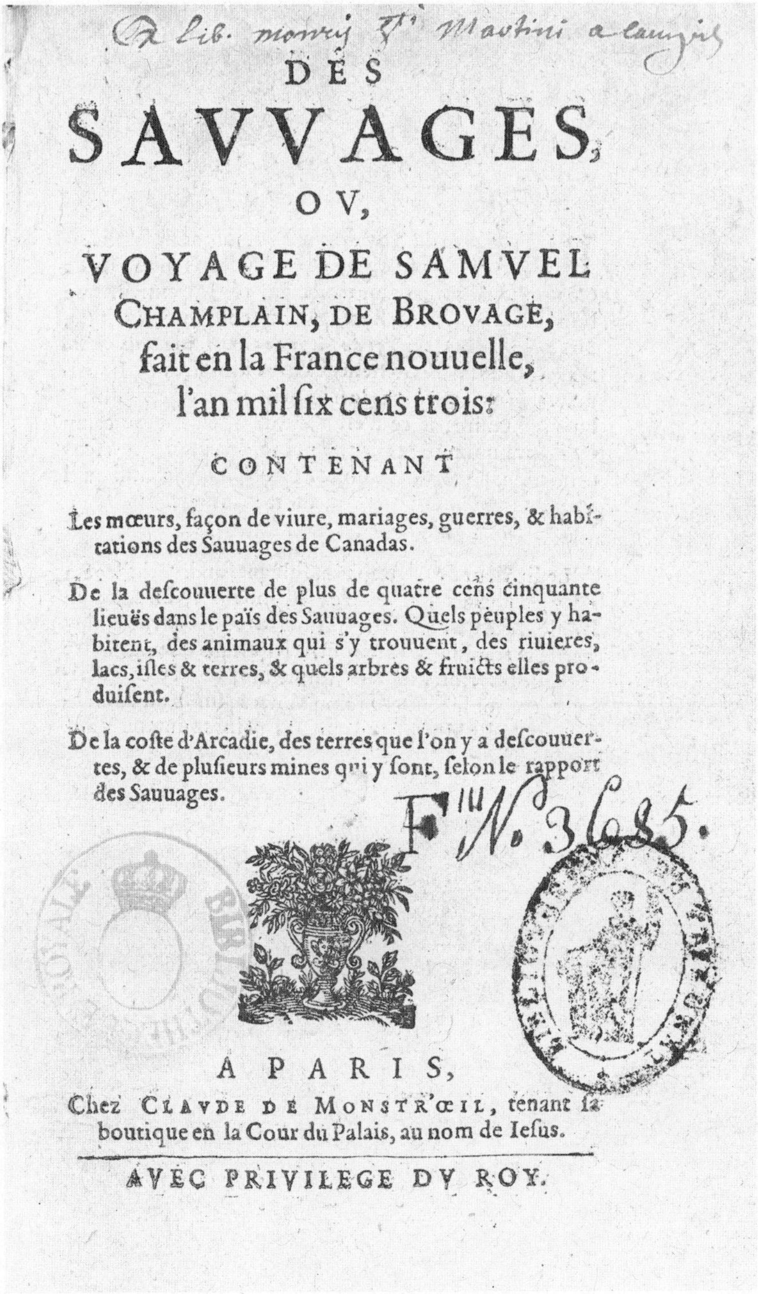

DES

SAVVAGES,

OV,

VOYAGE DE SAMVEL CHAMPLAIN, DE BROVAGE, fait en la France nouuelle, l'an mil six cens trois:

CONTENANT

Les mœurs, façon de viure, mariages, guerres, & habitations des Sauuages de Canadas.

De la descouuerte de plus de quatre cens cinquante lieuës dans le païs des Sauuages. Quels peuples y habitent, des animaux qui s'y trouuent, des riuieres, lacs, isles & terres, & quels arbres & fruicts elles produisent.

De la coste d'Arcadie, des terres que l'on y a descouuertes, & de plusieurs mines qui y sont, selon le rapport des Sauuages.

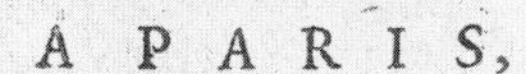

A PARIS,

Chez CLAVDE DE MONSTR'ŒIL, tenant sa boutique en la Cour du Palais, au nom de Iesus.

AVEC PRIVILEGE DV ROY.

Figure 1 Title page of *Des Sauvages,* [1603]. Bibliothèque nationale de France (BnF), RES-LK12-719, Tolbiac Magazin (TM)

DES

SAVVAGES,

OV,

VOYAGE DE SAMVEL

CHAMPLAIN, DE BROVAGE,

faict en la France nouuelle, l'an mil six cens trois.

CONTENANT,

Les mœurs, façon de viure, mariages, guerres & habitations des Sauuages de Canadas.

De la descouuerture de plus de quatre cens cinquante lieuës dans le pays des Sauuages, Quels peuples y habitent, des animaux qui s'y trouuent, des riuieres, lacs, isles, & terres, & quels arbres & fruicts elles produisent.

De la coste d'Arcadie, des terres que l'on y a descouuertes, & de plusieurs mines qui y sont, selon le rapport des Sauuages.

A PARIS,

Chez CLAVDE DE MONSTR'OEIL, tenant sa boutique en la Cour du Palais, au nom de Iesus. 1604.

Auec Priuilege du Roy.

Figure 2 Title page of *Des Sauvages,* 1604. BnF, RES-LK12-719(A), TM.

CHRONOLOGIE SEPTENAIRE

DE

L'HISTOIRE

DE LA PAIX ENTRE LES ROYS DE FRANCE ET D'ESPAGNE.

Contenant les choſes plus memorables aduenuës en France, Eſpagne, Allemagne, Italie, Angleterre, Eſcoſſe, Flandres, Hongrie, Pologne, Suece, Tranſſyluanie, & autres endroits de l'Europe: auec le ſuccez de pluſieurs nauigations faictes aux Indes Orientales, Occidentales & Septentrionales, depuis le commencement de l'an 1598. iuſques à la fin de l'an 1604.

DIVISEE EN SEPT LIVRES.

A PARIS.
Par IEAN RICHER, ruë S. Iean de Latran, à l'Arbre verdoyant: Et en ſa boutique au Palais, ſur le perron Royal, vis à vis de la galerie des priſonniers.

M. D. CV.

Auec Priuilege du Roy.

Figure 3 Title page of Pierre-Victor Cayet's *Chronologie septenaire.* Library and Archives Canada, RES-DC122-C38.

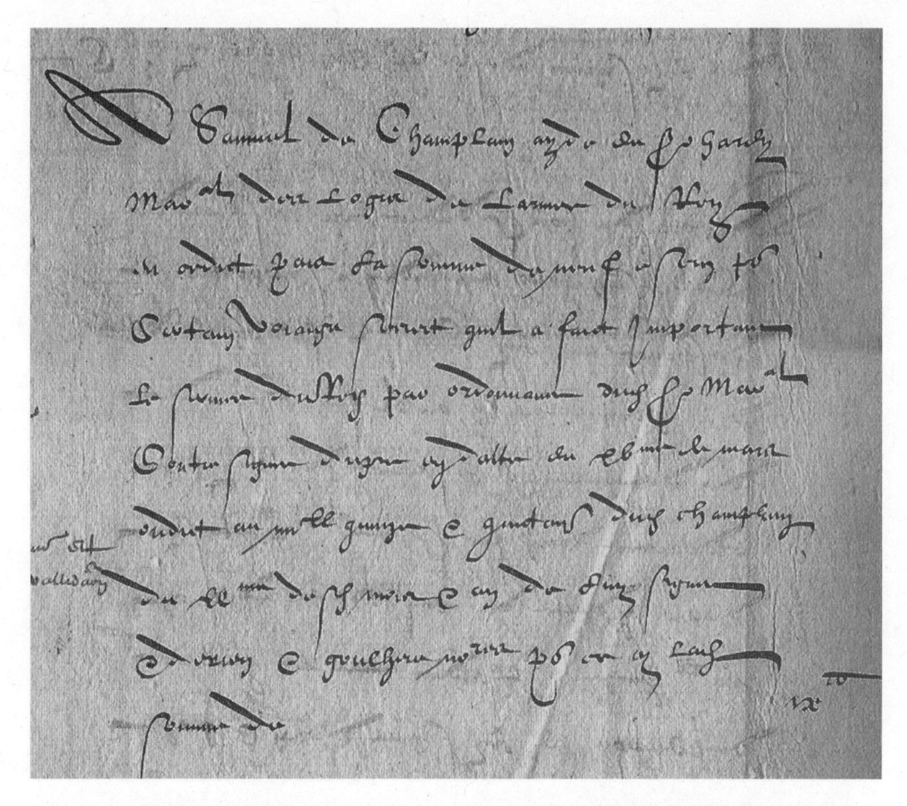

Figure 4 (*above*) First known documentary mention of Samuel de Champlain, 15 March 1595, naming him *ayde* to Sieur Hardy and principal on a secret voyage in the king's service. Département d'Ille-et-Vilaine, Archives régionales de Bretagne, Rennes, C2914, f. 229v. For a translation, see Document B, "Personnel and Pay Records," no. 2, f. 229v.

Figure 5 (*facing page, above*) "Duche de Bretaigne Dessigné par le Sieur Hardy Mareschal des logis du Roy," by Henricus Hondius, in Joannes Janssonius's *Theatrum Universae Galliae* (Amsterdam, 1631), map 27, 37.0 x 50.5 cm; scale 10 lieux françoises = 7.0 cm. British Library, map 15355(1). In view of the fact that Champlain was an aide to Jean Hardy when they were together in Brittany, it is highly probable that he worked on the compilation of this map.

Figure 6 (*facing page, below*) Cartouche and portion of the Hondius map of the Brittany coast, from Blavet (southeast), past Quimper (centre), to Brest (northwest).

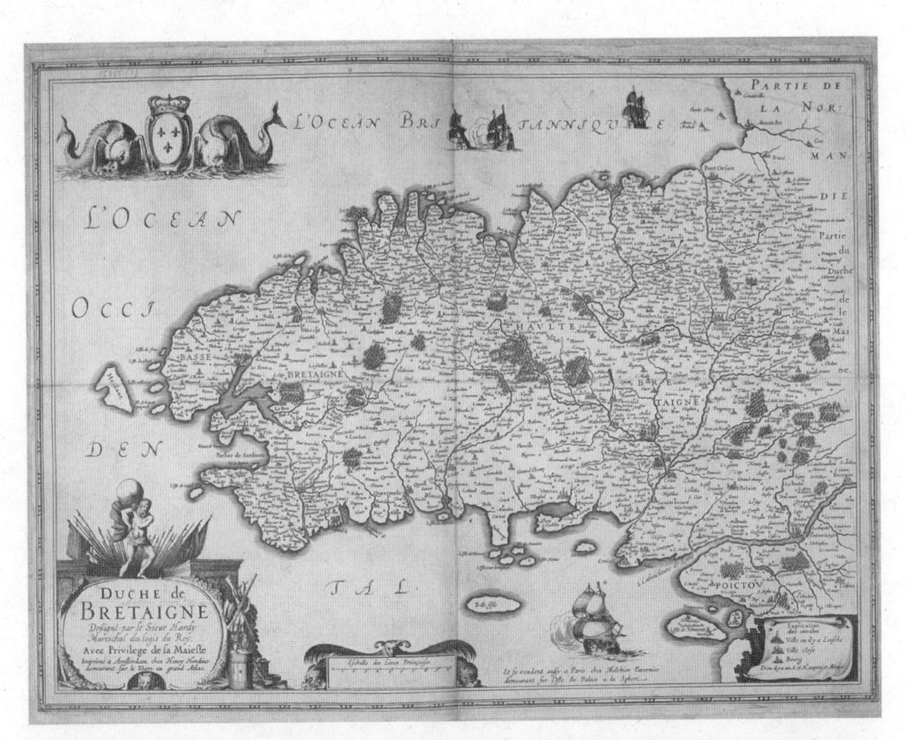
L'OCEAN BRI
TANNIQUE
PARTIE DE
LA NOR
MAN
DIE
Partie du Duche de le Maine
L'OCEAN
OCCI
DEN
TAL
BASSE
BRETAIGNE
HAULTE
BRE
TAIGNE
POICTOU
DUCHE de
BRETAIGNE
Desßigné par le Sieur Hardy
Mareschal des logis du Roy.
Avec Privilege de sa Maieste
Imprimé a Amsterdam chez Henry Hondius
demeurant sur le Dam au grand Atlas.
Eschelle des Lieux Françoises

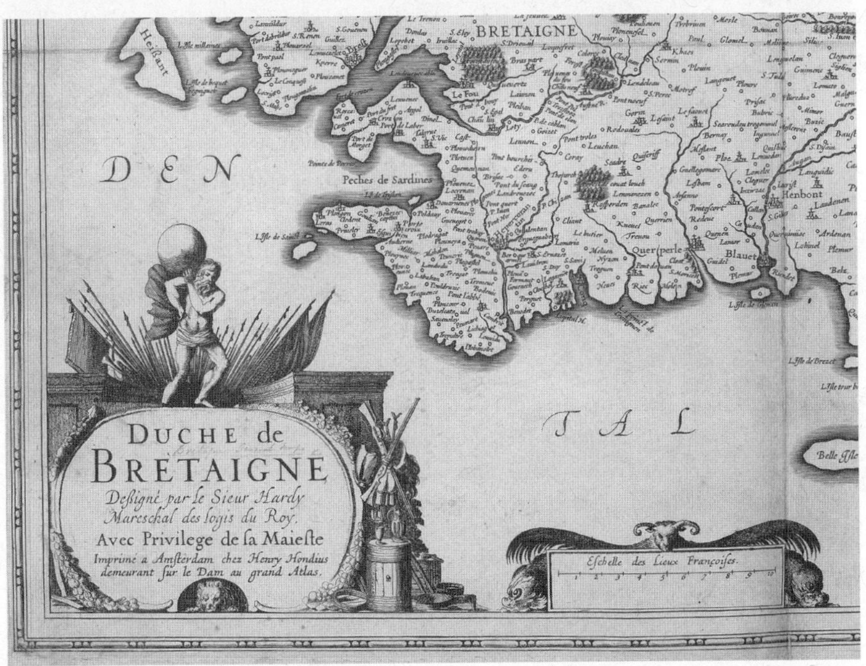
BRETAIGNE
Heissant
DEN
TAL
Peches de Sardines
Querperle
Blauet
Henbont
Belle Isle
DUCHE de
BRETAIGNE
Desßigné par le Sieur Hardy
Mareschal des logis du Roy.
Avec Privilege de sa Maieste
Imprimé a Amsterdam chez Henry Hondius
demeurant sur le Dam au grand Atlas.
Eschelle des Lieux Françoises.
1 2 3 4 5 6 7 8 9 10

CHAP. VI.

The Voyage of SAMVEL CHAMPLAINE *of* Brouage, *made vnto* Canada *in the yeere* 1603. *dedicated to* CHARLES de Montmorencie, *&c. High Admirall of* France.

Their Voyage to *Tadouſac.* *Chap.* 1.

WE departed from *Honfleur*, the fifteenth day of March 1603. This day we put into the Roade of New *Hauen*, becauſe the winde was contrary. The Sunday following being the ſixteenth of the ſaid moneth, we ſet ſaile to proceed on our Voyage. The ſeuenteenth day following, we had ſight of *Ierſey* and *Yarnſey*, which are Iles betweene the Coaſt of *Normandie* and *England*. The eighteenth of the ſaid moneth, wee diſcryed the Coaſt of *Britaine*. The nineteenth, at ſeuen of the clocke at night, we made account that we were thwart of *Vſhent*. The one and twentieth, at ſeuen of clocke in the morning, we met with ſeuen ſhips of *Hollanders*, which to our iudgement came from the *Indies*.

The firſt report of the Sauages touching the Head of the Riuer. A Riuer running 60. leags into the Countrie of the *Algoumequins*. A Lake of 15. leagues. Another Lake of 4. leagues. Fiue other Saults. A Lake of 80. leagues long. Brackiſh water * It ſeemeth hereby to trend ſouthward. The laſt Sault. Another Lake 60. leagues long, very brackiſh. A Strait of 2. leagues broad. Another mightie Lake.

the ſaid Sault, is a very thinne Wood, through which men with their Armes may march eaſily, without any trouble; the aire is there more gentle and temperate, and the ſoyle better then in any place that I had ſeene, where is ſtore of ſuch wood and fruits, as are in all other places before mentioned: and it is in the latitude of 45. degrees and certaine minutes.

When we ſaw that we could doe no more, we returned to our Pinnace; where we examined the Sauages which we had with vs, of the end of the Riuer, which I cauſed them to draw with their hand, and from what part the Head thereof came. They told vs, that beyond the firſt Sault that we had ſeene, they trauelled ſome ten or fifteene leagues with their Canoas in the Riuer, where there is a Riuer which runneth to the dwelling of the *Algoumequins*, which are ſome ſixty leagues diſtant from the great Riuer; and then they paſſed fiue Saults, which may containe from the firſt to the laſt eight leagues, whereof there are two where they carrie their Canoas to paſſe them: euery Sault may containe halfe a quarter or a quarter of a league at the moſt. And then they come into a Lake, which may be fifteene or ſixteene leagues long. From thence they enter againe into a Riuer which may be a league broad, and trauell ſome two leagues in the ſame; and then they enter into another Lake ſome foure or fiue leagues long: comming to the end thereof, they paſſe fiue other Saults, diſtant from the firſt to the laſt ſome fiue and twenty or thirty leagues; whereof there are three where they carrie their Canoas to paſſe them, and thorow the other two they doe but draw them in the water, becauſe the current is not there ſo ſtrong, nor ſo bad, as in the others. None of all theſe Saults is ſo hard to paſſe, as that which we ſaw. Then they come into a Lake, which may containe ſome eighty leagues in length, in which are many Ilands, and at the end of the ſame the water is brackiſh, and * the Winter gentle. At the end of the ſaid Lake they paſſe a Sault which is ſomewhat high, where little water deſcendeth: there they carrie their Canoas by land about a quarter of a league to paſſe this Sault. From thence they enter into another Lake, which may be ſome ſixty leagues long, and that the water thereof is very brackiſh: at the end thereof they come vnto a Strait which is two leagues broad, and it goeth farre into the Countrie. They told vs, that they themſelues had paſſed no farther; and that they had not ſeene the end of a Lake, which is within fifteene or ſixteene leagues

Figure 7a (*above*) Opening lines of Hakluyt's translation of *Des Sauvages,* in *Hakluyutus Posthumus, or Purchas His Pilgrimes …,* vol. 4 (London, 1625), p. 1605. Library of Congress, Rare Books and Special Collections Division, Hans P. Kraus Collection of Sir Francis Drake, no. 40, G159.P98.

Figure 7b (*below*) Portion of p. 1614 (in *Hakluyutus Posthumus,* ibid.), with the first Algonquin story about the river and lakes system west of the Lachine Rapids.

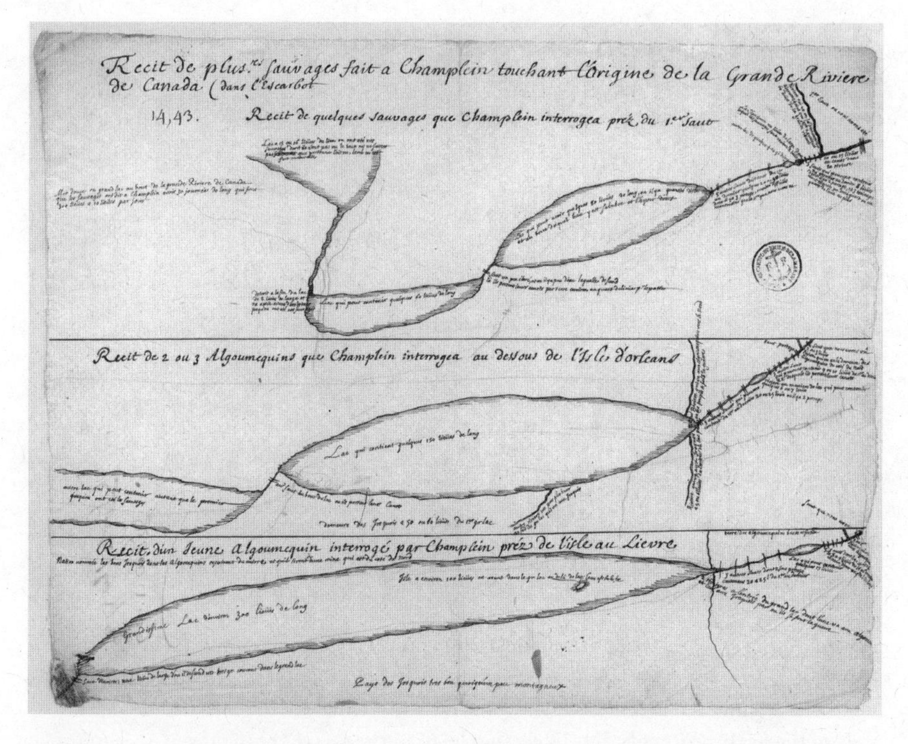

Figure 8 The three Algonquin accounts of the upper St. Lawrence and eastern Great Lakes in *Des Sauvages*, interpreted and drafted by Claude Delisle (c. 1695) from a reprint of Champlain's account in Marc Lescarbot, *Histoire de la Novvelle France contenant les navigations, découvertes, & habitations faites par les François …* (Paris: Iean Millot, 1609), pp. 378–84. Archives nationales de France, Marine, JJ/6/75, pièce 163.

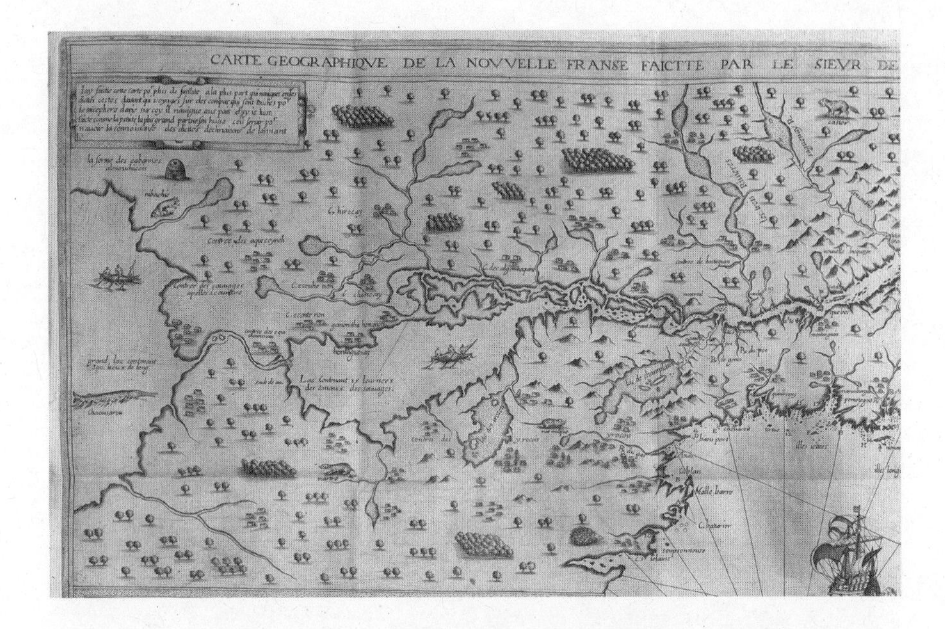

Figure 9 (*above*) Western portion of Champlain's "Carte Geographiqve De La Novvelle Franse …" (1612), in *Les Voyages dv Sievr de Champlain …* (Paris: Jean Berjon, 1613). Library and Archives Canada, C118494. By this time, Champlain had not been west of the Lachine Rapids. It is likely that he used the second of the three Algonquin verbal accounts and sketch maps he obtained in 1603.

Figure 10a (*facing page, above*) Champlain's first known signature as rendered on the "Donación de Guillermo Elena a Samuel de Champlain." Archivo Histórico Provincial de Cádiz, pr. 1512, notario Marcos de Rivera, 1601, f. 249^{v}.

Figure 10b (*facing page, below*) Champlain's last signature, on "Testament de Champlain," Quebec, 17 November 1635. Archives nationales, Paris, Minutier central, étude 62, vol. 138, f. 3.

a par grande difficulté en l'execution d'

et ce dixseptiesme de Novembre a Kebec mil

et trente cinq en presence de ceux qui son

signez. Champlain

PART ONE

INTRODUCTORY ESSAYS

A. Champlain and His Times to 1604: An Interpretive Essay

1 FRANCE AND SPAIN

Early life, relatives, and friends

There cannot be very many people who have made such an indelible imprint on the imagination and history of Canada as Samuel de Champlain, about whom so little personal information is known. Even though he wrote four substantial books about his activities,[1] he never mentioned the date of his birth, his parents, his education, his early life, his career in Henry IV's army, or anything personal of any consequence except one brief phrase. Not once did he record the name of his wife, Hélène Boullé, to whom he was married for twenty-five years, except to refer to her on a couple of occasions as *ma famille*. The little that is known about his wider family comes from a few manuscript legal papers. Unfortunately, this meagre record cannot be expanded very far because the early parish records of his birthplace, Brouage, no longer exist.[2]

Champlain was born at Brouage in the province of Saintonge,[3] France, sometime between 1567 and 1580. He died at Quebec, Canada, on 25 December 1635, after suffering a severe stroke early in October of that year.[4] The earliest and most often quoted date of his birth is 1567. Unfortunately, this date cannot be corroborated.[5] Instead, his biographers adopted a neutral "circa 1570" until recently, when the issue was raised again by Jean Liebel, who tried to build the case that

1 His four books contain 1,263 pages, 5 large maps, 22 plans, and 14 illustrations.

2 In 1867 the archivist of the Bibliothèque de La Rochelle, Leopold Delayant, and in 1875 the archivist Meschinet De Richemond of the Département Charente-Inférieure, wrote that the *registres de la paroisse de Brouage* had been deposited at Marennes, where they were destroyed by a fire in 1690. There are no longer any records relating to the population of Brouage earlier than August 1615 (Delayant, *Notice*, 2; Slafter and Otis, *Voyages of Champlain*, 1:206).

3 Champlain wrote on a number of occasions that he was "from Brouage." We do not know for certain that he was actually born there, although his parents and his maternal grandmother were residents. Saintonge became the Département Charente-Inférieure in 1790 and Charente-Maritime in 1941.

4 JR (*Jesuit Relations*), 9:201, 207–9.

5 See Document A, "Early Biographies," no. 6. Any archival papers that may have carried his birthdate no longer existed well before Rainguet wrote his biography.

Champlain was born "circa 1580."[6] We are certain however, that a birthdate in the early to mid 1570s is more realistic in view of Champlain's responsibilities by the mid 1590s.[7]

There has been much speculation on whether Champlain was born a Huguenot.[8] In every existing document in which he is mentioned or in which he mentions himself, he appears as a convinced and dedicated Catholic. His birth in Brouage, probably in the 1570s when it was Huguenot, and his first name Samuel, a Protestant name, may mean something, but there were also Catholics living in Brouage during the years it was Huguenot, and there were Huguenots there in the years it was Catholic. As for his given name, Samuel, there are at least two Catholic St. Samuels.[9] Although mainly used by Huguenots, the name Samuel could have been an acceptable name to some Catholics, perhaps signifying a birthdate on the feast day of the saint. Of the men to whom Champlain was responsible early in his Canadian career, Aymar de Chaste and Gravé Du Pont were Catholics, while Pierre Du Gua de Mons[10] was a Huguenot. All three had been early supporters of Henri IV, who was born a Huguenot in 1553 and switched to Catholicism in 1593, demonstrating that religion did not matter as much as complete loyalty to the king. In 1610 Champlain married a Huguenot from a Huguenot family, the twelve-year-old Hélène Boullé. This was an arranged marriage that carried a large dowry but could not be consummated until she was of age. The unhappy Hélène took instruction, albeit reluctantly, and converted to Catholicism by the time she was fourteen. The only known relative of Champlain's who was a Huguenot, at least early in his career, was his uncle by marriage, Guillaume Allene.[11] Champlain's birth during the 1570s when Brouage was Huguenot, his given name Samuel, and his marriage into a Huguenot family are all suggestive of a Huguenot origin. If he was baptized a Huguenot in Brouage, it may have been

6 Liebel, "On a vieilli Champlain," 236

7 For our interpretation of the evidence for an earlier birthdate, see appendix 1, "Champlain's Birthdate and Appearance."

8 DCB (*Dictionary of Canadian Biography*), vol. 1, "Champlain, Samuel de." This article by Trudel gives a good summary of the arguments.

9 Of the two St. Samuels, one, whose feast day is 16 February, was martyred at Caesarea in Palestine during the fourth century. The other martyr, St. Samuel of Edessa, has a feast day on 9 August. There is also a St. Samuel the Confessor in the Coptic Orthodox Church. His feast day is 18 December (Belèze, *Dictionnaire*, 386). See also: http://www.catholic.org/saints/.

10 See *DCB*, 1, "Du Gua de Monts, Pierre." We have adopted the spelling "Mons," which Champlain used throughout, rather than "Monts." In Biggar, *The Works*, Champlain's spelling of the name was changed to "Monts" in the translation.

11 Le Blant and Baudry, *Nouveaux documents*, 1:2–4

between 1572 and 1577, the only years that a Protestant pastor, Nicolas Folion (dit La Vallée), lived in the town.[12] A Huguenot origin and his conversion to Catholicism probably account for Champlain's explanation of his faith to the Montagnais headman Anadabijou at Tadoussac in June 1603.[13] It is the only case of such a treatise in his books. Coming as it does at the very beginning of his writings, one might ask oneself, Is it the *cri du cœur* of one who converted to Catholicism?

Champlain always signed himself "Champlain" on any documents he authored or witnessed. He never attached his given name or any of his titles to his signature, much as the royalty of his times wrote "Henri" or "Louis" in a hand at least twice the size of any other signatures on the page.[14] On documents authored by others, his name is sometimes spelled Champellain or Champelain. These variants in spelling may be variants in the pronunciation of the name, although at least once he signed his name "Champelain."[15] Research through French genealogies of this period, both on the Internet and in specialized libraries, has not produced any other people named Champlain, showing at the very least that the name was uncommon. In view of the spelling of his father's versions of the name (below) and the persistent use of "Champelain" or some variant in notarial documents, it may be that Samuel preferred "Champlain" at a very early period in his career.

According to the marriage contract between Samuel de Champlain and Helayne (Hélène) Boullé,[16] dated at Paris, 17 December 1610, his parents were Anthoine de Champlain, *cappitaine de la marine*, deceased at the time of the marriage, and *dame* Margueritte le Roy, still living but not present at the wedding.[17] The only other reference to what may have been Champlain's father is the sale of a half-share in the small 30-tun ship *Jeanne*, on 23 December 1573, at Brouage by Anthoyne Chappelin (also Anthine Chappelain),[18] *pilotte de navyres, demeurant à Jacopolis sur Brouage*. If this was indeed Champlain's father, his occupation was a pilot of ships in the harbour of Brouage as well as a captain in the navy. The Champlain family therefore had the status, respectability, and modest

12 Société de l'histoire du protestantisme, "Champlain," 275
13 See Document G, "*Des Sauvages*," ch. 3.
14 See appendix 2, "Champlain's Signature and Titles."
15 Delafosse, "Séjour de Champlain," 578
16 Although dated, the best biography of Hélène de Champlain is probably that by Bourde de la Rogerie, *Hélène Boullé*.
17 The reproduction of the marriage contract in Biggar, *The Works* (2:315), has some errors. The most accurate version seems to be "La minute notariée" by Cathelineau.
18 The document is signed "Anthine Chappelain." Note that the particle *de* is absent from early references to his name (Le Blant and Baudry, *Nouveaux documents*, 1:10–11).

wealth that comes from such an occupation in a small town where the harbour dominated the economy. We know nothing more about Anthoine de Champlain or his origins. Champlain's mother was listed as *dame*, meaning lady, or wife of middle rank, rather than the wife of a nobleman. Although Champlain appears not to have had any siblings,[19] a number of documents show that he had a first cousin, Marie Camaret (Cammaret), who lived in Paris in 1619 with her husband Jacques Hersan (Hersant, sieur d'Arsant), at the time a *picqueur* [piqueur] *des chiens de la Chambre du Roy.*[20] Two documents state that she was the daughter of George Camaret, a *capitaine*, and Françoise Le Roy, a sister of Champlain's mother. Champlain and Marie Camaret had jointly inherited a house from their mothers which had come down to them through a common grandmother.[21] We know nothing more about the Le Roy family except that there was a third sister to Margueritte and Françoise who was married to Guillaume Allene, also a pilot and captain. We do not know her given name. His father and two uncles were all senior officers in the navy. Champlain therefore came from a naval family. He was definitely not descended from "poor fisher folk," as many biographers described him.

There were at least three men with whom Champlain had close relationships or who had some degree of influence on him in his youth: his uncle Guillaume Allene; the geographer-engineer Charles Leber, sieur du Carlo; and, a bit later in his life, François Gravé Du Pont. We know nothing about Champlain's relationship with his father, if any, because he never mentioned him, nor did anyone else. One has to assume that because the father was a *cappitaine de la marine* and *pilotte de navyres* in Brouage, he must have had something to do with Champlain's statement that it was the art of navigation "which has drawn me to love it from a tender age and which, for almost my entire life, has stirred me to venture out upon the turbulent waves of the ocean."[22]

19 He is listed in a 1619 document as *filz et herittier seul de deffuncte Margueritte Le Roy sa mere* (son and only heir of the deceased Margueritte Le Roy his mother); see Le Blant and Baudry, *Nouveaux documents*, 1:376.

20 Le Blant and Baudry, *Nouveaux documents*, 1:374, 376, 397, 399. A *piqueur* is a pack huntsman, called a "whipper-in" in England. He handled a pack of dogs for the master huntsman by keeping them on the scent during the hunt and then bringing down the quarry (Du Fouilloux, *La vénerie*, chs. 38–40). By 1635 the Camarets lived in La Rochelle, and in 1637 Jacques Hersan was listed as *contrôleur des traites foraines et domaniales de La Rochelle* (controller of foreign and crown trade for La Rochelle); see Bourde de la Rogerie, *Hélène Boullé*, 14.

21 Le Blant and Baudry, *Nouveaux documents*, 1:399; Vigé and Vigé, *Brouage*, 2:288; Campeau, *Monumenta novae franciae*, 3:33.

22 See Document F, "Excerpts from Champlain's Works," no. 5.

Champlain's uncle by marriage, Guillaume Allene (Allenne, Allaine, Aleyne, Alayne, Arellane, Ellena, Hellaine, Heleyne), also widely known as the *capitaine provençal* (*provançal*), is one relative of Champlain's about whom more information is available.[23] In Spanish documents he is called Guillermo (Guillermon) Elena (Eleno) or *capitán provenzano*. A native of Marseille, son of Anthoine Allenne and Gassin Andriou, Guillaume married Guillemette Gousse, daughter of Nycolas Gousse and Collette François, in La Rochelle on 17 November 1563. At the time of his marriage Allene was a master pilot (*maistre pilote*) and both he and his wife were Huguenot.[24] Through the 1560s and 1570s he was listed as a *marchand et bourgeois* in La Rochelle, participating in voyages along the coast of Africa, South America, and Newfoundland. In 1569 he received *lettres de marque*[25] from the Huguenot queen of Navarre, Jeanne d'Albret, mother of Henri IV, while she and her son were in La Rochelle battling the Catholic army.[26] For the next few years he became a successful corsair, sailing out of La Rochelle against Spanish and *Flamans* shipping.[27] In 1579 Allene and his wife still lived in La Rochelle, but in 1584 he is listed as a captain in the navy living in Brouage.[28] There is no longer any mention of his wife Guillemette. When Allene revised parts of his will in 1601, making substantial gifts (*donación*) to Champlain, including a property near La Rochelle,[29] he wrote that he did this in part "for the love that I bear him, on account of having been married to an aunt of the aforesaid [Champlain], a sister of his mother, and for other grounds

23 Although a good biography of Allene has yet to be written, some of the essentials are in Delafosse, "L'oncle"; Vigneras, "Encore" and "Le voyage."

24 Ibid., 2–4.

25 A *lettre de marque* (letter of reprisal) was a warrant (commission) given by national authorities to private citizens (hence "privateer") to arm a ship for the purpose of capturing and plundering enemy merchant ships in time of war. In France this warrant was also called a *lettre de course*, and the person to whom it was issued a *corsaire*. Given the sense of "privateer" and "corsair," these terms should not be equated with the term "pirate."

26 Delafosse, "L'oncle," 211

27 Ibid. It is likely that the *Flamans* mentioned in the *lettres de marque* were the Dutch living in the Spanish-occupied Netherlands and not the Dutch in the United Provinces (Netherlands) to the north of them.

28 Le Blant and Baudry, *Nouveaux documents*, 1:7

29 The original document clearly identifies the property to be "near La Rochelle" (Document C, "Gift from Guillermo Elena"). Other writers have assumed Brouage was meant, since that town is "near La Rochelle" (Fiquet, "Brouage in the Time of Champlain," 37). The problem with this assumption is that Allene lived in La Rochelle from at least 1567 to 1584, when he moved to Brouage. Why would he write "near La Rochelle" when he meant Brouage?

and just considerations that endear him to me."[30] This aunt could not have been Guillemette Gousse of La Rochelle but must have been from the Le Roy family in Brouage. This implies that Guillemette was probably deceased by 1584 and Allene had married again. It also implies that Margueritte Le Roy, Champlain's mother, had at least two sisters: Françoise Le Roy, mother of Marie Camaret, and Allene's wife. It is probable that Allene and Champlain's father Anthoine knew each other; both were trained as pilots and naval captains, and both operated out of the port of Brouage. Perhaps it was Anthoine who introduced Guillaume Allene to one of his sisters-in-law. By the time Allene turned up in Brouage, the town was Catholic again, and in order to live there he must have undergone conversion, as perhaps did the Champlain family. We know little more about Allene until the late 1590s.

Another person who may have had contact with Allene and the Champlain family was Charles Leber (Le Bert), sieur du (de) Carlo (Carlot, Charlot). In 1625 he was *ingénieur et géographe ordinaire du Roy, et sargeant major en Brouage*. Like other sixteenth- to early-seventeenth-century citizens of Brouage, Charles Leber remains a shadowy person because of the paucity of documents regarding the town.[31] On 29 December 1625, Champlain made Charles Leber, sieur du Carlo, a gift of Allene's *donación* by transferring the document into his name "for the good and true, natural love which the said sieur de Champlain has always born and still bears for the said sieur du Carlot, and also that such is his good pleasure and will to do thus."[32] The document that described the gift was presented to Leber on 19 February 1626 and was accepted by him and registered on 25 February 1626.[33] A month later, on 23 March, Champlain sold Leber the house he had

30 Ibid.

31 Julien-Labruyere, *Dictionnaire biographique*, 801. Unfortunately, du Carlo's birthdate is not known. Consequently, we do not know if he was about the same age as Champlain or older.

32 Leymarie, "Inédit sur le fondateur," 84. French: "*pour la bonne et vray amour naturelle que a tousjours portée et porte le dict sieur de Champlain audict sieur du Carlot, et aussy que tel est bon plaisir et volonté d'ainsi le faire, transportant, dessaisissant, voullant, etc.*" The document of conveyance to Charles Leber makes explicit reference to the *donación* signed by Allene in 1601 in Cádiz. There is no doubt that this property was a gift, not a "sale," and that it was near La Rochelle and not "in Brouage" as suggested by Fiquet ("Brouage in the Time of Champlain," 37, 41n3).

33 Leymarie, "Inédit sur le fondateur," 85. Although both Champlain and du Carlo lived in Paris at the time, du Carlo was temporarily absent and could not see the documents until 19 February 1625. See also Société des archives historiques, *Bulletin*, 20:90.

inherited from his mother for 1,800 livres.[34] Who was Charles Leber du Carlo, and what was his relationship with Champlain?

The transfer of Allene's estate to Leber was an enormously generous gift by Champlain, suggesting that he and Leber had a long-standing relationship. It may also suggest that Allene and Leber had been friends, which would account for the fact that Champlain gave the estate to Leber much as he received it from his uncle Guillaume Allene. Could it be, as some have suggested, that Champlain had received training in surveying, draughtsmanship, and cartography from Charles Leber and that the gift of the estate "near La Rochelle" may have been that of a grateful former apprentice to a master who laid the groundwork for Champlain's future successful career?[35] Leber was an excellent surveyor and a fine cartographer, as seen in his manuscript maps of the coast of France from Cherbourg to the northern coast of Spain (c. 1625), Île de Ré (c. 1625), Brouage (c. 1627), and La Rochelle (c. 1628).[36] An interesting feature of these maps is that they are similar in cartographic style to Champlain's only surviving manuscript map, *descr[i]psion des costs p[or]ts rades Illes de la nouuelle france …* 1607, reinforcing the notion that Champlain may have been an apprentice of his[37] or perhaps trained under the same instructor. The earliest reference to Leber states that he was living in Paris by the early 1620s and that in January 1623 he sought permission to open a butcher's and fish shop in Brouage.[38] In 1625 he was still living in Paris when he began to draw the maps listed above, when he received from Champlain Allene's estate as a gift, and when on 23 March 1626 he bought the house Champlain inherited from his mother.[39] Finally, two documents state that Charles Leber died between 23 March and 26 November 1629, leaving underage

34 Vigé and Vigé, *Brouage*, 2:288

35 Champlain's relationship with du Carlo is stated as a hypothesis by Vigé and Vigé (*Brouage*, 2:280–2), which others have tentatively accepted (Glénisson, *La France d'Amérique*, 15; Fiquet, "Brouage in the Time of Champlain," 37). However, Fiquet added that there was probably a school in Brouage where navigation, surveying, and cartography were taught.

36 Buisseret, *Ingénieurs*, 110–12, and "French Cartography," 1517. These maps were produced while Leber du Carlo lived in Paris.

37 Heidenreich, "The Mapping of Samuel de Champlain," 1540

38 Online at www.HistoirePassion, AD33/1B21. The form reads "*Permission d'avoir une boucherie et poissonnerie à Brouage, en faveur de Charles Le Ber, Sr de Carlo.*"

39 In 1620 Champlain had purchased his cousin Marie Camaret's half of the house for 500 livres, and in 1626 he sold the house to Charles Leber for 1,800 livres (Vigé and Vigé, *Brouage*, 2:288).

children (*enfants mineurs*).[40] Unfortunately, we do not know when he was born. The fact that his earliest known maps date from 1625 and that he had underage children when he died in 1629 suggests that he may have been younger than Champlain, who must have been at least fifty at that time. If this reasoning is correct, then it is unlikely that Leber was Champlain's teacher. It may simply be that the two were close friends who had studied surveying and cartography under the same teacher or at the same school. This possibility will be explored below, under Champlain's education.

In 1604 Champlain began a correspondence with another *géographe du Roy*, Guillaume de Nautonier, author of a treatise of the earth's magnetic field and of a theory for determining longitude by observations of magnetic declination and latitude.[41] Champlain used de Nautonier's theoretical work and sent him observations of magnetic declination as early as 1604, from Acadia, for which de Nautonier was grateful, hoping "that on his [Champlain's] return, he will bring us a full account of that region, as he is extremely capable."[42] How the two came to know each other is not known, but it could have been through Leber sometime before the start of Champlain's 1604 voyage.

Early in 1603, Champlain met François Gravé Du Pont, a veteran of several voyages to Canada, who was to command Aymar de Chaste's ship, the *Bonne-Renommée*, on the voyage to explore the St. Lawrence River.[43] Over succeeding years the two became good friends to the point where Champlain declared in 1618, "I was his friend, and his years would lead me to respect him as I would my father."[44] There is little doubt that Champlain learned a great deal from Gravé, especially coastal navigation, Native relations, and trade. He became a trusted confidant, one of the few from whom Champlain sought advice.

40 Delafosse, "Séjour de Champlain," 572. Julien-Labruyere (*Dictionnaire biographique*, 801) contends that Charles Leber was still alive in 1653, but he appears not to have seen the reference to the notarial documents cited by Delafosse that Leber was dead in 1629.

41 Guillame de (le) Nautonier, sieur de Castelfranc, en Languedoc (1557–1620) was appointed *géographe ordinaire du Roy* by Henri IV in 1606. At one time he was also a Huguenot minister at the neighbouring towns of Castelfranc, Réalmont, Vénès, and Montredon. His major work was *Mecometrie de l'eymant*, published 1602–03, with revisions and additions in 1604. For a commentary on this work, see Mandea and Mayaud, "Guillaume Le Nautonier."

42 Nautonier, *Mecometrie de l'eymant*, 2nd edn., "Sixieme Livre ... 1604," 7^{v}

43 DCB, 1, "Gravé Du Pont, François." Also called Pont-Gravé, Pontgravé, and Du Pont-Gravé, he was born in Saint-Malo November 1560 and died sometime after 1629 (Le Blant and Baudry, *Nouveaux documents*, 1:1).

44 Biggar, *The Works*, 4:363

Like everything else about him, Champlain's education is a matter of speculation. There is every reason to suspect that he did not have much formal schooling, certainly not in what is called a classical education. His written French was serviceable, but without the classical allusions common to his time in the writings of educated people. After spending two and a half years in the company of Spaniards in the West Indies, and some time in Spain looking after the affairs of his uncle, it is almost certain that he spoke some Spanish. His writings on mapping and navigation, as well as his maps and charts, show that he used simple surveying techniques. He does not demonstrate that he had any training in basic geometry, let alone in the new mathematical developments used in navigation and surveying, such as trigonometry. He wrote late in life that he had learned navigation, instrumental observations, etc., "both by experience and by the teaching of many good navigators, as well as through the special pleasure I have derived from the perusal of books on this subject."[45] In other words, he learned by observing and listening to others, through his own experience, and through reading.[46] When Champlain is first mentioned, he is a young man – a *fourrier*[47] – in the land-based army of Henri IV during the religious wars in Brittany. In this role, he was required to have basic training in surveying and mapping, to make route maps, and to lay out camp sites for army units. There is no question that he had such surveying skills, otherwise he could not have carried out the tasks required of him. Judging from his large-scale mapping of harbours and bays after 1603, he learned those skills in France, because he used French measures consistently and with some skill. Where did he acquire such training before he arrived in Brittany in 1595?

The suggestion has been made that Champlain may have attended a "special academy" in Brouage, where he could have acquired his skills in surveying and cartography.[48] This academy is described in some detail by the Swiss physician and traveller Thomas Platter from a visit to Brouage between 5 and 6 May 1599.[49]

45 See Document F, "Excerpts from Champlain's Works," no. 8.

46 Heidenreich, *Explorations and Mapping*, 126; Heidenreich and Dahl, "Champlain's Cartography," 312–32; Heidenreich, "The Mapping of Samuel de Champlain," 1542–7.

47 When Champlain first surfaces in the documentary record in 1595, he is a *fourrier*, defined in 1690 as *Officier qui marque les logis pour le Roy, & toute sa Cour, quand il voyage* (officer who marks the lodgings for the king and all his court when he travels); see Furetière, *Dictionnaire universel*, "Fourrier." Champlain's rank and tasks will be examined in greater detail below.

48 Fiquet, "Brouage in the Time of Champlain," 37–9

49 Keiser, *Thomas Platter: Beschreibung der Reisen durch Frankreich*, 1:451–4. The volumes are written in the German of Platter's times before spelling was standardized. Thomas

This academy admitted only sons of noble and well-born men (*vom adell unndt wollgebornen*) who were between the ages of fourteen and twenty. Training consisted of knightly pursuits (*ritterspilen*), such as riding and horse jumping, dancing, fencing, and playing the mandolin. After eating breakfast each day, the students learned surveying and how to lay out fortresses and fortifications.[50] Once the "food was digested," the disciples, as Platter calls the students, partook in physically more rigorous training, especially horsemanship. Normally the program was two years, but students could shorten or lengthen their course of study depending on their aptitude. The students who were the most distinguished and accomplished horsemen, fencers, dancers, etc., were paid to teach the others. The headmaster (*rector*; modern: *Rektor*), a distinguished older man from the nobility,[51] was paid by the king but also received funding from each student. Graduates were suitable for careers in the army or in the service of a lord. In an academy of this kind, the *rector* would likely have made recommendations and looked after placements for his better students.

There were other academies for well-born young men in France, but they taught more scholarly subjects as well.[52] According to Platter, academic studies at Brouage went into a great decline (*mechtigen großen abgang*) because of the long wars of religion, to the point that the students were ashamed to utter Latin because, if they did so, they would be contemptuously called priests.[53] Platter concluded that these schools made training in similar Italian academies unnecessary.

It is tempting to link Leber and Champlain to this academy. It would explain much about their later careers. Champlain would have been acceptable to the school. He came from a respectable family that was prominent in Brouage, and showed an aptitude for surveying and drafting. The lack of polish and absence of classicisms in his written French may be a reflection of the academy's curricu-

Platter the younger (1574–1628), a Protestant and Swiss national, studied medicine in Montpellier. Like his father (1499–1582), he was a traveller and humanist interested in education.

50 *Nach dem essen gleich lehrnet man sie meßen, vestungen in grundt legen oder fortificieren* (ibid., 452). *Meßen* (*messen*) is measuring, mensuration, or surveying. Here, surveying is meant because the word is used to describe the act of laying out fortresses and fortifications (ibid., 452).

51 *ein stattlicher alter vom adel*, ibid.

52 There were other academies like the one in Brouage in such places as Paris and Orléans (ibid., 453).

53 *sagen gleich, es seye einer ein pfaff, so er latein redet* (ibid.). The German word *Pfaffe* was used in a contemptuous manner to denote a priest (*Priester*), especially by Protestants.

lum. Upon graduation, he was qualified and probably recommended to become a *fourrier* in the *maison du Roy*, where he was required to draft route maps and plans for the *maréchal des logis*, Jean Hardy. We know from Champlain's plan of Quebec that he understood the basic principles of fortification.[54] Leber may have been the more studious of the two, judging from his maps and other areas of qualifications, as reflected in his title, *ingénieur et géographe ordinaire du Roy, et sargeant major en Brouage*. He may have taught in the academy for a while before going on to be an engineer, geographer to the king, and sergeant major in the army. If so, it is probable that Leber took longer to graduate and/or acquired more training at a later date, as well as spending many years in the field surveying the French coastline. This could account for his late output of maps – but not for the underage children he left when he died in 1629.

In his dedicatory letter to Marie de Medici, mother of Louis XIII, Champlain wrote that it was the art of navigation that had "drawn me to love it from a tender age and which, for almost my entire life, has stirred me to venture out upon the turbulent waves of the ocean."[55] While this is entirely possible, we know nothing of his early youth at sea. The first evidence we have of Champlain on a ship is when he left Blavet[56] for Cádiz in 1598 on the *San Julián*, commanded by his uncle, Guillaume Allene. While his surveying skills show they were learned in France, not so his writings on navigation, which display methods he learned from the Spanish. Just as there is no evidence that Champlain achieved the rank of *maréchal de logis*, as stated in the manuscript *Brief discours* that has been attributed to him,[57] there is no evidence that he was ever a pilot, captain, or navigator on a ship of any consequence. When we read about him on a ship, he was either engaged in surveying a harbour or coastline or he was a passenger, with others in command. In his many descriptions of his Atlantic crossings, he never made any claims to have served in the capacity of a navigator or captain; only his later biographers did so. His titles, *capitaine ordinaire pour le Roy en la Marine du Ponant* and, after 1625, *capitaine pour le Roy en la Marine du Ponant*, may have been honorific. Perhaps it was necessary to hold the rank of at least a *capitaine* to become a lieutenant to a viceroy, or perhaps it permitted him to take command of a ship in the absence of qualified officers. His *Traitté de la marine*, the treatise

54 For example, on his plan of Quebec he wrote that he had constructed platforms in the style of *tenailles* for the placing of the cannon (Biggar, *The Works*, 2:39, symbol N). The term *tenailles* is also used in English (Muller, *A Treatise ... of Fortification*, 33–4).

55 See Document F, "Excerpts from Champlain's Works," no. 5

56 Blavet, a town on the south coast of Brittany, is now called Port-Louis.

57 Biggar, *The Works*, 3:3

on navigation he published in 1632,[58] is a compendium of sensible observations on how to operate and command a ship, which he had ample opportunity to observe on his twenty-three Atlantic crossings. The rules he gives for navigating are "cookbook" examples that involve no geometry or even arithmetic to apply or understand but, rather, the use of a series of tables from which variables such as course distance and direction could be taken to work out particular problems.[59] His real strength was that of a surveyor. He could correct his compass for magnetic declination but did not know about annual variation. He could calculate latitude but had to approximate longitude from estimated distances on particular parallels of latitude. He understood the principles of triangulation through observations with a compass – estimating or measuring distance and scaling the results on paper with a ruler and proportional divider – but not through geometry or trigonometry. He could do a stationary survey or a much more complicated running survey along a coastline. Finally, he shows that he had the skill and training to construct maps from his many observations. In the application of what he learned and within the limitations of his instruments, he shows great diligence and skill. His greatest weakness was in estimating distance, failing to convert these estimates to a common scale on his small-scale maps, and on occasions using more than one prime meridian on the same map without indicating he had done so.

In *Les Voyages, 1632*, Champlain makes a curious statement about his birth that is difficult to explain. When asked by Aymar de Chaste if he would like to join his expedition to the St. Lawrence in 1603, Champlain replied, "I could not do so but by the command of his said Majesty, to whom I was under an obligation both by birth and by a stipend [*pension*] with which he honoured me as a means to maintain me near him."[60] Champlain then asked de Chaste to request the king, on his behalf, to permit him to join the expedition. It is likely that he had been awarded a *pension de Roy*, which was a stipend (allowance) sometimes given by a monarch to an impecunious relative or to a person for services rendered in the past, with the hope of good advice in the future. We do not know when this stipend was assigned or for what reason, but it indicates that he was personally known to the king and that the king trusted him, found him reliable, and valued his advice enough to pay Champlain a stipend to live near him.[61]

58 Ibid., 6:257–346

59 Heidenreich, *Explorations and Mapping*, 115–26

60 See Document F, "Excerpts from Champlain's Works," no. 7.

61 This pension should be interpreted as a living allowance or stipend, not as a retirement pension. Without any documentary evidence, some writers have assumed that Champlain was given the stipend by Henri IV for his observations of Spanish colonial

What is less clear is what Champlain meant by writing that he was obliged to the king by birth (*de naissance*). The few writers who have commented on this statement felt that it was simply an allusion to the fact that both the king and Champlain were French[62] or that the Champlain family may have felt some "loyalty to a Protestant Henry."[63] The first of these explanations is unlikely in view of the number of French who journeyed out of the country without seeking the king's personal permission. The second is equally unlikely, in that the realm was full of people who formerly had, or still had, a Huguenot background. If Champlain had been a Huguenot, why would he draw attention to such a detail, since the king – and presumably Champlain – had become Catholics well before 1603? If the phrase *de naissance* is taken literally, it seems that Champlain suggested that he was somehow related to the Bourbon monarch; but in what way? His mother's maiden name, Le Roy, is suggestive, but we know nothing of her family. If Champlain had a family connection to the Bourbon monarchy, the *pension de Roy* to maintain him near the king and the obligation he felt to the king through his birth (*de naissance*) may be related. Is this why Champlain was attached to Jean Hardy, an officer in the king's household (*maison du Roy*) by 1595, when he is first mentioned in a document? Was Champlain related to the Bourbon monarchy and therefore given a stipend to maintain him in the king's household? Given the fact that in all his writings Champlain makes no specific comments about his origin except this one, such a suggestion is startling, to say the least. In the absence of any concrete evidence, it seems pointless to take this notion any further. However, if it were true, many questions about his future career and his relationship to his superiors would be answered:

- In what circumstances was he appointed to the king's household as a *fourrier*?
- Why did this particular *fourrier* run confidential messages for the king to the maréchal Jean d'Aumont, commander of the Brittany army?
- How did Champlain get to know so many influential people?
- Why did he feel he had to ask the king for permission to travel to Canada in 1603 and again in 1604?

activities. In his biography of Champlain, David Fischer wrote, "We have evidence that Champlain was receiving an annual pension from the king of 600 livres a year" (Fischer, *Champlain's Dream*, 658n12). This sum would have been about twice what Champlain was making as a *fourrier* in Brittany. Unfortunately, Fischer does not cite a reference for his "evidence."

62 Deschamps, *Les voyages*, 4n5, 8–9n1

63 Bishop, *Champlain*, 41

- Why did Henri IV feel it necessary to issue a letter, through his secretary of state Louis Potier de Gesvre, ordering Champlain to go on the 1603 voyage?
- Why did he report directly to the king every time he returned to France from an overseas voyage (1602, 1603, 1607, and 1609)?
- At a time when connections to the court meant everything to a successful career, how could Champlain, if he was a mere commoner from one of the outer provinces with no connections to the nobility, have risen through the ranks from *fourrier* to become de facto governor of New France? He served as lieutenant to a lieutenant general (Du Gua de Mons, 1608–12), lieutenant to five viceroys (comte de Soissons, 1612–13; prince de Condé, 1613–19; Condé and marquis de Thémines, 1619–20; duc de Montmorency, 1620–25; duc de Ventadour, 1625–28), and finished his career as lieutenant general to the most powerful person in France after the king, the cardinal de Richelieu (1638–35).[64]

Since the Bourbons were not in the habit of appointing commoners to represent them, were Champlain's natural abilities so remarkable to have achieved such successes? We have no answers to these and other question relating to him, his family, his marriage, and his career, but they would all be answered if he was connected by birth to the Bourbon monarchy or at least to nobility.

When Henri IV was assassinated on 14 May 1610, Champlain heard about it towards the end of June, causing him to write, "I was much pained to hear such bad news," and three years later in his dedicatory letter to the young Louis XIII he described the former king as *Henry le grand, d'heureuse memoire*.[65] At the very least, Champlain had lost his sponsor and most faithful supporter.

The war in Brittany, 1589–1595

On 1 August 1589, Henri III, king of France, was murdered by the fanatical monk Jacques Clément. As he was dying, he named Henri of Navarre, a Huguenot, as his legitimate heir and successor.[66] Since many in France could not accept an excommunicated Huguenot as king, the last phase in the religious wars that had plagued France during the sixteenth century was touched off.

In Brittany, opposition to Henri, in the form of the Catholic (Holy) League, rallied around Philippe-Emmanuel de Lorraine, duc de Mercœur et de Pen-

64 See appendix 2, "Champlain's Signature and Titles."

65 "Henry the Great, of happy memory" (Biggar, *The Works*, 2:145, 208).

66 Henri of Navarre had been excommunicated in 1585 by Pope Sixtus V when it appeared that he might become the successor to the throne of France.

thièvre (1558–1602), who had been appointed governor of Brittany by Henri III in 1582.[67] Upon Henri III's death, Mercœur declared independence for Brittany in the name of his underage son, Philippe Louis de Lorraine Mercœur. There were, however, two highly controversial candidates for the succession to the throne of France: the Huguenot Henri of Navarre, named successor by Henri III, and in theory at least if not in fact the Catholic infanta Isabella Clara Eugenia (1566–1633), daughter of Felipe II (1527–98), king of Spain, and his third wife, Elisabeth of Valois. Although Henri III was Isabella's uncle and she was Catholic, Salic law, which was in effect in France, forbade females from dynastic succession. Moreover, having the succession to the throne of France fall into the hands of Spain was anathema to the larger part of the French population. Within weeks after the assassination, as the country started to break into rival factions, Mercœur contacted Felipe II, as defender of the Catholic faith, for advice and help against Henri of Navarre. After Henri's success against the Catholic League at the village of Arques near Dieppe on 18 September 1589, help from Spain became more urgent. Spanish sources informed Felipe II that Mercœur claimed he had about five thousand soldiers, although others thought it was more like eight or nine thousand.[68] Following a number of rapid successes by Henri and a major victory close to Paris at Ivry on 14 March 1590, Felipe II had to act both to keep a Huguenot from the throne of France and to place his daughter Isabella on it. Spanish intentions and activities in France now began to worry Queen Elizabeth of England, who tried to help Henri and his cause by commissioning privateering activities under Drake, Raleigh, Frobisher, and others to harass and capture Spanish shipping wherever they could find it.

In June 1590, anticipating the Spaniards and to provide a base of operations for them, Mercœur laid siege to Blavet by land and sea, finally taking and burning it with enormous loss of life to the Blavétins.[69] Late in July, Felipe II ordered sixteen Spanish and two Italian companies, totalling 3,000 men, to Brittany under the command of the able Don Juan d'Aguila.[70] He also ordered the Duke of Parma to enter Brittany with troops from the Spanish-occupied territories in the southern Netherlands.[71] His ultimate object, in which he succeeded, was to break Henri's

67 The correspondence of Mercœur and documents relating to the activities of the League in Brittany are in Carné, *Documents sur la ligue de Bretagne*.

68 Ibid., 11:6. For details of Henri of Navarre's early campaigns to unite the country, see Buisseret, *Henry IV*, 28–44.

69 This event is still celebrated through the *gwerz* (lament) of *Lopéran* (*Locpézran*), a fifteenth-century name for Blavet. The modern name for Blavet is Port-Louis.

70 Carné, *Documents sur la ligue de Bretagne*, 11:11, 17

71 Buisseret, *Henry IV*, 36

siege of Paris. Harassed by Elizabeth's navy and by storms on the Bay of Biscay, the Spanish troops destined for Blavet were unloaded at Saint-Nazaire, near the mouth of the Loire River. Although some vessels commanded by Don Diego Brochero reached Blavet on 25 August, Aguila's troops had to march overland, and they reached Blavet and Vannes late in October. Along the way they lost about 250 men. Early in November, Jérôme d'Aradon[72] laid siege to Hennebont with Mercœur in order to return it to the League and secure Blavet from the land. The town fell on 22 December, opening that part of Brittany to the unhindered access of Spanish troops. By 25 November, Spain had 2,697 soldiers in place,[73] and during December, as soon as fighting had ceased, they began construction of the fortress Fuerte del Aguila at Blavet, which commanded the town and harbour. On 8 January, 2,000 more soldiers were promised, raising the total number of Spanish troops to 4,715 by 1 May 1591.[74] A breakdown of Spanish troops drawn up on 11 July 1591 lists a formidable force of 930 (*caballeria de coracas*) mounted men, 790 (*arcabuzeros de acaballo*) mounted men, each armed with an harquebus, and 3,000 (*infanteria a pie*) foot soldiers.[75]

The insertion of Spanish troops and its navy into ports along the Brittany coast was greeted with increasing alarm by the English. Elizabeth now saw that more help had to be given to Henri. Besides the usual raids on Spanish shipping, an expeditionary force of 2,400 men under the experienced officer John Norris (Norreys, Nourichs, Norrhis) was sent in May 1591 to land on the northern coast of Brittany at Paimpol and raid Île-de-Bréhat, where Mercœur had a fort. More significantly, later in the year about 4,000 soldiers from England under the Earl of Essex and 6,500 German horse and 10,000 foot were sent as reinforcements to Henri to aid in his siege of Rouen.[76] While there was considerable action in the rest of France, in Brittany each side seemed to avoid the other, probably because Henri's forces were not strong enough to take on the League and the newly arrived Spaniards. In fact, Henri had only fourteen garrisons in and near Brittany, eight around Nantes from the Rivière Vilaine south to a line between Pornic and Clisson, four in the western part of the Côtes-d'Armor between Île-de-Bréhat

72 Jérôme d'Aradon had been governor of Hennebont until it was taken by the king's army. He and his brother René d'Aradon, governor of Vannes, were strong Spanish and League sympathizers.

73 Carné, *Documents sur la ligue de Bretagne*, 11:17

74 Ibid., 25, 38

75 Ibid., 59–61

76 Ibid., 37–41

and Guingamp, one in central Brittany at Rostrenen, and an important one in the port of Brest.[77]

On 25 July 1593, Henri took the significant step of removing one of the most serious objections that most Catholics had to his succession to the throne by abjuring Protestantism, at the same time persuading his Huguenot allies that there would be no discrimination against them. Although some Catholics were not convinced of his sincerity and Pope Clement VIII did not lift his ban of excommunication until 17 September 1595, increasingly important families and towns came to his side, initiating a major shift in the war.

Sometime in September 1593, General John Norris returned to Île-de-Bréhat and Paimpol with 1,963 English soldiers and 24 officers.[78] Their objective seems to have been to winter in Brittany and in the new year make their way south to disrupt any attempt by Spain to take the port of Brest or to set up naval bases from which they could harass English shipping and launch a renewed attempt to invade England. Through a contract with Henri's authorities in Brittany, the English troops were quartered at Paimpol and Lanvollon,[79] and their ships berthed in harbours near Paimpol. Strict instructions were laid down regarding food procurement and the behaviour of the troops towards local citizens and their institutions. Nevertheless, during the winter Norris's army ravaged the countryside for food, and it moved to La Roche–Derrien by early May 1594, poised to march south.[80]

The year 1594 had a good beginning for Henri of Navarre. Since he had abjured Protestantism in July the previous year, Catholic support had increasingly been turning to him. On 27 February he was crowned King Henri IV at Chartres. With this act, resistance against him waned in Paris to the point that he could enter the city on 22 March. In late April, however, while the English forces were organizing themselves to move south, the Spanish, under Don Juan d'Aguila, began to make plans for a fort to blockade shipping into the harbour of Brest.[81] As early as November 1592, d'Aguila had advocated to Felipe II that Brest should be a principal Spanish objective because without Brest, nothing could be done against the English.[82] On 1 May, building material, cannons, and munitions were

77 Barthélemy, *Choix de documents*, 92–6

78 Ibid., 143–51

79 Lanvollon, a small town and castle, is about 14 km east-northeast of Guingamp in Brittany.

80 Barthélemy, *Choix de documents*, 164, 168

81 Carné, *Documents sur la ligue de Bretagne*, 12:19

82 Ibid., 11:160

received at the construction site of the fort, and by mid May construction was well on the way.[83] The new fort, Castillo del León, was built on a high cliff, 60 metres above sea level, on the northern point of the peninsula of Roscanvel, at a place now called Pointe des Espagnols. This was a perfect spot from which to blockade Brest, since it overlooked the *goulet de Brest*, a long narrow channel leading from the Atlantic into the harbour.[84] Of the 5,329 Spanish troops in Brittany, three companies, totalling 401 men, under the command of Don Thomas de Parredes (Praxède) were in the fort by 27 May.

In 1592, maréchal Jean d'Aumont[85] was appointed commander of Henri's army against the Catholic League and its Spanish allies in Brittany. By 1594, with the increase of Spanish activities, the time had come to put that army into action. Early in August 1594, d'Aumont left the vicinity of Rennes with approximately 3,000 soldiers. By late September he had taken Guingamp, Morlaix, Saint-Pol-de-Léon, and other places in northern Brittany. On 12 October he took Quimper, near the south coast, after inflicting severe artillery damage to the town and its fortifications from the heights of Mount Frugy.[86] From Quimper, d'Aumont's army quickly moved on to Crozon for an attack on Castillo del León. He arrived on 18 October and met up with Norris's English army of 2,000 men, which had arrived eight days earlier. On 2 November the combined armies began their assault with artillery on the Spanish fort. In this they were joined by a squadron of English ships with about 1,200 men commanded by Sir Martin Frobisher. Losses were heavy. Close to half the French and English soldiers were killed, and only eleven Spaniards survived. During the battle Frobisher was wounded, and he died at Plymouth on 22 November, five days after Castillo del León fell. Following the battle, d'Aumont's army settled into Quimper for the winter, while Norris went to Morlaix and from there to Saint Pol-de-Léon and Roscoff, on the north coast near his ships.[87] Before the king's troops could rest, one more operation had to take place. Early in January 1595, d'Aumont sent orders to capitaine Jean du Mats de Montmartin, who was stationed at Châtelaudren,[88] to join him in rooting out Guy Eder de la Fontenelle,[89] a League sympathizer, a notorious brigand and murderer, who used the turbulence of the war to raid and plunder

83 Ibid., 28–32; Barthélemy, *Choix de documents*, 154–5
84 At the site of the fort the channel is only 1.7 km wide.
85 Jean d'Aumont, baron d'Estrabonne, comte de Châteauroux (1522–95). He was made *maréchal de France* by Henri III in 1579.
86 Carné, *Documents sur la ligue de Bretagne*, 12:50
87 Ibid., 12:60
88 Châtelaudren is about 30 km north of Corlay.
89 For the life of Fontenelle and his role in the Brittany wars, see Baudry, *La Fontenelle*.

through Brittany. La Fontenelle had occupied Corlay since June 1594, using it as a base for his operations. D'Aumont left Quimper in the hands of Julien du Pou, its governor, and a large garrison of men, including some from Morlaix, with instructions to rebuild the fortifications. Corlay was captured on 10 February 1595, and in spite of Fontenelle's flight – only to cause future problems – the people of western Brittany began to see the beginning of the end of constant fighting. After Corlay, d'Aumont went on to the château de Comper near Rennes, where after the siege he died on 19 August from an harquebus shot.

With d'Aumont's "pacification" of large parts of Brittany, along with Henri IV's conversion to Catholicism and Spain's disaster at Crozon, support for the League was rapidly waning. In addition, there were serious strains between the Spanish occupation forces, Mercœur, and other members of the League. It was therefore opportune to declare war on Spain, which Henri IV did on 17 January 1595. Since Brittany was now relatively quiet, d'Aumont had the opportunity to deploy his forces into an occupation army, composed of 34 regional garrisons with some 7,400 men, a huge increase over the 14 garrisons Henri had late in 1591 (map 1). An extraordinarily detailed inventory was completed by 17 February 1595 for each garrison, giving the size and composition of the army units, the names of some senior officers, and the pay schedules.[90] Of these garrisons, *Kempercorantin* (Quimper)[91] had 481 men and officers, as well as 250 on loan from those stationed at Morlaix who had been sent to help rebuild the fortifications. It was at this garrison that Champlain appeared on 1 March 1595. A marginal note, written beside the entry for Quimper after the main report was finished on 17 February, reads: "*Ceste garnison a esté augmentée d'ung fourier, par moys, 20 écus*" ("This garrison has been increased by one *fourrier*, 20 écus per month").[92] This *fourrier* was probably Samuel de Champlain, who began to draw pay on 1 March at Quimper. There were no *fourriers* at Quimper until Champlain ar-

90 Barthélemy, *Choix de documents*, 179–97. Of the 7,414 officers and men in the royal army stationed in Brittany, 66% were *gens de pied* (common foot soldiers); 9.4% were *salades* (cavaliers); 10% *harquebusiers à pied* (foot soldiers armed with an harquebus); 7.4% *harquebusiers à cheval* (mounted soldiers armed with an harquebus); 6.5% *hommes de guerre montez et armez à la legère* (mounted and lightly armed soldiers); 0.7% were Swiss mercenaries. In February 1595, the wages of these soldiers and officers was given as 38,167 écus and 20 sols per month.

91 Quimper (Quimper-Corentin, *Kempercorantin*), a town in Brittany at the junction of the Rivers Odet and Stier. Quimper-Corentin was the ancient name for Quimper. The second part of the town's name was in honour of St. Corentin, first bishop (sixth century) of the bishopric of Cornouaille in Brittany, of which Quimper was the seat.

92 Barthélemy, *Choix de documents*, 193

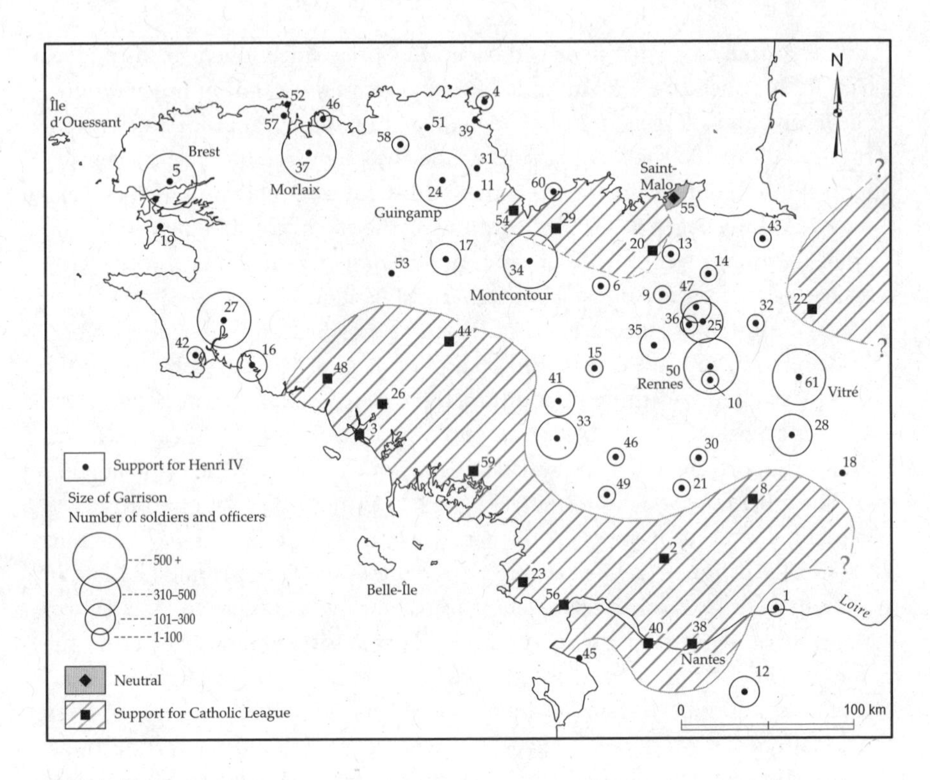

Map 1 Bretagne, 1595

In the late summer and fall of 1594, maréchal Jean d'Aumont made a circuit of Brittany at the head of Henri IV's army, reducing such important towns as Guingamp, Morlaix, Saint-Pol-de-Léon, and Quimper, and ending with the destruction of the Spanish fortress Castillo del León, which overlooked the maritime entrance to Brest. By March 1595, when Champlain arrived in Brittany, d'Aumont's army had garrisons in thirty-four towns, villages, and châteaux, with some 7,400 men. Information from documents written in 1595 show that the Catholic League held a few places on the north coast, all of the south coast from Quimperlé eastward, and both banks of the Loire River. The Spanish presence was confined to the south coast. Eastern, central, and western Brittany were largely in the hands of Henri IV. The largest garrisons – Quimper, Brest, Morlaix, Guingamp, Montcontour, Rennes, and Vitré – were roughly equidistant from each other at 40–50 km intervals in an east-west arc through Brittany, facilitating a rapid cavalry response to any problems. Although being wooed by both sides, the important port of Saint-Malo, remained neutral.

Among the many sources for this map, the most important were Barthélemy, *Choix de documents inédits sur l'histoire de la Ligue en Bretagne*, 149–204, and Carné, *Documents sur la ligue de Bretagne: correspondance du duc de Mercœur et des ligueurs bretons avec l'Espagne*, 12:8–100. See also figure 5.

PLACE NAMES: BRETAGNE, 1595

L = Catholic League, L/S = with Spanish Garrison, R = Royalist (number gives size of garrison)

Location on map	*Place names on documents 1595*	*Hardy – Hondius map, 1631*	*Modern name*	*Size of royal garrison*
1	Ancenis	Encenix	Ancenis	R <50
2	Blain	Belain	Blain	L
3	Blavet	Blavet	Port-Louis	L/S
4	Bréhat	Isle de Brehal	Île-de-Bréhat	R 30
5	Brest	Brest	Brest	R 633
6	Broon	Bron	Broons	R 60
7	Castillo de León	Fort de Crozon	Pointe des Espagnols	R
8	Chasteau Briant	Chāu Briant	Châteaubriant	L
9	Chasteau de la Latte	?	Château-de-la-Latte	R 30
10	Chastillon	Chastillon	Noyal-Châtillon	R 87
11	Chateaulaudren	Chāulāndra	Châtelaudren	R
12	Clisson	Clißon	Clisson	R 265
13	Coesquen	?	Coëtquen	R 32
14	Combourg	Conbourg	Combourg	R 70
15	Comper	Comper	Comper	R <50
16	Conquarneau	Conq.	Concarneau	R 210
17	Corlay	Corlay	Corlay	R 180
18	Craon	Craon	Craon	R
19	Crozon	Crozon	Crozon	R
20	Dinan	Dinan	Dinan	L
21	Fougeray	Fougeray	Grand Fougeray	R 100
22	Fougéres	Fougere	Fougères	L
23	Guerande	Gueraude	Guérande	L
24	Guingamp	Guingandt	Guingamp	R 700
25	Hédé	Hede	Hédé	R 380
26	Hennebont	Henbont	Hennebont	L/S
27	Kempercorantin	Hempercorantin	Quimper	R 731
28	La Guerche	La Guerche	La Guerche de Bretagne	R 430
29	Lambatte	Lamballe	Lamballe	L
30	La Marzelière	?	La Marzelière	R 40
31	Lanvollon	Lāolon	Lanvollon	R
32	Le Bordaige	Borduge	Château de Bordage	R 83
33	Malestroict	Malestroit	Malestroit	R 363
34	Montcontour	Montcōtour	Montcontour	R 530
35	Montfort	Montfort	Montfort	R 170
36	Montmuran	Mōmuran	Château Montmuran	R 70
37	Morlaix	Morlais	Morlaix	R 595

Location on map	*Place names on documents 1595*	*Hardy – Hondius map, 1631*	*Modern name*	*Size of royal garrison*
38	Nantes	Nantes	Nantes	L
39	Paimpool	Pampol	Paimpol	R
40	Pellerin	Le Pellerin	Le Pellerin	L/S
41	Ploërmel	Ploermel	Ploërmel	R 254
42	Pontlabé	Pont l'abbé	Pont l'abbé	R 30
43	Pontorson	Pont Orson	Pontorson	R 80
44	Pontivy	Pontivi	Pontivy	L
45	Pornic	Pornic	Pornic	R
46	Primel	Primel	(près de) Plougasnou	R 85
47	Québriac	Quebriac	Québriac	R 150
48	Quimperlé	Querperle	Quimperlé	L
49	Redon	Redon	Redon	R <50
50	Rennes	Rennes	Rennes	R 621
51	Roche Derrien	La Roche derriē	La Roche-Derrien	R
52	Roscoff	Roscof	Roscoff	R
53	Rostrenen	Rostrenen	Rostrenen	R
54	Saint-Brieu	S. Brieuc	St-Brieuc	L
55	Saint-Malo	S. Malo	Saint-Malo	Neutral
56	St. Nazare	S. Andre	Saint-Nazaire	L/S
57	Saint-Pol-de-Léon	s. Pol	Saint-Pol-de-Léon	R
58	Tonquenet	Lonquedec	Tonquédec	R 50
59	Vannes	Vannes	Vannes	L/S
60	Le Verdelet	La Roche de Verdelay	Le Verdelet (près de Pléneuf)	R 30
61	Vitré	Vitreÿ	Vitré	R 530

rived, although there were two *maréchaux des logis* who performed work in the cavalry units similar to that of *fourriers.*

Champlain in the army of Henri IV in Brittany, 1595

On two occasions Champlain wrote that he had been "employed in" and "served in"[93] the army of Henri IV under "maréchal d'Aumont, M. de St. Luc, and the maréchal de Brissac."[94] These statements are substantiated by the records of Gab-

93 Champlain used both the words *employé* (employed) and *seruy* (served).

94 See Document F, "Excerpts from Champlain's Works," nos. 1 and 7; maréchal Jean d'Aumont (1522–95); *maître de l'artillerie de France*, François d'Espinay de Saint-Luc (1554–97); and maréchal Charles II de Cossé, duc de Brissac (1550–1621).

riel Hus, *trésorier des États de Bretagne*, who recorded the pay issued to the royal troops stationed in Brittany.[95] As far as is presently known, these are the earliest records that have Champlain's name attached to them. It appears that Champlain first appeared in Brittany late in February or early in March 1595[96] as a *fourrier* and *aide* to the *maréchal des logis du Roy*, Jean Hardy, and that he was stationed for most of the time in the garrison at Quimper, commanded by the governor of the town, Julien du Pou, sieur de Kermouger. How long he held these two offices and stayed in Quimper is not known. Two years later he was an *enseigne* to the sieur de Milleaubourg, also in the garrison at Quimper. There is no evidence that Champlain was a maréchal des logis himself, as is stated in the *Brief discours*, although he may have served at some point as a fourrier in a cavalry unit, whose men were called *maréchaux des logis de l'armée*.[97] The rank of *maréchal des logis du Roy*, with a salary of 66⅔ écus per month, should not be confused with a *maréchal des logis de l'armée*, who was paid about 12 to 17 écus per month.

In order to appreciate the significance of Champlain's position, it must be seen in terms of Henri's court and the Brittany army. The *fourrière*, in which Champlain was a junior officer by virtue of being the aide to the maréchal des logis du Roy, Jean Hardy, was one of the seven offices of the *maison du Roy* (royal household). It was managed by the *grand maréchal des logis de la maison du Roy*,

95 See Document B, "Personnel and Pay Records," no. 1. Gabriel Hus, sieur de la Bouchetière, was *trésorier des États de Bretagne*, and mayor of Nantes from 1599 to 1601. He died in 1609. The fact that Champlain had served in the army in Brittany in 1595, based on the records of Hus, was noted as early as 1913; see *Bulletin et Mémoires de la Société Archéologie* 47–8.

96 Biographers of Champlain who dated his arrival in Brittany have invariably given the year as 1594 (Pocquet, *Histoire de Bretagne*, 5:260; Armstrong, *Champlain*, 21–2; Morison, *Champlain*, 17; Fischer, *Champlain's Dream*, 63). A few opted for dates as early as 1592 and 1593 (Deschamps, *Les voyages*, 5; Slafter, *Voyages*, 1:19). Bishop, who also opted for 1594 on the basis of Pocquet's statement that Champlain "fought bravely at Crozon" in 1594, is the only biographer who questioned the date by adding that he "regretted" that the author did "not give his authority" (Bishop, *Champlain*, 15–16). No biographer has ever given a documentary "authority" for a date earlier than 1595. All have assumed that Champlain arrived with d'Aumont in 1594. Champlain never once wrote that he "fought" in Brittany, let alone that he fought the Spaniards, only that he "served" (*servy*) and was "employed" (*employé*) in the army of d'Aumont. At present there is not a scintilla of evidence that he arrived in Brittany before March 1595.

97 Biggar, *The Works*, 1:3. For the differences between the various ranks, see the next paragraph.

who was directly responsible to the king.[98] Under him were twelve maréchaux des logis du Roy, each of whom was responsible for a district or province within France and each of whom commanded four fourriers des logis du Roy. Hardy was one of the twelve maréchaux des logis, and Champlain one of his four fourriers.[99]

Under the direction of the grand maréchal, the maréchaux des logis and fourriers planned the route and organized the lodgings of the king and his court while they were travelling. The fourriers proceeded well ahead of the king and his entourage, marking the doors of potential lodgings with chalk according to the number of "fires," or hearths (*feux*), available in the buildings. The maréchaux des logis and their fourriers were also responsible for surveying and laying out temporary camps and barracks when regular lodgings were not available. As well, the maréchaux des logis du Roy were responsible for the lodgings of the army. Under their command were a number of maréchaux des logis de l'armée, who were in fact fourriers responsible for securing lodgings for cavalry units, while the fourriers de l'armée looked after the requirements of foot soldiers. Their field procedures were similar to those of the fourriers des logis du Roy, in that they marked the capacity of the buildings to hold men according to the number of "fires" they contained. In the thirty-four regional garrisons of Brittany in early 1595, there were twenty-two maréchaux des logis de l'armée, all of whom were assigned to cavalry units, and only six fourriers, all assigned to companies of foot soldiers. The maréchaux des logis and fourriers were not responsible for food procurement, which was in the hands of a *commissaire général des vivres* (commissioner general of provisions), who had his own assistants, including a fourrier who looked after their lodgings.[100]

In order to record their work properly, the *fourriers* were required to make maps containing roads, villages, and the chalk-marked buildings, roughly to scale and direction. They also had to be able to make maps of the surveyed camps and barracks. It is certain that Champlain had these skills before he joined the fourrière, perhaps through the tutelage he had received in the academy he attended. It may have been through his surveying skills and his court connections

98 The organization of the *fourrière* and functions of its officers are described in Guyot, *Traité des droits*, 1:612–17. See also entries in Ganeau, *Dictionnaire universel*, editions of 1721 and 1732. The grand maréchal in the 1590s seems to have been Charles de Neufville de Villery, marquis d'Alincourt, comte de Bury, *grand maréchal des logis de la maison du Roy et gouverneur du Lyonnais*. He was made *chevalier* in the Ordre du Saint Esprit in 1597.

99 In October 1595 two more of Jean Hardy's fourriers are named, Paul Hubert and René du Boys. Both received a monthly pay of 33⅓ écus (*Bulletin du Comité de la langue*, 481).

100 *Bulletin du Comité de la langue*, 482

de naissance that he was recommended and accepted as a fourrier in the fourrière, rather than just in the army. The sketch maps compiled by the fourriers and maréchaux des logis were later used to produce detailed maps of some parts of France.[101] A map that grew out of the Brittany campaign was the *Duche de Bretaigne, Dessigné par le Sieur Hardy Mareschal des logis du Roy*[102] (figs. 5 and 6). Since Champlain was a fourrier and aide to Jean Hardy, it is probable that he compiled some of the material for the map and aided in its execution. During Henri IV's reign and that of his successors, maps were considered to be secret documents, which accounts perhaps for the late publication date (1631) of the *Duche de Bretaigne*.[103]

The difference in pay between the different ranks in the fourrière is striking. According to the records of Gabriel Hus, Jean Hardy,[104] maréchal des logis du Roy, was stationed in Brittany by at least the beginning of January 1595 with a basic salary of 66⅔ écus per month.[105] On 3 October 1595 he was still in Brittany with the same pay.[106] A fourrier des logis du Roy such as Champlain received from 20 to 33⅓ écus per month, while the maréchaux des logis de l'armée, who did the work of a fourrier in a cavalry unit, each received 12 to 17 écus, and a fourrier de l'armée only four écus monthly. Considering that the maréchal des logis du Roy and his fourriers were non-combatants, their pay was high compared with that of men in the fighting forces, who were in considerable danger.

101 Buisseret, "Monarchs, Ministers, and Maps," 109–11

102 The map was published by Henricus Hondius as map 27 in the *Theatrum universae galliae*, printed in Amsterdam by Joannis Janssonius, 1631 (Koeman, "The Theatrum").

103 The link between Champlain, Jean Hardy, and the map *Duche de Bretaigne* was an exciting discovery, which we shared with David Buisseret in an e-mail of 15 April 2007. After viewing our supporting documentary evidence, he agreed with us that in his role as fourrier and aide to Jean Hardy, maréchal de logis du Roy, Champlain must have worked on the compilation and drafting of the base map from which the *Duche de Bretaigne* was engraved. Further, that in order to be entrusted with the responsibilities Champlain had at the time, it is likely that he was older that the fifteen years proposed by Jean Liebel ("On a vieilli Champlain," 236). Since the map is the work of an engraver as part of a series of maps in an atlas, it cannot be used as an indication of Champlain's drafting techniques or skills.

104 See standard genealogical tables for the Hardy family in http://perso.orange.fr/bernard.linais/index.htm and http://benoit.maury.geneal.free.fr/page001.htm, where he is listed as "HARDY Jean (né 1549 à Etampes, mort 13/07/1617 à Etampes à l'âge de 68 ans) Sieur de la Guinette, Maréchal des Logis de la maison du Roy, enterré dans l'église de Saint Basile d'Etampes. Marié ca 1580 à Etampes, avec Anne Naudet."

105 See: Document B, "Personnel and Pay Records," no. 2 (f. 185, Jan.–June 1595).

106 *Bulletin du Comité de la langue*, 2:481–2

The average monthly pay at Quimper in February 1595 for a member in a standard company of 50 *hommes de guerre montez et armez à la legère* (mounted and lightly armed soldiers), including the pay of four officers, was 10 to 12 écus per month, while a member of a standard company of 50 *gens de pied* (foot soldiers), including four officers, was about five écus per month.[107] Examining the entire pay record of the 34 garrisons listed in February 1595 where it is possible to ascertain the monthly pay issued to junior officers and soldiers, the pay of a fourrier de l'armée, at 4 écus per month, was the same as that of a *thambourin* (drummer) or a *phifre* (piper).[108] An *enseigne* (ensign) was paid 10 écus; a *sergent* (sergeant) 6 to 8 écus; a *corporal* (corporal) 6 écus; a *harquebusier à pied* (foot soldier armed with an harquebus) 4 écus, the same as a *salade* (cavalier); while a *harquebusier à cheval* (mounted soldier armed with an harquebus) received about 10 écus. More senior officers earned anywhere from 20 to 33 écus per month; for example, a *lieutenant* received 18 écus or more, and a *capitaine* anywhere up to 33 écus per month.[109] Champlain's pay as a fourrier des logis du Roy was therefore as high as that of a lieutenant and on occasions as high as that of a captain.

A closer examination of Champlain's wages (regular pay) shows that his monthly salary varied from 25 to 33⅓ écus per month over the time he was in Brittany in 1595:[110]

- 1 March to 30 April, a total of 66⅔ écus at 33⅓ écus per month; the same monthly pay as Paul Hubert and René du Bois, two of Hardy's other fourriers;[111]
- 1 May to 30 November, a total of 175 écus at 25 écus per month;
- On 28 December he received 25 écus for the month.

From March 1 to the end of December he thus received a total of 266⅔ écus for his wages, for an average of 26⅔ écus monthly. One of the more interesting aspects of Champlain's record is the extra pay he drew, totalling 42 écus:

107 Barthélemy, *Choix de documents*, 193–5

108 In 1562 the pay for a drummer (*thambourin*; modern: tambour de basque), piper (*phifre*), and *fourrier* (*fourryer*) was 12 livres each. Since one écu was worth three livres, the pay schedule had not changed in this instance between the armies of Charles IX and Henri IV (*Bulletin du Comité de la langue*, 3:544).

109 For the details of these generalizations, see Barthélemy, *Choix de documents*, 179–93. For an extract describing Quimper, see Document B, "Personnel and Pay Records," no. 1, "The King's Garrisons in Brittany, 1595, *Kempercorantin*."

110 See Document B, "Personnel and Pay Records," no. 2 (1 March to 28 December).

111 *Bulletin du Comité de la langue*, 481

- On 15 March he received 9 écus, for "a certain secret voyage which he made involving the king's service, by order of sieur mareshal [Jean d'Aumont]."
- On 31 March, he was paid 3 écus "to find Monsieur le mareshal [Jean d'Aumont] and for representing to him some matter involving the service of the king."
- Between 6 and 11 July he received 30 écus for "fire duties" – marking buildings according to the number of "fires" (hearths).[112]

Not only was Champlain a fourrier and aide to Jean Hardy, the maréchal des logis du Roy for Brittany, he was also delivering sensitive messages between the court of Henri IV and maréchal Jean d'Aumont, who was the senior commander of the Brittany operations.

What is clear from the pay records of the royal army in Brittany is that the army was paid from provincial taxes (*fouages*) raised by the États de Bretagne.[113] Unlike most of the French provinces, Brittany and a few other provinces were not taxed directly by the crown but maintained local autonomy in taxation. In the case of Champlain, the entry of the accounts under which his name appears makes it clear that he was not paid from the royal accounts but, like most of the army, from taxes raised by the États de Bretagne, whose treasurer, Gabriel Hus, issued Champlain's pay.[114] The États de Bretagne was the provincial government of Brittany, composed of members from the three États (estates): clergy (*l'église: premier état*), nobles (*noblesse: deuxième état*), and commoners (*tiers état*), the latter being people not from the clergy or aristocracy. Normally it met late in a year[115] to discuss upcoming expenses and to make estimates of the revenues needed to cover them through the tax system by assessing a tax (*fouage*) on some lands and every hearth (*feu*), a crude surrogate measure for population. Most of the fouages were collected from the commoners.[116] The États de Bretagne was in an unenviable position in that the duc de Mercœur and his agents were trying to raise money from it at the same time as the financial agents of his enemy Henri IV.

112 See, Document B, "Personnel and Pay Records," no. 2 (15, 31 March and 6–11 July).

113 See introduction to Document B, "Personnel and Pay Records," no. 2.

114 Ibid., see /[f. 192ᵛ].

115 The minutes of the États de Bretagne from 1491 to 1589 have been assembled and published by La Lande de Calan, *Documents inédits*. They are an interesting view into the operations of this system of government.

116 Chateaubriand, *Œuvres completes*, 247–8. This is a short but clear exposition of the hated *fouage* system.

The war in Brittany, 1595–1598

As 1595 progressed, Henri IV had further significant victories. In June he managed to defeat a much larger Spanish force at Fontaine-Française,[117] and on 18 September the Pope lifted Henri's excommunication. Both were blows to the Spanish and League causes.

Nothing is known about Champlain's activities in 1596. It seems likely that he was working for Jean Hardy on the map of Brittany – a major task of collating sketches to a common design, scale, and orientation. In the spring of 1597, Champlain was still present at the garrison in Quimper but was not listed as a fourrier. It is probable that the fourriers were no longer necessary, because the troops were stationary in their garrisons. Champlain was now listed as an ensign attached to the sieur de Milleaubourg[118] in the garrison at Quimper under the command of the governor, Julien du Pou, sieur de Kermouger.[119] Milleaubourg had been an officer in the Catholic garrison at Soubise[120] in 1585, when it was overwhelmed by the Huguenot army of Henri de Bourbon, prince de Condé. Subsequently, Milleaubourg joined the Catholic forces of François d'Espinay de Saint-Luc,[121] then governor of Brouage, and participated in the successful defence of that town in 1586.[122] When Saint-Luc took over the forces in Brittany from Jean d'Aumont after the latter died in August 1595, he may have brought some of his old comrades with him. It seems reasonable to suppose that Champlain, who had been born and reared in Brouage, was assigned to an officer who had served in the defence of that town. If Champlain was old enough in 1585–86 to have participated in the defence of Brouage, he would have known Milleaubourg and Saint-Luc. Unfortunately, we do not know what kind of army unit Milleaubourg commanded. If we did, we would know whether Champlain was in an infantry or

117 A small village with a château northeast of Dijon in the department of Côte-d'Or, Burgundy. For the significance of the battle, see Buisseret, *Henry IV*, 57–8.

118 Also Millaubourg and Millambourg. See Document B, "Personnel and Pay Records," no. 2 (2 April 1597).

119 Julien du Pou was governor of Quimper from 1592 to 1610.

120 Louis de Milleaubourg and his supposed relationship to Champlain is briefly mentioned in Vigé and Vigé, *Brouage*, 282. Soubise is a small town about 4 km southwest of Rochefort on the south side of the Rivière Charente, 8 km northeast of Brouage.

121 François d'Espinay, seigneur de Saint-Luc (1554–97), was an early follower of Henri IV. In 1592 he was appointed lieutenant general under maréchal Jean d'Aumont in Brittany. He became a chevalier du Saint-Esprit in 1595 and succeeded d'Aumont later in the year when the latter died at Comper. In 1596 Saint-Luc was appointed *grand maître de l'artillerie de France*, which ended his tour in Brittany.

122 Beauchesne, "Brouage," 59

a cavalry unit, since the rank of *enseigne* existed in both. Of the other senior officers mentioned at Quimper in 1597, both Kermouger and Champfleury had also been there in 1595. The sieur de Kermouger, besides being governor of Quimper and commandant of the garrison, commanded a company of cavalry, while the sieur de Champfleury commanded foot soldiers. Champlain's role at the garrison is therefore not clear. As ensign to Milleaubourg, he would have carried the standard of his military unit and served in a similar role as he had for Jean Hardy. The reference to Champlain's presence at Quimper in 1597 seems to imply that all the men listed were "*capitaines* and *commandants* in the companies established in the garrison."[123] The only commandant was Kermouger, and it is unlikely that an ensign could simultaneously hold the rank of captain.

Sometime during 1597 or earlier, Champlain may have met his *oncle provençal*, Guillaume Allene, who mentioned in his bequest (*donación*) to Champlain that he was in Quimper about that time to make a will.[124] If the two met, it is probable that through this visit Champlain learned that his uncle would be with the Spanish garrison in Blavet (Port-Louis). We know nothing more of Champlain and his activities until he surfaces the following year in Blavet. It may be that he participated in the successful defence of Quimper under a renewed attack by Guy Eder de la Fontenelle in early May 1597. A number of officers, such as capitaines Magence and du Clou, led by the commandant of Quimper, gouverneur Kermouger, were there with Champlain at the time and fought in this engagement.[125]

The year 1598 saw the end of the long series of religious wars that had plagued France since 1562. With most of the country now under Henri IV's control, the king's armies began to concentrate on the forces of the duc de Mercœur and the Catholic League in Brittany. On 18 February the king's army, estimated at about 14,000 men, left Paris for Brittany, while at the same time the chancellor of France, Pomponne de Bellièvre, and Marquis Nicolas Brûlart de Sillery began negotiations for a peace with Spain and the withdrawal of the remaining Spanish troops that supported the League.[126] Seeing that his position was hopeless, Mercœur submitted to Henri at Angers on 20 March and went into exile in Hungary. On 13 April, with Mercœur out of the way, Henri IV signed the Edict of Nantes, making Catholicism the official state religion but granting a large measure of religious freedom to the Huguenots, including the right to hold public office. The only choice now remaining to Spain was to enter into an agreement with Henri

123 See Document B, "Personnel and Pay Records," no. 2: Champlain *enseigne* (2 April 1597).

124 Written in 1601, Allene actually stated that "it could have been about four years ago." See Document C, "Gift from Guillermo Elena," last paragraph.

125 Baudry, *La Fontenelle*, 262–71

126 Knecht, *Rise and Fall*, 470; Buisseret, *Henry IV*, 69–76

that permitted it to withdraw its troops in peace. On 2 May 1598, Henri IV officially ended the war with Spain by signing the Treaty of Vervins. At that time, about 5,000 Spanish troops were still in the fortified city of Blavet, and about another 2,000 camped at the shipyards of Le Pellerin.[127] Under clauses XV and XVII of the treaty, the troops at Blavet were permitted to return to Spain with their arms and munitions within three months after the treaty was signed.[128] Peace having been declared, Henri IV's army could be scaled down, and some of the troops were disbanded.

Blavet to Cádiz and the West Indies, 1598–1601

The principal document that has hitherto been used to trace Champlain's life from the end of the religious wars in 1598 to his first journey to Canada in 1603 is the manuscript *Brief discours*, which is attributed to him.[129] This document, detailing his voyage to the West Indies on a Spanish ship, with accompanying illustrations, exists in three manuscript copies of an unknown original. These three manuscripts are located in Providence, Turin, and Bologna, all written in different hands, none of which resembles Champlain's.[130] Although the *Brief discours* was known to contain factual errors, no one seriously doubted that Champlain was the author until about the middle of the twentieth century. Increasingly, however, scholars have had serious reservations about the accuracy of some of the facts stated in the document.[131] By now it is generally assumed that Champlain did not write or author the *Brief discours*, but it is thought that it may have been written by a scribe who had access to notes made by Champlain after his return from the voyage or notes made from a verbal report.[132] It is doubtful that Champlain would have taken the risk of keeping a diary and making drawings and maps while on a Spanish ship, when it was well known that Spain liked

127 Blavet is now Port-Louis in the municipality of Morbihan, Brittany. Le Pellerin is a small town on the south bank of the Loire, about 15 km west of Nantes.

128 *Articles Accordés*, 15:13, 17:14

129 Biggar, *The Works*, 1:3–80

130 For an analysis of the three variant texts of *Brief discours*, see Giraudo, "The Manuscripts" and "Research Report."

131 For critical reviews that range from politely pointing out errors in the document to calling it a fake and questioning Champlain's veracity, see Bishop, *Champlain*; Bruchesi, "Champlain"; Vigneras, "Le voyage"; and especially Codigliola, "Le prétendu voyage."

132 For a recent, more sympathetic analysis of *Brief discours*, see Gagnon, "Is the *Brief Discours* by Champlain?" As this book was going to press, David Hackett Fischer published a fairly convincing reconciliation between the details from the *Brief discours* and the evidence from Spain (Fischer, *Champlain's Dream*, 74–100).

to keep these sensitive voyages secret. It is more likely that the notes were made when he returned from the voyage, before he gave a verbal report to Henri IV. Perhaps such a set of notes was used to construct the *Brief discours*, and missing details were then filled in from other accounts or from the scribe's imagination. Whatever the case, caution must be exercised in using any information given in this manuscript unless it can be corroborated by other documents. There is no reason to doubt that Champlain had undertaken such a trip, because he mentioned having been to the West Indies on several occasions; the question is whether the contents of the *Brief discours* wholly reflect his experiences.[133] What is of importance here is not the account of his voyage through the West Indies but his trip from Blavet to Spain, his sojourn in Cádiz before and after he returned from the West Indies, and his return to France.

As some of the French troops were being demobilized, Champlain was among them, and he made his way from Quimper to the Spanish garrison in Blavet to meet his uncle, Guillaume Allene. On 3 July, General Pedro de Zubiaur[134] arrived at Blavet from Spain as commander of three ships. Since this was not nearly enough to complete the evacuation, fifteen additional ships had to be leased locally. Through the maréchal de Brissac, who was in charge of the French forces supervising the evacuation, two ships were leased from Julien de Montigny de la Hautière after an inspection of ships at Blavet by a Spanish captain, Domingo Martin, and *l'alfere Arellane*.[135] The ships chosen were the *San Julián*, a *hourgue* of 500 tuns,[136] and a second ship, *Le Jacques*, of 100 tuns. The circumstances in

133 For example, the dedicatory poem by De la Franchise in *Des Sauvages*, which boldly states, "He saw Peru, Mexico, and the marvel / of infernal Vulcan." See Document G, "*Des Sauvages*." Also Document F, "Excerpts from Champlain's Works," no. 7.

134 Don Pedro de Zubiaur (c. 1541–1605) (Çubiaur, Soubriago, Subiaure), *General de las Armadas del Océano* of Spain, also *Almirante de la Mar Océano* in 1597 (Garcias Rivas, "En el IV centenario," 157–71). The scribe of the Providence manuscript spells the name "Soubriago" (Biggar, *The Works*, 1:10). The date of arrival at Blavet is given in Vigneras, "Le voyage," 166.

135 *Alfere* (alférez) is an ensign or second lieutenant, the lowest rank of commissioned officers. In view of the many different ways Champlain's uncle Guillaume Allene's name has been spelled in contemporary documents, this is probably another reference to him. At the time he was captain of the *San Julián* and therefore familiar with her capabilities. Julien (Jullian) de Montigny, sieur de la Hautière (Hottière, Haultière), then living in Vannes, was an ardent supporter of the Spanish and the League (Carné, *Documents sur la ligue de Bretagne*, 12:159, 177).

136 The *San Julián*, is described in the document as a *hourgue* (*houcre*) or "hooker" in English. The sixteenth-century Dutch hooker was a small, open, merchant or fishing vessel used in coastal waters. It had a main and mizzen mast and ranged from 60 to 200 tuns

which La Hautière acquired the *San Julián* are not known. On 15 July de Brissac entered into an agreement with La Hautière to lease the two ships from him as troop transports for 300 Spaniards, with their armaments, for the sum of 40 réaux per tun daily, but he himself had to assume responsibility for handling the crew and equipment.[137] Once the lease was signed, in the knowledge that the *San Julián* was going to be captained by his uncle and proceed to Spain, Champlain most likely conceived the idea of going along and using the opportunity to gather intelligence for Henri IV. It must have taken some time to load this large ship. The manifest shows that besides the 300 men, their officers, and their equipment, the *San Julián* loaded munitions and 65 cannons.[138] On 23 August, Zubiaur's fleet of eighteen ships set sail from Blavet.[139] After rounding Cape Finisterre and a stopover at Vigo, the fleet reached Cádiz on 14 September.[140] Upon arrival, Zubiaur learned that the English had taken the Spanish fortress at Puerto Rico and that he was appointed to command an armada to retake it.[141] Now rapid plans had to be made to put a fleet together and dispatch it to the island. While the rest of the French ships were unloaded and returned to Blavet, the *San Julián* was requisi-

burthen. In view of the stated capacity of the *San Julián* (500 tuns) and the task it was hired to perform, this ship could not possibly have been a *hourgue*. Perhaps the smaller *Le Jacques* was meant or the term *hourgue* had a wider meaning. A Spanish document gave Allene as the captain of *felibote San Julián* (Giraudo, "Research Report," 96). *Felibote* is the Spanish term for the *fluyt* (flûte), a type of ship of Dutch origin but widely used in other navies. The flûte was a cargo vessel, ship-rigged on three masts, with a rounded stern and lightly armed to create more room for cargo. If the *San Julián* was a flûte, it was a very large one, because these vessels rarely had capacities greater than 300 tuns. Aubin mentioned that the flûte was used in the fleets to the Indies (Kemp, *Oxford Companion*, 318, 395; Aubin, *Dictionnaire*, "Flûtes," "Hourque").

137 The author of the *Brief discours* stated that de Brissac seized the ship (*pris et arresté*), but this is incorrect; the ship was seized once it reached Spain. He also incorrectly stated that the Spanish transports were commanded by Champlain's uncle; Allene commanded only the *San Julián*, General Zubiaur commanded the Spanish fleet (Biggar, *The Works*, 1:5). See also Document F, "Excerpts from Champlain's Works," no. 1.

138 Carné, *Documents sur la ligue de Bretagne*, 12:162–3. In the *Brief discours*, it is stated that 46 cannons from Blavet were delivered to Puerto Rico (Biggar, *The Works*, 1:22).

139 Vigneras, "Le voyage," 168. The author of *Brief discours* gave the date of departure as *commencement du moys d'aoust*, the beginning of August (Biggar, *The Works*, 1:5). A letter written by Julien de Montigny, however, put the date of departure of "his ships" as 9 September, 17 days later (Carné, *Documents sur la ligue de Bretagne*, 12:162n2).

140 Vigneras, "Le voyage," 168

141 Giraudo, "Research Report," 95

tioned by Zubiaur for the trip to the West Indies, in violation of the contract with La Hautière.[142] The ship was well built and already had on board the arms, munitions, and cannons necessary to re-equip the fortress. On 12 October the *San Julián* was sent from Cádiz up the Guadalquivir River to Sanlúcar to be careened and outfitted for the journey.[143] This "seizure" resulted in a lawsuit by La Hautière against the Spanish government. Zubiaur however, claimed that La Hautière was not the rightful owner of the ship, that in fact the *San Julián* belonged to a man called Landricart, in Brittany, from whom La Hautière had taken it during the recent war.[144] Although Zubiaur described the ship as having been "taken from the French," he may have felt justified in seizing the ship, or simply did not care, because Allene claimed that he had full powers to act for Landricart and that by virtue of these powers the ship could be freighted for service to the West Indies. The ownership of the *San Julián* is mentioned in the 1601 *donación* from Allene to Champlain, in which Allene wrote that he owned one-eighth of the ship and "*Senor de Landricart Endebanes en Bretania*" the remainder.[145]

One cannot help but wonder about the role of Allene in all this. He must have known the ownership complications of the *San Julián* when he helped select the ship with Captain Martin in Blavet, because he already owned part of it. He must also have had the power to act for his partner Landricart to reassert the ownership of "their" ship when he sailed it into Spanish waters. It is probable that Allene told Zubiaur about its ownership and therefore facilitated the seizure of the valuable 500-tun vessel in the hope that it would be turned over to him and Landricart at a later date. It could only have been after the Spanish seizure of the ship was completed and the orders given to set sail for Puerto Rico that Champlain found out that the ship was actually going to be sailing to the West Indies.

When confirmation came on 20 November that the English fleet had left Puerto Rico, General Zubiaur was no longer necessary as a commander at the head of an armada. He was replaced by Don Francisco Coloma, who was charged with the task of replenishing Puerto Rico and making the annual circuit of the

142 Ibid., 170. The scribe of the *Brief discours* wrote that Zubiaur had "engaged [the ship] for the service of the king of Spain" (Biggar, *The Works*, 1:7).

143 Giraudo, "Research Report," 170

144 Ibid. Repeated attempts to identify Landricart Endebanes of Brittany have failed. It is therefore difficult to say why his ship was confiscated by La Hautière during the war and why Zubiaur took it from La Hautière, a League and Spanish sympathizer.

145 Document C, "Gift from Guillermo Elena"

West Indies to transport gold and silver bullion to Spain.[146] A Spanish captain, Yeronimo de Vallebrera, was appointed to the *San Julián* to replace Allene.[147] On 3 February 1599 the *San Julián* left Sanlúcar as a part of Coloma's fleet on its way to the West Indies, presumably with Champlain on board.[148] Allene was paid 300 réaux on 26 February 1599 for having sailed the *San Julián* from Blavet to Cádiz,[149] and later in the year when Zubiaur was named commander of a Spanish fleet headed for Gibraltar, he chose Allene as captain of one of his ships.[150]

The details of the voyage to the Indies, especially the known facts from Spanish sources as compared with the *Brief discours*, have been discussed by others in great detail.[151] The role of Champlain on the voyage is not clear. The author of the *Brief discours* wrote that Champlain's uncle had "committed the charge of the said ship to me [Champlain] to watch over it" (*me commist la charge dudict vasseau pour avoir esgard à iceluy*).[152] This does not imply, as some have suggested, that Champlain commanded the *San Julián*; he simply did not have the training for such a task, nor would Spain have permitted it. The French wording suggests that if he had any role at all, it was looking after his uncle's interests in the ship.[153]

Although one cannot be certain about any aspects of Champlain's early life, one thing that seems clear is that this young man, with a good background and an interest in geography and field surveying, also acquired some knowledge of navigation while on Spanish ships. There is no other way to explain his consistent use of the Spanish marine league on his small-scale maps in the navigational examples in his *Traitté de la marine* (1632) and in references in his other books, where Frenchmen would have used French measures.[154] The only text on navigation he referred to was Pedro de Medina's *Arte de navigar*,[155] though this may

146 Vigneras, "Le voyage," 171. The circumstances surrounding the involvement of Spain in the taking, outfitting, and departure of the *San Julián* for the West Indies is different in the *Brief discours*.

147 Ibid.

148 Ibid., 174. The author of the *Brief discours* wrote that the fleet left at the beginning of January.

149 Ibid., 172; Giraudo, "Research Report," 95–6

150 Vigneras, "Le voyage," 172

151 See especially Vigneras, "Le voyage," and Gagnon, "Is the *Brief Discours* by Champlain?"

152 See Document F, "Excerpts from Champlain's Works," no. 2.

153 Nicot, *Thresor de la Langue francoyse*, 251a. This is also the opinion of Gagnon, "Is the *Brief Discours* by Champlain?" 87.

154 There are 15.5 Spanish marine leagues to a degree of latitude, while the French marine league consisted of 20 to a degree (Heidenreich, "Measures of distance," 122, 133; Heidenreich, *Explorations and Mapping*).

155 Biggar, *The Works*, 6:322

not be significant, because the text appeared in five French translations between 1554 and 1579, as well as in Italian, Dutch, and English.[156] Champlain's persistent use of the Spanish marine league to the exclusion of the French marine league is significant, however, because it suggests that he became used to estimating distances on the high seas in the Spanish marine league and could not or would not readjust to the shorter French league. It is probable, therefore, that by the time he returned to France from the West Indies, he had taken the opportunity to learn the basic principles of navigation and had learned them by observing Spanish navigators and reading Spanish texts.

Like almost everything about Champlain's West Indian voyage, the date of his return is controversial. The author of *Brief discours* stated that the *San Julián* returned two years and two months after leaving Sanlúcar, while Champlain wrote in 1632 that he returned "nearly two and a half years after the Spaniards left Blavet."[157] These references would put his return sometime between early February and early April 1601.[158] He was definitely back before 26 June 1601, when he signed his uncle's *donación* as a witness.[159] Previous writers on the subject have assumed that the *San Julián* returned with one of the Spanish fleets because it is so stated in the *Brief discours*.[160] Because Francisco Coloma's fleet returned to Sanlúcar on 23 February 1600[161] and other Spanish convoys returned on 11 August 1600 and 4 January 1601,[162] the dates given by Champlain and the author of the *Brief discours* were discounted as errors. Like Morris Bishop, most historians argued for the return of the *San Julián* in August 1600, but this seems unlikely. More recent research shows that the ship had to undergo extensive repairs at Veracruz on the coast of Mexico in June 1599 and in so doing lost contact with Coloma's fleet. The *San Julián* may even have been sold, the captain and crew being transferred to other ships.[163] What happened after that is contradictory, but early in 1601 the ship was back in Spain with Champlain on board.

156 Waters, *The Art of Navigation*, 82, 163nn2, 3
157 Document F, "Excerpts from Champlain's Works," nos. 3 and 7
158 Given three dates of departure from Blavet – early August, 28 August, and 9 September (see note 139 above) and two dates for the departure from Sanlúcar, beginning of January and 3 February (see note 148 above), and a return to Spain after 2 years 6 months from Blavet and 2 years 2 months from Spain, it seems that the *San Julián* returned sometime between early February and early April 1601.
159 Document C, "Gift from Guillermo Elena"
160 Biggar, *The Works*, 1:76
161 Vigneras, "Le voyage," 192
162 Bishop, *Champlain*, 341
163 Vigneras, "Le voyage," 188; Giraudo, ""Research Report," 95. In the *Brief discours* the author wrote that the ships were careened at Saint-Jean de Luz (San Juan de Ulúa) near

When Champlain returned to Spain he made his way to his uncle Guillaume Allene, who was living in Cádiz, and on 26 June 1601, Allene thanked Champlain for looking after him during his illness by presenting him with the *donación*, and also charged him with selling Allene's share of the *San Julián* and its cargo in order to pay off the sailors.[164] On 3 July Champlain gave a power of attorney (*poder*) to the lawyer who had witnessed the *donación* a week earlier to effect the sale of his uncle's interests in the ship.[165] Although Allene was in the Madeira Islands in August 1600,[166] it is likely that if Champlain had returned about that time, as others have proposed, the order to sell Allene's share of the ship and the cargo to pay off the sailors would have occurred much earlier than 3 July 1601. It seems, therefore, that Champlain returned to Sanlúcar, at the mouth of the "River of Seville,"[167] and made his way to Cádiz to visit his uncle sometime between late February and early April 1601. He then looked after his ill uncle and his affairs, and on 26 June he received the *donación*, a substantial gift that included Allene's estate in the neighbourhood of La Rochelle, his one-eighth interest in the *San Julián* and its cargo from the West Indies voyage, and whatever Champlain could retrieve from the lawsuit his uncle had relating to a ship in San Sebastián.[168] Although he could not take possession of these gifts until his uncle died, he was charged with the immediate sale of his uncle's interest in the *San Julián* and its cargo to pay the sailors. Since Allene had also charged Champlain to take over the responsibility from his former hostess (innkeeper), Maria Augustín, for settling the lawsuit with Ojelde Zalan (Ogero Challa?), it is probable that Champlain travelled to San Sebastián in order to do so.

It is generally assumed that Allene succumbed to his illness in 1601. The last reference to him, however, is a command by Henri IV, written on 20 September

Veracruz. He says nothing about the *San Julián*. After being careened, these ships rejoined the rest of the fleet at Havana, in Cuba, for the journey to Spain (Biggar, *The Works*, 1:70–1).

164 Document C, "Gift from Guillermo Elena"

165 Ibid., Supplement to the *donaçión*, 3 July 1601

166 Vigneras, "Encore," 547

167 Biggar, *The Works*, 1:80. The "River of Seville" is the Guadalquivir River.

168 Ibid.; also Document C, "Gift from Guillermo Elena," paras. 1–3. The reference to the ship in San Sebastián tied up in a lawsuit may be Allene's share of the 150-tun hooker *Adventureuse*, which he sold to Ogero (Ojelde?) de Challa (Zalan?) de Byarriz (of Biarritz) in 1572 at La Rochelle (Le Blant and Baudry, *Nouveaux documents*, 1:7). San Sebastián is a fishing port in the Spanish Basque country, about 10 km from the French border and 50 km from Biarritz, in the French part of the Basque country.

1604 to the citizens of Metz to permit the bailiff, sieur de Vitry,[169] to confiscate "the goods of the late *capitaine provençal*," who had been "condemned to death by sentence of our bailiff de Vitry the previous year [1603]."[170] There is no indication of the nature of the offence, nor that he was executed. It is doubtful that Allene ever returned to France after he left Blavet in 1598. His involvement in the alienation of the *San Julián* from France to Spain, his residence in Spain after 1598, and his service under Pedro de Zubiaur after 1599 were probably enough to merit a death warrant and confiscation of his property. If Zubiaur returned to one of his earlier occupations, as a corsair who harassed English, Dutch, and French Huguenot shipping, Allene's participation as captain on one of Zubiaur's ships would alone have been enough to have him condemned to death.[171] It may also be that a death warrant was issued in order to carry out the confiscation of his property in return for the "theft" of the *San Julián*. Allene's conversion to Spain seems to have been complete by the time he dictated and signed the *donación*, in which he even refers to the Spanish monarch Felipe II as "our lord the king."[172] One wonders if the wily old veteran of many intrigues and shifting allegiances gave his estate near La Rochelle to Champlain in order to keep it safe from probable confiscation for his misdemeanours. Certainly, by 1601, Allene must have been aware of his nephew's relationship to Henri IV and probable immunity from prosecution. On 29 December 1625, Champlain gave Allene's estate near La Rochelle to Charles Leber, sieur du Carlo, a gift that du Carlo accepted on 19 February 1626.[173]

Exactly when Champlain returned to France from Spain is not known. His last documented act in Spain took place on 3 July 1601, when he signed a power of attorney to order the sale of his uncle's assets in the *San Julián*. In view of Allene's illness, it is probable that Champlain remained in Spain to settle other aspects of his uncle's estate before returning to France. In contrast to Allene, Champlain seems to have circulated among the French and Spanish with impunity. It is likely that he was conversant enough with Henri IV and the French court

169 This is probably Louis de l'Hôpital, sieur de Vitry, an early supporter of the League and governor of Meaux, who changed sides in January 1594 by surrendering Meaux to Henri IV. He became *capitaine des gardes du corps du Roy*, charged with executing the king's orders (La Force, *Mémoires authentiques*, 1:143).

170 Henri IV, *Recueil des letters*, 6:294. There is no reason to suppose that the *capitaine provençal* named in this letter is someone other than Champlain's uncle, Guillaume Allene.

171 Vigneras, "Le voyage," 171

172 Document C, "Gift from Guillermo Elena," note 23

173 Leymarie, "Inédit," 83–5

to enjoy their protection or at least to remain free of suspicion of any defection to the Spanish.

We do not know whether Champlain did anything between the time he dealt with his uncle's affairs and his reappearance at the court in Paris. One of his biographers wrote that he worked as an *armateur*, an outfitter of ships, and another that he was a *géographe du roi* working as a surveyor at the Louvre, but there is no proof for these statements.[174]

2 LA NOUVELLE FRANCE, OU CANADA

The St. Lawrence Valley in the early sixteenth century

Before European contact, the St. Lawrence Valley was occupied by two Iroquoian-speaking groups named by modern ethnologists and archaeologists after their principal villages, Hochelagans and Stadaconans. The names they called themselves are not known. Collectively they are now termed the St. Lawrence Iroquoians.[175] The heavily fortified village of Hochelaga, with about 50 large longhouses composed of perhaps 1,000 people, stood on Montreal Island.[176] It is likely that their territory stretched from the mouth of the Ottawa River to the entrance of Lac Saint-Pierre. Although they practised agriculture, the Hochelagans were also heavily involved in fishing and to a lesser degree in winter hunting. On the north side of the river between Portneuf and Cap Tourmente were seven small unfortified villages, with a population of perhaps 3,000, whose main village was Stadacona, at the present site of Quebec City. Cartier named this area the "province of Canada" after the St. Lawrence Iroquoian word for "village."[177] These people ranged approximately from Lac Saint-Pierre to the Saguenay River, with some fishing expeditions as far as Gaspé and the north shore of the St. Lawrence east of

174 Document A, "Early Biographies of Champlain," no. 7; and Fischer, *Champlain's Dream*, 108. For the argument that Champlain was not a *géographe du roi*, see appendix 2, especially notes 9 to 12.

175 This section is based to a large extent on Heidenreich, "History of the St. Lawrence–Great Lakes Area." The term "Iroquoian" refers to the broad linguistic family of Iroquoian speakers. This term should not be confused with "Iroquois," which is the collective name for the Iroquoian-speaking Five Nations. The St. Lawrence Iroquoians – *Ouendat* (Huron), Five Nations Iroquois, and others – were separate political and territorial entities, although they had similar cultures, a common origin, and spoke similar but not identical languages.

176 Biggar, *Jacques Cartier*, 156

177 Ibid., 245

Anticosti Island. Their subsistence economy was based less on horticulture than on fishing, sea mammal hunting, and the winter hunting of land mammals. It is probable that these villages were largely independent, though Cartier implied that Stadacona exercised some influence on the four villages downstream.[178] When the French arrived in the mid 1530s, the Stadaconans and Hochelagans were at peace with each other but were fighting with some of their neighbours: the Hochelagans with the *Agojuda*,[179] who attacked down the Ottawa River, and the Stadaconans with the *Toudaman*, who were probably the Mi'kmaq.[180] Further warring was described up the Richelieu River and south of Lake Champlain.

The intertribal situation gleaned from Cartier's writings and from archaeological evidence suggests that well before the sixteenth century, warfare had broken out in the Lower Great Lakes area, producing a movement of people from the eastern end of Lake Ontario into the St. Lawrence Valley, from the north shore of Lake Ontario into the southern Georgian Bay–Lake Simcoe area, and along the north shore of Lake Erie towards Lake St. Clair and the west end of Lake Ontario, creating large no man's lands between the warring groups.[181] These wars were primarily between the various Iroquoian-speaking agricultural communities in the Lower Great Lakes area and between some of them and groups in the Upper Susquehanna River area and the agriculturally based Algonquian speakers of the Michigan peninsula. The Algonquin and Montagnais bands of hunters and fishers who later allied themselves with the French were probably involved on the side of the Iroquoians north of the Lake Erie, Lake Ontario, and St. Lawrence waterway, but the evidence is not conclusive. By the late sixteenth century, alliances and enemies had hardened into the patterns so well described by Champlain and his successors.[182]

Jacques Cartier in the St. Lawrence Valley

Direct contact between the Native peoples west of the Atlantic coast began in 1534 with the first voyage of Jacques Cartier. The Montagnais in the Strait of Belle

178 Ibid., 161

179 The *Agojuda* (*mauvais gens*) were likely the *Ouendat* (Huron), judging from the slat armour they wore. Both Champlain and Sagard describe Huron armour in identical terms to Cartier's for the *Agojuda* (Biggar, *The Works*, 3:77, 135; Wrong, *Long Journey*, 154).

180 On his voyage of 1534, Cartier was greeted by the Mi'kmaq with the words "*Napou tou daman asurtat*" (Biggar, *Jacques Cartier*, 50). It is likely that the French called them *Toudaman* from part of the greeting.

181 Heidenreich, "History of the St. Lawrence–Great Lakes Area," 476–8

182 Ibid.

Isle were familiar with French fishermen, but the Mi'kmaq of the southern Gulf of St. Lawrence and the interior Stadaconans and Hochelagans were not.[183] On rounding Miscou Island and entering Chaleur Bay, Cartier encountered several hundred enthusiastic Mi'kmaq in their canoes, waving skins at the ends of their paddles. When Cartier eventually landed, their behaviour suggests that they had heard of Europeans, but most, if any, had never seen them before. As in other instances of first contact, the Mi'kmaq tried to barter whatever they had, mainly food and clothing, for whatever the French were willing to exchange. Cartier, however, had come not to trade but to find a route westward to the Orient. He bartered and gave gifts to establish friendly relations but regarded the entire transaction as *peu de valleur* (of small value).[184] At Gaspé he met a different group of people, the Stadaconans who had gone there to fish. They were very apprehensive of the Europeans, indicating no prior knowledge of them. Cartier gave some gifts, raised a cross (to the annoyance of their headman Donnacona), "detained" his two sons, and sailed away.[185] After leaving the Gaspé, he noted the mouth of the St. Lawrence north of Anticosti Island, which he thought might be the sought-after passage westward.[186] Eventually, Cartier circumnavigated the Gulf of St. Lawrence and returned to France.

On his second voyage the following year, Cartier increasingly became obsessed with discovering a route westward and finding precious minerals. When he arrived at Stadacona and returned Donnacona's sons, he was offered a closer relationship with the villages in the area. The Stadaconans were impressed by French arms, and it is probable that they were seeking an alliance. Deeply distrustful of anything his hosts proposed, Cartier failed to see the importance of this gesture. Consequently, his subsequent visit to Hochelaga without the involvement of the Stadaconans deepened their distrust towards the French, which not even gift giving could overcome. The winter at Cartier's settlement Sainte-Croix on the Saint-Charles River became a gruesome experience when almost all of the crew became deathly ill and 25 of the 110 men died of scurvy before the Stadaconans

183 The behaviour of these groups when they first met Cartier's expeditions were quite different. While the Montagnais stepped on board the ships with a great deal of familiarity, the Mi'kmaq greeted them with expressions of joy, dancing, singing, and touching. The Iroquoians at the Gaspé, however, greeted Cartier's crew initially with deep suspicion; but when the French visited them a year later, they were greeted in a similar manner as they had been greeted by the Mi'kmaq (Biggar, *Jacques Cartier*, 53, 55–6, 60, 76, 125, 150–1, 162–3).

184 Ibid., 52, 56

185 Ibid., 64–7

186 Now called Détroit de Jacques-Cartier

could show Cartier a cure. Cartier attributed the cure not to the Stadaconans but to "God in His divine Grace who had pity on us."[187] Unfortunately, he never described the tree *annedda*, whose branches had cured them.[188] This was one of his many failings as an explorer, for had he done so he could have saved hundreds of lives during the early settlement of Canada.

Once spring arrived, the Stadaconans began to fuel Cartier's obsession about precious minerals by elaborating earlier stories of a wealthy "Kingdom of Saguenay" in the northern interior, about one month's journey west-northwest of the mouth of the Saguenay River, but also approachable up the Ottawa River.[189] It is probable that they concocted the stories to get rid of the French menace in their midst – the heavily armed strangers who kidnapped people and could not be secured by an alliance. Unfortunately, these stories led Cartier to kidnap Donnacona, his two sons, and seven others before departing for France, where the captives were expected to repeat their knowledge of Saguenay to King François I.

Upon his return to France, Cartier related his discoveries, and Donnacona relayed to the king his stories of the wealthy Kingdom of Saguenay. João Lagarto, a Portuguese navigator who was present at the interview, was skeptical of Donnacona, saying that the "Indian King" was only telling these stories "so as to return to his own land."[190] Cartier, François I, and the rest of their entourage, however, believed Donnacona. None of the Stadaconans ever returned to Canada.

War between France and Spain (1536–38) delayed the next voyage until 1541. Cartier was ordered to explore the route to the Kingdom of Saguenay and act as a guide for Jean-François de La Rocque de Roberval, his superior in the venture, who was to found a colony, begin missionary work, and effect the conquest of Saguenay.[191] Unlike the previous voyages, this one was organized like those of the Spanish conquistadores and included 300 soldiers as well as cannons.[192] Cartier left Saint-Malo with 5 ships and 400 men on 23 May 1541 and reached Stadacona on 23 August. Because of organizational difficulties, Roberval was delayed until 1542. From Stadacona, Cartier moved upstream to the Rivière du Cap

187 Biggar, *Jacques Cartier*, 204–15

188 *Annedda* was probably an antiscorbutant, such as the eastern hemlock (*Tsuga canadensis*) or white cedar (*Thuja occidentalis*).

189 Biggar, *Jacques Cartier*, 170, 200–1, 221–2

190 Biggar, *A Collection of Documents*, 79

191 Ibid., 178–85

192 Ibid., 275–9. The sheer incompetence of Cartier is breathtaking. After two previous voyages to Canada, how could he have thought of moving an army and siege cannons through the rugged, heavily forested lakes and river system of the Laurentian Shield, north of the St. Lawrence River?

Rouge, where he built a new settlement, Charlesbourg-Royal. During his stay, he was plagued by increasingly deteriorating Native relations and a severe winter. Basque fishermen, who came in contact with some St. Lawrence Iroquoians in the Gulf of St. Lawrence late in 1542, claimed that in open fighting with Cartier's men the Natives had killed 35 of them.[193] Cartier departed for France in the spring with a load of "diamonds" and "gold" he had found near his settlement. In the harbour of St. John's, Newfoundland, he met up with Roberval, but Cartier refused to return to Canada. Roberval continued in his three ships with 200 to 300 people, ranging from released prisoners to courtiers, including women.[194] Upon reaching Charlesbourg-Royal, he enlarged it and built a second fort at the waterfront; collectively, these forts were called France-Roy.[195] After a disastrous winter, during which he had to quell uprisings and suffered the loss of a quarter of his men to scurvy,[196] Roberval decided to head home. On the way, he tried to explore north on the Saguenay River, but this effort came to naught. Both Roberval and Cartier discovered, when they returned to France, that their "diamonds" and "gold" were quartz crystals and iron pyrite.

It is clear that the Cartier-Roberval expeditions returned to France with a poor image of Canada. The country had terrible winters that could kill you. There was no easy wealth in minerals, and not only were the Natives hostile, but they were impoverished and had nothing worthwhile to trade. In short, unlike Mexico or Peru, this was a land from which little profit could be derived.[197]

Free trade versus monopolies under Henri III and IV

Following the Cartier-Roberval fiasco, France lost interest in Canada except as a place for fishing and for hunting sea mammals. As a consequence of the Cartier-Roberval voyages, the St. Lawrence Iroquoians were openly hostile to European visitors. According to André Thevet, some of Cartier's men had wilfully and cruelly murdered a number of Natives, an act that had resulted in retaliation.[198]

193 Ibid., 463

194 Biggar, *Jacques Cartier*, 263

195 Ibid., 266

196 Ibid., 267–8

197 These sentiments are paraphrased from captions on a map by Pierre Desceliers, "[World Map] Faicte a Arques, l'an 1550." Biggar (*Jacques Cartier*, plate XIV, 224) reproduced the part with the captions.

198 Schlesinger and Stabler, *André Thevet's North America*, 15, 42. Thevet claimed that he had visited Cartier for five months in his house in Saint-Malo and called him "my good friend" and "my great and intimate friend" (ibid., 48, 75, 82). If true, it may be that on

The long-term result of these murders was that the St. Lawrence Iroquoians, claimed Thevet, "have not allowed any Christian to approach and set foot on the ground, on their shores and boundaries, or do any business with them as we have since found out from experience"[199] ... "And you must not go to trade there except in good numbers since they still remember this outrage, and once offended it is impossible to conciliate them."[200]

With the virtual closure of the St. Lawrence by the Native people and the unfavourable image of Canada, European attention shifted to the maritime resources of the Gulf area. Spanish and French Basques opened up summer whaling stations and processing plants along the north shore of the Strait of Belle Isle. Others, mainly French Basques, Normans, and Bretons, engaged in fishing and sealing throughout the eastern reaches of the Gulf, especially its islands.[201] Although a small trade in hides and furs had probably been an incidental activity of some fishermen and whalers since early in the century, it was not until the late 1560s that a few ships were returning with profitable cargoes of fur, mainly from the Atlantic coast and eastern parts of the Gulf.[202] The initial trickle of furs seems to have led some merchants to take a renewed interest in Canada. In 1577 Henri III decided to explore the possibilities of a Canadian trade and the construction of settlements when he gave Troilus de La Roche de Mesgouez,[203] governor of Morlaix in Brittany, a monopoly on the trade and appointed him viceroy of New France. La Roche's plans, however, were put on hold after a couple of unsuccessful expeditions in the late 1570s and early 1580s, followed by the religious wars and his seven-year stint (1589–97) as prisoner of the duc de Mercœur.

Early in the 1580s, the St. Lawrence was again opened to trade "the olde mater beginning to grow out of minde ... and being drawen by gyfts ... they [the Natives] are againe to admit a traffique [trade]."[204] Prior to that time the Native groups in the St. Lawrence Valley had little direct contact with Europeans, which

these occasions he obtained additional information directly from Cartier that is not in the explorer's texts (ibid., 103). For a critical study of Thevet's veracity, see Lestringant, "Nouvelle-France et fiction."

199 Schlesinger and Stabler, *André Thevet's North America*, 15

200 Ibid., 42

201 Harris and Mathews, *Historical Atlas of Canada*, vol. 1, plate 25

202 The earliest reference to a trade in skins was written in 1539. It was not to furs but to "hides," which fetched ten cruzados each in Portugal (Biggar, *A Collection of Documents*, 78). For early trading, see Trudel, *The Beginnings of New France*, 57.

203 *DCB*, 1, "La Roche de Mesgouez, Troilus de"

204 Hakluyt, *Principal Navigations*, 8:145–6. This is no doubt a reference to the problems created by the Cartier expeditions.

accounts for the paucity of European trade goods on archaeological sites belonging to those people.[205] With La Roche safely out of the way in the duc de Mercœur's prison, the traders from Saint-Malo, Rouen, and Dieppe had a more or less free rein on the infant fur trade. In 1581 a small Breton ship made a profitable voyage to the Gaspé, followed by a similar voyage to the coast of Acadia in 1582.[206] In 1583 three ships from Saint-Malo traded at the Lachine Rapids, the first known Europeans in the area since Roberval departed.[207] Unfortunately, Cartier's nephew, Jacques Noël, who was among those traders, did not say who the Natives were or how much was traded. The same year, Étienne Bellenger, out of Le Havre, traded successfully between Cape Breton and the Penobscot River, though he experienced some Native problems along the coast of what is now Nova Scotia.[208] In 1584 five ships returned to Saint-Malo from "the Contries up the Bay of St. Lawrence" with such a large profit that ten ships were outfitted the following year, though it is not known how many actually sailed.[209] In 1587 rival traders fought each other, culminating in the destruction of three ships owned by Jacques Noël and his partners. On Noël's return to France, he petitioned Henri III to grant him a monopoly of the fur trade for twelve years in memory of his uncle, who had pioneered the connection to Canada, and in return for the 8,630 livres the king still owed to the Cartier family.[210] Unable to pay Noël the debt, Henri III granted him a twelve-year monopoly on 14 January 1588.[211] Seeing their trade threatened, the merchants of Saint-Malo put up a series of protests that lasted until the king revoked the monopoly on 9 July of the same year.[212] The free traders had won the round, and colonization was shoved aside for the time being.

It is probable that there was little trade on the St. Lawrence during this time. Most of the trade seems to have been in the Gulf area and upstream as far as Tadoussac. Indeed, Marc Lescarbot, writing in 1610, stated that before the voyages of the late 1590s, "Tadoussac had hardly been heard of," the Natives sought cod fishermen near Newfoundland to trade with them, and "none of our merchants had passed Tadoussac save Jacques Cartier."[213] Sometime before the 1580s the St. Lawrence Iroquoians who had populated the banks of the river from Quebec to

205 Jamieson, "The Archaeology of the St. Lawrence Iroquoians," 403

206 Quinn, *Voyages of Sir Humphrey Gilbert*, 2:363, 467; also Taylor, *The Original Writings*, 1:205–6

207 Taylor, *The Original Writings*, 1:205–6, 2:288

208 Quinn, "The Voyage of Étienne Bellenger," 328–43

209 Taylor, *The Original Writings*, 2:278

210 Trudel, *The Beginnings of New France*, 59

211 Ramé, *Documents inédits* (1865), 34–44

212 Biggar, *Early Trading Companies*, 185–6; Ramé, *Documents inédits* (1865), 48–51

213 Grant, *The History of New France by Lescarbot*, 3:4, 25

the Lachine Rapids had disappeared. Their former village sites, some of which have been excavated, do not contain European trade material, which suggest that their disappearance occurred before the beginning of trade in the early 1580s.[214] Champlain's observations in 1611 confirm this. When he was on Montreal Island, he noted that the land he saw had been cleared at one time but was now abandoned.[215] Remnant groups of the St. Lawrence Iroquoians have been traced by archaeologists to the Huron, Mohawk, Onondaga, and Abenaki.[216] It is not known why they were dispersed. Champlain and his contemporaries suspected intertribal warfare,[217] some of them blaming the Iroquois.[218] Some modern writers have suspected epidemic diseases, brought either through Cartier's voyages or by later traders.[219] This is unlikely since none of the St. Lawrence–Great Lakes Native groups had any collective memory of such diseases when they eventually broke out in 1634. Whatever the case may have been, the St. Lawrence Valley was depopulated west of Tadoussac by the early 1580s, forcing the traders to operate from Tadoussac eastward along the coast of the Gulf of St. Lawrence.

With the end of the religious wars and the accession of Henry IV, colonization took on a higher priority. Although commerce with the Natives and their conversion to Christianity were important, it is clear that the king had a long-standing interest in overseas settlement and was determined to erect colonies in Canada. Not only had he kept himself informed of overseas activities, but to him colonization became a priority that took precedence over trade. In his commission to Pierre Du Gua de Mons, issued on 8 November 1603, Henri, using the "royal we," made statements such as these:

> [W]e therefore, being long since informed of the situation and condition of the countries and territory of La Cadie …
>
> [H]aving long since seen, by the report of the ship-captains, pilots, merchants, and others who for many years have visited, frequented, and trafficked with various peoples of these parts, how fruitful, advantageous, and useful to us, our estates, and subjects would be the occupation, possession, and colonisation thereof …

214 Jamieson, "The Archaeology of the St. Lawrence Iroquoians," 403

215 Biggar, *The Works*, 2:176

216 Jamieson, "The Archaeology of the St. Lawrence Iroquoians," 402–4; Bradley, *Evolution of the Onondaga*, 83–9; Kuhn et al., "The Evidence for a Saint Lawrence Iroquois Presence," 77–86

217 Biggar, *The Works*, 2:176; 3:263

218 Grant, 3:117, 267–8; Jamet, "Relation du Père Denis Jamet," 350

219 Jamieson, "The Archaeology of the St. Lawrence Iroquoians," 402

You shall especially people, cultivate, and settle the said lands as promptly, carefully, and wisely as time, place, and circumstances will permit …

You shall cause to be built and constructed one or more forts, public places, towns, and all such houses, dwellings, and places of habitation, ports, harbours, shelters, and barracks as you shall deem fit, useful, and necessary for the execution of the said enterprise …[220]

In these endeavours, Henri IV had little support from his *conseil des affaires*, the notable exception being councillor Pierre Jeannin.[221] In fact, some of his chief ministers, such as the duc de Sully[222] and the sieur de Villeroy,[223] were openly critical. In his *Mémoires*, Sully mentions his opposition to the thinking and edicts late in 1603 that led to the colonizing efforts of de Mons: "… *la navigation du sieur de Monts pour aller faire des peuplades en Canada, du tout contre vostre advis, d'autant, disiez-vous, que l'on ne tire jamais de grandes richesses des lieux au dessus de quarante degrez.*"[224, 225] ("… the voyages of the sieur de Monts to go

220 Grant, 2:211–16; French text, ibid., 490–2

221 Barbiche, "Henri IV and the World Overseas," 30–2; Buisseret, *Henry IV*, 138–9. Pierre Jeannin (1540–1622) was an important jurist and diplomat who served under Charles IX, Henri III, Henri IV, and Louis XIII. A Catholic, he supported Henri IV in 1595 and was appointed by him *conseiller d'État* in 1601 and *intendant des finances* in 1602. After Henri IV's death and Sully's forced retirement, Marie de Medici, regent for the young Louis XIII, appointed Jeannin *surintendant des finances*. Champlain mentions him on several occasions as an important supporter of his plans for colonizing Canada. Lescarbot dedicated his 1618 edition of the *Histoire* in part to Jeannin (Grant, *History of New France*, 1:9–11).

222 Maximilien de Béthune, baron de Rosny, duc de Sully (1559–1641), a lifelong Huguenot and friend of the king, was a gifted administrator who held many offices under Henri IV, but is best known as his *surintendant des finances*.

223 Nicolas de Neufville, sieur de Villeroy (1542–1617), a Catholic and perhaps Henri's most experienced councillor, had been secretary of state under Charles IX and Henri III, as well as under Henri IV.

224 Sully, *Mémoires des sages*, 1:516. Sometime after his forced retirement following Henri IVs assassination, Sully wrote his memoirs and published them privately at Château de Sully in 1638. What was unusual about them is that they were written in the second person. In this quotation, Sully makes it seem that de Villeroy had written him a letter or memo reminding him of his (Sully's) policies regarding colonization. The version reproduced here is in a nineteenth-century edition of the original. We have emended *au dessous de quarante degrez* (below forty degrees) in the nineteenth-century copy, to *au dessus* (above). The letter *o* must have been added by mistake, because it makes no sense.

225 In 1745 the abbé Pierre Mathurin de l'Écluse des Loges (1716–83) published Sully's memoirs after he had rewritten them into ordinary narrative form. To the rewritten passage

and make settlements in Canada, totally against your advice, for as much, you were saying, as one never draws great wealth from lands located above forty degrees.") The king, however, prevailed and made colonization a personal priority in spite of the indifference or resistance of his advisers and the open hostility of merchants who insisted on free trade.

In 1596 La Roche was released from prison, and the following year he approached Henri for a renewal of his monopoly. The monopoly was approved on 12 January 1598, making him lieutenant general of New France and charging him with the erection of settlements.[226] For settlers, he was given permission to take "criminals, malefactors and paupers of both sexes."[227] After selecting 250 potential settlers, including 50 women, the total number was reduced to 40. Their destination was Sable Island, a lonely and windswept crescent of sand in the Atlantic, some 110 km from the coast of Nova Scotia, totally unsuitable for a colony. In 1603 a pathetic remnant of 11 colonists was rescued.[228]

With La Roche busy trying to colonize Sable Island, the free traders still had the fur trade to themselves. This changed dramatically when, on 22 November 1599, Henri IV issued an exclusive ten-year monopoly for the trade in Canada to his loyal comrade-in-arms during the religious wars, the Huguenot Pierre Chauvin de Tonnetuit. According to Champlain, the intermediary between Chauvin and the king in the acquisition of the monopoly was François Gravé Du Pont, who persuaded Henri that Chauvin would begin the colonization of Canada.[229] Since Chauvin belonged to the Huguenot denomination, which had just been placed on an equal footing with Catholics through the Edict of Nantes, Henri must have felt that it was time he did something for his loyal Huguenot supporters by permitting them to participate in the overseas trade. This was a serious threat to the free traders and anti-Protestants. The merchants of Saint-Malo especially had been opposed to any monopolies in the St. Lawrence fur trade for some time, and they reacted strongly to the one just granted to Chauvin. On 3 January 1600 they began to register their protests.[230] In this they were joined by

of the original cited here, he inserted a footnote referring the reader not to *Des Sauvages* by Champlain but to Pierre-Victor Cayet's abridged version in *Chronologie septenaire*; see, for example, Sully, *Mémoires*, 3:398–9. See also Document H, "Of the French Who Have Become Accustomed to Being in Canada."

226 La Roche's commission is reproduced by Lescarbot in Grant, *History of New France*, 2:196–204.

227 Ibid., 204–7

228 *DCB*, 1, "La Roche de Mesgouez"; Trudel, *The Beginnings of New France*, 61–5

229 Biggar, *The Works*, 3:305–6

230 Ramé, *Documents inédits* (1867), 12; ibid. (1865), 51–2

La Roche de Mesgouez, who felt he was being squeezed out of the monopoly he had just been granted. In response, the wily monarch made Chauvin a lieutenant of La Roche on 15 January 1600 and stipulated that Chauvin's monopoly was to cover only 100 leagues along the St. Lawrence east of Tadoussac.[231]

In the summer of 1600, Chauvin traded in Canada with four ships. Before leaving, he built an establishment at Tadoussac and left sixteen men to winter there. Champlain, who saw this post in 1603, was severely critical of Chauvin.[232] He thought that Tadoussac was the wrong place for a settlement because it had an inadequate harbour and poor sandy soil. In this observation, he was supported by Gravé Du Pont, who had apparently suggested to Chauvin that he explore and place his settlement much farther upstream. In fairness to Chauvin, his commission covered only the St. Lawrence as far as Tadoussac, not west of it. To Champlain, apparently, what was worse was that the appointment went to a Huguenot, which was to him, according to a variant in the 1632 edition of *Les Voyages*, a *horreur & abhomination* because it would introduce *les heretiques* into Canada.[233]

Champlain was also critical of Chauvin's building, which he sarcastically called *une maison de plaisance* (country cottage). This structure was four by three *toises* (about 7 by 5 metres) and 2.5 metres high. It had a central fireplace and was supposed to house sixteen men[234] (frontispiece). The winter was a disaster; some men survived into the spring of 1601, but only because local Natives had taken care of them.[235] It is probable that none would have survived if the settlement

231 Trudel, *The Beginnings of New France*, 66

232 Biggar, *The Works*, 3:305–11

233 It is entirely possible that these sentiments were introduced into some copies of *Les Voyages, 1632*, by a censor. For example, in the 1632 edition of *Les Voyages*, printed by Louis Sevestre, on page 35 there is a sentence that describes the kinds of people Chauvin was planning to bring to Tadoussac (edition in the National Library, Ottawa, 971.01/13 20, Canadiana 896029522; also Biggar, *The Works*, 3:307). This sentence reads "*Plusieurs personnes d'arts & de mestiers s'acheminent & se rendent au lieu de Hondefleur lieu de l'embarquement*" ("Many artisans and mechanics make their way and repair to Honfleur, the place of embarkation"). In *Les Voyages, 1632*, on the same page, printed by Claude Collet, this sentence was replaced by "*Tout ira assez bien, horsmis qu'il n'y aura que des Ministres & Pasteurs Calvinistes*" ("Everything will go well enough, except that there will be none but Calvinist Ministers and Pastors"); edition in the Bibliothèque nationale de France, Paris, RES-LK12-722(A).

234 Biggar, *The Works*, 3:309. Champlain drew a small picture of the *abitasion du Capp^tt chauuain de lan 1600* on his chart of Tadoussac (ibid., 2, facing p. 19).

235 DCB, 1, "Chauvin de Tonnetuit, Pierre de." In this biography, it is stated that eleven of the sixteen men died over the winter, leaving five survivors. This interpretation is based on the statement "*les unze moururent*" (the eleven died) in the Sevestre printing

had been farther upstream, since this had been an unoccupied shoreline after the disappearance of the St. Lawrence Iroquoians.

Amid continuing complaints from the free traders, Chauvin traded during 1601 and 1602 but did nothing with his settlement. Henvi IV, concerned with the lack of progress in settling Canada, authorized the merchants of Rouen in the fall of 1602 to join forces with Chauvin and press forward.[236] On 21 December 1602 the merchants of Saint-Malo protested this decision and applied for the trade themselves.[237] Henri's response seven days later was to set up a commission under the governor of Dieppe, Vice-Admiral Aymar de Chaste, and the sieur de la Cour, president of the parliament of Normandy, to look into the matter and put forth a series of recommendations that would lead to equitable trade and colonization.[238] On 2 January 1603, in order to give the commission time to complete its study, an ordinance was issued to the "*habitants des villes maritimes, portz et havres des provinces de Normandye, Bretaigne, Picardye, Guyenne, Biscaye, pays Boullonoys, Calais et autres costes de la mer océanne*" to desist from further trading beyond the Gaspé until the commission submitted its recommendations.[239] The following day, the Saint-Malo merchants were informed of this decision and invited to go to Rouen to consult with Chauvin and other interested merchants in discussions on how to regulate the fur trade and how to explore and settle Canada.[240] These discussions could potentially lead to a large undertaking that would involve many merchants, who would share the trade and the expense of colonization. The question that must have been foremost in the minds of those present was whether the profits from the trade would be great enough to offset the expense of building settlements, bringing colonists to Canada, and taking care of them until they were self-sufficient. The merchants of Saint-Malo, probably the most experienced at the conference, decided on 26 January 1603 to withdraw from any further discussions.[241]

The end of the negotiations came early in February with the death of Chauvin.[242] Rather than begin all over again, Aymar de Chaste, who was a Catholic

of *Les Voyages, 1632*, 37. On the same page of the printings of *Les Voyages* by Le Mur and Collet, it states, "*les uns moururent*" (some died). It is probable that *unze* is a misprint for *uns* and that the number of survivors is therefore unknown.

236 Trudel, *The Beginnings of New France*, 68

237 Ramé, *Documents inédits*, (1867), 12–14

238 Ibid. 15–17

239 Ramé, *Documents inédits* (1867), 17–18

240 Ibid., 19–21

241 Ibid., 14–15

242 *DCB*, 1, "Chauvin de Tonnetuit"

and had been another early and consistent supporter of Henri during the wars, was made successor to Chauvin's monopoly. As chief commissioner of the inquiry he probably had a good idea of the dilemma of free trade versus monopoly and the determination of the monarch to push colonization. It is likely that the failure of Chauvin's men to winter at Tadoussac in 1600–01, along with La Roche's disaster at Sable Island and knowledge of the winters suffered by Cartier, Roberval, and their crews, led in part to plans for an expedition in 1603 that was designed "to make the voyage in order to see this land, and what the associates would do there."[243] Could Canada be settled by the French? This was a serious question that had to be answered, as was the question whether it was worthwhile to explore the interior west of the Lachine Rapids. If the answer was yes to these two questions, monopoly control of the trade was necessary to generate the revenue needed to accomplish both ends. If the answer was no, the free traders could be given clearance to do as they liked. It is probable that two Montagnais brought to Paris by Gravé Du Pont in the fall of 1602 (see below) were there to provide some of the answers.

Probably acting on his knowledge of the debates and deliberations of the commission,[244] as well as the experiences of Chauvin's men and the reports of the two Montagnais, de Chaste decided to authorize three ships that represented his own interests and those of the other merchants.

Preparations for the voyage to Canada, 1603

In the fall of 1602, François Gravé Du Pont had returned from trading on the St. Lawrence and brought two Montagnais men with him from the Tadoussac area. He presented them to Henri IV, and they told the king that "the big river[245] (which we once thought to be only a bay or gulf, because it was eighteen leagues across its mouth at the sea) was more than 400 leagues long and cut across an infinity of fine lands and lakes, into which a great number of beautiful rivers also flowed together; and because one could go there with the canoes which the *Sauvages* use to navigate by way of this great river."[246] The two Montagnais offered to make a presentation to their people from the king – that if they allowed

243 Biggar, *The Works*, 3:315
244 As yet no report of the commission of inquiry has been found.
245 The "big river" is the St. Lawrence.
246 Document H, "Of the French Who Have Become Accustomed to Being in Canada," by Pierre-Victor Cayet. As *chronographe du Roy* (royal historian of Henri IV's reign), former tutor of the king, professor at the Collège de Navarre, and supporter of overseas expansion, Cayet was probably present when these discussions took place.

the French to settle on the St. Lawrence, the French would help negotiate a peace with their enemies, the Iroquois, or help defeat them.[247] These offers were enthusiastically received by both parties, with the result that Gravé Du Pont, "along with a few other sea captains (with the good wishes of the King), resolved to return there and see the interior of the land with the help of the *Sauvages* in the same way as they had seen the coast alongside the ocean."[248]

With this meeting, some of the motives for returning to Canada fell into place. A crucial briefing for the voyage must have occurred sometime late in 1602 after reports of an English voyage to Norumbega[249] had been published, or early in 1603 after Chauvin had died.

From 14 May to 18 June 1602, while Gravé Du Pont and his group were trading on the St. Lawrence, an English expedition captained by Bartholomew Gosnold had been active on the coast of Norumbega.[250] The Gosnold expedition arrived back in England on 23 July, and by the end of October an account of its voyage had been published under the authorship of John Brereton, together with a "treatise," by Edward Hayes, urging the exploration and "planting" of English settlements on the coast of Norumbega.[251] This little publication was so popular that a new, enlarged edition was published in late 1602 or early 1603.[252] From entries in Champlain's *Des Sauvages*, it is fairly certain that this publication, especially the

247 Document G, "*Des Sauvages*," ch. 2. It is probable that a preliminary agreement was made between Chauvin and the Montagnais that led to the construction of a post at Tadoussac in 1600. In order to finalize this agreement, Gravé Du Pont brought the two Montagnais "ambassadors" to Henri IV in 1602 (Thierry, *La France de Henri IV en Amérique*, 71). If this was the case, Henri IV recognized the significance of the potential benefits that such an "agreement" could bring and worked it into a more comprehensive and binding one.

248 This is an important addition by Cayet not mentioned in *Des Sauvages*. Cayet had in fact summarized one of the aims entrusted by Henri IV to Gravé Du Pont for the 1603 voyage, namely, the "discovery of lands in Canada and the adjacent countryside" (Document E, "Decrees and Commissions"). It was Champlain who was eventually assigned to the completion of this aim.

249 Norumbega was the maritime coast of today's Nova Scotia, New Brunswick, Maine, and Massachusetts, roughly what the French called Acadia.

250 Quinn and Quinn, *English New England Voyages*, 143–65

251 Brereton, *A Briefe and True Relation*; Quinn and Quinn, *English New England Voyages*, 36

252 Quinn and Quinn, *English New England Voyages*, 37. The second, enlarged edition had the words "With divers instructions of speciall moment newly added in this second impression" placed on the title page above the place of publication and the publisher's name.

Hayes treatise, was used in planning the 1603 expedition under Gravé Du Pont[253] and led to the 1604 expedition to Acadia.

Edward Hayes, like Richard Hakluyt, was a promoter of English westward expansion.[254] The short treatise attached to the Gosnold voyage was not his first attempt to goad England into "planting in these parts" of North America, but it was important because it came at a time of renewed French interest in the same area. His contention was that England should "plant" settlements of "Christian people" in the temperate parts of northeastern North America between forty and forty-four degrees north latitude.[255] These areas, he insisted, had not been claimed by any "Christian prince," but they would have been if the French had not been preoccupied with their religious wars. The lands, he argued, were "the rightful inheritance" of Queen Elizabeth I, because they were first discovered by John Cabot and his sons during the reign of Henry VII.

Hayes suggested that the coast of Norumbega should be explored and the "naturall inhabitants" questioned about the interior geography, because it was likely that one would learn as much from them as from "our owne navigations." In practical terms, he suggested that like the French, the English should be looking for mighty rivers that flowed from the interior to the coast and that settlements should be made at the mouths of these rivers. He pointed out that the French, especially Jacques Noël, had followed such a course by questioning the Natives while travelling up the St. Lawrence and across a number of rapids to a huge lake, whose entrance had fresh water, but its western end was salt. This lake, he said, was at forty-four degrees and was called *Tadouac* by the "Salvages" (Natives) and most likely had a river connection to the coast of Norumbega. Besides exploring rivers that led to the interior, settling the coast of Norumbega would lead to the discovery of many resources such as those in the Bay of Menan (Fundy), including copper and silver deposits and "Salt as good as that of Buruage" (Brouage) in France.[256]

253 For an examination of the French and their relationship to the Hayes treatise, see Hunter, "Was New France Born in England?" Also Hunter, *God's Mercies*, 197–9. As far as we have been able to determine, Hunter was the first to explore this relationship. For an elaboration of the impact of the Gosnold/Hayes documents on planning the 1603 French expedition, see appendix 5, "Champlain's *Des Sauvages* and Edward Hayes's *Treatise*."

254 For a brief biography of Hayes, see DCB, 1, "Hayes, Edward."

255 What follows is based on Hayes's statements in Quinn and Quinn, *English New England Voyages*, 168–203.

256 The Bay of Menan (Menon, Menim) was the Bay of Fundy, so called by Étienne Bellenger in 1583 (Ganong, *Crucial Maps*, 458–9; Quinn, "The Voyage of Étienne Bellenger").

These objectives must have been greeted with some alarm by Henri IV and the supporters of colonization in his council: England was seriously interested in lands to which the French believed they had a right and in which they were in the process of reactivating their trade and planning possible settlements. Not only were the English, like the French, interested in colonization, but they obviously considered northeastern North America to be "rightfully" theirs. Their plans for settlement made good sense. Although the French attempt to settle at Tadoussac by Chauvin and on Sable Island by La Roche had just failed, colonization was still a high priority with Henri IV. Another sensible suggestion was that of questioning the Natives about the geography of the interior. Was Champlain told to follow such a procedure in a briefing session prior to the voyage of 1603? Within a few days of his arrival at Tadoussac he used this technique to good advantage. The activities of Cartier and Noël must have been well known to Henri's court, except that Hayes claimed that Noël had actually been to the great western lake called *Tadouac* and that it was salt water at its western extremity.[257] Edward Wright's map, published by Hakluyt, on which the lake called *Tadouac* and the river connection to the Atlantic coast (Norumbega) are delineated, must also have been known to the French court by this time, but who would get there first?[258] Hayes's treatise was rife with references to the wealth that was obtainable in northeastern North America, but the statement that there were copper and silver deposits as close to the St. Lawrence as the Bay of Fundy, as well as salt deposits as good as those from Brouage, must have struck the French as especially worthy of note.

The parallels between *Des Sauvages* and the Hayes treatise in some of the information and the conjectures expressed do not seem accidental; both England and France had residents in each other's countries who gathered information for them. Another incident in the rapid dissemination of information that involved Champlain was Henry Hudson's map. It was in Champlain's hands late in 1612, only a few months after it had been engraved and printed by Hessel Gerritsz.[259]

257 The voyages of Jacques Cartier and intelligence from Jacques Noël were published by Richard Hakluyt in the third volume of his *Principal Navigations* (1598–1600).

258 The magnificent world map by Edward Wright appeared in the second volume of *Principal Navigations* (1599). On it, "The Lake of Tadouac the boundes wherof are unknowne" is the source of the St. Lawrence River, with a connecting river "Gamas" branching off to the Atlantic coast. Earlier maps clearly identify the River Gamas as the Penobscot River. For even earlier maps that show a river connection between the Atlantic coast and the St. Lawrence, see Ganong, *Crucial Maps*, 117–22.

259 Heidenreich and Dahl, "The Two States," 11–12

Champlain prepares to join the voyage

Of the sequence of events that led to his participation in the 1603 voyage, Champlain wrote: "I found myself at court, newly arrived from the West Indies ... in order to make a true report of them [the West Indies] to his Majesty upon my return."[260] This statement implies that he went straight to Paris after settling matters for his uncle. At some point he had received "a stipend with which he [Henri IV] honoured me as a means to maintain me near him,"[261] which obliged him to return to Paris. Whatever else he may have had in mind, the report to the king was a compelling reason for Champlain to return to the court rather than lingering somewhere outfitting ships[262] or perhaps exhibiting his pictures.[263]

Once he had returned to France, Champlain "from time to time" visited Aymar de Chaste, who was now governor of Dieppe,[264] to see if he could be helpful to him in whatever enterprise he was planning. An impressive person with great influence at court, de Chaste had become interested in the possibility of trading and colonizing in Canada, a fact that he had communicated to Champlain.[265] Probably early in 1603, after Chauvin had died and de Chaste was fairly certain that he would be permitted to outfit a ship for Canada, he asked Champlain if he "would like to make the voyage in order to see this land, and what the associates would do there."[266] Champlain was, of course, enthusiastic about such an undertaking but felt that he "could not do so but by the command of his said Majesty." Champlain then asked de Chaste if he would speak to the king on his behalf. De Chaste honoured this request, and the result was "an order from his Majesty for me to take this voyage and make a faithful report of it to him."[267] As paraphrased by Champlain, this "order" (*commandement*) came in the form of a letter written

260 Document F, "Excerpts from Champlain's Works," no. 7.

261 Ibid. This may have been *une pension du Roy*, given to those whom a monarch wanted to reward for services (Ganeau, *Dictionnaire universel*, 4:701).

262 Document A, "Early Biographies of Champlain," no. 6

263 Document D, "Factum of the Merchants of Saint-Malo"

264 For biographies of de Chaste, see Michaud, *Biographie universelle*, vol. 8, "Chaste"; and Glénisson, *La France d'Amérique*, 352.

265 Document F, "Excerpts from Champlain's Works," no. 7

266 Ibid.; also the other quotations in this paragraph

267 Champlain wrote on at least three occasions that permission from the king to join the 1603 expedition came in the form of an order *par commandement de sa Maiesté* (Document F, "Excerpts from Champlain's Works," nos. 6, 7, 9). He followed the same procedure for the 1604 voyage to Acadia, receiving permission from the king "on condition I should always make a faithful report to him" (Biggar, *The Works*, 3:321).

by Henri's secretary of state, Louis Potier de Gesvre, to François Gravé Du Pont, the captain of de Chaste's ship *Bonne-Renommée*, to permit Champlain to "see and explore all that which could be seen and explored in those parts," assisting him "as much as possible in this enterprise."[268]

From the foregoing it is clear that Champlain was eager to join de Chaste's expedition, that de Chaste wanted to have him along, that Champlain felt he was under obligation to the king and would have to seek his permission to go, and that the permission came in the form of an order that involved a report to the king on his return. In other words, Champlain went under orders from Henri IV to report to him on conditions in Canada that were conducive to settlement and to follow up the leads on the western river and lakes system given by the two Montagnais men. This is also the view of some of the early biographers of Champlain.[269] In view of the sequence of events outlined by Champlain, it is difficult to understand the contention of some writers, notably Marcel Trudel, who see Champlain going as "a private passenger … a mere observer" and again "simply as an observer"[270] – unless, of course, they agree with the merchants of Saint-Malo, who made a similar statement early in 1613. In this factum, they sought to discredit Champlain and his 1603 tour of the St. Lawrence by stating that he and others were there "simply as passengers"[271] and for the money they were making. But if Champlain was simply going as an observer on the request of de Chaste, why would the king request that his secretary of state write a letter to the captain of de Chaste's ship and order Champlain to report to the king on his return?

How well was Champlain prepared for the tasks he was to accomplish? Namely, to assess the potential of the banks of the St. Lawrence for settlement and to determine the possibilities for westward exploration. It is probable that he knew nothing about New France; at least, he does not show prior knowledge in *Des Sauvages*. Of the published material that might have been available, he had not even read Jacques Cartier's accounts.[272] He referred to Cartier only once, when he wrote that *Jacques Quartier* had not travelled farther up the St. Lawrence than the little river now called Rivière Jacques Cartier just west of Quebec.

268 Louis Potier, baron de Gesvres, comte de Tresmes (c. 1542–1630), was secretary of state (*secrétaire d'État*) under Henri III and Henri IV. For a paraphrase of Potier's letter, see Biggar, *The Works*, 3:315–16.

269 Document A, "Early Biographies of Champlain," nos. 1 and 5

270 *DCB*, 1, "Champlain, Samuel de"; Trudel, *The Beginnings*, 74

271 Document D, "Factum of the Merchants of Saint-Malo"

272 For a discussion of Cartier's published voyages that might have been accessible to Champlain, see Hoffman, *Cabot to Cartier*, 131–2; Biggar, *Jacques Cartier*, ix–xii.

This, of course, is an error and was corrected by Champlain in his later books after he had read Cartier.[273] He probably had not read Brereton's little book with the Hayes treatise, because there is no evidence that he could read English. But since he expressed a number of Hayes's ideas in *Des Sauvages*, it is likely he was briefed on them or was instructed by Gravé Du Pont, to whom he probably owes other bits of prior knowledge about New France.

After Champlain arrived in Canada and departed from Tadoussac on his survey up the St. Lawrence, it becomes evident to the reader that he had neither a compass nor a device for taking latitude, such as an astrolabe or a cross-staff. His journal of 1603 is unlike his later ones, especially his surveys of the Acadian coast from 1604 to 1607, where he gave his courses by compass readings, calculated the compass declination and latitude of potential harbours, and prepared maps from the results of triangulating coastal features by means of a compass and estimated distances. Throughout the entire voyage of 1603 he refers to the left bank of the St. Lawrence River as "the north side" and the right bank as "the south side." By the time he reaches Montreal Island, he calls the east side of the island the north side of the river and describes the Monteregian Hills as being to the south of the island, when they are to the east. In fact, what he demonstrates is that he had no inkling of the major changes in the course of the river. Indeed, on his map of 1612, which appears to be based on the 1603 survey, the St. Lawrence runs in a constant direction from west to east, or from the southwest to northeast, if his oblique meridian can be trusted and was not a later addition to the map, which was first drafted in 1603.[274] As well, he made no calculations of compass declination. He gave only four readings of latitude. Of these, three were in the Gulf of St. Lawrence, probably taken by the pilot to whom Champlain refers, and the other at the Lachine Rapids, which may have been based on a reading that predated Champlain. His diligence in making instrumental recordings changed so much between 1603 and 1604 that one is forced to believe that he had different orders for the 1604 Acadian voyage and had done a crash course on making instrumental observations. His estimates of distance are discussed elsewhere.[275] Suffice it to say that on the 1603 voyage they were indifferent to poor, but they improved

273 Champlain, *Les Voyages, 1613*, 184–91; Biggar, *The Works*, 2:35–44

274 In *Les Voyages, 1613*, 320–5, Champlain wrote that he placed an oblique meridian on the 1612 map for French navigators who could not, or would not, adjust the declination of their compasses to the Grand Banks, where the compass needle pointed towards the northeast, while in France it pointed to the northwest. It is likely, however, that the oblique meridian was intended specifically for the 1612 map and was not on the sketch map he prepared for Henri IV.

275 Appendix 4, "French Measures of Distance, Weight, and Coinage"

on his later voyages. His description of places was very good. In spite of the poor distance estimates, one can usually find the islands, rivers, and other physical features he mentioned with a high degree of reliability.[276]

It is likely, therefore, that the only preparation Champlain had for the 1603 voyage was his earlier training as a fourrier, whatever he had learned on Spanish ships, a thorough briefing for the voyage in early 1603 that included some ideas from Edward Hayes's treatise, and his conversations with Gravé Du Pont.

To Canada and return, 1603

Aymar de Chaste, who had been granted Pierre de Chauvin's monopoly on Chauvin's death, lived long enough to launch the 1603 expedition to the St. Lawrence but died just before his ships returned from Canada. The monopoly was then transferred to Pierre Du Gua de Mons, another faithful supporter of Henri IV.

On 13 March 1603, Henry IV tried to put an end to the eternal feuding between the free traders, such as the merchants from Saint-Malo, and the monopolists represented by de Chaste and his supporters, by issuing a decree followed by a commission for the trade in Canada.[277] Both documents clearly state that the three parties were to engage in and share in any expenses related to "the discovery of lands in Canada and the adjacent countryside."

On 15 March, two days after the decree and the commission were posted, three ships set off for New France, all captained by men originally from Saint-Malo. The first, the *Bonne-Renommée*, representing the interests of de Chaste and the merchants of Rouen, was captained by François Gravé Du Pont. It departed from the harbour of Honfleur for the St. Lawrence River with Champlain on board. Their tasks were to seal the treaty with the Montagnais, to engage in trade, to explore the settlement potential of the banks of the river, and to check on the Native stories regarding the possibility of exploring westward to the great interior lakes or seas. The second ship, captained by an associate of Gravé's, Jean Sarcel de Prévert, was to sail to the vicinity of Acadia to trade and to search for copper and other mineral deposits. It is probable that the impetus for this task came from the Hayes treatise. A third ship, representing the interests of the Saint-Malo merchants, captained by Gilles Éberard du Colombier, was sent out, but unfortunately we do not know its destination. It probably engaged only in trading, which is all the Saint-Malo merchants expressed an interest in.

276 Besides topographic maps (scale 1:50,000), the internet program Google Earth was of great use in following Champlain's course, even though the physiography of the St. Lawrence Valley has changed considerably since his time.

277 Document E, "Decrees and Commissions"

Some 44 days into their voyage, after unfavourable winds in the Seine estuary and extensive storms in mid-Atlantic, the first icebergs were sighted at about 45°40'N, some 100 to 120 leagues off Newfoundland. To avoid the pack ice, the ships sailed southwest and began to traverse the southern Grand Bank at about 44°N. On 6 May, engulfed in fog, the crews heard the waves beat on a shore, and it was decided to head out to sea. The following day they emerged from the fog within sight of Cape St. Mary's. Following the south shore of Newfoundland, they encountered heavy ice again off Cape Ray and through the western reaches of the Cabot Strait.[278] It is likely that once the Gulf of St. Lawrence was reached, the three ships separated to proceed to their destinations. On 20 and 21 May, the crew of the *Bonne-Renommée* sighted Anticosti Island and the Gaspé Peninsula and entered the passage now called Détroit d'Honguedo, which was then regarded as the entrance to the St. Lawrence River (map 2). The ship continued to skirt the south shore of the river as far as Bic, the landmark which the captain knew signalled the place where he had to cross the river to Tadoussac, which he did on 24 May. After waiting in the outer harbour for favourable tides and currents the ship entered the small harbour of Tadoussac on 26 May. The entire journey had taken seventy days, only three days shorter than the voyage in 1611, which was the longest of Champlain's twenty-three Atlantic crossings.[279] The course the ship had taken from France across the Grand Banks to Tadoussac followed the standard route, demonstrating that the captain and navigator had made this crossing many times before. Unlike others who crossed the Atlantic for the first time – for example, the Récollet brother Gabriel Sagard in 1623[280] – Champlain made no mention of the abundance of fish, birds, and sea mammals between the Grand Banks and Tadoussac.

On Tuesday the twenty-seventh, the day after the French landed, the Montagnais called a *tabagie* (meeting) at Pointe aux Alouettes across the river from Tadoussac to hear what intelligence their two men had brought from France. The convener of the meeting was the respected headman and orator Anadabijou. In an egalitarian society such as the Montagnais, it would be his task to present a summary and analysis of the substance of the meeting and perhaps give his

278 The areas where Champlain reported heavy ice are still areas of drifting ice floes and icebergs during spring ice conditions (Scarratt, *Canadian Atlantic Offshore Fishery Atlas*, 20–1).

279 Of the 23 times Champlain crossed the Atlantic, the 12 journeys westward from France took an average of 52 days (max. 73 days, min. 18 days), while the 11 going eastward to France took 36 days (max. 49 days, min. 25 days). Adverse winds and ice conditions were the main factors in the crossings to Canada.

280 Wrong and Langton, *The Long Journey*, 30–43

opinion on any action to be taken, but he did not have the power to make decisions for those who were present. It is likely that his "80 or 100 companions" were the informal spokesmen of the winter hunting groups, each of which consisted of a number of closely related Montagnais families.[281] If it was necessary to come to a decision, these "headmen" would consult with their family members and then meet again as a group to discuss the matter in order to reach a measure of consensus. At this meeting, Gravé Du Pont and Champlain were present but without a specific role, since neither said a word.[282] They were seated next to Anadabijou, probably in order to symbolize the importance of the meeting. Although other Frenchmen may have been present to lend gravity to the discussions, none is mentioned. Nor did Champlain indicate the presence of any Algonquins or Etechemins[283] at this *tabagie*. It is unlikely that they were present, because the meeting concerned the news which the two Montagnais "ambassadors" had brought back from France.

Fittingly, the *tabagie* opened with an oration by one of the two "ambassadors" setting out the subject matter of the meeting. If one of the French had been able to speak Montagnais, it is most likely that he would have been the one who presented the French case as spokesman for the king. The speech of the Montagnais man followed a three-point pattern:

- He stated that the king had given them a good reception in France, had entertained them well, and sent his greetings to the assembled group. This part of the oration established that the king respected them.
- He said that the king wished to settle in their country. In return, he would make peace between them and their enemies, the Iroquois, or send troops to defeat them. This was the heart of the matter, which the assembled group would have to discuss and render a decision about.
- Finally, the "ambassadors" entertained their listeners with vignettes of what they had seen in France, all of which pointed out to them that the French were a powerful group of people worth making an agreement with.

281 For a discussion of leadership structure among the Montagnais, see Leacock, "Seventeenth-Century Montagnais Social Relations," 191.

282 Some writers have misrepresented Champlain in this affair, thinking that he "sealed" this agreement. See, for example, Dickason, *Canada's First Nations*, 103. As far as we can tell from Champlain's brief report, neither he nor Gravé Du Pont uttered a word at the *tabagie*. The entire "agreement" had been "negotiated" by the Montagnais and Henri IV before the voyage to Canada. Their presence seems to have been symbolic.

283 The Algonquins were from the Ottawa River area and the *Estechemins* (Etechemins) were various groups that lived on the Atlantic coast between the St. John and Kenebec Rivers.

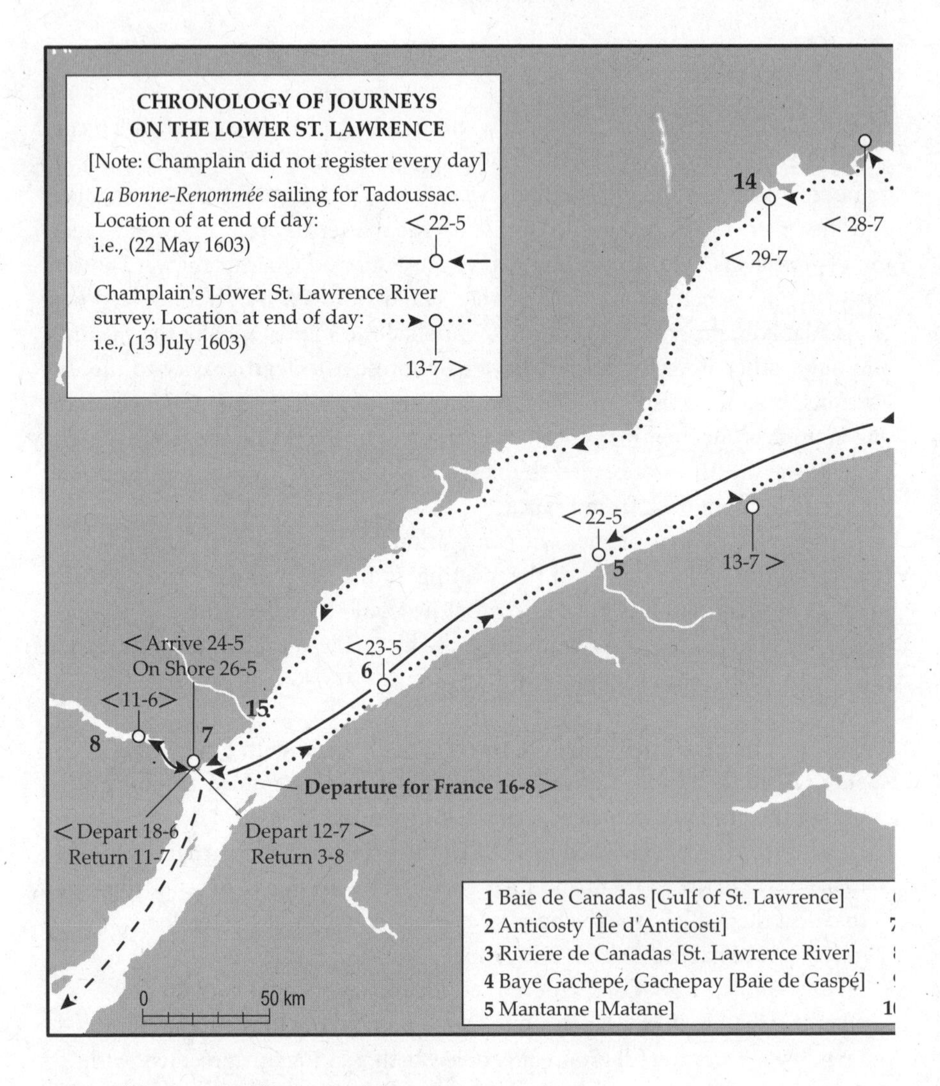

Map 2 The Lower St. Lawrence River: A Resource Survey

On its way from Honfleur to Tadoussac, the *Bonne-Renommée* sailed past Anticosti Island on 20 May and made anchor at Tadoussac on the twenty-fourth. On 11 June, Champlain explored up the Rivière Saguenay to about Île Saint-Louis, 35 km from Tadoussac. On 18 June he departed on his survey up the St. Lawrence River to *Le Grand Sault*, returning on 11 July (see map 3). The next day he began his survey of the lower St. Lawrence River as far as Île Percé on the south shore. He then crossed to the north shore near Baie de Moisie and was back in Tadoussac on 3 August. He concluded that "all these lands are very poor and replete with fir trees … they are neither as pleasant

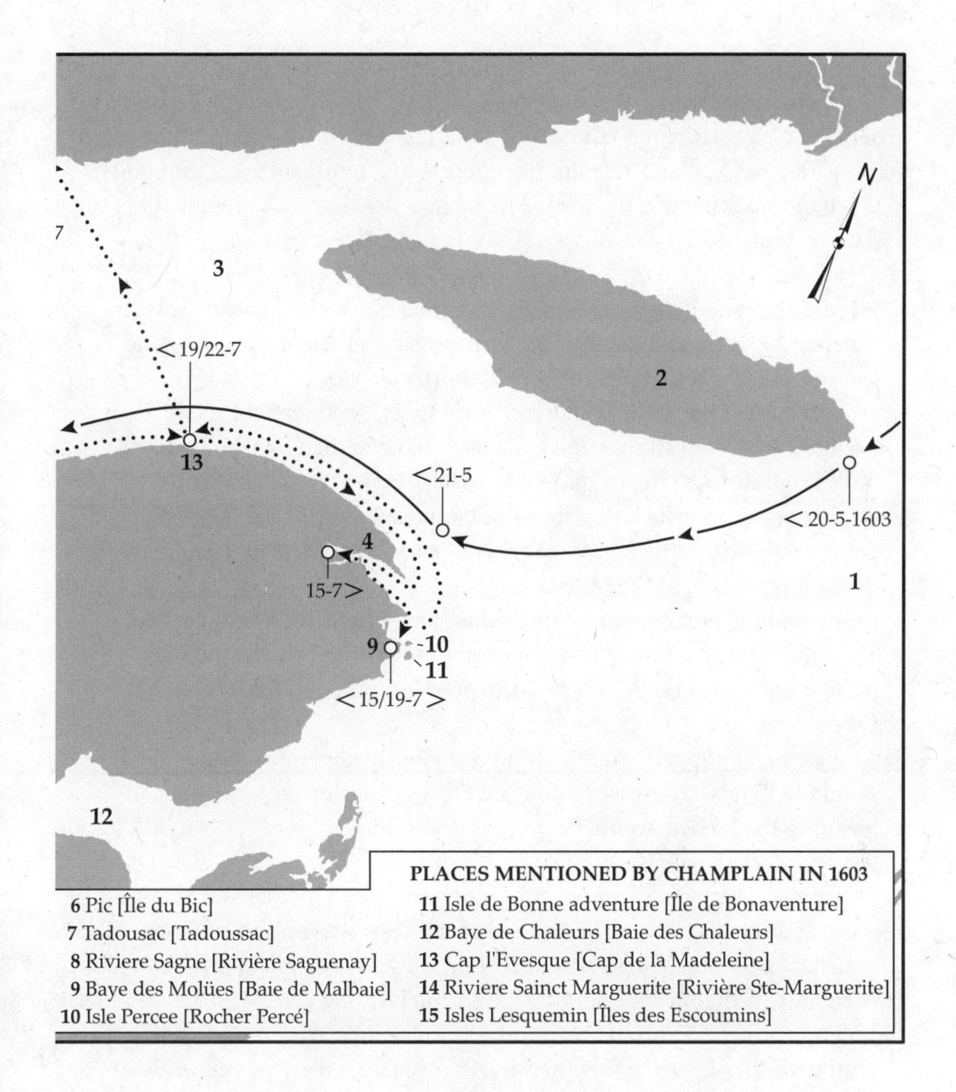

nor as fertile as those of the south." The expedition began its return to France on 16 August, stayed at Île Percé from 18 to 24 August, and made Le Havre on 20 September.

Following the opening oration setting out the subject matter of the *tabagie*, it was customary for the participants to join in a ritualized smoke by passing a pipe around to all participants. The act of smoking united the group in a common activity and gave each participant the opportunity to organize his thoughts.

Once the smoking was finished, Anadabijou stood up and presented his summary and analysis.

- He said they should be very glad to have the king as their friend. This statement told those assembled in what direction their spokesman was leaning. It was greeted positively by a chorus of "yes."
- He agreed that the French could settle on their land and fight their enemies. It is interesting that Anadabijou avoided the French offer to broker a peace with the Iroquois and instead linked settlement with waging war. To the listeners, it was obvious that in return for the French being allowed to settle on Montagnais territory they were obliged to help the Montagnais fight the Iroquois.
- He wished the French well, more so than any other nation in the world. This statement was probably directed not only at the French who were present but also at the Montagnais, expressing the new relationship between them.
- At the end, he slipped in a matter not previously mentioned but on the mind of all those assembled – that a close relationship with the king, meaning the French, would be profitable to them.

Seeing general agreement among those present, further discussion was not necessary, and the participants reconvened for a feast that symbolized the successful conclusion of the *tabagie*. It is likely that the feast was only for the "80 to 100" participants in the meeting, not for all in the hunting groups, since there were only "eight to ten cauldrons" of food and only men were mentioned as being present with bark bowls in their hands. After the feast there was general rejoicing and dancing to celebrate a victory over the Iroquois in which the Montagnais had participated.

The importance of this *tabagie* has been pointed out by many writers.[284] What occurred here was a reciprocal agreement between a European nation represented by its king and a number of autonomous bands composed of interrelated families who shared a common territory, who were collectively called Montagnais by the French.[285] This was not a treaty in the modern accepted sense of the

284 See, for example, Beaulieu, "Birth of the Franco-American Alliance," 153–61.

285 The term *Montagnais* (mountaineers, or people of the mountains) was coined by the French to describe the people who lived in the mountainous interior north of the St.

word, because any agreement made by the spokesmen of the various bands was not binding on the individuals that composed the bands. It was simply an agreement to which the majority of the Montagnais would adhere as long as the individuals who composed the bands could see the agreement as an advantage to them. This agreement gave the French access to settle in the St. Lawrence Valley, but in return they would have to help fight the Iroquois Confederacy and give the Montagnais access to the French trade. After Quebec was built in 1608, it fell to Champlain to honour this agreement by accompanying his new Native "allies" on a raid up the Richelieu River (*Rivière des Iroquois*) against a Mohawk encampment. Although historians have pilloried Champlain for "starting" the long-drawn-out Iroquois wars, the Montagnais and their various allies were in conflict with the Iroquois well before the French arrived, and because of the agreement Champlain had no choice in the matter.

Following the day of the *tabagie*, Anadabijou suggested to all the Montagnais families that they move back to Tadoussac, "where their friends were." As the thousand individuals, in two hundred canoes, made their way across the river, Champlain had his first encounter with the birch bark canoe, finding them more manoeuvrable and faster than the shallop he was in. By the time he reached the Lachine Rapids a month later, he had grasped the full significance of the canoe to exploration, the first European to do so.

We do not know what happened between 28 May and 9 June when Champlain's narrative resumes. Presumably, it was during these days that Gravé Du Pont and his men engaged in trading. Throughout his journal Champlain makes no comments about the fur trade, probably because this was not his business but that of Gravé Du Pont, who would be delivering his own report to Commander Aymar de Chaste.

On 9 June the Montagnais again celebrated their victory over the Iroquois, this time along with their allies in the affair, the Algonquins and some Etechemins. The one hundred Iroquois heads they had taken were displayed, speeches were given, gifts exchanged, and dances performed.[286] Champlain took the occasion to make some cursory observations of Native customs. These were augmented no doubt by reports from Frenchmen who had been to Canada before,

Lawrence. Like other Native groups, the Montagnais had a name for themselves. If it was necessary to distinguish themselves from other groups of people that were linguistically and culturally different, they called themselves *Innu*, meaning "human beings." They spoke dialects of the Cree language (Rogers and Leacock, "Montagnais-Naskapi," 186).

286 This seems to have been a particularly difficult time for the Iroquois. In the late 1650s, the Jesuits wrote that according to Mohawk elders, "toward the end of the last century, the Agnieronnons [Mohawk] were reduced so low by the Algonkins that there seemed to be scarcely any more of them left on the earth" (*JR*, 65:205).

and they are so general and distorted that they have little ethnographic value. Champlain was not long enough at Tadoussac to observe marriages or burials, and he accepted uncritically what other Frenchmen told him. Even in later years among the Huron and others, Champlain's ethnological observations are of limited value compared to those of the Récollets and Jesuits, who could speak the various Native languages. Anything he could see, such as aspects of material culture, were generally well described, but not aspects of non-material culture, such as religion or social and political organization, which depended on language for an adequate description and interpretation.

On or about this time, Champlain appears to have questioned Anadabijou about his religion and supplied him with a formulaic sketch of Christianity. In a recent analysis of the topic, Dominique Deslandres summarized Champlain's recitation of the Creed as follows: "A single God, the Trinity, the Incarnation, the Passion, the Resurrection, the virginity of the mother of Christ, the communion of saints, miracles … which seems more like a recitation of the Credo than the expression of a desire to teach the doctrine to non-Christians."[287]

Champlain makes it appear that he spoke directly with Anadabijou, when it is certain that neither could speak the other's language. If interpreters were present, they are not mentioned. Why would Champlain engage in such a discourse? There is no independent evidence that this conversation actually took place, nor is there any indication that Champlain had any concern for the salvation of Native souls until 1615, when he brought the Récollets to Canada. In his summary of *Des Sauvages*, Pierre-Victor Cayet makes no mention of Champlain's discourse with Anadabijou, even though Cayet was himself a promoter of Catholic causes,[288] and Champlain never referred to this incident in his later writings. In fact, were it not for Anadabijou's myths, which sound so plausible that it is doubtful Champlain invented them, this recitation of the Creed looks like a literary device of Champlain's to make a public declaration of Catholicism at a time when such declarations were becoming increasingly politic. Other than this, it served no useful purpose, neither to Champlain nor to Anadabijou, who would have understood little of the philosophical concepts he was being offered even if an interpreter had been able to convey them correctly. We should not view Champlain's effort at proselytizing, if indeed it took place, as a precursor of the well-organized missionary onslaught of the Récollets and Jesuits, who learned to speak to their Native listeners in their own languages.

287 Deslandres, "Champlain and Religion," 192

288 See Document H, "Of the French Who Have Become Accustomed to Being in Canada," ch. 3.

On 11 June, two days after the *tabagie* of the Montagnais and their allies, Champlain began his explorations by travelling "some 12 or 15 leagues" up the Saguenay (map 2).[289] His description of the river and its surroundings is essentially accurate. From a methodological point of view, the importance of this trip was that he developed a way of obtaining geographical information from the local Native people; he simply asked them, presumably through an interpreter, to give him a verbal description and sketch map of the rivers, lakes, and lands beyond. This was a significant information-gathering phase before actual exploration could take place.[290] In this manner, Champlain obtained from his Native companions a good description of the route up the Saguenay River to a northern salt sea (Hudson Bay), which he accurately surmised to be "a great gulf of this our sea," the Atlantic Ocean.

On Wednesday, 18 June, Champlain, Gravé Du Pont, and an unknown number of other men began their journey up the St. Lawrence to the Lachine Rapids in what Champlain called a *barque* (map 3).[291] From the entries in *Des Sauvages*, it is evident that most of his observations of soils, vegetation, animal life, and the suitability of land for agriculture were made on the outward journey to the rapids, which took sixteen days. By contrast, his return journey to Tadoussac took eight days and involved no further descriptions of land quality.

The expedition followed the left bank of the St. Lawrence River and on 22 June reached the future site of Quebec. Although Champlain considered some of the larger islands along the way to be "somewhat pleasant," it was not until he

289 For Champlain's estimates of distance, see appendix 4, "French Measures of Distance, Weight, and Coinage." A distance of 12 to 15 leagues would take him to about Île Saint-Louis, 35 km from Tadoussac.

290 Heidenreich, "The Beginning of French Exploration," 238–41

291 Champlain was probably using the term *barque* in a non-technical manner, such as "boat." He was consistently vague about describing the vessels on which he was sailing. Technically a *barque* (also "barque" in English) is a small three-masted sailing vessel, square-rigged on the foremast and mainmast, and fore and aft rigged on the mizzen. Sometimes the mizzen was rigged for a lateen sail (Kemp, *Oxford Companion*, 61–2). Most of the early barques in Canadian waters had capacities of 25 to 75 tuns. It is doubtful that anyone would have taken a true barque up the unknown St. Lawrence River as an exploration vessel. In a later account of this trip, Champlain wrote that they had travelled in *moyennes barques de 12 à 15 tonneaux* (Champlain, *Les Voyages, 1632*, pt. 1, 40). This type of *barque* (boat) may have been a two-masted pinnace, often used for coastal exploration, although Champlain never mentioned a pinnace on any of his journeys. In Cayet's summary of *Des Sauvages*, he assumed that the entire voyage from Tadoussac to the Lachine Rapids and return was made in a shallop. See Document H, "Of the French Who Have Become Accustomed to Being in Canada."

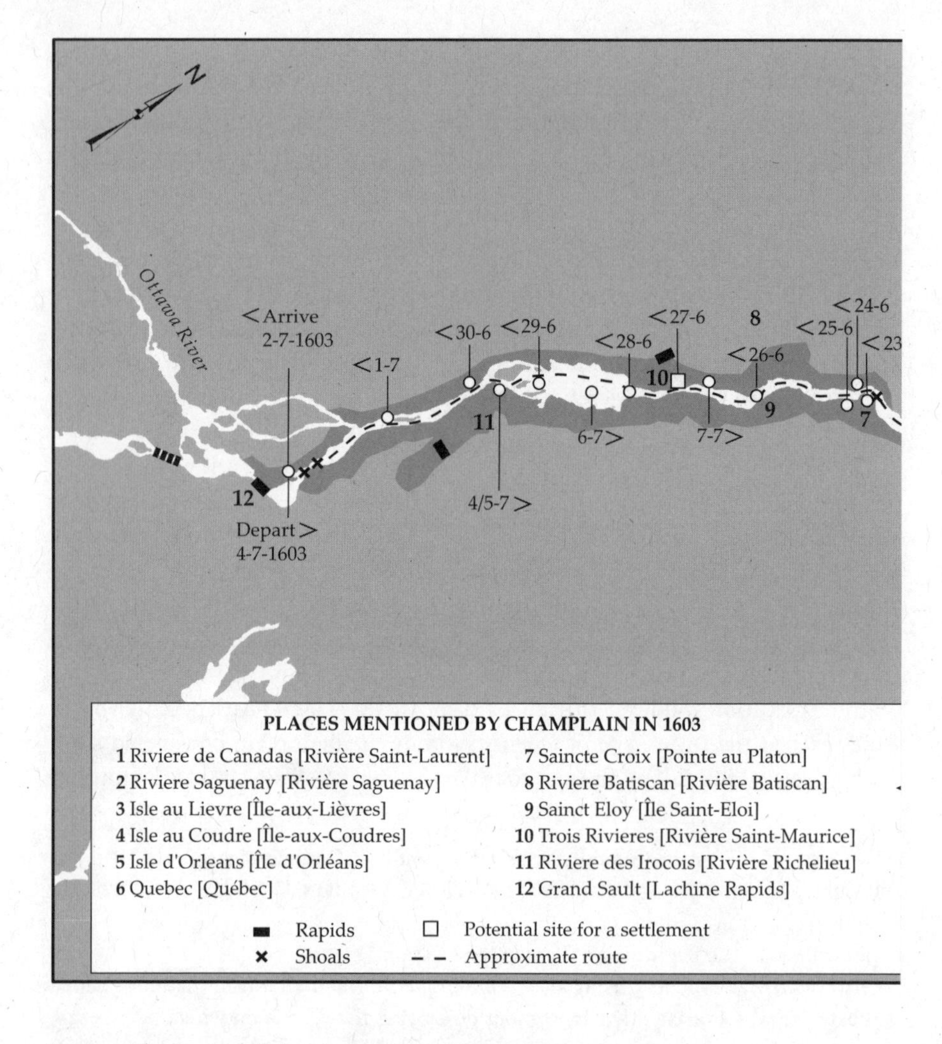

Map 3 The Central St. Lawrence River: A Resource Survey

On 18 June 1603, Champlain left Tadoussac in a ship of 12–15 tuns on his way to assess the banks of the river for their agricultural capability and to investigate the possibility of westward exploration across the Lachine Rapids. Although the ship sailed only along the "north shore" (left bank), he was not impressed by what he saw until he passed the south shore of the Île d'Orléans, which he called "very pleasant and level." As he sailed upstream from Quebec, he became increasingly enthusiastic about the quality of the land even though some of it was a little "sandy" near the river. He judged the vegetation, fish, and animal life to be plentiful and similar to that of France. Arriving near Rivière Saint-Maurice, he thought the climate to be a bit warmer and the vegetation

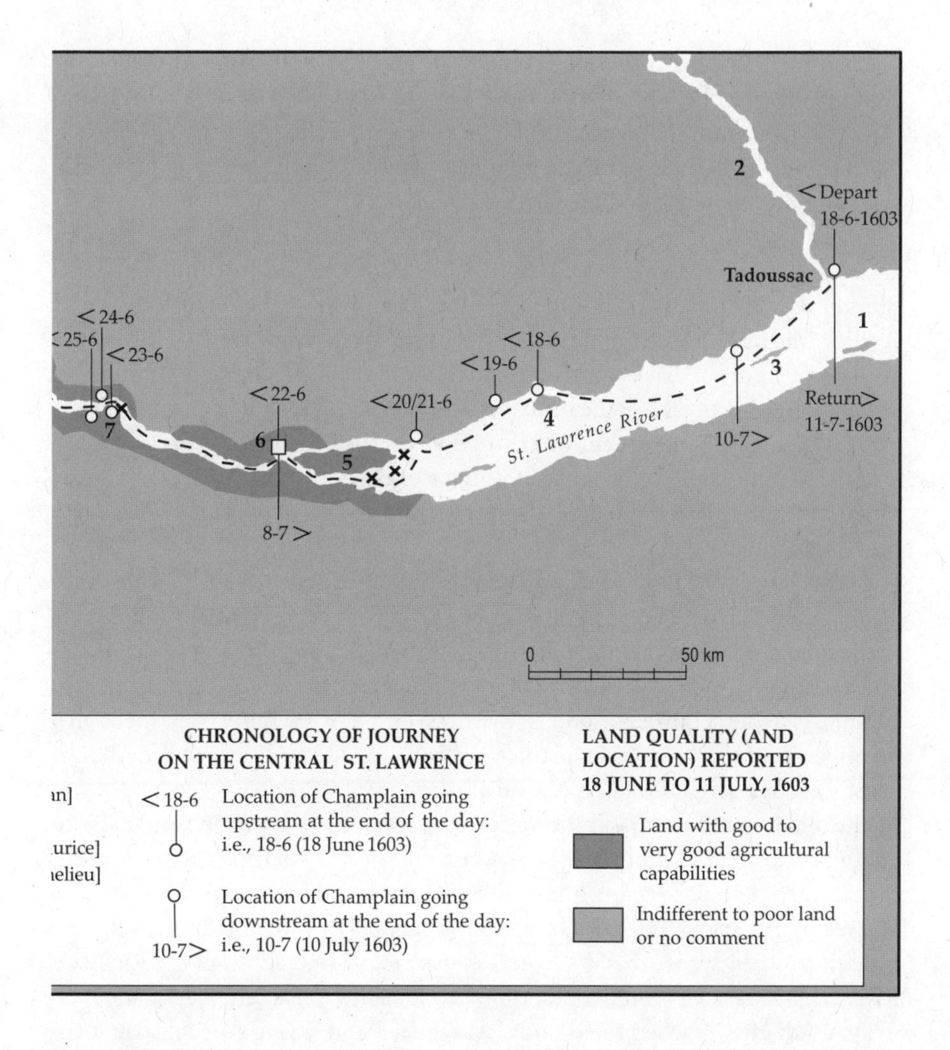

more advanced. As he continued, the quality of the land steadily improved. When he attempted to investigate Rivière Richelieu, the ship was stopped by the rapids at Saint-Ours. He also commented on dangerous shoals around Île d'Orléans, Pointe au Platon, and between Île Sainte-Hélène and Montreal Island. On 2 July he arrived at the Lachine Rapids. After walking the length of the rapids, he collected a map and a verbal description of the upper St. Lawrence to Lake Huron from a party of Algonquins (see map 4 and figs. 7 and 8). On 4 July he began his return journey to Tadoussac, which proceeded quickly and without further comment. Along the way, he collected two more Algonquin accounts of the geography west of the Lachine Rapids.

reached Île d'Orléans that he saw land that was "very pleasant." At Quebec he commented that the land, if cultivated, "would be as good as ours." To Champlain, the agricultural potential of the future colony definitely began at Quebec, all the lands he had passed since leaving Tadoussac being "nothing but mountains on both the south and the north shores."

On setting out from Quebec the following day, he noted that "the country keeps on getting more and more beautiful." At Pointe Platon, he commented that the soil was "better than in any place that I had seen." The farther they proceeded, "the more beautiful the country became." When they reached the islands at the mouth of the Rivière Saint-Maurice on 27 June, Champlain commented that not only was the land "the most pleasant we had yet seen" but that the largest of the islands, Île Saint-Quentin, was a good site for a fortified settlement. Such a settlement, he thought, would not only be of benefit to the Natives who were afraid of the Iroquois, but could also serve the purpose of making friends with the Iroquois.

On the last day of June the expedition reached the Richelieu River, where they encountered the fortified encampment of a group of Natives on their way to raid the Iroquois. Making their way up the Richelieu, Champlain and his companions eventually reached the Saint-Ours Rapids, which prevented them from continuing. Here he acquired a good Native description of the route from the Richelieu River through Lake Champlain and down the Hudson River to the Atlantic coast. He considered the soil in the area to be good but "somewhat sandy."

The quality of the soil and the variety of vegetation, wild fruits, and animals continued to impress him as they proceeded from the Richelieu River towards the Lachine Rapids. Meanwhile, Champlain was becoming increasingly impressed by the speed and portability of the canoe. On 1 July he decided to have himself paddled across the St. Lawrence just west of the mouth of the Richelieu River. This may have been the first time a European sat in a birchbark canoe.

On 2 July they reached the islands, gravel bars, and strong currents below the Lachine Rapids. The skiff they had brought with them, built expressly for crossing the rapids, turned out to be inadequate, whereas the canoes seemed to have fewer problems. After he saw the rapids and walked their length, Champlain realized that they were impossible to negotiate with any kind of ship. Instead, he reasoned, one must use the canoe, which had the advantage of being portable. He then made the most insightful statement of the journey, noting that "with the canoes of the *Sauvages*, one may travel freely and quickly throughout all the lands, as much in the little rivers as the big ones, so well that by directing one's course with the help of the said *Sauvages* and their canoes, one would be able to see all that can be seen, good and bad."[292] In a day's observation of the largest ob-

292 Document G, "*Des Sauvages*," ch.8, fol. 23^{v}.

stacle to inland exploration, Champlain had solved the problem. With the help of the Natives and the use of their technology, the interior of Canada could become accessible. What this implied, of course, was that peaceful relations had to be established with those Natives who could act as guides and were willing to conduct the Europeans in their canoes. When Champlain returned to the St. Lawrence in 1608, he worked diligently to bring these principles to reality.

Having solved the problem of how to conduct exploration, he attempted to learn what was beyond the rapids. By now he must have had some confidence in the geographical reports that his interpreters were able to elicit from their Native companions. He had tried this method on the Saguenay, at the mouth of the Saint-Maurice, and on the Richelieu River and had received seemingly satisfactory reports. This time, at the Lachine Rapids, he also asked that a map be drawn for him to accompany the verbal report of his informants. Unlike the two Montagnais who had reported on the river systems west of the Lachine Rapids to Henri IV, the ones questioned by Champlain were Algonquins who probably lived up the Ottawa River. What he received was a good account of the waterways from the Lachine Rapids up the St. Lawrence River, including all major rapids and portages along the way, across Lake Ontario, over Niagara Falls, and the length of Lake Erie to the St. Clair River, where it flowed out of a huge lake (Lake Huron) (map 4; fig. 8, top diagram). Although all the water was said to be drinkable and his informants had not been to this huge lake, they told him that "the water there is very bad, like that of this sea." Champlain thought the lake to be a reference to the "South Sea" (Pacific Ocean) and what they described as "bad" water to be "salt" water.[293] The similarity between the Algonquin words for "salt water" and "bad water," used to describe the Great Lakes beyond Lake Erie and Georgian Bay, was to confuse the French from the time of Champlain's journey in 1603 until well after 1648, the year Father Ragueneau, SJ, solved the linguistic problem.[294]

On his return journey, at the Île d'Orléans on 9 July, Champlain questioned "two or three" Algonquins, who drew a map for him and described the route as far as the entrance to Lake Erie in almost identical detail to that of the first group of Algonquins. All along the way the water was drinkable. In passing, they also mentioned the Oswego River, which was the main route south to the Iroquois country and the Cataraqui–Rideau river system that led north to the Ottawa

293 Unfortunately, Champlain could not record the Algonquins' description in their own language. The words for salt water, muddy, or stinking water are virtually the same – they describe bad or undrinkable water. See Cuoq, *Lexique Algonquine*, 439–40 (*Winickwaw*, "salis-le"; *Winidjickiwaka*, "c'est salipar la boue, c'est bourbeux, fangeux"; *Winipik*, "eau sale").

294 Heidenreich, "An Analysis," 50–3

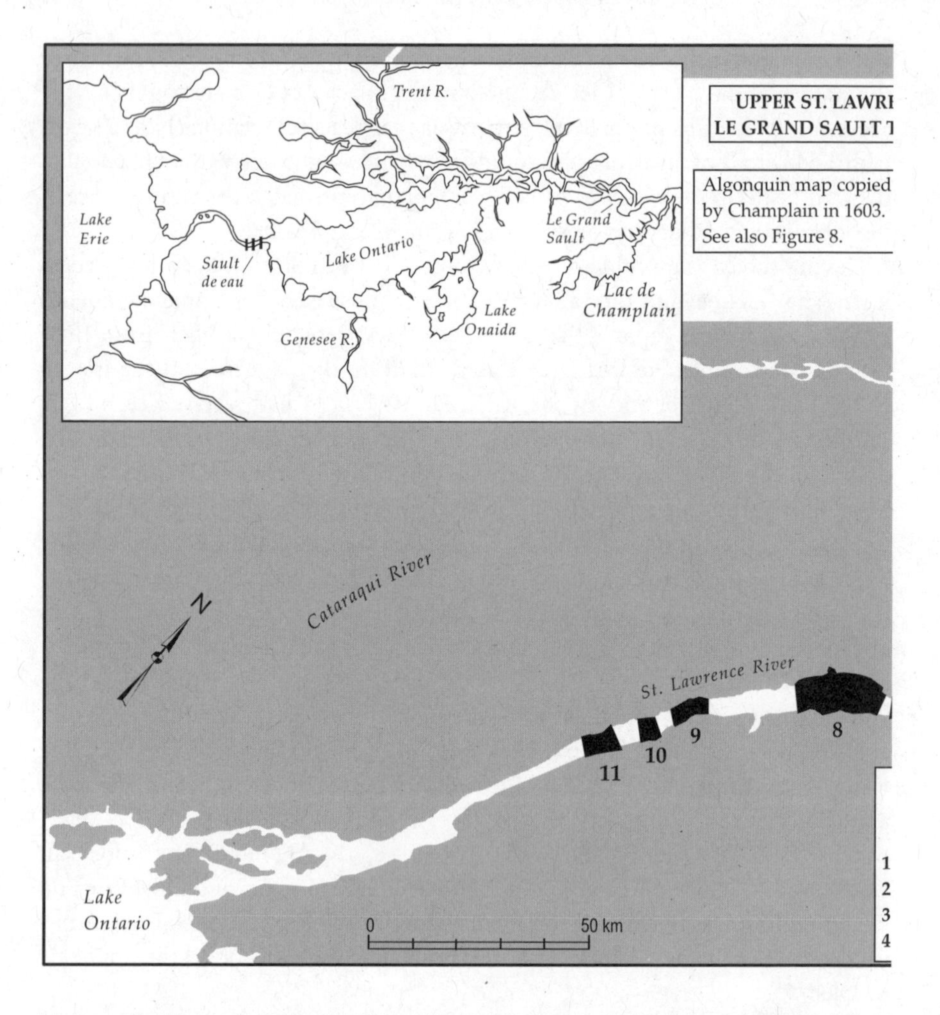

Map 4 The Upper St. Lawrence River: Algonquin Reports of Rapids and Distant Lakes

In addition to a resource survey of the banks of the St. Lawrence, Champlain was to find out more about the water route leading west from *Le Grand Sault*. He obtained three verbal descriptions with sketch maps from Algonquins in July 1603, which outlined an identical geography of the upper St. Lawrence but differed somewhat regarding the western lakes (see fig. 7 and chs. 8 and 9 of *Des Sauvages*). The length and difficulty of the rapids were briefly described by the Algonquins; although the rapids have largely disappeared through the construction of the St. Lawrence Seaway, they were named and placed on French maps by the early eighteenth century. In addition to these rapids, the Algonquins mentioned Niagara Falls as the final *sault de eau* that had to be overcome

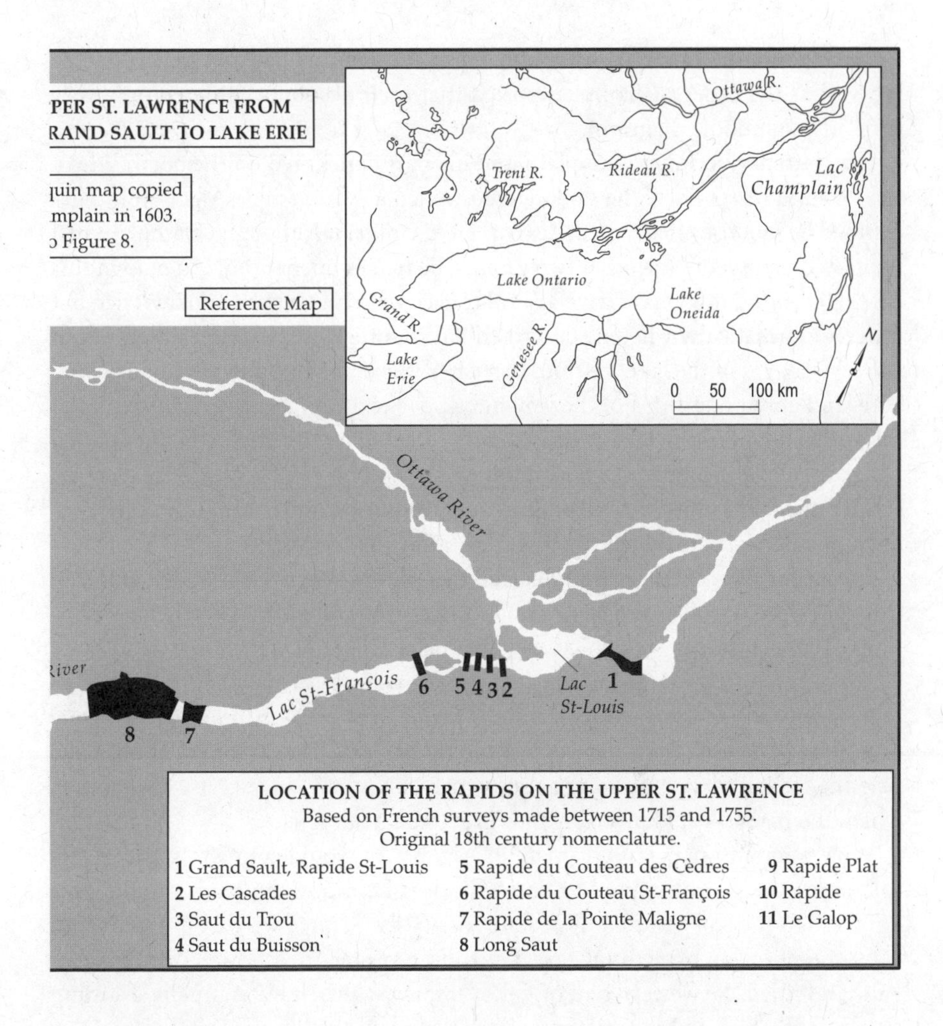

to reach the interior lakes. Of the sketch maps Champlain collected, he seems to have placed more faith in the second one, which he reproduced on his "Carte Geographiqve" of 1612 (see above and fig. 8). These reports confirmed what he had suspected earlier – that the interior of Canada could be explored only with canoes.

River Algonquins (map 4). Although these Algonquins had not penetrated as far west as the others, Champlain noted that their accounts "differed only very little" (fig. 8, middle diagram).

On Thursday, 10 July, Champlain and his party anchored on the north side of the Île aux Lièvres. Here, he said, he questioned a "young male Algonquin" who claimed he had travelled a great deal on Lake Ontario. Although Champlain had requested maps from the other two groups, he makes no mention of a map in this case. His young informant gave a similar account of the upper St. Lawrence but differed dramatically when he described Lake Ontario (fig. 8, lower diagram). In the eastern end of the lake, he said, the water was drinkable, but after one passed an island about one-third of the way into the lake, the water deteriorated and at the western end of the lake it was "totally salty." Like the others, he mentioned Niagara Falls and a huge sea beyond it, whose end no one had seen. He also described the routes into the Algonquin and Iroquois countries and the existence of copper deposits, to which the "good Iroquois" had access.[295]

Except for two sentences that referred to a salty western part of Lake Ontario, this description made perfect sense and is corroborated by the other descriptions. The two sentences (which we have placed here in italics) are as follows: "… they enter a very large lake … [where] the water is drinkable. *But he told us that, continuing on some hundred leagues farther, the water is still worse. Arriving at the end of the said lake, the water is totally salty.*"[296] The statement about Lake Erie that followed said nothing about water quality except that it was *une mer*, which the reader would probably interpret to be a salt sea.

What is one to make of all this? No Algonquin would spin a false story like this unless he was prompted to do so. Even if he did, Champlain should have known that the story was a physical impossibility. A huge sea that drained from west to east in "an exceedingly great current of water" that flowed over Niagara Falls and filled the western half of Lake Ontario with salt water, finally draining out of the eastern end of the lake as fresh water, was a silly fabrication. It was as absurd as claiming that water could flow uphill. If the two sentences about salt water are removed, the whole statement becomes reasonable. Champlain concluded his thinking about these three stories with the statement that the sea the Algonquins had mentioned might be the Pacific Ocean ("sea of the South") some four hundred leagues from the Lachine Rapids. Had Champlain lost his critical faculties or was he trying to give credence to the salty western end of Lake On-

295 "The good Iroquois" were the Iroquoian speaking *Ouendat* (Wendat), later called "Huron" by the French.

296 Document G, "*Des Sauvages*," ch. 9, fol. 27v.

tario and the sea beyond it that had been postulated by Edward Hayes?[297] For his published map (fig. 9), it seems that Champlain opted for the second of the two Algonquin accounts.

On 11 July the expedition reached Tadoussac and immediately set off again, making Île Percé four days later (map 2). This was the farthest point Champlain reached along the south shore of the Gulf of St. Lawrence. The journey seems to have been uneventful, and neither on this trip nor on his other two passages along this section of the St. Lawrence did he make any observations on the physical landscape. At Percé he commented that the entire coast they had passed from the Gaspé to this island was used for the dry and green fishery. During his brief stay at Percé he collected more geographical accounts about the interior river systems from a Native group he called the *Canadiens*,[298] as well as the bare outlines of the coast eastward, including the Baie des Chaleurs, probably from the fishermen who were stationed at Percé.

Before returning to Tadoussac, Champlain mused on the importance of finding a route connecting possible rivers on the coast of Florida (Acadia/Norumbega) to the interior salt sea. What he wrote was in fact a well-constructed summary of those parts from Hayes's treatise.[299] He stated, further, that some Natives had told him of copper and perhaps silver deposits in the interior, upstream in a river that flowed into the Baie des Chaleurs.

The return journey from Percé to Tadoussac began on 19 July. After being delayed by storms and buffeted about on the St. Lawrence, the ship reached the north shore somewhere east of the Rivière Sainte-Marguerite nine days later. Champlain now began his survey of the north coast towards Tadoussac. This survey was completed when he reached Tadoussac on 3 August, having seen nothing but a rugged shoreline, shifting sands, shoals, small islands, and "poor lands … neither as pleasant, nor as fertile as those of the south." In the mouth of Rivière des Escoumins he reported a Basque whaling station but regarded the harbour as "not worth anything at all."

Back at Tadoussac he observed the return of another successful Montagnais war party against the Iroquois, and gave a brief description of their customs before embarking on a campaign. He also noted, for the second time on the 1603 voyage, the importance which the Natives of northeastern North America gave

297 Appendix 5, "Champlain's *Des Sauvages* and Edward Hayes's *Treatise*": "A mighty lake"
298 The *Gaspesiens*, who are the western branch of the Mi'kmaq
299 Appendix 5, "Champlain's *Des Sauvages* and Edward Hayes's *Treatise*": "A river connection"

to dream interpretation. In later years he made use of this on several occasions by telling the Natives his dreams in order to influence certain decisions.[300]

On 16 August the *Bonne-Renommée* began its return voyage to France. Before leaving, it took on board the son of Bechourat (Begourat),[301] the Montagnais headman, and an Iroquois woman – the former so that he could see what the previous year's two "ambassadors" had seen, and the latter to save her from being "eaten." Two days later, at Percé, Champlain met up with Jean Sarcel de Prévert, who had been exploring for minerals. He, too, had Natives on board: a man from coastal Acadia and a woman and two children of the *Canadiens*.[302] How much Prévert actually saw of the mineral deposits he was supposed to be searching for and how much he was told by his Mi'kmaq guides is impossible to say. Drawing on the precedent set by the Montagnais who were promised French aid in fighting the Iroquois in return for permitting the French to settle on the St. Lawrence, the Mi'kmaq offered to take the Frenchmen to rich mineral deposits in return for helping them fight their *Armouchiquois* enemies on the coast of Acadia.[303]

Unfortunately, Champlain repeated two stories that he was told by Prévert. The first was a "description" of the "monstrous … unnatural" *Armouchiquois*; the other of the "frightful monster which the Natives call the *Gougou*." These earned him the ridicule of Marc Lescarbot and other writers. It is interesting to note that although Hakluyt included the *Armouchiquois* description in his trans-

300 See, for example, the dream two nights before the attack on the Mohawk on 30 July 1609, in which Champlain "dreamt" that the enemies would be defeated (Biggar, *The Works*, 3:95; Heidenreich, "The Skirmish," 18). A second incidence occurred on 8 June 1613, when Champlain told the Algonquin headman Tessoüat that he had dreamed that Tessoüat was going to send a canoe to the Nippissing after having told Champlain that he could not take him there because the Nippissings were on unfriendly terms with his Algonquins (Biggar, *The Works*, 3:290–1). Champlain had obtained this information from his interpreter, and it made Tessoüat look deceitful.

301 Gravé Du Pont brought Bechourat's infant son to the court of Henri IV, who decided to have him raised with his own children. He was baptized on 9 May 1604, his godparents being two of Henri's legitimized children by his mistress Gabrielle d'Estrées (1573–99): Catherine-Henriette (b. 1596) and Alexandre (b. 1598). Henri's son, the future Louis XIII (b. 1601), named the baby *Petit Canada* and was very upset when Little Canada died 18 June, aged 17 months (Thierry, *La France de Henri IV en Amérique*, 107).

302 The purpose of these Native "passengers" is not known. Neither do we know whether any of them returned to North America on a subsequent voyage. The *Canadiens* are the western Mi'kmaq.

303 Document G, "*Des Sauvages*," ch. 12. Also appendix 5, "Champlain's *Des Sauvages* and Edward Hayes's *Treatise*": "Prévert's 'copper mines.'"

lation of *Des Sauvages*, he omitted any mention of the *Gougou*. As with the huge lake that was saltwater at one end and freshwater at the other, one can reproach Champlain for being overly credulous; or perhaps, as Charlevoix suggested, it was "the defect of upright souls" who "could not understand that a man who would gain nothing by lying, would do it of his own free will, and believed everything that Prévert said to him in good faith."[304]

On 24 August the ships sailed for France. Twenty-seven days later they arrived at Le Havre and from there proceeded across the mouth of the Seine to Honfleur.

While the French expedition of Gravé, Prévert, and Colombier had been in the Gulf of St. Lawrence and up the river to the Lachine Rapids, the English had again been on the coast of Acadia. A trading expedition of two ships, organized by the merchants of Bristol under the command of Martin Pring, had left England on 29 March, reached New England on 7 June, and explored the coast from about Cape Elizabeth to Cape Cod.[305] This voyage was a direct follow-up to the Gosnold voyage of the previous year. Its objectives were to become familiar with the coast, to find a good harbour where trading could be carried out, and to establish what kind of goods the Natives were interested in. Trading took place along the coast and in Cape Cod Bay, where a post was erected. After two months the two ships headed home, reaching England on 2 October. Among the goods they had traded for were some Native artifacts, including a seventeen-foot birchbark canoe; but the English did not draw the same conclusion as Champlain about its usefulness as a tool for exploration.[306] It is not known if the French were aware of the Pring expedition, because an account of it was not published until 1625, by Samuel Purchas. However, in view of the flow of information between the two countries, it is likely that the French were aware of what the English were up to.

Return and preparations for the 1604 voyage

Upon landing at Honfleur, the participants in the voyage were told that Aymar de Chaste had died during their absence. This was a shock, because further expeditions depended on the transfer of de Chaste's monopoly to an equally well-connected and capable "nobleman whose authority was apt to thrust envy aside"; otherwise, the free traders would emerge again and no progress would be made

304 Document A, "Early Biographies of Champlain," no. 2

305 The details of this voyage are given in Quinn and Quinn, *English New England Voyages*, 212–30.

306 Ibid., 222–3

in settlement.[307] In Champlain's words, some "merchants in France who were interested in this trade began to complain that they were forbidden to traffic in furs, by the trade having been given to one alone."[308]

Champlain set off immediately to find the king, who was at the time in Saint-Germain. There he showed the king the map he had made of the areas he had explored, and gave him "a very special account" that he had made of it.[309] Neither the map nor the account have been found. It is likely that the "very special account" was a verbal commentary on the map, dealing specifically with the possibility of settlement and westward exploration. It is unlikely that this "special account" was *Des Sauvages,* which is too long to have interested the king and contained observations only and not the kind of analysis and recommendations that the king and his council would have wanted. Apparently, the king was "pleased" with what Champlain told and showed him. By now, Champlain had met Henri IV on several occasions, and it is probable that he knew that the king had a strong visual sense, could draw his own maps, and was a great enthusiast for cartographic depictions.[310] A well-drawn map with a commentary related directly to the map would have been the best way to appeal to Henri's interest. After speaking about his explorations, Champlain must have broached the topic of a replacement for Aymar de Chaste, because the king promised "not to give up this project, but to have it pursued and facilitated."[311] In Champlain's opinion, there was no "defect" in de Chaste's enterprise, it "having been well begun." The expedition of 1603 was to be the foundation on which a broader enterprise could be built.

Pierre Du Gua de Mons, the person who emerged to replace de Chaste, had impeccable credentials. A Protestant nobleman, de Mons had been a strong supporter of the king through the religious wars and for his service had been awarded a substantial pension and the governorship of Pons, a town in his native province of Saintonge – the same province in which Champlain was born.[312] Moreover, he had become familiar with Canada when he travelled there with Gravé Du Pont in 1600 during Chauvin's monopoly.[313]

307 Aymar de Chaste died on 13 May 1603. For Champlain's opinions of de Chaste, see Document F, "Excerpts from Champlain's Works," no. 7.

308 Biggar, *The Works*, 3:318

309 Ibid.

310 Buisseret, "Monarchs, Ministers, and Maps," 107

311 Biggar, *The Works*, 3:318

312 *DCB*, 1, "Du Gua de Monts, Pierre"

313 Biggar, *Early Trading Companies*, 42

After preliminary discussions with the king and his council, de Mons was given a commission on 31 October 1603 by Charles de Montmorency. Among Montmorency's other major positions, he was a councillor in His Majesty's Council of State and Privy Council, as well as being admiral of France and of Brittany.[314] In the commission, de Mons was named vice-admiral and lieutenant general of Acadia from forty to forty-six degrees north latitude.[315] He was charged with the exploration of the coast and interior of Acadia, which was believed to have good soil and a good climate; he was to construct forts and establish settlements; to promote Christianity and pursue commerce with the Natives; to get the Natives to submit to the rule of the French crown; to work the mines "recently reported"; and to trade in furs under monopoly conditions for the next ten years in the Bay of Fundy, the St. Lawrence River, and the Bay of St. Clair.[316] Significantly, this expedition was considered particularly important because "certain strangers design to go to set up colonies and plantations in and about the said country of La Cadie, should it remain much longer as it has hitherto remained, deserted and abandoned."[317]

One week later, on 6 November, de Mons submitted to the king a petition composed of seven articles outlining his proposal for the exploration, settlement, and commercial exploitation of Acadia.[318] In substance, it was similar to the commission of 31 October, except that de Mons also petitioned to be able to levy fines of 30,000 francs from anyone caught trying to break the monopoly. This petition was followed two days later by a commission from the king naming de Mons his lieutenant general of *l'Acadie, autrefois appellé Norembèque*, with powers to make war and peace, and affirming all the clauses in the previous

314 The commission is reproduced in Grant, *History of New France by Lescarbot*, 2:217–20.

315 "*Vic'-Admiral & Lieutenant general en toutes les mers, côtes, iles, raddes & contrées maritimes qui se trouveront vers ladite province & region de la Cadie, depuis les quarantiéme degrez, jusques au quarantesixiéme, & si avant dans les terres qu'il pourra découvrir & habiter*" (Grant, *History of New France by Lescarbot*, 2:493).

316 On Champlain's map of 1612, *baye S[t] teclaire* [sic] is the present Grand Bay, about 4 km around the headland to the west of Channel–Port aux Basques, Newfoundland.

317 Grant, *History of New France by Lescarbot*, 218. This is, of course, a reference to the Gosnold expedition and to the exhortation in the Hayes *Treatise* for the English to begin settlement of the Acadian coast between 40 and 44 degrees latitude (Quinn and Quinn, *English New England Voyages*, 168). See appendix 5, "Champlain's *Des Sauvages* and Edward Hayes's *Treatise*."

318 [Blanchet], *Collection de manuscrits*, 40–3

documents.[319] The mention of Norumbega, the English name for Acadia, cannot be accidental. It is probably another reference to the Hayes proposal. On 18 December, these documents were followed by two others: one by de Mons outlining in greater detail his plans to bring settlers and artisans to Acadia, the other by the king to his subjects, confirming the articles given to de Mons and outlining in some detail the areas where his monopoly applied: "from Cape Race to the 40th degree, including the whole confines of La Cadie, Cape Breton, and the land adjacent, the Bays of Saint-Clair and Chaleur, Île Percée, Gaspé, Chischedec, Miramichi, Lesquemin, Tadousac, and the River of Canada, upon either bank, and all the bays and rivers which have their mouths within the said confines."[320]

Disobedience of this prohibition could lead to the confiscation of ships, arms, provisions, and merchandise and a fine of 30,000 francs, all to the profit of de Mons's company. It was a solid monopoly, designed for the maximum benefit of de Mons and his partners and backed by the king, who hoped that this gesture would finally lead to French settlement and commerce ahead of whatever the English were up to. When all these arrangements were in place, Champlain was approached: "The said Sieur de Mons asked me [Champlain] if I would care to take this voyage with him. The desire I had on the last voyage had increased so that I agreed to his request, providing I had his Majesty's permission, which I received on condition I should always make him a faithful report of all I saw and discovered."[321] As with the voyage of the previous year, Champlain felt he had to obtain the king's permission to go, and again the king honoured the request on condition that Champlain give him a full report. This report was presented to de Mons around October 1607 and to the king early in 1608.[322]

As mentioned earlier, not all the members of the king's council were in favour of Henri's colonization projects, especially not the powerful superintendent of finance, the duc de Sully.[323]

Accomplishments of the 1603 voyage

Henri IV and Du Gua de Mons as successor to de Chaste's monopoly both got what they wanted: a survey of the banks of the St. Lawrence River regarding

319 Ibid., 44–5. The document in Blanchet is a *résumé*. The complete document is in Grant, *History of New France by Lescarbot*, 2:211–16.

320 Ibid., 222. A somewhat different version is in [Blanchet], *Collection de manuscrits*, 47. The places mentioned are all fishing areas, ports, and trading stations.

321 Biggar, *The Works*, 3:321–2

322 Report to de Mons (ibid., 2:3) and to the king, (ibid., 4:37)

323 See notes 221–5.

their suitability for settlement and enough information about the rivers to the west of the Lachine Rapids to decide with some degree of certainty whether they connected with a salt sea. What they were given was a report and map from the king's trusted informant, Champlain, and probably another report from Gravé Du Pont, with the assurance that the St. Lawrence Valley was habitable, at least above Quebec, and that the river system west of the Lachine Rapids, according to Native reports, ultimately led to a salt sea. Significant for the French was that this large river valley with agricultural capabilities was currently uninhabited, except during the summer and fall, when small Native groups were engaged in hunting and fishing there. The more permanent villages of the St. Lawrence Iroquoians reported by Jacques Cartier had disappeared. What was disturbing, however, was that the English were also gathering information on the suitability of northeastern North America for settlement and were contemplating the discovery of a navigable connection from the Atlantic coast to the same interior river and lakes system. Having satisfied themselves that it was in France's interests to secure northeastern North America, the king and his agent Du Gua de Mons had to decide how to keep the English out. The only logical solution to this dilemma was to claim the coast of Acadia through exploration and settlement before the English could do the same. Since there was some urgency in the matter, the decision was made to shift operations involving exploration and settlement in the following year, 1604, to Acadia, while continuing the fur trade on the St. Lawrence.

These were the immediate accomplishments of the 1603 voyage. Champlain, however, also observed and favourably commented on the growing French practice of seeking amicable relations with the Native groups they contacted. In later years, he continued and strengthened this practice in order to lay secure foundations for the exploration and settlement of Canada and for trade with its Native inhabitants. In his first voyage, he also laid the methodological foundations of how Canada should be explored: by binding the Natives into French allies, by accepting them as geographical informants, by travelling with them in their canoes, and by adapting to Canadian environmental conditions. These were all observations that led to the rapid French exploration of the North American interior. The first English expeditions reached Lake Ontario between 1685 and 1687. By that time the French had explored and mapped the entire Great Lakes system, the Mississippi River to the Gulf of Mexico, and the Albany, Abitibi, and Saguenay–Lac Mistassini–Rupert river systems to James Bay. It was Champlain's observations and instructions on how exploration should be carried out that laid the foundations for successful French exploration and secured France's position in North America.

With the voyage to the St. Lawrence in 1603, Champlain had come a long way from being a fourrier in Henri's Brittany army. He had become a published author and authority on the suitability of Nouvelle-France for settlement. He had found a future course for his life that gradually moved him from obscurity to undeniable prominence. What got him there were not only his natural and acquired abilities – tenacity, honesty, a capacity for hard work, a desire to improve himself, an uncomplaining nature, skill in surveying, and simple description of observable facts – but also complete loyalty to his superiors and the trusting, active support of his monarch. It is a certainty that without Henri's support, Champlain would never have risen from obscurity and made it to Canada. Fortunately, Henri IV lived long enough as Champlain's supporter that others in authority could take the measure of the man and continue to employ him in a productive capacity. What we do not know is the answer to the intriguing question regarding the relationship between Champlain and his monarch. Was Henri le Grand, *le bon roi*, more than simply the supporter of an able person? Or was there also a relationship between them that was *de naissance*? It is likely that we will never know the answer to that question.

B. Textual Introduction to *Des Sauvages*

1 DEALING WITH CHAMPLAIN'S FRENCH

Champlain's French

Seventeenth-century France is known as the Golden Age of French Literature. It is the age of Corneille, Racine, Molière, Boileau, and La Fontaine – the French classical canon, equivalent to William Shakespeare of England or Miguel de Cervantes of Spain. Both of the latter died in 1616. Five years earlier, in 1611, Shakespeare had concluded his career with the New World theme of *The Tempest*, the same year that the King James Version of the Bible was first published.

Within such a grandiose context, Champlain's *Des Sauvages* of 1603–04 suits a Renaissance classification better than any association with the Golden Age. For convenience, 1610 marks the turn of period to the Golden Age at the assassination of the French king, Henri IV, on 14 May of that year.[1] Yet every age is an age of transition, wherein the old European languages tenaciously cling to their medieval roots even while branching out into new arenas like living organisms. The exuberant burgeoning of the French language would not be clipped and pruned under the strict control of the Académie française until at least 1635, the year of both the academy's foundation and Champlain's death.

In such fertile yet untamed soil, Champlain launched his first journal with no literary pretensions. He did not even sprinkle his text with classical rhetoric, allusions, or comparisons, as did most of his contemporaries, notably Marc Lescarbot, who has achieved more fame in France for his republication of *Des Sauvages* than Champlain seems to have attracted for the original. Unlike Lescarbot, Champlain did not receive a classical education – or ignored it if he did. As Morris Bishop has commented, Champlain "wrote good serviceable French, with none of the graces of scholastic rhetoric. He never ornamented his pages with Latin tags and classical allusions, in an age when the sorriest author cited his school texts for decorum's sake."[2] English Canadian historians tend to have low expectations of the literary merit of the early explorers,[3] but some authors

1 Atkinson, *La littérature géographique*, 9
2 Bishop, *Champlain*, 7
3 Warwick, "Humanisme chrétien," 26

have been particularly scathing in their condemnation of Champlain's French, especially Henry Biggar, the Champlain Society's general editor for the first edition of these voyages: "I never before in my life dealt with so disorderly a writer as Champlain"; and William Grant, who pronounced an even more severe judgment on Champlain's "jog-trot pedestrian style, for the most part as simple and undistinguished as that of an Ontario schoolboy."[4] Grant, as a teacher, may have been voicing his frustration as much with "schoolboy" carelessness in Ontario as with Champlain. Such complaints are not entirely unwarranted from a modern perspective. Nonetheless, *Des Sauvages* compares favourably with the adaptation which the official chorographer (geographical historian) of Henri IV, Pierre-Victor Cayet, made of it in his *Chronologie septenaire* of 1605.[5]

All dismissals aside, we could classify Champlain's Renaissance style within his own historical context as a *style nu* (naked style).[6] The style was perhaps unofficially launched by Michel de Montaigne in his preface *Au lecteur* (To the reader) to the *Essais* that he wrote at intervals from c. 1571 until his death in 1592. The fifth and last posthumous edition was published in 1604 by Abel L'Angelier, an associate of Champlain's first publishers, Claude de Monstr'œil and Richer III.[7] Here Montaigne states clearly:

> *Que si j'eusse esté parmy ces nations qu'on dict vivre encore sous la douce liberté des premieres loix de nature, je t'asseure que je m'y fusse tres-volontiers peint tout entier, et tout nud.*[8]
> (For if I had been among these nations said still to live in the sweet freedom of nature's primal laws, I assure you that I would very willingly have depicted myself whole, and wholly naked.)[9]

Montaigne's metaphor, comparing his own naked honesty to a natural state devoid of clothes, is clearly inspired by his idealized view of the New World Native. Is it a coincidence that in the 1615 edition of *Le Mercure François*, Estienne Richer's *avis au lecteur* (notice to the reader) reflects the same attitude?

4 Heidenreich, *Champlain and The Champlain Society*, 36
5 See Document H in both French and English.
6 Holtz, *Pierre Bergeron et l'écriture du voyage à la fin de la Renaissance*, 176, citing Estienne Richer's "avis au lecteur" of 1615
7 Balsamo et al., *Essais*, lxxv–xc. See also "Printing History," below.
8 Montaigne, *Essais*, f. 2
9 The English words "primal" and "whole, and wholly naked" have been adopted from the translation by M.A. Screech, *The Essays: A Selection*, 3.

Quand au stile de ce livre, tu ne le trouveras pas grave, ou enrichy de fleurs de bien dire; ains seulement un recit bref, simple & nud des choses comme elles sont advenuës, ou comme elles ont esté escrites & publiees.[10]
(As for the style of this book, you will not find it to be heavy, or enriched with flowers of rhetoric; rather, only a brief, simple and naked account of affairs as they happened, or as they have been recorded and published.)

It is this tradition of naked honesty and style to which Champlain belongs. His dedication of *Des Sauvages* to Charles de Montmorency is admittedly less self-conscious from a literary point of view: "*je n'ay voulu pourtant m'arrester à leur dire, & ay expreßément esté sur les lieux pour pouvoir rendre fidelle tesmoignage de la verité.*"[11] Yet by recording the essentials of what he personally experienced on the 1603 voyage in a straightforward, empirical report, Champlain places himself squarely in the Renaissance tradition.[12]

It may surprise North American readers to learn that there is little evidence of a local dialect in Champlain's French. On the other hand, there is no reason to doubt Champlain's own testimony that he came from Brouage, near Rochefort, some 35 kilometres south of La Rochelle and 113 kilometres north of Bordeaux in the Charente-Maritime area of Saintonge. The linguistic region has been named Franco-Provençal, Middle-Rhône, and, more recently, Saintongeais-Poitevin in the French southwest, an area intermediate between southern and northern France.[13] As popularized by Dante Alighieri, the southern region has been called the *langue d'oc* (the language of *oc*), *oc* having developed from the Latin *hoc* (this) into "yes," and the northern region the *langue d'oïl* from *hoc ille* (that), another form of "yes." The traditional boundary between these two principal regions was the Loire Valley. From the conquest of Poitou and Aquitaine by Louis IX in 1242 until the Ordinance of Villers-Cotterêts of 1539 – which made French the only of-

10 Estienne Richer, *Le Mercure François*, 1615, f. ãiij^{v}. University of Toronto electronic resources: 546160.

11 Letter to Charles de Montmorency, *Des Sauvages*, ãij. Montaigne's diatribe against travellers who invent fictions to amplify their own experience, in his essay "On the Cannibals," is well known: "*Je voudrroye que chacun escrivist ce qu'il sçait, et autant qu'il en sçait*" ("I would wish that everyone would write what he knows, and as much as he knows of it"), implying not more than that (Balsamo et al., *Essais*, 211).

12 Although Lescarbot criticized Champlain for including an account of the *Gougou* in the last chapter of *Des Sauvages*, Champlain does set a limit to his credulity: "*si je mettois tout ce qu'ils en disent, l'on le tiendroit pour fables*" (f. 36); see Thierry, "Champlain et Lescarbot," 121.

13 Pope, *From Latin to Modern French*, 17–18, 500; Gautier, *Grammaire*, 6; Huchon, *Le français*, 39–40.

ficial language of the realm, with its obligatory use in legal documents (clause 111) – the crown set about legislating the French standard from its seat of power in Paris. The gradual centralization of power influenced the southwestern regions to the extent that the northern border of the *langue d'oc* gradually slipped south to the Gironde, today an estuary of the Garonne River emptying into the Bay of Biscay. Consequently, Saintongeais is now identified with Poitou, between the Loire and the Gironde, but retains vestiges of what is called an *oc* substratum, the older *occitan* or *provençal* tendencies.[14]

Only four words within the thirteen short chapters of *Des Sauvages* exhibit traces of the Saintongeais dialect of western France. All are nouns and are given here in order of convenience: *treffes* (truffles, f. 16), *groizelles* (currants, ff. 16, 21), *esbouillonnement* (from *un esbouillonnement estrange*, a strange bubbling forth, f. 23), and *pible* (poplar tree), for modern French *peuplier*, in chapter 8 (f. 21).[15] The modern French form for *treffes* would be *truffes*, for *groizelles* it would be *groseilles*, and for *esbouillonnement* it would be *bouillonnement* (n.m. bubbling or boiling). *Treffes* does not appear in any of the regional dictionaries as "truffles," but Agrippa d'Aubigné, who was born just south of Saintes and Brouage in 1552 (new style) and was educated in Saintonge until the age of ten, used *treffles* to describe the truffles of his cultural heritage.[16] This suggests that he pronounced the medial vowel as Champlain did. In the *Dictionnaire poitevin-saintongeais*, the remaining two regionalisms are written *gruséle*, *groeséle*, or *grouséle*, and *éboujhement*, the latter defined as an action of setting in motion, literally "a budging."[17]

One of the characteristics of Saintongeais is the ellipsis of the subject pronoun, but more often in the first person singular or plural (*jhe* or *nous*), or in the neuter pronoun (*ol*), which Champlain does not use.[18] Champlain tends to include both *je* and *nous*, the former to express his own opinion and the latter when referring

14 Gautier, *Grammaire*, 7; Knecht, *Renaissance Warrior*, 353. We are indebted to Professor Emeritus Rober Taylor, specialist of *occitan*, for guidance in this respect. See also Eric Nowak, http://www.troospeanet.com/article.php3?id_article=148. The necessity for one common language in France would not override the local dialects until 1680–1700 (Huchon, *Le français*, 42).

15 We are indebted to Mme. Simoni-Aurembou for her confirmation of three of these western regionalisms: *groizelles*, *esbouillonnement*, and *pible* (private correspondence, 17 February 2008). See also Simoni-Aurembou, "En quelle langue a écrit Samuel de Champlain," 1–7.

16 Lazard, *Agrippa d'Aubigné*, 13–24; see Document G, "*Des Sauvages*," note 113.

17 *éboujhement* (note m.) 1. action de se mettre en mouvement (Pivetea, *Dictionnaire poitevin-saintongeais*, 119, 376). See G, "*Des Sauvages*," notes 157, 159, and 177.

18 Doussinet, *Grammaire saintongeaise*, 125–6

to movements that the entire expedition is making. The subject pronouns that Champlain drops are usually in the third person: *il*, *ils*, and the *il* of *il y a*. This is not so exceptional for the period. We find similar tendencies, most notably in Jean Nicot's *Thresor de la Langue francoyse* of 1606.[19] To a reader today, Champlain's use of disjunctive pronouns as personal pronouns in the nominative case may be just as disorienting. At the conclusion of Anadabijou's speech in chapter 2 of *Des Sauvages*, for example, Champlain writes: "*nous sortismes de sa Cabanne, & eux commencerent à faire leur Tabagie*" ("we went out of his lodge, and they began to make their *Tabagie*," f. 4^{v}; our underlining). Here, the *eux* (disjunctive pronoun) functions as *ils*. Likewise, one finds *luy* (disjunctive pronoun) for *il* in the Brittany records for 1595: "*… lequel Rouxeau auroit depuis baillé audict Loppin pour deniers contans les susd. acquictz & paiement des sommes que luy a fournies*" ("… which Rouxeau would have handed over since then to the said Loppin for cash, counting the above-said receipts and payment of sums which he furnished," f. 525; our underlining).[20] In both cases, the function of *eux* and *luy* do not appear to be particularly emphatic, as one would expect.

What makes Champlain's habits more difficult to follow are the sudden changes he makes in subject without notifying the reader. For example, "*Dieu luy donna de la viande, & en mangea*" (f. 9^{v}) in chapter 3 of *Des Sauvages* leads the reader to assume that *Dieu* (God) *donna* (gave) and *mangea* (ate), which cannot be the case. Similarly, a little farther on in the same chapter, there is an abrupt switch in subject from the *je* of "I believe" to the third person plural *sçavent* of "they know," by way of *il y a*, as in "there is no law": "*Voilà pourquoy je croy qu'il n'y a aucune loy parmy eux, ne sçavent que c'est d'adorer & prier Dieu*" ("That is why, I believe, they have no law among them, nor know what it is to worship and pray to God," f. 11). The suppression of *il y a* allows "they" to be introduced without disruption. In chapter 8 again, it is the third person plural pronoun that is missing: "*Nous veismes quantité d'isles, la terre y est fort basse, & sont couvertes de bois*" (f. 21^{v}). In this case, the subject of *sont* is reiterated for clarity: "We saw a great many islands. The land there is very low and these islands are covered with trees." It is as if Champlain was relying on the inflections of his verbs to represent their subjects. His childhood introduction to Saintongeais and his adult exposure to Spanish may have exaggerated a tendency that even mainstream French was reluctant to abandon. In comparison with other romance languages, such as Spanish, French is the only one to have developed, like English, obligatory subject pronouns for all forms of the verb apart from the imperative.

19 Nevertheless, by about 1550 Ronsard's appeal for the use of French personal pronouns before verbs was taking effect in France (Gougenheim, *Grammaire*, 68).

20 Document B, "Personnel and Pay Records"

In other respects, Champlain's orthographic tendencies reflect the odd and oscillating variety that was common in an age before dictionaries. The use of the apostrophe to mark the feminine ending of the adjective *grand'* (as in *grand' riviere*) was commonly found in *grand' mère* until 1935, when the Académie française modified the usage to *grand-mère*.[21] This apostrophe could also have been added by the printer, since the forms *grand Riviere* (f.), *grand Baye* (f.), and *grand poche* (f.) without the apostrophe coexist with *grand'* in *Des Sauvages*. Extra letters from a twenty-first-century point of view are often indicative of the Latin etymological origin of the French words; for example, the etymological *s* in *isle* (from the Latin *insula*) and the etymological *c* in *faict* (from the Latin *factum*). Often the loss of these letters will be signalled by a circumflex; e.g., *îles*, and *côte* for *coste* (coast), etc., but the modern *chaque* for *chasque* is an exception. Such Latin remnants have also been called "dead consonants" (*consonnes mortes*) since they were no longer pronounced; e.g., the *l* in *sault* (*saut*: rapids) and *c* in *dict* (*dit*: said).[22] Extra vowels may reflect either the same phenomenon, the growth of the vernacular language out of Latin, or differences in pronunciation; for example, the *i* in *montaignes* (*montagnes*: mountains) may reflect the pronunciation, whereas the extra *e* in the principal parts of verbs, such as the past participles *veu* (*vu*: seen) and *peu* (*pu*: been able), or literary past forms of the infinitive *veoir* (*voir*: to see), *veis* (*vis*: I saw) and *veismes* (*vîmes*: we saw); of *être* (to be or to go), i.e., *feusmes* (*fûmes*: we were or we went); and of *faire* (to do), i.e., *feismes* (*fîmes*: we did), are merely orthographic. As for accents, the conventions governing their usage were likewise still developing. The unsteady application of accents to the monosyllabic homophones *a* (verb) and *à* (preposition), and *ou* (or) and *où* (where) is a common feature of the time. Champlain's use of the acute accent on a final *-é* when it is masculine, but not for the feminine *-ee*, was equally conventional in that period. Its sporadic appearance in a medial position, in *expreßément* (f. ãij), for example, and the lack of the medial grave accent on *e* (*è*) in *Des Sauvages* also were common currency for the time.[23] In all such cases, the typographical habits of the compositor who set the text for Claude Monstr'œil were likely to interfere with Champlain's actual orthography. The orthographic

21 Catach, *Histoire de l'orthographe française*, 349

22 Geofroy Tory named superfluous consonants *consonnes mortes* in his *Champ fleury* of 1524–29 (Catach, *Histoire*, 108).

23 Such orthographic tendencies persisted until the end of the seventeenth century and the grave accent was only introduced very slowly in the eighteenth century (Catach, *Histoire*, 146–7).

variation between the two editions of *Des Sauvages* attests to the whims of the compositors.[24]

The sea change in French attitudes towards Champlain is apparent in the work of Marie-Rose Simoni-Aurembou, director of research emeritas of the Institut de linguistique française at the Centre national de la recherche scientifique (CNRS). Mme Simoni-Aurembou notes that Champlain wrote naturally and spontaneously in French but must have been writing in haste and discomfort, with little chance to revise his work before publication.[25] We can see signs of the immediacy of his reportage in his frequent use of verbs in the present tense, even when the literary past is his preferred tense. The same oscillation between the present and the literary past can be seen in the journals attributed to Jacques Cartier. For example, after Champlain and Gravé Du Pont have disembarked into their skiff with a few Natives in order to explore the St. Lawrence River from the area around present-day Montreal to the Lachine Rapids (ch. 8), Champlain reports what *on void* (one sees). On the other hand, when describing the landscape or Native customs, he seems to switch to the present tense in recognition that these lands and customs were ongoing realities, whether or not the French were there to observe them. In a subtle way, this habit suggests a certain respect for Native autonomy. The latter case accounts for the oscillation between past and present in chapters 2 and 3, as well as in Champlain's descriptions of the *Armouchiquois* on the New England coast in chapter 12. Most of the time, Champlain seems to be drafting his log just after the event to which he refers in the past. His last chance for revisions, or for transforming foul papers into a fair copy, would most likely have occurred on board the ship during the home crossing.

Occasionally, Champlain's descriptions of the Natives become judgmental or sententious. For example, he twice describes them as wicked because of their perceived propensity to lie and seek revenge for damages (ch. 3, f. 8; ch. 12, f. 33). His sententious "*promettent assez & tiennent peu*" ("they promise much and perform little," f. 8) seems to repeat a standard judgment like the French proverb: "*De grand langage / peu de fruict, grand dommage*" ("Little fruit from big words, great harm") or, even closer, like that of the langue d'oc: "*Toujours promettre et ne jamais tenir, c'est conforter l'extravagant*" ("Always to promise and never perform is to confirm extravagance").[26] Although such harsh judgments seem

24 See "Establishing the French Text," below.

25 Simoni-Aurembou, "En quelle langue a écrit Samuel de Champlain," 1–7

26 Le Roux de Lincy, *Le livre des proverbes français*, 2:282; "*Toujhour proumëtrë é noun tënë, ës lou fat ëntrëtënë*" (Trinquier, *Proverbes et dictons de la langue d'oc*, 189, no. 1220).

uncharacteristic of Champlain, there is no linguistic or textual evidence in *Des Sauvages* for the interpolation of any material from a source outside his own experience. At the same time, Champlain's judgments could easily have been influenced by his more experienced mentor, François Gravé, sieur Du Pont, who had experienced life at the trading post in Tadoussac since 1600.

Champlain exhibits a certain eclecticism in adopting foreign vocabulary into his French. On his first trip to Canada he may have introduced the Native words *matachia* and *tabagie* into the French written tradition. One nautical term found in *Des Sauvages* is likewise unattested before Champlain's usage: the verb *variser* (to tack);[27] also *petunner* (to smoke).[28] In this his first journal as well, Champlain uses the term for "corn" that was in common use among the Spanish: *bled d'Inde* (or *blé d'Inde*, s.m., ch. 9, f. 28), literally "Indian corn," rather than that preferred in his native Brouage. In Brouage, they called this same corn the *blé d'Espagne*,[29] thereby dissociating themselves from the Spanish encounter with maize in the West Indies. Champlain's usage may have been influenced by the two years he spent among the Spanish on the *San Julián* in the West Indies.

It is also interesting to note that Champlain does not name the places he comes across in Canada but repeats toponyms that had become customary since Cartier. It was on one of Jacques Cartier's expeditions that *Saincte Croix* (Holy Cross, ch. 6) was named. Most often, the name had been chosen according to the saint's day upon which the place was discovered.[30] Although Champlain repeats such European place names as *cap de Saincte Marie* (Cape St. Mary's, f. 2ᵛ), *poincte de Sainct Mathieu* (St. Mathew Point, ch. 2), the islands of *sainct Pierre, sainct Paul* (f. 2ᵛ), *sainct Eloy* (f. 17), and *sainct Jean,* along with the river *saincte Marguerite* (f. 31), he is not responsible for naming them. On the other hand, what we know today as the St. Lawrence River he calls *la riviere de Canadas.* The latter is most remarkable for three reasons: the consistency of its usage in *Des Sauvages*, the Native perspective that it reflects, and the plural inflection of *Canadas.*

The use of *Canadas* in French literature is attested to at least twice before *Des Sauvages* was published. Marguerite de Navarre named the northern continent the *isle de Canadas* (the island of Canada) in her sixty-seventh story of the *Heptaméron*, reflecting the French hope around 1550 that Canada was only a minor

27 Huguet's *Dictionnaire de la langue française du seizième siècle* fallaciously attributes the first usage of *variser* (mod. Fr. *louvoyer*) to Marc Lescarbot, but Lescarbot was merely republishing Champlain in his *Histoire de la Nouvelle France* of 1609 (HU, 7:405).

28 See Document G, "*Des Sauvages,*" notes 35–6.

29 Simoni-Aurembou, "En quelle langue," 4

30 Huchon, *Le français*, 57

obstacle en route to the Orient.[31] Cartier's first and second journals, which were published in Rouen by Raphaël du Petit Val in 1598, attest to the second occurrence of *Canadas* in the title and preliminaries: *Discours du voyage fait par le capitaine Jaques Cartier aux Terres-neufves de Canadas, Norembergue, Hochelage, Labrador, & pays adjacens, dite Nouvelle France* [*etc.*], in the "Printer to the Readers" and an opening poem by an unidentified "C.B." Nevertheless, the body of the texts, translated from Ramusio's Italian, exhibits our fixed convention: Canada.[32] The initials C.B. may refer to Claude Brissard, but Raphaël du Petit Val is known to have spoken Italian, and consequently it is likely that he himself did the French translation from Giovanni Battista Ramusio's Italian version.[33] It is all the more curious, if such was the case, that this publisher allowed such a discrepency between the *Canadas* of the preliminaries and Ramusio's "Canada."[34]

Notwithstanding these precedents, Champlain's usage of *Canadas* is almost entirely consistent. At the first printing of *Des Sauvages*, there are sixteen occurrences of *Canadas* against three of "Canada." Moreover, two of the latter are corrected to *Canadas* for the second edition, suggesting that it was always Champlain's intention to write "*Canadas*."[35] In stark contrast to the proliferation of *la riviere de Canadas, la grande Riviere de Canadas, l'entree de la grande baie de Canadas,* and *la terre de Canadas*, the absence of *Nouvelle France* (New France) in *Des Sauvages* is remarkable. The only instance of the latter European perspective arises in the title: *Des Sauvages ou, Voyage de Samuel Champlain, de Brouage, fait en la France nouvelle,* where the adjective "new" (*nouvelle*) simply extends the realm of France into a new region. Clearly, such a claim is aimed at a French audience. As for *Canadas,* it is unlikely that the final *s* would have been pronounced in French, when final consonants had generally become mute by the mid-sixteenth century.[36] Furthermore, the *terre de Canadas* (f. 1^{v}) is a significant indicator that Champlain conceived the country as being one land, not in the plural. Moreover, he always uses *de Canadas* and *en Canadas* instead of *des Can-*

31 Claude de La Charité's preface, in Huchon, *Le français,* 12

32 Grégoire Holtz, working at the Bibliothèque nationale de France (BnF) kindly verified that *Canadas* appears only in the title in *L'Imprimeur aux Lecteurs* (the Printer to the Readers, f. aij), and in the opening poem by C.B., "Sur le voyage de Canadas."

33 C.B. is tentatively identified by Michel Bideaux as Claude Brissard, active in Orléans at the time (Bideaux, *Relations*, 46; Arbour, "Raphaël du Petit Val, de Rouen," 88).

34 Ramusio, *Terzo volvme*, ff. 441^{v}–453^{v}

35 "Canada" is corrected to *Canadas* in the chapter heading to chapter 2 and in the listing for it in the Table of Chapters at the beginning.

36 Huchon, *Le français*, 47

adas or *aux Canadas*. This suggests that *Canadas* was singular for him, just as it was for Jacques Cartier (or his Italian translator), for whom *le pays de Canada* and *la prouvynce de Canada* meant the Laurentian region.[37]

Champlain clearly uses *Canadas* as a feminine noun (*en Canadas*) in the singular, possibly by analogy with *la Nouvelle France*.[38] Another possible influence might be a Spanish place name. Of the two Cañadas in Spain,[39] Champlain would certainly have encountered the volcanic caldera (basin-shaped crater) on Mount Teide (*El Teide*) in Tenerife, Canary Islands, called Las Cañadas in the feminine plural. The Canary Islands, now actually part of Africa, were a Spanish colony from 1479 to 1978. They were known to Antiquity as the Fortunate Isles, though Tenerife developed a reputation as the *Isola dell'Inferno* (the Island from Hell).[40] The caldera now measures 16 by 9 kilometres in length and breadth, while Mount Teide itself is "the third largest volcano on a volcanic ocean island" at 3,718 metres above sea level.[41] Champlain spent six days in the Canaries after his departure on the *San Julián* from Sanlúcar de Barrameda for the West Indies in January 1599, according to the *Brief discours*. Although his map of the area does not mention Las Cañadas, in common with other maps of the period,[42] it is quite likely that Spanish sailors en route to the West Indies took their bearings from the complex system of volcanoes and caldera, subsumed under the name Las Cañadas. If the hypothesis is correct that Champlain had not read Jacques Cartier's journals before his first trip to Canada,[43] he may, consciously or unconsciously, have transferred the Spanish toponyme to the New World when he heard an almost analogous place name being used there by the French. Because of his Spanish experience, he may have assumed the orthographic existence of

37 Morissonneau, *Langage géographique*, 33

38 The tendency to think of Canada as feminine by analogy with France did not cease upon the defeat of the French by the English, since private family papers (*inédits*), for example, dating from c. 1890 and linking the Canadian family Fournier dit Préfontaine with a Parisian (Fornerius), who died in 1534, oscillate indiscriminately between a masculine and feminine gender for Canada.

39 The other Cañadas in Spain is present-day Medina de las Torres, about 10 km south of Zafra or 40 km south of Almendralejo.

40 It is first named the *Isola dell'Inferno* on the Medici Portolan of 1351 (Broekema, *Maps*, 3).

41 http://volcano.und.edu/vwdocs/volcimages/africa/tenerife.html; *Wikipedia*, http://wikipedia.org/wiki/Teide and http://fr.wikipedia.org/wiki/%C3%8Eles_Canaries

42 See plate 6, attributed to Champlain (Biggar, *The Works*, 5 portfolio), and that of Willem Barentsz of 1595 (Broekema, *Maps*, plate 2).

43 The distinction between the two explorers' allusions to Canada extends to maps. No original map representing Cartier's voyage uses the name *Canadas*, yet this form appears on all three of Champlain's earliest maps, those of 1612, 1613, and 1616.

the final *s* in the mouths of his French compatriots, even though it was unpronounced. He could have forgotten the Spanish *ñ* sound and accentuation on the second syllable when he was beyond the oral range of Spanish speakers. On the other hand, he may have been aware of whatever French tradition influenced Marguerite de Navarre and Raphaël du Petit Val.[44]

Spanish influence in sixteenth- and early-seventeenth-century France is not so surprising. Yet Champlain was directly engaged in the war against Spain until the Treaty of Vervins and his voyage to the West Indies on a Spanish ship. In this light, his readiness to adopt new vocabulary from the potentially hostile foreigner – both the Spanish and the Canadian Native – reveals an open and eclectic tolerance in his character. His ability to convert adversaries into allies by adopting their customs, vocabulary, and possibly even linguistic habits ought to be appreciated, rather than denigrated as inferior to the florid rhetoric of contemporary humanists. In any case, his journal fulfilled its primary function by communicating his direct experience in an empirical *style nu*.

Principles of translation

As a general rule, the translation is literal. The meaning of "literal" is not as simple as it seems, however, since a word-for-word translation (x = x) is impossible between any two languages in the best of circumstances. One advantage that the French language has over English is gender. The distinction between two genders helps to identify antecedents to pronouns, adjectives, and past participles in a way that cannot be replicated in English. Consequently, for clarity in English, we have sometimes had to supply the noun to which the French pronoun refers. Champlain had a habit of reporting all that the Natives told him in one long run-on sentence – all of it dependent on "he said" (*il dit*), as evidenced by the repetition of "that" (*Que*) before each new clause. In order to translate these sentences into manageable units for the twenty-first-century reader, we have brought the clauses to full stops with periods and have repeated the implicated "he said" before the "that." Conversely, on occasion, we have trimmed off Champlain's circumlocution as being redundant in English. Finally, there are many instances of an inversion in Champlain's French of what would be a natural English word order. In all such occurrences, we have not hesitated to adjust

44 No historian or linguist, to our knowledge, has yet adequately explained the sporadic occurrence of *Canadas* in sixteenth-century French literature and its explosive finale in *Des Sauvages*. There have been other whimsical attempts to link Canada to a Spanish etymology, however. See Carpin, *Histoire d'un mot*, 49–52, for a brief summary.

his syntax to a natural flow in English, just as any translator would reposition the French adjective, following its substantive, to precede the English noun.

More significantly, the first principle of this translation is fidelity to the full meaning of the French text and, whenever possible, the authorial intention. If the meaning is unclear, the diverse possibilities are discussed in the footnotes. This fidelity also connotes an underlying respect for both Champlain's intentions and the dignity of the First Nations. Past translations have occasionally misrepresented Champlain when they have assumed him to be at fault, instead of making more of an effort to fathom his true meaning. Similarly, the choice of descriptors applied to the Natives can carry pejorative overtones to which we are more sensitive in 2010. Consequently, an avid attention to Champlain's vocabulary and its etymological development has allowed us to translate it in fresh ways that – we hope – avoid any pejorative resonance. Although we, too, most certainly reflect the biases of our times in ways that we cannot fully control, we have made every effort to avoid projecting our own assumptions onto *Des Sauvages*.

Such projections slipped all too easily into the first Champlain Society translation. The earlier edition, for example, assumed that the Natives always took the scalps of their enemies as trophies, when the text actually reads *testes* (*têtes*).[45] Re-reading the evidence in its historical context, we call a *tête* a *tête*.[46] Similarly, the French adjective *salubre* from the Latin *salubris*, which gave rise to "salubrious" in English as well as "salut" (hail, health, salvation) in French, was mistranslanted as "brackish" (slightly salty) in the previous Champlain Society edition, following the London edition of *Purchas His Pilgrimes* of 1625, which was reprinted by Edward Gaylord Bourne in 1906. The Bourne edition explains that "Champlain's use of the word seems to imply that he associated *salubre* with 'sel,' salt, and 'salé,' salty."[47] We feel that from a Native perspective on healthy water, "wholesome" and "drinkable" are more faithful to Champlain and his native tongue.[48] Similarly, we have dared to modernize the traditional translation of *chaudière* and *chaudron* as "kettle" and replace it with "cauldron." The Latin root *calidus* also gave rise to *chaud* (hot) and *chauffé* (heated) in French. Consequently, *chaudière* and *chaudron* designate heating receptacles. Since the *chaudière* in French has been associated with the "steam boiler" since the nineteenth

45 See Document G, "*Des Sauvages*," note 42.

46 According to William Barker, the English expression "to call a spade a spade" descends from a misunderstanding by humanist Desiderius Erasmus over classical Greek. The equivalent in French is *appeler un chat un chat* (Barker, "Pandora's Box: To Call a Spade a Spade," presented at the Canadian Society of Renaissance Studies, Ottawa, 1998).

47 Bourne, *Algonquians, Hurons, and Iroquois*, 224n20

48 See Document G, "*Des Sauvages*," note 232.

century, and the English word "kettle" suggests an electric receptacle in which to boil water for tea, we have sought to replace the terms in their proper context by using "cauldron" (< *calidarius*) for both. *Chaudron*, in the strict French sense of the term, with its diminutive suffix *-eron*, is merely a smaller form of *chaudière*.

Champlain's use of *descouverte* or *descouverture* (discovery) is also problematic, not just for us but also for him in his own time. When the merchants of Saint-Malo asserted in 1613 that Champlain could not claim to have discovered Canada, they were attempting to expose a lie and thereby discredit him. But Champlain used the two French nouns in a broader sense. First of all, he clearly recognizes Cartier's previous achievements when he credits him with the discovery of a river near the Batiscan in chapter 6 of *Des Sauvages*: "*Jacques Quartier au commencement de la descouverture qu'il en fit*" (f. 16). This usage of *descouverture* is very close to what we would call exploration, but the ambiguity between exploring the unknown and discovering for the first time remains. In 1603 Champlain aimed to discover the source of the St. Lawrence River and new possibilities for permanent settlement. Although we might call this exploration, we have chosen to preserve the ambiguity of the term "discovery" in order to account for the complaints of the merchants of Saint-Malo.

Champlain's tendency to oscillate back and forth between the present and the past sometimes makes the verbs tricky to translate. For example, the first folio starts out entirely in the literary past until one verb in the present tense pops up: "*nous vinsmes si proche de terre que nous oyons la mer à la coste*" (f. 2). We have translated this sentence as if it read "*nous vînmes si proche de terre que nous oyions la mer battre à la coste*" ("we came so close to land that we heard the sea beating against the shore"), instead of "we hear the sea." In another passage, on ff. 21 to 21^{v}, beginning "*Nous passasmes contre une isle qui est fort aggreable*" and continuing to "*Nous veismes quantité d'isles, la terre y est fort basse*," he twice passes from the literary past to the present in interrelated clauses. The two main verbs in the first person plural, *passasmes* and *veismes* (*passâmes* and *vîmes* in the modern French literary past) are followed in both cases by *est* in the present tense. Nevertheless, we have had to translate them all in the past to conform to twenty-first-century expectations. Since the prevalence of such verbs in the present tense are significant indicators of the immediacy of Champlain's reportage or his awareness of an ongoing reality, as argued above, we resort to such minor adjustments as seldom as possible.

When checking vocabulary whose meaning has changed since the seventeenth century, we have consulted the *Dictionnaire historique de la langue française*, Huguet's *Dictionnaire de la langue française du seizième siècle*, and the *Trésor de la langue française* as starting points and then Beaulieu and Ouellet's French edition of *Des Sauvages*, which includes a glossary, for comparison. All these sources rely on early dictionaries of the Académie françaíse and Furetière's dictionary of 1690. If a term remains ambiguous or still presents difficulties, we have consulted Jean Nicot's *Thresor de la langue francoyse* of 1606 and, more rarely, Wartburg's *Französisches Etymologisches Wörterbuch*.

2 PRINTING HISTORY OF *DES SAUVAGES*

The following is a discussion of the major printings of *Des Sauvages* in French and in English translation. The section is divided into original printings that bear Champlain's name and were prepared by him; reprints, translations, and abstracts printed during his lifetime; and later reprints, abstracts, and translated versions of his original texts. Every effort was made to discover the number and location of remaining original printings. Not included in this discussion are microform versions of the original and subsequent texts, although it should be noted that all the texts discussed here that were printed before 1970 exist in microform versions.

Original printings

n.d. [1603]

DES / SAVVAGES, / OV, / VOYAGE DE SAMVEL / CHAMPLAIN, DE BROVAGE, / fait en la France nouuelle, / l'an mil six cens trois: / CONTENANT / Les mœrs, façon de viure, mariages, guerres, & habi-/tations des Sauuages de Canadas. /

De la descouuerte de plus de quatre cens cinquante / lieuës dans le païs des Sauuages. Quels peoples y ha-/bitent, des animaux qui s'y trouuent, des riuieres, / lacs, isles & terres, & quels arbres & fruicts elles pro-/duisent.

De la coste d'Arcadie, des terres quel'on y a descouuer-/tes, & de plusieurs mines qui y sont, selon le rapport / des Sauuages.

A PARIS, / Chez CLAVDE DE MONSTR'ŒIL, tenant sa / boutique en la Cour du Palais, au nom de Iesus. / AVEC PRIVILEGE DV ROY

1604

DES / SAVVAGES, / OV, / VOYAGE DE SAMVEL / CHAMPLAIN, DE BROVAGE, / *faict en la France nouuelle, l'an / mil six cens trois.* / CONTENANT, / Les mœrs, façon de viure, mariages, guerres, & habi-/tations des Sauuages de Canadas.

De la descouuerture de plus de quatre cens cinquante / lieuës dans la pays des Sauuages, Quels peoples y ha-/bitent, des animaux qui s'y trouuent, des riuieres, lacs, / isles, & terres, & quels arbres & fruicts elles produi-/sent.

De la coste d'Arcadie, des terres que l'on y a descouuer-/tes, & de plusieurs mines qui y sont, selon le rapport / des Sauuages.

A PARIS, / Chez CLAVDE DE MONSTR'ŒIL, tenant sa bou-/tique en la Cour du Palais, au nom de Iesus. 1604 / *Auec Priuilege du Roy*

Champlain departed from Honfleur on 15 March 1603, reaching Tadoussac on 24 May. By 2 July he was at the Lachine Rapids, whence he departed for Tadoussac on the fourth. In recounting some of the circumstances of the 1603 expedition for *Les Voyages, 1632*, he mentioned that on leaving the rapids he "wrote a short account with a correct map of all I had seen and discovered" before he reached Tadoussac.[49] It is therefore likely that a substantial part of the material that went into *Des Sauvages* was written by 23 August, when the expedition left Gaspé for France. On 20 September the ships reached *Havre de Grace*,[50] and a few days later, Honfleur.[51] On learning of Aymar de Chaste's death, Champlain went immediately to see King Henri IV, who was at Saint-Germain-en-Laye near Paris.[52] There he had an audience with the king and showed him "*la carte dudit pays, avec le discours fort particulier que je luys en fis*."[53] It may be that the *discours*, as the word suggests, was a verbal rendering of Champlain's most important findings. The map has never been found.

Claude de Monstr'œil (also Montr'œil, Demontre'œil, Monstrœil, Montreuil, and Monstreul), Champlain's alleged printer, was a *libraire*[54] living in Paris, with a shop in the Cour du Palais from 1576 to 1604.[55] Denis Pallier has assembled a number of official documents involving minor disputes or sales that clearly identify this Claude de Monstr'œil as one of the *marchands libraires* (merchant sellers) occasionally associated with the university.[56] Similarly, the extract from the licence to print and distribute *Des Sauvages* bestows exclusive printing rights upon the same Claude de Monstr'œil in the University of Paris (*Libraire en l'Université de Paris*) "so that no other person may print it or have it printed, sold and distributed over the course of five years, except with the consent of the said

49 Champlain, *Les Voyages, 1632*, pt. 1, 41
50 *Havre de Grace* is on the north shore of the estuary of the Seine River, now the city of Le Havre.
51 Honfleur is southeast of Le Havre on the south shore of the Seine estuary.
52 Cuignet, *L'Itinéraire*, 117
53 Champlain, *Les Voyages, 1632*, pt. 1, 41
54 The title *libraire* is ambiguous; it indicates variable combinations of bookseller and publisher.
55 Renouard, *Répertoire*, 312; Muller, *Répertoire*, 9, 113
56 Pallier, *Recherches*, 115n33, 528n309, 550nn457, 459

Monstr'œil." If appearances can be trusted, Champlain's first journal figured among Monstr'œil's last books.[57]

Having arrived back in France from Canada on 20 September 1603, Champlain wasted no time in having his journal published. Within two months – by 15 November – he had obtained the king's *privilège* (licence). Normally, the process of procuring a licence involved a review of the book's contents by a *conseiller référendaire* (a referee), the king's representative in the royal chancellery, who would then write a report to the *Maître des Requêtes* (Master of Requests) recommending the work for publication with more or less favourable conditions. Part secretary, part judge, the *référendaire* could delay publication or could hasten it by putting authors in contact with publishers and printers. He also sold the *privilège* to the publisher for anywhere from five to twenty years. The price range of such sales is not known today.[58] Claude de Monstr'œil therefore held the monopoly on *Des Sauvages* for the minimum number of years, but the astonishing rapidity with which the licence was conferred suggests some institutional or personal influence in support of Champlain's exploration journal.[59]

Ironically, the French Wars of Religion have contributed to our knowledge of Monstr'œil's probable production habits. Under threat from the ultra-Catholic League, Henri III had fled Paris on 13 May 1588 and organized the Estates General in Blois from 16 October 1588 to 16 January 1589. During this political turmoil, three Parisian *libraires-imprimeurs* moved with the royal suite to Blois: Jamet Mettayer, the royal printer, Jean Richer, and Claude de Monstr'œil. After the king had arranged for the double assassination of the duc de Guise on 23 December 1588 and his uncle the cardinal on the following day, the three *libraires-imprimeurs* moved to Tours, where Henri III entered the city on 6 March 1589. When Henri III in turn was ambushed on 1 August 1589, dying of his wound the following day,[60] the exiled Parisian publishers turned to his unlikely succes-

57 Monstr'œil's widow succeeded her husband under the name Catherine Nyverd (d. 1625), although his supposed son, Michel de Monstr'œil, *libraire* since 1593, shared the same address (Renouard, *Répertoire*, 312).

58 We are indebted to Grégoire Holtz for sharing parts of his unpublished thesis that describe the *référendaire*'s function and discuss the collaboration between Monstr'œil and Richer (Holtz, "Pierre Bergeron").

59 Private correspondence with Grégoire Holtz, 13 May 2008. A Jean Richer published extracts from Champlain's *Des Sauvages* for Pierre-Victor Cayet the following year in 1605. See below and Document H, "Of the French Who Have Become Accustomed to Being in Canada."

60 We are indebted to Laurence Augereau for her help in unravelling this sequence of details in private correspondence, 10 June 2008. See Jouanna, *Histoire et dictionnaire des guerres de religion*.

sor, Henri de Navarre, the future Henri IV, for hope and protection. After the regicide, these *libraires-imprimeurs*, with a growing number of refugees of the same profession, continued to publish pamphlets in support of Henri de Navarre's claim to the throne. In 1591 six of them, naming themselves booksellers (*marchands libraires*), signed a contract to establish a company in Tours for two years, from October 1591 to 1593. Curiously, Abel L'Angelier's signature is also appended to this contract, although he did not invest in the company or share the profits.[61] The contract lists the investment of the members at around 75 écus each as well as their stock of four different kinds of paper (*papier bâtard, petit bâtard, pappier au petit pot*, and a majority 94 reams of *pappier espagnol gros uny*), the room they rented for storage and administration, and their manner of drawing lots to choose one of themselves to administer the operation for two months each over the two years. The assigned administrator supervised the printing, kept the books – notably, the *Journal des affaires de ladicte Companye* (the business journal of the said company) – apportioned the paper to the printers every Saturday, displayed the printed folios to the company when ready, and then assembled the books before distributing 50 copies of each book to each member, to a total of 300. The contract committed the company to reprint a list of eight classic or contemporary humanist texts and share the profits or losses.[62] Most significantly, only two of the members promised actually to print the books: Jamet Mettayer, the royal printer, and Jean Richer. As a rule, they were not permitted to reprint any book of more than six folios without consulting the entire company.[63] Consequently, only Jean Richer (and not Claude de Monstr'œil) actually operated a press, at least in Tours, in league with the royal printer.

The relationships and patterns of production established by this association were not without precedent in previous years. The company's first administrator, Jean Richer, had collaborated with Claude de Monstr'œil since they had arrived at Blois together in the train of Henri III. From 1588 to 1589, they published six books or pamphlets in Blois, and they continued the collaboration when they followed the court to Tours, with another twenty-nine publications between 1589

61 Giraudet, *Une association*, 19–60

62 Surviving records indicate that only three of the eight titles were actually published: *La Diane* by Montemayor (1592), *Imitations* by Bonnefons (1593), and Seneca's *Epistres* (1594) (Augereau, *La vie intellectuelle*, ch. 3).

63 As for the printing, "*lesdicts Mettayer & Richer cy dessus dicts nommez promectent ensemblement les imprimer*" (Giraudet, *Une association*, 24–33, esp. 31). The reference to the above-named (*cy dessus dicts nommez*) alludes to Jamet Mettayer, *imprimeur du Roy*, and Jean Richer, named in this way at the beginning of the contract (Giraudet, *Une association*, 19).

and 1594. In all these cases, the names of both Richer and Monstr'œil appear on all the title pages.[64] In fact, they did not publish independently throughout this period, and a notaried act of 4 November 1591 names them *libraires et imprimeurs associés* (associated printers and booksellers).[65] Their publications featured a broad selection of humanist and religious literature, with some royalist propaganda.[66] Furthermore, the two men settled in the parish of Saint-Pierre-des-Corps outside the walls of Tours, apart from the other booksellers in the association, probably because they had arrived sooner. There they participated in parish life together, to the extent that Richer became the godfather of Monstr'œil's son. Previous to this period, Richer had collaborated with Abel L'Angelier in 1584, 1585, and in 1587 when Richer published an edition of Montaigne's *Essais*, thanks to L'Angelier's mediation. A recent doctoral thesis by Laurence Augereau suggests that the two-year association of 1591 was a late manifestation in Tours for protection against the growing number of exiled publishers there and the rising cost of paper. The more common way of collaborating was by private agreement.[67]

Samuel de Champlain's *Des Sauvages* provides further evidence for the continued collaboration of Claude de Monstr'œil and Jean Richer at the turn of 1604. Although the title page mentions only Monstr'œil, one of Richer's smaller devices, the *Fleuron et l'un des bandeaux à leur chiffre* (floral ornament and one of the bands with their monogram) decorates it as well.[68] Yet Richer's monogram, "I" (= J) and "R" have been removed from the two sides of the depicted urn (see figure 1). Perhaps Richer was reluctant to put his name to *Des Sauvages* because of the financial risk of investing in an unknown author. If the two men followed their pattern of publishing in Tours, one can assume that they first launched a relatively cautious print run of about 300 copies, with a readiness to reprint upon demand. When these copies sold out, the ten quires (ã, A–I) that comprise *Des Sauvages* would have required a consultation between them to determine the economic feasibility of resetting the thin volume in 1604.[69]

64 Labarre, *Répertoire bibliographique*, 6–23

65 Archives départementales 37, Charles Bertrand 3 E 5/251, 4 November 1591 (Augereau, *La vie intellectuelle*, 1477).

66 Labarre, *Répertoire bibliographique*, 18–19, 126–33; Augereau, *La vie intellectuelle*, 669; Arbour, "Raphaël du Petit Val," 130–8

67 Augereau, *La vie intellectuelle*, 611 and ch. 3. Jean Balsamo and Michel Simonin are more skeptical of the relationship between Richer and L'Angelier, noting only two editions whose publication they shared, those of 1584 and 1587 (Balsamo and Simonin, *Abel L'Angelier*, 84n107).

68 Renouard, *Les marques*, 318, fig. 991

69 See this chapter, "Establishing the French text," for a closer examination of the textual evidence (pp. 110–19).

For both printings of *Des Sauvages* during the winter of 1603–04, Claude de Monstr'œil was of a venerable age compared with Jean Richer *le jeune*. The elder Richer, *un des plus savants imprimeurs de son temps* (one of the most learned printers of his time),[70] had died sometime before 1599, when his second wife was named a widow in a contemporary document. His son, Jean III, had been seventeen when he was apprenticed to another printer on 3 November 1584, so his career overlapped that of his father. Both father and son operated at the same address on rüe Saint-Jean-de-Latran *à l'enseigne de l'Arbre verdoyant* (at the sign of the verdant tree) and used the same devices.[71] Whereas the Cour du Palais was famous for specializing in novelties, including travel literature to the New and Old World, the placement of the Richer shop on Saint-Jean-de-Latran meant that this family dynasty also operated in the university milieu, specifically in the student quarter of the Sorbonne.[72] Several generations of the two dynasties of Richer and Monstr'œil published about forty books in collaboration up to the 1620s.[73]

In France, the division of labour between *libraires* (booksellers) and *imprimeurs* (printers) did not begin to be codified systematically until 1618, under Louis XIII, so in 1604 it is still difficult to distinguish them.[74] In major publishing centres such as Paris, many combined the two tasks as *imprimeurs-libraires* (or *libraires-imprimeurs*) in the manner of the Richers, both father and son. Although the licence to print *Des Sauvages* was awarded to Monstr'œil as *libraire* only, records from the association in Tours, 1591–92, list him with Jean Richer as *libraires et imprimeurs*.[75] If Monstr'œil was only the bookseller in Champlain's case, Richer would have printed *Des Sauvages* for him, with Monstr'œil possibly bearing the brunt of the financial investment. At the same time, there was a hierarchy of *libraires*, involving *libraires-jurés* who had taken the oath of the

70 "JEAN RICHER (*Johannes Richerius*), un des plus savants imprimeurs de son temps, fut reçu libraire-juré par l'Université de Paris, en l'année 1573" (Giraudet, *Une association*, 44nn3, 4).

71 Renouard, *Répertoire*, 371

72 Holtz, "Pierre Bergeron," 146–7, 152

73 Ibid., 174 and note 118. Philippe Renouard has acknowledged that almost all the editions published in the name of Jean Richer between 1588 and 1597 were printed in collaboration with Claude de Monstr'œil and wondered to what degree Jean III was involved, even in these earlier years (Renouard, *Répertoire*, 371). The younger Richer and either Monstr'œil or his widow held the monopoly on the favourite *Voyages* of Villamont, which they published seven times from 1595 to 1609 (Holtz, "Pierre Bergeron," 175).

74 Mellottée, *Histoire économique*, 40 ff

75 Laurence Augereau cites documents from the Archives nationales de France, the Archives départementales d'Indre-et-Loire, and the Archives municipales de Tours, which list Monstr'œil as *imprimeur* from 1589 to 1594 (Augereau, *La vie intellectuelle*, 1477–9).

university. The simple *libraires* who had sworn no oath were the lowest on the scale.[76] Although there is no evidence to indicate that Monstr'œil was a *libraire-juré*, he clearly enjoyed, or at least claimed, the privileges and protection of the university.[77]

Further evidence of collaboration between the Tours associates at a later time is furnished by the woodcuts they shared. For example, the ornament entitled *bandeau aux oiseaux* by Jean Balsamo and Michel Simonin, used in their book on Abel L'Angelier, decorates both the title to chapter 1 of *Des Sauvages* and the *Recueil des Harangues* of Guillaume Du Vair, published by L'Angelier in 1610. More significantly, Jean Richer III reused this same ornament for the upper border of his preface to *Le Mercure François ov, la Svitte de L'histoire de la Paix* of 1610.[78] Considered the first French periodical, the *Mercure François* may have been used as propaganda for the monarchy.[79] The two sons of Jean Richer the elder, Jean III and Estienne, founded this annual publication as a continuation of the *Chronologie septenaire* of Pierre-Victor Cayet (1525–1610), the official historiographer of Henri IV.[80] Since the younger Jean Richer printed these three works, *Des Sauvages*, the *Chronologie septenaire*, and *Le Mercure François*, it is not surprising that the *Chronologie septenaire* of 1605 includes a summary of *Des Sauvages*.[81] Claude de Monstr'œil, the licensed publisher of Champlain's slender volume had just died. In any case, Monstr'œil's close association with the Richer family, and at least indirectly with Cayet, suggests that he was a moderate Catholic (open to Calvinism) who was loyal, above all, to the monarchy.

The book *Des Sauvages* is 16 centimetres high and has 80 printed pages (40 leaves), of which 8 are preliminary matter and 72 contain the text. The first printing was probably finished late in 1603. Internal evidence from the 1632 edition of his *Voyages* indicates that Champlain claimed to have published his first journal

76 Runnalls, *Les mystères*, 25

77 Pallier, *Recherches sur l'imprimerie à Paris*, 556–7, lists Jean Richer as a royalist *libraire-juré* during the Catholic League, but not Monstr'œil; Mellottée, *Histoire économique*, 138–9.

78 Richer, *Le Mercure François*, f. ãij

79 Litalien, "Historiography of Samuel Champlain," 16n1; Glénisson, "Champlain's Voyage Accounts," 283

80 Holtz, "Pierre Bergeron," 174 and note 123; also Document H, "Of the French Who Have Become Accustomed to Being in Canada," note 3. The official catalogue of the Bibliothèque nationale de Paris lists seven distinct variations of Cayet's name, including two in Latin (Cajetanus, PVP; and Palma, Petrus Victor Caietanus), along with Palma Cayet, Pierre Victor Palma-Cayet, and a pseudonym P.V.P.C. (Catalogue BN-Opale Plus, notice d'autorité personne no. FRBNF11895643, rev. 18 March 2005).

81 See Document H, "Of the French Who Have Become Accustomed to Being in Canada," and below, "Establishing the French Text," esp. note 109.

TABLE 1 COMPARISON OF [N.D.] AND 1604 PRINTINGS

Folio	*[n.d.] printing*	*1604 printing*
4	*reprenoit sa parolle*	*reprenant sa parolle*
7v	*sur nos ennemies*	*de nos ennemies*
7v	*dās leurs cabannes*	*à leurs cabannes*
12v	*d'où il descēd*	*d'où descend*
13	*lesquelles viennent*	*lesquels y viennent*
17	*qu'il porte son fruict*	*qu'ils portēt leurs fruits*
27	*l'isle aux Coudres*	*l'isle au Coudre*
31	*ponts*	*saults*
34v	*il nous en a monstré*	*il nous a esté monstré*
36v	*pour l'emmener en France*	*pour l'amener en France*

in 1603: "*comme il se voit par ses relations cy dessus imprimées depuis l'an 1603, iusqu'à present 1631.*"[82] The date of the licence corroborates this claim and could be considered sufficient evidence without the inclusion of the date 1603 on the title page. Nevertheless, since the date of the licence (15 November) was so close to the new year, the undated first edition will be referred to here as the "[n.d.] printing." In modern bibliographies and reference works it is usually presumed to have been printed in 1603 or is described more cautiously as "[1603]." In 1604 the entire book was newly typeset and reprinted in the same size and length.[83] Many minor differences distinguish the two printings. Setting aside obvious errors on the part of the compositors, there are only ten changes of any significance (see table 1).

A search made for existing copies of the two books resulted in ten copies of the [n.d.] printing and five of the 1604 printing (see tables 2 and 3). It is probable that more copies of the book exist in European and private collections. There is no record of any originals having been sold since early in the twentieth century.

A title page of *Des Sauvages*, reproduced in the Champlain Society edition of *The Works of Samuel de Champlain*, edited by Henry Percival Biggar,[84] actually

82 Champlain, *Les Voyages, 1632*, pt. 2, 295

83 See below, "Establishing the French Text," for a more detailed examination of the two editions.

84 Biggar, *The Works*, vol. 1, facing p. 83

TABLE 2 *DES SAUVAGES*, [N.D.] PRINTING, LIBRARY LOCATIONS

Library	*Catalogue number*
Bibliothèque nationale de France, Paris, France	RES-LK12-719 Tolbiac
Bibliothèque nationale de France, Paris, France	8 H 1520 (Ex. 1) and 8 H 1521 (Ex. 2) Arsenal
Bibliothèque Sainte-Geneviève, Paris, France	8 G 334 INV 3160 RES
British Library, London, England	Grenville Library, shelf G.7268
Huntington Library, San Marino, CA, USA	Rare Books, 18726
James Ford Bell Library, Minneapolis, MN, USA	1603 Ch
John Carter Brown Library, Providence, RI, USA	JCB E603 .C453d
New York Public Library, New York, NY, USA	Lenox Collection, *KB 1604
University of Pennsylvania, Philadelphia, PA, USA	Rare Bk. & Ms. FC6 C3587 603d
Yale University, New Haven, CT, USA	Beinecke, Taylor 231

TABLE 3 *DES SAUVAGES*, 1604, LIBRARY LOCATIONS

Library	*Catalogue number*
Bibliothèque nationale de France, Paris, France	RES-LK12-719(A) Tolbiac
CNAM, Bibliothèque centrale, Paris, France	12 Y 8 (P.1)†
Harvard University Library, Cambridge, MA, USA	Houghton, Can 205.3
Newberry Library, Chicago, IL, USA	Ayer 121 .C6 1604
Pierpont Morgan Library, New York, NY, USA	E2 60 C

† As this publication was in the final stages of preparation, a copy of the 1604 edition was located in the Bibliothèque centrale but not in time to be examined by us.

has the date 1603 below the last line on the page. This title page created some confusion among cataloguers. Knowing that Biggar had obtained the photograph of that title page from the British Museum in 1917 led to an inquiry to John Goldfinch at the British Library, who replied that "the date is unmistakably handwritten, very neatly but not with any intent to deceive, and done I should guess before the acquisition of the book by Thomas Grenville, who left it to the Museum in

1846, from where it passed to the BL in 1973."[85] One cannot help but wish that Biggar had inserted a note that the date on the photograph was spurious.

Reprints and summaries during Champlain's lifetime (to 1635)

1605
[Champlain, Samuel de] "Des François qui se sont habituez en Canada." In Pierre-Victor Cayet, *Chronologie septenaire de l'histoire de la paix entre les roys de France et d'Espagne ... divisee en sept livres.* Paris: Par Jean Richer, ruë S. Iean de Latran, à l'Arbre verdoyant: Et en sa boutique au Palais, sur perron Royal, vis a vis de la galerie des prisoniers, book 7, item 22: 415–24
REPRINTED 1605. 2nd edn., book 7: 416–25
REPRINTED 1607. 3rd edn., book 7: 416–25

See introduction to Document H, "Of the French Who Have Become Accustomed to Being in Canada: Summary of *Des Sauvages* by Pierre-Victor Cayet, 1605."

1609
Lescarbot, Marc. *Histoire de la Novvelle France contenant les navigations, découvertes, & habitations faites par les François ...* Paris: Chez Iean Millot, tenant sa boutique sur les degrez de la grand sale du Palais, book 2: 305–25, 341–5, 365–85, 415–19
1611
Lescarbot, Marc. *Histoire de la Novvelle France ...* Paris: Chez Iean Millot
REPRINTED 1612. Paris: Chez Iean Millot
REPRINTED 1866. *Histoire de la Nouvelle-France par Marc Lescarbot.* 3 vols. Paris: Librarie Tross
1617
Lescarbot, Marc. *Histoire de la Nouvelle France contenant les navigations, découvertes, & habitations faites par les François ...* Paris: Chez Adrian Perier, Ruë Saint Jacques, au Compas D'Or, book 3: 280–98, 313–17, 335–53, 381–5
REPRINTED 1618. Paris: Chez Adrian Perier.

Of Lescarbot's *Histoire*, the 1611 and 1612 editions are essentially corrected reprints of his 1609 book, with a few additions. In the 1609 edition, about 42 printed pages of the 888 are devoted to Champlain's *Des Sauvages.* The 1617 edition, reprinted in 1618, continues the history begun in the 1609 edition to the end of 1615 and includes a long section on the Native inhabitants of New France.

85 Communication by John Goldfinch, British Library, to Conrad Heidenreich, 21 February 2005

About 48 printed pages of the 970 are a reprint of *Des Sauvages*, interspersed with comments and selections from Jacques Cartier's second voyage of 1535–36.

Much of the historical narrative in Lescarbot's writings was compiled from secondary sources. The significant exception is his description of de Mons's activities on the coast of Acadia from 1606 to 1607, activities that were witnessed by him. His rendition of Champlain's writings is not completely faithful to the original.

Lescarbot was one of the few contemporaries of Champlain who was openly critical of him, particularly where Champlain's observations were at odds with those of Cartier and where Champlain uncritically accepted information from others, as in the case of the *Gougou*. The most puzzling aspect of Lescarbot's rendition of *Des Sauvages* is that it contains nothing that is not in the original. Lescarbot arrived in Acadia in July 1606 and departed in August 1607. One would think that he would have read *Des Sauvages* before he came to New France, or read it once he got there, or at least discussed the voyage with Champlain during the winter of 1606–07. Yet there is no new information in any of the editions of the *Histoire* that sheds any additional light on the 1603 voyage.

1625
Purchas, Samuel. *Hakluytus Posthumus or Purchas His Pilgrimes. Contayning a History of the World, in Sea voyages & lande Travells, by Englishmen & others*, The Fourth Part, Book VIII, Chap. VI: *1605–1619*, "The Voyage of Samuel Champlaine of Brouage, made unto Canada in the yeere 1603, dedicated to Charles de Montmorencie, &c. High Admirall of France." London: Printed by *William Stansby* for *Henrie Fetherstone*, and are to be sold at his shop in *Pauls* Churchyard at the signe of the Rose
REPRINTED 1906. 20 vols. Glasgow: MacLehose and Sons, 1905–06. "The Voyage of Samuel Champlaine of Brouage ..." 18:188–226

See introduction to Document I, "The Voyage of Samvel Champlaine of Brouage, made unto Canada in the yeere 1603."

Subsequent printings

1870
Laverdière, Charles-Honoré, ed. *Œuvres de Champlain publiés sous le patronage de l'Université Laval*. 2nd edn., 6 vols. Quebec: Imprimé au Séminaire par Geo.-E. Desbarats. "Des Savvages," 2:i–viii, 1–63
REPRINTED 1973. Montreal: Éditions du Jour

The first attempt to bring together and collate all of Champlain's books, pamphlets, and letters was completed in 1870 under the auspices of Université Laval and under the direction of abbé Charles-Honoré Laverdière. It is called the second edition because the first burned as it was going to press. The second edition was printed from a proof of the first edition that had escaped the fire. Laverdière worked from the [n.d.] version of *Des Sauvages* which his friend abbé Verreau had copied for him from the original in the Bibliothèque royale, now Bibliothèque nationale de France, Paris.[86] Laverdière believed this book to be exceedingly rare; in fact, he called it the only copy of *Des Sauvages* in existence. The 1604 printing, also in the Bibliothèque royale, was either missed or acquired later. After 1908, when the Champlain Society editors began work on their French/English edition, they discovered that the Verreau copy which Laverdière had edited and printed was untrustworthy, and a new French text had to be established from the originals. The Laverdière edition of the *Œuvres* was the first great Canadian venture into the editing and publishing of historical documents.

1880
Slafter, Edmund, ed. and intro. Trans. Charles Pomeroy Otis. *Voyages of Samuel De Champlain*. 3 vols. Boston: The Prince Society. "The Savages or Voyage of Samuel de Champlain of Brouage, Made in New France in the Year 1603," 1:225–91
REPRINTED 1966–67. New York: Burt Franklin

The Otis translation into English was the first after Hakluyt's version printed by Purchas. Otis used the original of the 1604 printing in the Harvard University Library as his base text and collated it with the [n.d.] printing owned by Mrs. John Carter Brown.[87] Although he noted the existence of the translation published by Purchas, he avoided its major errors. It is unfortunate that the Otis translation never had the impact of some of the later translations. The reason for this is that the Prince Society, founded in 1858, published a limited edition of its books.[88] Only one set of the three Champlain volumes was published for each of the two hundred members of the Society. The Prince Society, which had a similar organization to the Champlain Society, no longer exists. It published its last book in 1920. The scholarly editing and long introduction by Slafter was a model for its time.

1906
Bourne, Edward Gaylord, ed. Trans. Annie Nettleton Bourne. *The Voyages and Explorations of Samuel de Champlain (1604–1616) Narrated by Himself … Together with the Voyage of 1603 Reprinted from Purchas His Pilgrimes*. 2 vols. New York: A.S. Barnes. "The Second Voyage of Samuel Champlaine," 2:149–229
REPRINTED 1911. Toronto: Courier Press

86 Laverdière, *Œuvres*, 2:iii
87 Slafter and Otis, *Voyages of Champlain*, 1:215
88 Ibid., 1:306–18

REPRINTED 1922. New York: Allerton Book Co.
REPRINTED 1977. New York: AMS Press
REPRINTED 2000. Dartmouth, NS: Brook House Press

As the title of the Bourne edition indicates, Champlain's "Second Voyage" (*Des Sauvages*), was reprinted from the Hakluyt/Purchas text. The only difference was that the Bournes supplied footnotes and that they inserted the paragraphs about the *Gougou* which had been omitted by Hakluyt/Purchas. No effort was made to check and correct the Hakluyt/Purchas translation. One suspects that this book owes its popularity, as reflected by its many reprints, to the fact that a commercial press produced it, rather than it being limited to the restricted editions printed by the Prince (1880) and Champlain Societies (1922). The Courier (1911) printing contains a brief "Special Introduction" by William L. Grant that adds nothing of any consequence. What is puzzling is that Grant made no comments on the errors in the Hakluyt/Purchas translation of *Des Sauvages* used by the Bournes although he avoided these errors for his translation of Lescarbot (1911), which he was doing at the same time.

1911

Grant, William L., trans. and ed. Intro by Henry P. Biggar. *The History of New France by Marc Lescarbot.* 3 vols. Toronto: The Champlain Society, 1907, 1911, 1914, 2:77–84, 85–7, 88–94, 125–30, 131–6, 137–40, 168–74
REPRINTED 1968. New York: Greenwood Press

This is the first completely bilingual edition of Lescarbot's *Histoire, 1618;* in fact, it is the first text by any of the early writers on Canada given in both languages. Like Otis, Grant noted the Hakluyt/Purchas version of *Des Sauvages,* but he made his own translation of Lescarbot's transcription, which he checked against Champlain's original *Des Sauvages.*

1922

Biggar, Henry P., gen. ed. *The Works of Samuel de Champlain reprinted, translated, and annotated by six Canadian scholars under the general editorship of H.P. Biggar.* 6 vols. Toronto: The Champlain Society, 1922–36, 1:81–189
REPRINTED 1971. Toronto: University of Toronto Press

The French text of this edition containing *Des Sauvages* was collated by J. Home Cameron, chair of the French Department at the University of Toronto, and translated by Hugh H. Langton, chief librarian at the same university.[89] For his texts, Cameron used the [n.d.] printing in the British Museum (now British

89 For a lengthy discussion of the editorial proceedings involving *Des Sauvages*, see Heidenreich, *Champlain and the Champlain Society*, 35–55.

Library) and those in the Bibliothèque Sainte-Geneviève and Bibliothèque nationale, both in Paris, France. He also found what he considered to be "a unique copy bearing the date 1604" in the Bibliothèque nationale de France, Paris.[90]

The editing and production of the first volume containing *Des Sauvages* took from 1908 to 1922, presenting an interesting study in editorial procedures. Although Cameron reliably collated the French text, the translation has some errors that should not have occurred. As with most of the other translations, the word *salubre* was given as "salty" and "brackish," and new errors were introduced. For example, where Champlain wrote that the Natives cut the "heads" (*testes*) off some captives, "heads" was rendered as "scalps," probably on the assumption that Natives only scalped.[91] Some of these errors, notably *salubre*, were made not by Langton but by Biggar in his attempt to bring the Champlain Society edition in line with the popular Hakluyt/Purchas version.

What made this edition valuable for researchers is that for the first time a reliable, collated French text was made available, and the juxtaposition of the French and English texts made it possible to check the translation. The editorial work on place names and personalities is good, but is either dated or non-existent for Champlain's ethnographic observations. The work needed a full scholarly discussion of the original texts and a good introduction to Champlain's life and times. These were planned but never written.

1951

Deschamps, Hubert Jules, ed. *Les voyages de Samuel de Champlain, saintongeais, père du Canada*. Paris: Presses universitaires de France. "Des Sauvages," 53–97

Contained in this volume is a reprint of the [n.d.] edition with the date 1603 inserted on the title page. The *privilège*, the dedication to Charles de Montmorency, the poem, and the table of contents have been omitted. Some of the French has been slightly altered, marginalia have been retained, and the original folio pagination is given but not the *verso* pages. The editor's footnotes are pertinent and well researched.

1973

Macklem, Michael, trans. Intro. by Edward Miles. *Voyages to New France: Being a Narrative of the Many Remarkable Things … 1599 to 1601, with an Account … in the Year 1603*. [Ottawa]: Oberon. "Savages," 63–111

90 Biggar, *The Works*, 1:xvii

91 Ibid., 102. Where Champlain wrote *prenant les testes de leurs enemis* (taking the heads of their enemies), the translation makes *testes* into "scalps," and on page 103, *ausquels ils couperent les testes* (whose heads they cut off), is given as "whose scalps they cut off."

This is a very free translation of the original [n.d.] text in the Bibliothèque nationale de France.[92] It contains the same errors found in the Champlain Society edition where fresh water (*salubre*) is rendered as "salty" and "cut off heads" are given as "cut off scalps."[93] In addition, it introduces new errors by converting distances given by Champlain in leagues into miles and paces into feet by multiplying them by a factor of three.

1978
Champlain, Samuel de. *Des Sauvages: A facsimile of the Paris, 1603, Edition, Made from the Copy at the John Carter Brown Library, Brown University, Providence Rhode Island. With an Introduction by Marcel Trudel*. Montreal: Designed at the Stinehour Press, printed at the Meriden Gravure Co. for G. Javitch

This is a beautifully designed and printed photo-facsimile of the [n.d.] printing in the John Carter Brown Library. The book is limited to 100 copies.

1993
Champlain, Samuel de. *Des Sauvages: texte établi, présenté et annoté par Alain Beaulieu et Réal Ouellet*. Montreal: Éditions Typo

This is the latest scholarly treatment of *Des Sauvages* in French by two of Quebec's foremost scholars on the subject. The text used by them was the [n.d.] original in the John Carter Brown Library, given here as having been printed in 1603.

3 ESTABLISHING THE FRENCH TEXT

The difficulty in establishing an authoritative French text lies in the ambiguity between one set of extant copies of *Des Sauvages* with no date on the title page and another dated 1604. The possibility that the 1604 date was added as a stop-press correction has been seriously entertained; this would mean that there was only one edition of *Des Sauvages*, in 1604, though the Bibliothèque nationale in Paris has always catalogued them separately, as have the libraries in Britain and the United States with this volume in their holdings. The confusion was further complicated when the Champlain Society's 1922 edition published the 1603 date on the title page, as explained above.[94] Only the Grenville volume in the British Library (G.7268) had accreted the hand-written date before it was "heavily

92 Macklem and Miles, *Voyages to New France*, 17

93 Compare the references to scalps in Biggar, *The Works*, 1:102–3 (Champlain Society edn.) to the same references in Macklem and Miles, *Voyages to New France*, 68–9.

94 See note 85 above, Communication from John Goldfinch.

trimmed."[95] The trimming accounts for the cut-off look of the date "1603" in the facsimile published in the 1922 edition.[96] Consequently, there are still only two versions of the title page: [n.d.] and 1604.

The possibility of one edition with stop-press alterations is further enhanced by the fact that the same Parisian bookseller, Claude de Monstr'œil, claims responsibility for all the states, issues, and editions of *Des Sauvages* after Champlain received the licence on 15 November to have it printed any time in the next five years. Since by this time the Edict of Paris of 1564 had established the beginning of the legal year in France on the first of January, and since the application of this licence may easily have taken a few months, the date of the first publication could possibly straddle 1603–04.[97] Champlain himself wintered in France from 20 September 1603 to 7 April 1604, which would have allowed him to supervise the printing if he was so inclined. We must also keep in mind that the French citizen from Brouage, Samuel de Champlain, was still relatively unknown as both an explorer and an author. This was his first publication.

In the first and final analysis, the most concrete source of evidence derives from the textual variants themselves. In order to set aside, once and for all, all skepticism concerning the two distinct settings of *Des Sauvages*, we have considered it worthwhile to lay out the variants between the [n.d.] and the 1604 volumes in the most transparent and comprehensive manner possible. In the case of a stop-press correction to add the date 1604 to the title page, this would normally be the only correction on the page. Our collation, however, reveals nine variants on the title page alone. The substantial number of variations that arise in the following pages has led us to conclude that the entire book was truly reset in 1604. This thorough resetting of the text allows us to maintain the concept of two separate editions.[98]

95 Communication by Germaine Warkentin after an *in situ* examination of G. 7268 in the British Library (21 October 2007). See also note 85 above, Communication from John Goldfinch.

96 Biggar, *The Works*, vol. 1, facing p. 83

97 Barbiche and Chatenet, *L'édition des textes anciens*, 19

98 "An EDITION is the whole number of copies of a book printed at any time or times from substantially the same setting of type-pages" including all variant states and impressions within "its basic type-setting" (Bowers, *Principles of Bibliographic Description*, 39). Changes to title pages are particularly significant, since publishers might add a date to give a false impression that the book was "newly published" (ibid., 41). In this case, however, no printer of a stop-press correction would have bothered to make the other eight modifications on the title page, of which the most significant replaces *descouuerte* with *descouuerture*. For a discussion, see Heidenreich, *Champlain and The Champlain Society*, 52.

Nevertheless, the qualification of a substantial resetting is called into question by the repetition of a few more major errors in the second edition. Apart from the unreliable misnumbering of folios in both editions, the title for chapter 7 is repeated and that of chapter 8 omitted in both. Minor details, like the impossible hour of 17 (for 5 PM) in the morning mentioned in chapter 1 and the displacement of *le* at the bottom of f. 24^{v} are repeated without attentive correction. These errors could easily have been introduced by a compositor in the first place, then faithfully repeated. It is more difficult to imagine that the second compositor would not have noticed, in both editions, the repetition of three lines: "*laquelle peut tenir quelque demie lieue de long, & un quart de large: & une autre petite isle qui est entre celle du Nort*" (f. 22^{v}). Failure to correct this error is best explained by a reluctance to disturb the substantial layout of the text when the type was reset.

Since Champlain was a French citizen at the time, we have chosen to respect this French heritage by comparing RES-LK12-719 in the Bibliothèque nationale de France (BNF) in Paris, as the no-date edition [1603],[99] with the contiguous book on the shelf, RES-LK12-719(A), representative of the second edition, clearly dated 1604.[100] The 1604 CNAM edition in Paris was discovered only during the last stage of editing this book and consequently was never examined *in situ*. The other three extant volumes of the 1604 edition are found in the United States. It is worth noting that Canada has not one original copy of either edition in any holdings across the entire country! Of the five extant editions in Paris that have been examined *in situ*, only RES-LK12-719(A) is clearly dated 1604 on the title page. The others are undated, apart from the royal imprimatur of 15 November 1603. At the same time, because the variants are generally not "substantial," we have labelled the two editions by the same letter: A1 and A2.[101] Since a fastidious concern for the finer points of the French language does not seem to have been characteristic of Champlain, we have chosen as our base text the earlier edition, which was closer to the authorial production and intention. The *sigla* adopted can be summarized in this way:

99 This first original publication is also available on the BNF website, text no. 105065.

100 Besides conducting an *in situ* examination of this volume and the others in Paris, the facsimile of this 1604 volume has been consulted, which Professor J. Home Cameron ordered for use during his preparation of the first Champlain Society edition (1922). It is now located in the Fischer Rare Books Library at the University of Toronto (E10450).

101 See Marie-Christine Pioffet's choice of C1 and C2 for the last two editions of 1617 and 1618 of Marc Lescarbot's *Voyages en Acadie*. Pioffet chose C2 as her base text because she was convinced that the author, Lescarbot himself, was involved in ongoing modifications and additions to the text (*Voyages en Acadie (1604–1607)*, 55–6).

A1 = [1603] Paris: BnF, RES-LK12-719[102]
A2 = 1604 Paris: BnF, RES-LK12-719(A)[103]

A more technical report of the ideal copy and accidentals of the other copies in Paris concludes this section.

Table 4, below, lays out the most significant variants between the two editions. Listed in numerical order by quire and folio, the left column catalogues the minor errors that have been introduced into A2 in comparison with the corrections enumerated on the right. The difficulty arises in trying to assess errors from mere variations in orthography and punctuation. We have excluded here all orthographic variants and diverse punctuation as long as they do not interfere with comprehension, assuming that such eccentricities, from our point of view, will be anticipated by the reader. The majority of variants are of minimal significance; some inconsistencies between the two texts involve the use of capitals for such words as *ancre/Ancre* or *isle/Isle*, and orthographic variants such as *milieu/millieu*, *encore/encores*, *saut/sault*, and *dit/dict*, including the substitution of *y* for *i*: e.g., *ensuyvant/ensuivant*. Other interchangeable letters are used with less consistency: e.g., the medial and final *s/z*.[104]

Fairly consistent preferences in the second typesetting include addition of the cedilla to ç on such words as *Perçee* (especially in quire H, where it was forgotten the first time), the final *t* in *Laurent* (as opposed to *Laurens*), an apostrophe after *grand'* to mark the lost *e* when the noun modified is feminine, and either *i* or *y* for *ï* in variations of *resiouïr*, *resiouïssance*, and *païs*. The second edition also tends to drop the apostrophe in *n'y*, which introduces a confusion with *ni* (or *ny*: neither, nor). It is interesting to note that the preponderance of *Canadas* with a final *s* is considered the norm in both settings; in fact, the occurrence of *Canada* in the title of chapter 2, in both the table of contents and the actual heading, is corrected to *Canadas* for the second edition! After this adjustment, only one occurrence of *Canada* remains (ch. 8, f. 22) in the 1604 edition.

Otherwise, the 1604 compositor seems at times to be inspired by a contrary spirit. For example, if the previous edition had *lieüe* (place), the new compositor would either drop the diaeresis altogether (*lieue*) or switch it to the *e*, as in *lieuë*, as if asserting a spirit of individualism.[105] Similarly, the modification of

102 See BnF website, text no. 105065.
103 University of Toronto, Fischer Rare Book Library facs, E10450 (1604).
104 See above, section on Champlain's French.
105 There is no significant difference in the adoption or omission of the diaeresis in *lieue*, since the diaeresis in such cases *est utilisé un peu au hasard, à de multiples usages, avant de se fixer*. One of the most significant uses of *ü* helps to distinguish *v* from *u*, but

abbreviations in the second setting is almost humorous. The later edition, in contrast with the original, might shift the abbreviation mark, as in [n.d.] *viēt tomber*, to *vient tōber* [1604]. In both cases, the meaning is literally "comes to fall."[106] The compositor in the second instance appears to want to omit the abbreviation, only to reinstate it later when there is not sufficient space for the whole word. There are other instances of this tendency.

Consequently, we have taken the printer's usage as our norm, assuming that both *Canots* and *Canos*, *voir* and *veoir*, *dit* and *dict*, etc. were acceptable in the early seventeenth century, because neither the meaning nor the pronunciation vary with the orthography. In the case of the final *s*, which was not pronounced, we have ignored all the variations in such words as *encore*, *vingt*, *cent*, and *Laurent* but have recorded preferences for *quelques* over *quelque* in the plural. Conversely, if consistency establishes a norm, we have corrected the variations by the seventeenth-century standard; for example, the *l'* is applied fairly uniformly, so the lack of an apostrophe, as in the unique instance of *Lon* (f. 2ᵛ) is considered an error. Similarly, Champlain used the word *cyprez* for cedar tree so consistently that the one instance of *cypres* is modified to this norm, rather than conforming to our established orthography: *cyprès*.

The haphazard application of accents over *à* (to, at, in) and *où* (where) is more difficult to judge, since there does seem to be an effort made to distinguish them from their homonyms *a* (has) and *ou* (or) by the compositor of 1604, especially in quires F and G. When the accents are either omitted or applied in the appropriate places, we prefer the more modern usage and have finally decided to include the slightly confusing misapplications in table 4, below.

For this table, the reading on the left of the colon in both columns is that found in the 1604 edition. The reading on its right is what has been replaced from the earlier setting. To save space, three dots either before or after indicate the omission of exactly the same words as on the left. Similarly, we have tried to provide the literary context that justifies the entry wherever this is possible within the spatial limitations of the table. Based upon the footnotes of the French edition, the French has not been modernized.

Most of these so-called errors (either introduced or corrected) are questionable from the early-seventeenth-century perspective. The increased use of accents on

this is not the case here (Catach, *Histoire de l'orthographe française*, 133). Some printing houses used it throughout the seventeenth century to distinguish the vowel *u*, as in *boüe* (mud) or *devoüement* (devotion). This was the usage recommended by the Académie française in 1694 (ibid., 135). With the equally whimsical appearance of *lieuë* here, none of these variants is considered significant.

106 The abbreviation mark over the vowel represents a missing nasal consonant: *n* or *m*.

TABLE 4 ERRORS AND CORRECTIONS INTRODUCED INTO *DES SAUVAGES*, 1604 (A2)

Quire	*Errors introduced into* A2	*Corrections introduced into* A2
ã	a peine [for]: à peine (f. ãj^{v}) trs-humble: tres-humble (f. ãij^{v}) ãij: ãiij Lees: Les (f. ãiv^{v})	Sainct Mathieu [for]: Saincte Mathieu (f. ãiij^{v})
A	a la mer: à la mer (f. 2) l'entre: [noun] l'entree (f. 2^{v}) à: [verb] a (f. 2^{v}) – *hypercorrection*	à: a *in* à l'entree (f. 2^{v}) auec: a-/auec (f. 4) qu'ils ["which they"]: qui ils (f. 4^{v}) à qu'ils volussent plus de bien: a qu'ils … (f. 4^{v})
B	guerre.: guerre, (f. 5) cabano: cabane (f. 5^{v}) de ["by"]: sur [over] (f. 7^{v}) chair, d'Orignac: chair d'Orignac (f. 7^{v}) Algoumequins, Apres: Algoumequins. Apres (f. 7^{v})	quelques fois: quelques-fos (f. 6) toutes: toute leurs femmes (f. 7) Montaignez: Montaignes (f. 7^{v}) s'il ne croyoit: s'ils ne croyoit (f. 8^{v}) la la croyance: la là croyance (f. 10^{v})
C	volonte: volonté (f. 10^{v}) chose qui aduiennent: choses … (f. 11^{v}) 45. où 50.: 45. ou 50. (f. 12^{v}) mōtaignes, de rochers: mōtaignes de rochers (f. 12)	créez: crees (f. 10) songé: songè (f. 11^{v}) leurs parens: leur parens (f. 12^{v}) d'où descēd vn torrent: d'où il … (f. 12^{v}) (*either is possible, but* A2 *slightly better*)
D	ny: n'y (f. 13) quelque sept: quelques sept lieues (f. 13^{v}) a quelques 20. ou 25. lieuës: à … (f. 15)	lesquels y viennent: lesquelles viennent (f. 13) (*A1 limits small birds to swallows*) 5. lieues: 5. lieüe (f. 16^{v}) à Quebec: a Quebec (f. 15)
E	les terres estants: les terres estãt (f. 17) ou peuuent aller les Canos: où … (f. 17) lentree: l'entree (f. 19) terre basse, remplies: …, remplie (f. 20^{v}) touste les sortes: toutes les sortes (f. 20^{v}) ou le bois y est fort clair: où … (f. 20^{v})	qu'ils portēt leurs fruits: qu'il porte son fruict (f. 17) (*A2 is more consistent*) quelques trois ou quatre: quelque … (f. 19) la grand' riuiere: la grand riuiere (f. 19^{v})

Quire	*Errors introduced into* A2	*Corrections introduced into* A2
F	quelque deux lieues: quelques … (f. 24) ny: n'y (f. 24^{v}) quelque 80. lieues: quelques 80. (f. 24^{v})	viuent: vinent (f. 21) par où: par ou (f. 22^{v}) saults: saultr (f. 23) où l'on peut aller: ou … (f. 23^{v}) où il y a quantité de bois: ou … (f. 23^{v}) tous les autres: tout les autres (f. 23^{v}) à quelque quinze … : a … (f. 24^{v})
G	Iroquois,: Iroquois. (f. 25^{v}) ny ont esté que: n'y … (f. 26^{v}) qu,ils: qu'ils (f. 27^{v}) quelque trois cents: quelques … (f. 27^{v}) de Chaleurs: des Chaleurs (f. 28^{v}) du coste du Su: du costé … (f. 28^{v})	grand' riuiere: grand riuiere (f. 26) où ils les portent: ou … (f. 26) f. 27: f. 29 l'isle au Coudre: l'isle aux Coudres (f. 27) où ils sont cabannez: ou … (f. 27) grand' riuiere: grand riuiere (f. 27) vingt à vingt cinq: vingt a … (f. 27^{v}) f. 28: f. 27 qu'à son entree: qu'a son entree (f. 28)
H	Montaignz: Mōtaignez (f. 29) quelque quatre-vingt: quelques … (f. 29^{v}) iusques a: iusques à (f. 29^{v}) f. 29: f. 30 riuiere.: riuiere, (f. 30) ne seroient subiects: … subiect (f. 30) 15. degré: 51. degré (f. 31) deldeux lieuës: de deux lieues (f. 32)	grand' baye: grand baye (f. 30^{v}) à quelques six lieues: à quelque … (f. 32) à quelque 4. brasses: quelque … (f. 32) où entre vne Riuiere: ou … (f. 32) où il y a vne ance: ou … (f. 32) où il y a peu d'eau: ou … (f. 32) à l'abry: a l'abry (f. 32)
I	Lees: Les (f. 33) deuant que, d'aller: deuant que … (f. 33) trouuasnes: trouuasmes (f. 33) ny: n'y (f. 33) tontes: toutes (f. 34) canciēnement: anciennement (f. 34^{v}) a vn pied ou deux: à … (f. 34^{v}) toutes les sortes d'arbres que nous avons veus: … veues (f. 35^{v}) precedēt.: precedēt, m'a dit (f. 36) pou fables: pour fables (f. 36) [femme …] amenee: … amenée (f. 36^{v}) Canadiens, Le 24 …: Canadiens. (f. 36^{v})	Passant: Passans (f. 34^{v}) grand' Baye: grand Baye (f. 35) grande poche: grand poche (f. 36) vn des Sagamoz: vn des Sagamo (f. 36^{v})

à, *là*, and *où*, for example, demonstrates the uncertain evolution of usage towards our fixed modern conventions. One may note that the French *j* was just coming into usage; A2 introduces it into *jours* (f. 29v) and *jambe* (f. 34), but cancels it in *Majesté* (A1, f. 4–4v; 10v); e.g., *Maiesté*. The feminine ending *-ee* commonly has no accent, although one instance pops up at the end of A2: *amenée* (f. 36v). It is clear that the conventions were still unstable.

With this qualification, then, the table highlights the strength of the original edition. The recomposition in 1604 introduces slightly more errors than corrections: 54 errors to 47 corrections, a difference of 7. Almost as many grave accents on *à*, *où*, *là* are misapplied in A2 as are added correctly; there are 9 misapplications and 15 welcome additions. Most of the latter occur in quires F and G, the only quires where the so-called corrections outnumber the errors. Furthermore, the new setting in 1604 gives us insight into the amount of intervention into any given original text that a compositor might feel inclined to make. We cannot assume, therefore, that Champlain was responsible for every accidental, even in the first setting. Above all, these findings justify the decision to use the *editio princeps* [n.d.] as our base text.

One minor error in foliation between folios 26 and 31 remains to be examined. This confusion reveals the compositor's attempts to self-correct, which is never fully successful. A summary of the differences in foliation of the various copies is as follows:

Foliation: G26, G29, 27, H29, H29, H31
John Carter Brown Library: JCB E603 .C453d
Huntingdon Library: R. Hoe Collection, Rare Books, 18726
Yale University: Beinecke, Taylor 231
University of Pennsylvania: Annenberg Rare Books and Manuscript Library: FC6 C3587 603d
Bibliothèque Sainte-Geneviève: 8 G 334 INV 3160 RES
Arsenal site (BnF) 8 H 1520 (Ex. 1) (H29[1] corrected by hand to 28)

Foliation: G26, G29, 27, H29, H30, H31
Bibliothèque nationale de France: RES-LK12-719, Tolbiac
Arsenal site (BnF) 8 H 1521 (Ex. 2)
James Ford Bell Library: 1603 Ch
New York Public Library: Lenox Collection, *KB 1604
British Library: G.7268 (G29 corrected by hand to 27; 27 corrected by hand to 28)

Foliation: G26, G27, 28, H29, H29, H31
Bibliothèque nationale de France: RES-LK12-719(A), Tolbiac

From the outset, the compositor introduced three errors into the numbering of consecutive folios of the *editio princeps*: two in quire G (26, 29, 27) and one in quire H (29, 29, 31). In all cases, it is merely the numbering of the folios, and not the text or catchword between quires, that is affected by the errors. During the first print run, the compositor stopped the press to correct H29 to H30 but missed the previous confusion in quire G. This one consistent variation allows us to identify at least two distinct states of the first edition and suggests the following hypothesis.

In modern terms, we would label the 1604 edition a reprint. A perfect reprint or reissue was well nigh impossible for a hand-printed book at that time.[107] It is likely that Claude de Monstr'œil had a small to modest number printed of this first publication by an unknown author.[108] A small print run at that time would have been around 300 copies per edition, which was the norm for Monstr'œil's company in Tours, 1591–93.[109] It is also quite possible that the slim volume proved to be more popular than anticipated. In response to the market demand for more copies, the printer must have reset the old edition following the pattern already established.[110] By imitating the previous work, he would not have had to work out the details of composition all over again. If the repeated lines on folio 22^{v} were noticed in the course of the second typesetting, there would have been an understandable reluctance to make the correction, since that would have disrupted the entire layout for the rest of the book. The copy chosen as the model for this second typesetting must have been one from the first state of the original printing, since the error of foliation in quire G is finally corrected (26, 27, 28), but the one in quire H (29, 29) is reintroduced. The latter is unlikely to have been newly introduced, given its previous appearance. For convenience, the term "edi-

107 "Since with early books the printer ordinarily distributed his type for each gathering very shortly after the gathering or any of its formes had been printed, any reprint of the book normally called for complete, or almost complete, resetting and thus for a new edition" (Bowers, *Principles*, p. 37n).

108 Small print runs at 300 copies and medium ones at about 700, recorded by Graham Runnalls for the early sixteenth century in France (Runnalls, *Études sur les mystères*, 439), are still reflected by evidence from eighteenth-century England (McKenzie, *Making Meaning*, 62–72).

109 See above, "Printing History," for further details concerning the company of Tours.

110 "If after the book had been placed on sale, more copies were required than had originally been printed, the book was perforce completely reset and a new edition created" (Bowers, *Principles*, 108), "since it would be too expensive to keep a book in standing type until it was determined whether a new impression were needed" (Greetham, *Textual Scholarship*, 167).

tion" is maintained here for the [n.d.] and 1604 settings, though we have identified at least two variant states for the original edition. A further examination of the volumes found in Britain and the United States might uncover more states within either edition. At the least, the detailed examination presented here lays a foundation for further comparisons.

Description of A1

DES | SAVVAGES, | OV, | VOYAGE DE SAMVEL | CHAMPLAIN, DE BROVAGE, | fait en la France nouuelle, | l'an mil six cens trois: | CONTENANT [&c.] [Pot of flowers; Renouard 991] || A PARIS, | Chez Clavde de Monstr'œil, tenant sa | boutique en la Cour du Palais, au nom de Iesus. | AVEC PRIVILEGE DV ROY. ||| [ãj^v] *Extraict du Priuilege.* ||| [ãij] [upper border: Cupid between two birds among foliage] A TRES-NOBLE, HAVT | ET PVISSANT SEIGNEUR, | Messire Charles de Montmo-| rency, [&c.] ||| [ãiij] LE SIEVR DE LA FRAN-| CHISE AV DISCOVRS DV | Sieur de Champlain. ||| [ãiij^v] [floral, geometrical upper border] TABLE DES CHA-| PITRES. [&c.] ||| [ãiv^v] FIN.
Head Title] [Upper border: three birds in foliage][111] DES SAVVAGES, | OV | VOYAGE DV SIEUR DE | Champlain, faict en l'an 1603. ||| 13 chapters
Running Titles] Des Sauuages, ou Voyage [ff. 1^v–4^v, verso only]; Des Sauuages, ou, Voyage [ff. 5^v–36^v, verso only]; du Sieur de Champlain. [ff. 2–36, recto only].
Explicit] FIN.

Collation: 8° [ã^4, A–I^4], half-sheet imposition with no visible watermarks;[112] 40 leaves: signatures. Foliation, excluding the preliminaries [iv, 1–36 ff.]; variable errors in foliation from G27 to H30; no pagination; 28 lines per page, 160 x 110 mm. Type: roman. Preliminaries in italics; four ornate initials with floral decorations, M (ãij), M (ãiij), B (ãiij^v), N (f. 1), to mark the first word of all the preliminary material up to the beginning of the first chapter. Three ornamental upper borders [as above]. No maps. Catchwords: *que, croyance, lieue, sont, encores, quel-, -pelle, Les.*

111 *Bandeau aux oiseaux* (Band with birds), so named in its use by L'Angelier in Balsamo and Simonin, *Abel L'Angelier*, 377 and plate 5, 543. Jean Richer and Abel L'Angelier must have shared this woodcut in a collaboration that is documented by at least six editions of humanist texts between 1584 and 1600 (ibid., 84, entries 117, 143, 181, 330, 349). See also "Printing History," above, 100–2.

112 Frequently, half-sheet imposition in smaller formats causes the watermark to disappear "altogether into the binding" (Greetham, *Textual Scholarship*, 132).

Description of A2

DES | SAVVAGES; | [&c. as in A1] | *faict en la France nouuelle, l'an* | *mil six cens trois:* || [&c. as in A1] | au nom de Iesus. 1604. | Auec Priuilege du Roy. || [&c. as in A1] ||| [ãij] [upper border: eight naked figures: first on left seated; second holding a torch; third with horns like a goat; fourth is riding a goat; fifth with a devil's face, pointed ears and protuding tongue …] | [&c. as in A1].

Examination of five copies in Paris

BnF: RES-LK12-719 [1603] = A1
Paper: 170 x 105 mm
Vellum binding with stains; no inscriptions on the cover or spine; no marble ends; slightly gold-tipped edges to the paper. Two flyleaves at the back. There may have been two at the front as well, since one may have been glued into the front cover, which displays an irregular surface. Evidence of a provenance written in handwriting at the top of the title page: *Ex lib. Mons*[eigne]*ur Martini a* [*Lavinga*?], translated as "From the library of my lord Martini from [Lauingen sur Donau?]).[113]

BnF: RES-LK12-719(A) 1604 = A2
Paper (text): 160 x 103 mm (134 x 70 mm)[114]
Marble binding with brown binding tape of 25 mm; one blank paper flyleaf at both ends; well preserved.

Arsenal site (BnF): 8 H1520 [1603]
Paper: 155 x 93 mm
Deluxe binding with gold tooling on spine, reading *Voyage de Champlain au Canada*; red-tipped paper. Previous owner's name at the top of the title page is illegible and possibly truncated by the trimming of the pages.

Arsenal site (BnF): 8 H1521 [1603]
Paper: 158 x 97 mm
Brown paper wrapping covering marble ends similar to those found in 8 H1520 above; ff. 1–4 were clipped back to 88 mm from the 97 mm. Second number-

113 Here and below, Latin place names have been identified by consulting Benedict, *Graesse, Orbis latinus.*

114 Atkinson, *La littérature géographique de la Renaissance*, 349

ing of foliation added in ink above the printed version, without fault from f. 228 to f. 267, suggests that this exemplar of *Des Sauvages* was formerly bound in a collection.

Bibliothèque Sainte-Geneviève: 8 G 334 INV 3160 [1603]
Paper: 160 x 100 mm
Vellum binding in poor condition; pages tipped golden brown at the end. Provenance typed onto a label pasted into the front cover: *Ex Bibliothecâ / quam 16000. Voll. constantem / huic Abbatiæ S. Genovesæ Paris./ Testamento Legavit Car. Maurit./* LE TELLIER *Archiep. Remensis* (From a library which, containing 16,000 volumes, Car*dinal* Maur*ice* Le Tellier, Archbishop of Rheims, bequeathed in his will to this Abbey of Sainte-Geneviève, Paris). In a tiny hand in black ink on the upper middle page of the flyleaf was added: *édition originale introuvable (valeur vénale* ...), unobtainable original edition (market price ...?).

The volume in the Bibliothèque Sainte-Geneviève indicates that *Des Sauvages* belonged to the "renowned" library of Charles-Maurice Le Tellier, archbishop of Rheims (1642–1710), who received his doctorate of theology from the Sorbonne before being ordained priest in 1666. Intimate with Jacques-Bénigne Bossuet, he convinced the latter to write *Oraison funèbre de Michel Le Tellier* at the death of his father, Michel. Charles-Maurice became titular of the see of Rheims in 1671 and presided over the general assembly of the French clergy in 1700. After his death, his manuscripts were given to the king's library by his nephew in 1718. These are now held in the BnF, while the 50,000 volumes of his printed books are at the Bibliothèque Sainte-Geneviève,[115] of which 16,000 must constitute one category or subsection. Although we cannot read the actual market price, the indication that the original undated volume was rare (*introuvable*), along with this prestigious provenance, strengthens the hypothesis that the slim volume was a sell-out at first printing.

4 EDITORIAL PRINCIPLES AND PROCEDURES

Having decided upon a facing-page translation, we necessarily have two sets of concerns when establishing our text: one set for the original edition and another for the English translation.

115 *New Catholic Encyclopedia*, 8:519; J.F. Sollier, *Catholic Encyclopedia*, 2008, www.newadvent.org/cathen/09200c.htm

The French text of 1603

For a critical edition of what should be considered a standard French Canadian text, current conventions in editing French texts have been adopted from Bernard Barbiche and Monique Chatenet, *L'édition des textes anciens XVI–XVIII siècle* (Paris: Inventaire Général, 1990), with consideration given to Yvan G. Lepage, *Guide de l'édition de textes en ancien français* (Paris: Honoré Champion, 2001) and current publication practices in Quebec and France.[116]

For this new edition, the *editio princeps* of 1603 in the Bibliothèque nationale de France (BnF, RES-LK12-719) constitutes the base text. This emission of the first original publication is also available on the BnF website, text no. 105065. The 1604 edition with which the base text has been collated is the contiguous copy in the BnF in Paris: RES-LK12-719(A). In an effort to resist anachronistic standardization, the French text is respected as much as possible *tel quel* with only a few modifications.

1 We have restored the original style of the table of contents (*Table des Chapitres*), which was modernized and adapted to the 1922 edition.
2 We have added almost all the paragraphs in order to orient the reader and facilitate comparison between the original French text and the English translation. There is only one original paragraph in the French text, at *Le Mecredy 24* in chapter 6 (f. 16), which is noted in our edition by a paragraph sign.
3 We have placed the marginalia in the footnotes, keyed into the text by an asterisk. The marginalia replicates that found in the [1603] edition at the Bibliothèque nationale (RES-LK12-719) with no variants or line breaks recorded and no modernizations. Notes from the right and left margins are distinguished by RM] and LM], respectively.
4 Except when creating new paragraphs, we have maintained the original punctuation, even when it does not conform to modern usage; for example, commas and colons are often used as periods.
5 We have normalized the letters *i/j* and *u/v* to distinguish the vowels from the consonants, following the French convention in both Quebec and France.

116 We have consulted, most notably, texts published by the Presses de l'Université Laval, Ville-Marie Littérature, *Tangence* éditeur of the Université du Québec à Rimouski, and Septentrion in Canada, as well as editions from Librairie Droz, Switzerland, and Champion, France.

6 The orthography, syntax, and punctuation of the original text have been rigorously respected, except in the case of abnormalities that disrupt the flow of reading for the twenty-first century. In the rare cases when minor modifications are deemed necessary, they are signalled in textual notes at the foot of the page, along with variants and other necessary corrections.

7 The foliation in square brackets represents an idealized numbering that does not take into account the confusion between ff. 26–31, as discussed above. The editorial insertion of square brackets enclosing identification of the foliation is made as unobtrusively as possible; thus [f. 1], [f. 1^{v}], [f. 2], etc., where the verso is marked by a superscript "v" but the recto is implied by default. We have omitted the printer's catchwords and signatures for binding, except the initial letter in the latter case. The letter of the signature marks the change to a new quire. The printer's signatures are fully noted only in the preliminaries where there is no foliation.

8 We have expanded abbreviations by restoring the lost letters in italics that were originally signalled by a tilde over nasalized vowels (i.e., *ã*, *ẽ*, and *õ*, representing *n* or *m*) or, more rarely, over the consonant (i.e., over *q̃* for *que* in ch. 13, f. 36^{v}). Another abbreviation adopted from medieval Latin usage, which likewise occurs only once, is *pl'* for *plus* (ch. 9, f. 27^{v}, corrected from f. 29^{v}). All other italics, expecially for the preliminaries and marginalia, are preserved. Whenever the norm is in italics, the expanded abbreviations are removed from them in order to keep the expansions distinct.

9 We have maintained the ampersand (&), both for the italics and for the main text.

10 Accents and diacritic marks are reproduced as they appear in the text. The usage of grave and acute accents, for *à* and *-é*, respectively, excluding the feminine ending *-ee*, conforms to the contemporary convention. The greatest potential for confusion lies in the homonyms *a* and *à*, and *ou* and *où*, whose accents are applied inconsistently. The variants that apply them correctly have been preferred.

11 The spacing of words is not reproduced, except in the case of word divisions that might differ from modern usage but were used at that time and are clearly demarcated or amalgamated in the original text. Judgment on this issue is not always easy, since the printer often condensed the text in one instance and then spread it out in another in order to justify the text in its layout on the page. Sometimes the final punctuation is divided from the sentence by a space, but because the spacing is haphazard and inconsistent, we have settled upon the omission of spaces between the last letter and the punctuation, in the English style. Similarly, word divisions at line

breaks in the original edition are not maintained, since the justification alone accounts for them. Only word divisions between the recto and verso (and vice versa) of folios are preserved.

12 Original capitalization, numbers, and the spelling of proper nouns have all been respected without any effort to regularize them beyond the usage of the seventeenth century. The English translation and notes will clarify any ambiguities.

All textual notes record the original readings with no editorial interventions at all (i.e., *u/v*, *i/j* are not modified and abbreviations are not expanded there). The reading given on the left before the square bracket represents the preferred one adopted in the text above, while the rejected variant to the right is always labelled "A1" [no date] or "A2" [1604]. If A2 is not explicitly recorded on the right, this means that its variant has been adopted as the preferred reading, and vice versa. In many cases, there is no need to prefer A2 over A1, since the variants are too minor to hold any significance. Any editorial intervention into the main text with no witness in the original editions is indicated by the original readings for both A1 and A2 on the right of the square brackets, in conformity with the principles laid out above.

The English translation

The principles applied to the English translation of *Des Sauvages* conform to those expressed in the preface under "Notes on Spelling and Usage," with the following qualifications:

1 Since all unintended repetitions or errors in the original editions are noted in the French text published alongside the English, the English translation will not take them into account. Difficult passages, however, are clarified in the notes.

2 There is no regularized English-language orthography of French place names, Native names, or Native words. Any such names and words that are not recognizable in modern Canadian English are placed in italics in the main text, just as foreign-language terms are habitually italicized.

3 Square brackets are occasionally added to signal missing words; they appear only when additions were necessary to clarify the French text. Extra words required for the clarity of the translation are implied by customary usage or by the French grammatical agreements (*les accords*), involving inflections and gender unknown to English.

4 Occasionally, the marginalia found in the French text are translated into English and placed among the footnotes. Only those marginal notes that add an interpretation to the main text have been translated. All the others merely repeat the text in substance or form.
5 In some instances, we changed the capitalization in a manner not normally done in English. For example, it was necessary to capitalize the "He" representing God (*il*) in chapter 3, because it is easily confused with human beings referred to by the same masculine pronoun. Conversely, "devil" is occasionally capitalized, but the first instance prevails, where the *d* is not capitalized. The reader can easily check the French on the opposite page to view the original if desired.

PART TWO

DOCUMENTS

DOCUMENT A

Early Biographies of Champlain: Extracts / *Anciennes Biographies de Champlain: Extraits*

INTRODUCTION TO THE DOCUMENTS

As far as we can determine, the earliest biographical statement about Champlain, including the first review of one of his books, was written by Father Pierre-François-Xavier de Charlevoix, SJ, for his *Histoire et description generale de la Nouvelle France*, published in Paris in 1744 by five different printers, each producing a three- and six-volume edition. All the material used by Charlevoix was taken from scattered statements about Champlain in the *Jesuit Relations*, Lescarbot's *Histoire*, and Champlain's own writings. The Jesuits eulogized Champlain as a devout Christian and capable administrator who laid the foundations of "civilization" in Canada. The statements from the *Jesuit Relations* and Charlevoix's brief biography did much to lay the groundwork for mythologizing Champlain after the mid-nineteenth century.

Champlain was relatively unknown until the first third of the nineteenth century, when brief biographies of him began to appear in encyclopedia entries. These coincided in 1830 with the first reprint of any of his books (*Les Voyages, 1632*) since the original editions.[1] We have reproduced three of these encyclopedia entries as examples. For its time, the biography by Rossel, in the *Biographie universelle* (1813), which we have given only up to the end of 1603, is astonishingly good, mainly because he actually read Champlain's writings and thought about them. By contrast, the one from the *Nouveau dictionnaire biographique* (1837) and the entry in the *Encyclopédie du dix-neuvième siècle* of 1845 are more typical of the century. The latter gives the occupation of Champlain's father as an *armateur* (ship's manager), which the next author, Rainguet, transferred to Champlain himself. The biographical statement published in 1851 by the *notaire* Pierre-Damien Rainguet (1803–75) made a much more lasting impact than earlier or later biographies because he was the first and only writer to give Champlain a specific birthdate. Almost immediately (1854), the same date appeared on the

1 Champlain, *Voyages du Sieur de Champlain*, 1830. The maps and *Traitté de la marine* were not reprinted.

phony but influential portrait of Champlain by Louis-César-Joseph Ducornet.[2] We do not know why Rainguet chose 1567, nor has anyone found any documentation that would support it. Compared to the Rossel biography that preceded it, the one by Rainguet is full of obvious errors. Unfortunately, it became a model for later short biographies, such as that by Trousset (1885).

Space does not permit more biographical excerpts. Instead, we have compiled a list of standard biographies of Champlain, his wife Hélène Boullé, some novels and stories for young people, an epic poem, and two plays. These are organized by date at the end of the References.[3] In the first two, Faillon (1863) and Parkman (1865), Champlain's life is discussed within the context of the history of New France, which became a standard way of handling his life in nineteenth- and twentieth-century histories, such as Trudel (1973). The first true biographies we have been able to find are those by Margry (1865–67)[4] and Delayant (1867). The Margry biography was written for the Commission des arts et monuments de la Charente-Inférieure, which was lobbying for the erection of a statue of Champlain in Brouage. Both Pierre Rainguet and his brother l'abbé Augustin Rainguet were members of that commission, and, significantly, Margry accepted the date 1567 for Champlain's birth. Delayant (1867) was the first to question Rainguet's birthdate for Champlain as well as his assertion that the Champlain family were poor fisherfolk, calling both conjectural.

The publication of Laverdière's (1870) and Slafter's (1880) compilations of Champlain's writings and related documents made these rare works available to French- and English-speaking readers. Both contain serviceable biographies of the man, set in his milieu. With the availability of Laverdière's and Slafter's compilations, both limited editions, other compilations quickly followed – by Bourne (1906) and Biggar (1922–36). Of these two, only the Bourne edition has a brief biographical statement, and unfortunately the Biggar edition was limited to Champlain Society members. All of these compilations have now been reprinted.[5] With the original documents available and the tercentenary of Quebec in 1908, Champlain biographies became a minor industry. Of the early ones, Audiat (1893), Casgrain (1898), Dionne (1891, 1906), and especially Gravier (1900) are still worth reading, although most of them are unabashed hagiographies.

2 See appendix 1, "Champlain's Birthdate and Appearance."

3 See references, "Biographies of Champlain."

4 The author of the biography is not given. The authority for the author is Charavay, *Samuel de Champlain*, 1. At the time, Pierre Margry was *conservateur des Archives de la marine* in Paris. The date given by Charavay was *vers le milieu de l'année 1870*, but the work it was published in makes it clear that it was before 1867.

5 See part 1, B, "Printing History"

Of the more recent biographies, we prefer Bishop (1949) and Deschamps (1951). The brief biographies in Campeau (1967) and Trudel (1973) are reliable guides to Champlain and his role in the early history of New France, and Moore's (2004) little book is the best of its type for youthful readers. The plays about Champlain (Harper, 1908) and his wife Hélène (Hooke, 1942) can best be described as works of fiction.

By the 1970s a reaction had set in to Champlain as "hero" in the writings of iconoclasts such as the anthropologist Bruce Trigger, and the popular historian Pierre Berton. While Trigger consistently portrayed Champlain in the worst possible light,[6] with little attempt to understand the man, Berton, with no understanding at all, called him an "assassin" whose "victims" were "unsuspecting Indians."[7]

Champlain was not perfect, especially by modern politically correct standards. Like most of his contemporaries, he had little understanding or respect for Native government and religion, but he saw the people as human beings with whom he could work, from whom he could learn, and with whom the French could intermarry. In these respects he was very different from his contemporaries, the Puritans of New England and the Dutch of the Hudson River Valley, who regarded Natives as obstacles that stood in the way of progress and civilization, and who were still doubtful of their humanity more than one hundred years after Champlain's death.[8]

Recently, more balanced biographies have been published by French authors: Glénisson (1994), Legaré (2004), Capella (2004), Montel-Glénisson (2004), and by the French Canadian editors, Litalien and Vaugeois (2004). As this volume was going to press, *Champlain's Dream* (2008), a new biography by David Hackett Fischer, was published. This book will likely replace those published previously. Except for a number of important details, it reflects most of our major findings and conclusions.

Our list of biographies does not pretend to be comprehensive. We have presented it as an indicator of the continuing fascination writers and readers have had with Samuel de Champlain over the last 150 years.

6 See, for example, Trigger, *The Children*, 1:274–5, 329–30.

7 Berton, *My Country*, 65

8 See, for example, the classic studies on this topic: Axtell, *The Invasion Within*, ch. 7; Jennings, *The Invasion of America*, chs. 4 and 5.

1 Charlevoix, 1744. *Les voyages de la Nouvelle France Occidentale, 1632*

/[p. xlvij] *Les voyages de la Nouvelle France Occidentale,*[a] *ditte Canada, faits par le Sieur* de Champlain, *Xaintongeois, Capitaine pour le Roy en la Marine du Ponent, & toutes les découvertes, qu'il a faites en ce Pays depuis l'an 1603. jusqu'à l'an 1629. où se voit comme ce Pays a été premierement découvert par les François l'authorité de nos Rois Trés-Chrétiens jusque'à ce regne de Sa Majesté à present Regnante* Louis XIII. *Roy de France & de Navarre, avec un traité des qualités & conditions requises à un bon & parfait Navigateur, pour connoître la diversité des estimes, qui se sont en la navigation, les marques & enseignemens, que la Providence de Dieu a mises dans la Mer pour redresser les Mariniers en leurs routes, sans lesquelles ils tomberoient en de grands dangers, & la maniere de bien dessiner les Cartes Marines, avec leurs ports, rades, Isles, sondes, & autres choses nécessaries à la navigation. Ensemble une Carte générale de la descripsion dudit Pays en son Méridien, selon la délinaison de la Guide Ayman, & un Catechisme ou Instruction traduite du François en langage des Peuples Sauvages de quelque contrée, avec ce qui s'est passé en ladite Nouvelle France en l'année 1631. à Monseigneur lé Cardinal* Duc de Richelieu. In-quarto. *A Paris chez Pierre le Mur dans la Grand' Sale du Palais,* 1632.

M. de Champlain est proprement le fondateur de la Nouvelle France; c'est lui, qui a bâti la Ville de Quebec. Il a été le premier Gouverneur de cette Colonie, pour l'établissement de laquelle il s'est donné des peines infinies. Il étoit habile Navigateur, homme de tête & de resolution, désinteressé, plein de zéle pour la Religion & pour l'Etat. On ne peut lui reprocher qu'un peu trop de credulité pour des contes, qu'on lui faisoit; ce qui ne l'a pourtant jetté dans aucune erreur impor-

a LM] Champlain, 1613, 1620, 1632

1 Charlevoix, 1744. First Review of Champlain's *Les Voyages, 1632*

Charlevoix, Pierre-François-Xavier de. *Histoire et description generale de la Nouvelle France* … 3 vols. Paris: Rolin fils, 1744

EDITION USED: Toronto: Royal Ontario Museum (ROM), Canadiana Collection, FC305 (paginated), 2:xlvij–xlviij

ONLINE: Gallica. Bibliothèque nationale de France. Giffart, FRBNF30224082, 3-vol. edn., 2:xlvij–xlviij

ONLINE: Early Canadiana Online. Rolin fils, 6-vol. edn., 6:391–3 (Microform: CIHM, 33217–22)

ONLINE: Early Canadiana Online. *La Veuve Ganeau*, 6-vol. edn., 6: 391–3 (Microform: CIHM, 90691–6)

[p. xlvij] *The Voyages of Western New France, called Canada, made by the Sieur* de Champlain, of Saintonge, *Captain for the King in the Western Marines, and all the discoveries that he has made in this country from 1603 until 1629, in which is seen how this land was first discovered by the French under the authority of our very Christian Kings up to the reign of His Majesty now ruling, Louis XIII, King of France and Navarre; with a treatise on the qualities and conditions required for a good and perfect navigator to know the diversity of reckonings which are made in navigation, the signs and teachings that the Providence of God has set in the sea in order to correct the sailors in their courses, without which they would fall into great danger, and the way to draw charts correctly, with their ports, harbours, islands, soundings and other items necessary for navigation; together with a general map of the description of the said land in its meridian, according to the declination of the compass needle; and a Catechism or Instruction translated from French into the language of one of the* Peuples Sauvages *of the country; with what happened in the said New France in the year 1631. To Monseigneur le Cardinal Duc de Richelieu. In quarto. Published in Paris by Pierre le Mur in the Grand' Salle du Palais, 1632.*

M. de Champlain is, properly speaking, the founder of New France. It is he who built Quebec City. He was the first governor of this colony,[1] for the establishment of which he took infinite pains. He was a skilled navigator, a man of leadership and determination with no selfish motives, full of zeal for religion and the state. One might only reproach him for being a bit too overly credulous concerning the stories he was told, which, nevertheless, did not lead him into any significant

1 Champlain never had the title of governor. His last title was lieutenant governor to Cardinal Richelieu. See appendix 2.

tante. D'ailleurs ses Memoires sont excellens pour le fond des choses, & pour la maniere simple & naturelle, dont ils sont écrits. Il n'a presque rien dit, qu'il n'ait vû par lui-même, ou que /[p. xlviij] sur des relations originales de personnes sûres; comme ce qu'il a rapporté d'une maniere plus abregée que Lescarbot, des expéditions de MM. de Ribaut, de Laudonniere, & du Chevalier de Gourgues dans la Floride Françoise.

Dès l'année 1613. il publia ses premiers voyages en un volume *in-quarto*, divisé en deux livres, & imprimé à Paris chez Jean Berjon. En 1620. il en donna la continuation en un petit volume *in-octavo*, imprimé à Paris chez C. Collet. Enfin dans l'édition, dont je viens de donner le titre, il reprend toute l'Histoire depuis les premieres découvertes de Verazani, jusqu'à l'an 1631. Il y a joint un Traité de la navigation & du devoir d'un bon Marinier, & un abregé de la Doctrine Chrétienne du P. Ledesma Jesuite, traduit en Huron par le P. Jean de Brebeuf, avec le François à côté.

2 Charlevoix, 1744. Entreprise du Commandeur de Chatte, 1603

/[p. 111] Le Commandeur de CHATTE, Gouverneur de Dieppe,[a] lui succéda, forma une Compagnie de Marchands de Roüen, avec lesquels plusieurs Personnes de condition entrerent en societé, & fit un Armement, dont il confia la conduite à Pontgravé, à qui le Roy avoit donné des Lettres Patentes pour continuer les décou-

a RM] Entreprise du Commandeur de Chatte, 1603

error. Furthermore, his memoirs are excellent in their essentials and for the simple and natural way in which they are written. He said almost nothing that he had not seen with his own eyes or that /[p. xlviij] he had not heard first-hand from trustworthy people, such as that which he reported, more briefly than Lescarbot, of the expeditions to French Florida of MM[2] de Ribaut, de Laudonnière, and du Chevalier de Gourgues.

Since 1613, he has published his first voyages in a quarto volume divided into two books and printed in Paris by Jean Berjon. In 1620 he presented the continuation of it in a small octavo volume, printed in Paris by C. Collet. Finally, in the edition whose title I have just mentioned, he takes up the whole history again, from the first discoveries of Verazani [Verrazano] up to the year 1631. Attached to it is a treatise on navigation and the duty of a good sailor, and an abridged version of the *Christian Doctrine* by the Jesuit Father Ledesma, translated into Huron by Father Jean de Brébeuf, with the French to one side.

2 Charlevoix, 1744. Enterprise of Commander de Chatte, 1603

Charlevoix, Pierre-François-Xavier de. *Histoire et description generale de la Nouvelle France* ... 3 vols. Paris: Rolin fils, 1744

EDITION USED: Toronto: ROM, Canadiana Collection, FC305, 1:111

ONLINE: Gallica. Bibliothèque nationale de France. Giffart, FRBNF30224082, 3-vol. edn., 1:111

ONLINE: Early Canadiana Online. Rolin fils, 6-vol. edn., 1:172–3 (Microform: CIHM, 33217–22)

ONLINE: Early Canadiana Online. *La Veuve Ganeau*, 6-vol. edn., 1:172–3 (Microform: CIHM, 90691–6)

Marginal note: Enterprise of Commander de Chatte,[3] *1603*

/[p. 111] The commander de Chatte, governor of Dieppe, succeeded him [M. Chauvin] and formed a company of Rouen merchants, with whom some people of quality entered into a society and made a shipping company, whose leadership he conferred upon Pontgravé, to whom the king had given Letters Patent to

2 MM could be expanded to *messieurs* (their lordships, *lit.* my lordships), the plural of *monsieur.*

3 The orthographic variation here between *Chatte* and *Chaste* indicates, not unexpectedly, that the *s* was no longer pronounced by the eighteenth century. Pontgravé is François Gravé Du Pont.

vertes dans le Fleuve du Canada, & pour y faire des Etablissemens. Dans le même tems Samuël de CHAMPLAIN, Gentilhomme Saintongeois, Capitaine de Vaisseaux, & en réputation d'Officier brave, habile & expérimenté, arriva des Indes Occidentales, où il avoit passé deux ans & demi. Le Commandeur de Chatte lui proposa de faire le voyage de Canada, & il y consentit avec l'agrément du Roy.

Il partit avec Pontgravé en 1603. Ils s'arrêterent peu à Tadoussac,[a] où ils laiserent leurs Vaisseaux, & s'étant mis dans un Batteau leger avec cinq Matelots, ils remonterent le Fleuve jusqu'au Sault S. Loüis, c'est-à-dire, jusqu'où Jacques Cartier étoit allé; mais il paroît que la Bourgade d'Hochelaga ne subsistoit plus dès-lors, ou étoit reduite à très-peu de chose, puisque M. de Champlain, dont les Mémoires sont extrêmement détaillés, n'en dit pas un seul mot. A leur retour en France, ils trouverent le Commandeur de Chatte mort, & sa Commission donnée à Pierre du Guast, Sieur de MONTS, Saintongeois, Gentilhomme Ordinaire de la Chambre, & Gouverneur de Pons, lequel avoit encore obtenu le commerce exclusif des Pelleteries, depuis les quarante dégrés de Latitude – Nord, jusqu'aux cinquante-quatre, le droit de conceder des Terres jusqu'aux quarante-six, & des Lettres Patentes de Vice-Amiral, & de Lieutenant Général dans toute cette étenduë de Pays.

a RM] Premier voyage de Champlain

continue exploration in the River of Canada and build establishments there. At the same time, Samuël de CHAMPLAIN, a gentleman from Saintonge and ships' captain, with a reputation for being a brave, skilful, and experienced officer, arrived home from the West Indies, where he had spent two and a half years. Commander de Chatte asked him to make the trip to Canada, and he consented to it at the king's pleasure.[4]

Marginal note: 1603, Champlain's first voyage
He departed with Pontgravé in 1603. They hardly stopped in Tadoussac, where they left their ships, and having transferred to a lighter boat with five sailors, they went up the River again as far as the St. Louis rapids,[5] that is to say as far as Jacques Cartier had gone. But it appears that the village of Hochelaga from that period no longer existed, or that it was reduced to very little, since M. de Champlain, whose memoirs are extremely detailed, does not utter a single word about it. Upon their return to France, they discovered that Commander de Chatte had died and his commission had been given to Pierre du Guast, Sieur de MONTS, a native of Saintonge, a Gentleman Ordinary of the Chamber and Governor of Pons, who had again[6] obtained the exclusive fur trade from forty to fifty-four degrees of latitude north, the right to concede lands up to the forty-sixth, and letters patent from the vice-admiral and lieutenant general throughout this entire stretch of land.

4 "At the king's pleasure" (with the king's approval) may be slightly archaic but is still comprehensible and closer to the literal meaning of the French.

5 Champlain does not say how many men went with him to the Saint-Louis (Lachine) Rapids. When he got there, five sailors transferred from the larger vessel to a skiff in order to explore the rapids.

6 Again (*encore*), assuming that De Chaste had the same monopoly.

3 Charlevoix, 1744. Champlain mourut à Québec

/[p. 197] ... au mois de Decembre 1635 [... la joye ...] fut bientôt troublée par la perte, que fit peu de jours après la Colonie Françoise de son Gouverneur. Il mourut à Quebec vers la fin de cette même année, généralement regretté, & avec raison. M. de Champlain fut sans contredit un Homme de mérite, & peut être à bon titre appellé le Pere de la Nouvelle France. Il avoit un grand sens, beaucoup de pénétration, des vûës fort droittes, & personne ne sçut jamais mieux prendre son parti dans les affaires les plus épineuses. Ce qu'on admira le plus en lui, ce fut sa constance à suivre ses entreprises, sa fermeté dans les plus grands dangers, un courage à l'épreuve des contretems les plus imprevus, un zéle ardent & désintéressé pour la Patrie, un cœur tendre & compatissant pour les Malheureux, & plus attentive aux intérêts de ses amis, qu'aux siens propres, & un grand fond d'honneur & de probité. On voit en lisant ses Mémoires, qu'il n'ignoroit rien de ce que doit sçavoir un Homme de sa profession: on y trouve un Historien fidéle & sincere, un Voyageur, qui observe tout avec attention, un Ecrivain judicieux, un bon Géometre, & un habile Homme de Mer.

Mais ce qui met le comble à tant de bonnes qualités, c'est que dans sa conduite, comme dans ses Ecrits, il parut toujours un Homme véritablement Chrétien, zélé pour le service de Dieu, plein de candeur & de Religion. Il avoit accoûmé de dire, ce qu'on lit dans ses Mémoires, "Que le salut d'une seule Ame, valoit mieux que la conquête d'un Empire" ... Il parloit ainsi surtout pour fermer la bouche à ceux, qui prévenus mal-à-propos contre le Canada, demandoient de quelle utilité seroit à la France, d'y faire un Etablissement? ... /[p. 198] Il ne manqua à M. de Champlain, pour lui donner des fondemens plus solides, que d'être plus écouté de ceux,

3 Charlevoix, 1744. The Death of Champlain at Quebec, 1635

Charlevoix, Pierre-François-Xavier de. *Histoire et description generale de la Nouvelle France* ... 3 vols. Paris: Rolin fils, 1744

EDITION USED: Toronto: ROM, Canadiana Collection, FC305, 1:197–8

ONLINE: Gallica. Bibliothèque nationale de France. Giffart, FRBNF30224082, 3-vol. edn., 1:197–8

ONLINE: Early Canadiana Online. Rolin fils, 6-vol. edn., 1:306–8 (Microform: CIHM, 33217–22)

ONLINE: Early Canadiana Online. *La Veuve Ganeau*, 6-vol. edn., 1:306–8 (Microform: CIHM, 90691–6)

/[p. 197] ... in the month of December 1635, [the joy] was soon troubled by the French colony's loss of its governor a few days later. He died in Quebec towards the end of this same year, missed by just about everyone and with cause. M. de Champlain was, without question, a man of talent and perhaps rightly called the Father of New France.[7] He had great judgment, much insight, very upright views, and no one ever knew better how to come to a decision in the most ticklish matters. What one admired the most in him was his constancy in executing his ventures, his steadfastness in the greatest dangers, a courage put to the test of the most unforeseen mishaps, an ardent and disinterested zeal for the homeland, with a heart tender and compassionate for the unfortunate, more attentive to the interests of his friends than to his own, from a great depth of honour and integrity. Reading his memoirs, one sees that he knew everything that a man of his profession ought to know. We find him to be a faithful and sincere historian, a *voyageur* [traveller] who observes everything attentively, a judicious writer, a good surveyor, and a skilful *homme de mer* [seaman].

But the fulfilment of so many good qualities lies in the fact that in his conduct, as in his writings, he always appeared to be a truly Christian man, zealous for the service of God, filled with candour and religion. He was accustomed to say what one reads in his memoirs, "That the salvation of a single soul was worth more than the conquest of an Empire" ... He spoke in this way, above all, to prevent those people from talking who, ineptly forewarned against Canada, were questioning how useful it would be for France to build an establishment there ... /[p. 198] The only thing lacking to M. de Champlain, to give it more solid foundations, was to be more "listened to"[8] by those who got it operating and to be aided

7 The title "Father of New France" was bestowed on Champlain by Charlevoix.

8 Awkward but accurate; possibly "respected."

qui le mettoient en œuvre, & d'être secouru à propos. La maniere, dont il vouloit s'y prendre, n'a été que trop justifiée par le peu de succès, qu'ont eu des maximes & une conduite contraires.

Lescarbot lui a reproché d'avoir été trop credule; c'est le défaut des ames droittes, & on ne sçauroit en effet lui passer ce qu'il dit du *Gourou*, & de la figure monstrueuse des Sauvages *Armouchiquois*. Il avoit été trompé par un Malouin, nommé PREVERT, lequel prenoit souvent plaisir à inventer de pareils contes, qu'il débitoit avec beaucoup d'assûrance; comme quand il protesta un jour en présence de M. de Poutrincourt qu'il avoit vû un Sauvage joüer à la crosse avec le diable ... Champlain ne pouvoit pas comprendre qu'un Homme, qui n'avoit aucun intérêt à mentir, le fît de gayeté de cœur, & crut de bonne foi tout ce que lui disoit Prevert. Dans l'impossibilité d'être sans défaut, il est beau de n'avoir que ceux, qui seroient des vertus, si tous les Hommes étoient ce qu'ils doivent être.

4 Rossel, 1813. "Champlain," *Biographie universelle*

/[p. 28] CHAMPLAIN (SAMUEL), premier gouverneur de la Nouvelle-France, ou Canada, né à Brouage, se distingua de bonne heure dans la marine, et servit, pendant la guerre de 1595, sur les côtes de Bretagne contre les Espagnols. Immédiatement après la conclusion de la paix, il fit un voyage aux Indes occidentales, où il resta deux ans et demi. Sa fortune était vraisemblablement très-modique; car Henri IV, voulant se l'attacher, lui fit à son retour une pension qui lui donna les moyens de se maintenir honorablement auprès de sa personne. Le commandeur de Chaste, gouverneur de Dieppe, obtint du roi, peu de temps après, la commission de faire de nouveaux établissements dans l'Amérique septentrionale, et eut le désir d'engager un homme du mérite de Champlain dans cette grande enterprise; celui-ci y consentit très volontiers. Henri IV lui permit de faire ce voyage, et le chargea de lui en rendre directement un compte fidèle. Champlain s'embarqua à Honfleur sur le vaisseau de Pont-Gravé, marin très expérimenté de Saint-Malo, avec lequel il fit par la suite beaucoup d'autres voyages, et se lia d'une étroite amitié. Leur vaisseau partit le 15 mars 1603, et mouilla le 24 mai dans le fleuve Saint-Laurent. Ils s'embarquèrent ensuite dans de petits bâtiments, et remontèrent le fleuve jusqu'au Saut Saint-Louis, où Jacques Cartier s'était également arrêté en 1535, pendant son second voyage. (*Voy.* CARTIER). Champlain, après avoir visité les rives du fleuve, revint en France, et présenta au roi le récit de son voyage. La narration en a été publiée à Paris en 1603, in-8°, sous ce titre: *Des sauvages,* ou

appropriately. The way in which he wanted to go about it was only too justified by the lack of success reaped by [those with] opposite principles and conduct.

Lescarbot reproached him with having been too credulous; this is the defect of upright souls, and one could not in fact let him get away with what he said about the *Gourou* and the monstrous form of the *Armouchiquois Sauvages*. He had been deceived by a native of Saint-Malo named PREVERT, who often took pleasure in inventing similar tall tales, which he recited with great assurance, as when he protested one day, in the presence of M. de Poutrincourt, that he had seen a *Sauvage* playing lacrosse with the Devil … Champlain could not understand that a man who could gain nothing by lying would do it of his own free will, and believed everything that Prevert said to him in good faith. In the impossibility of being faultless, it is fine to have only what would be virtues, if only all men were what they ought to be.

4 Rossel, 1813. "Champlain," *Biographie universelle*

Rossel, L. "Champlain (Samuel)," *Biographie universelle ancienne et moderne*. 45 vols. Paris: L[ouis]-G[abriel] Michaud, 1813, 8:28–9

EDITION USED: University of Toronto, St. Michael's College, Kelly Library, CT143 M5. Paris: A. Thoisnier Desplaces, 1811. Reprint 1844, 7:461–2 [Paris: BnF microform; Ledien: IDC, 1983]

ONLINE: Gallica. Bibliothèque nationale de France, FRBNF37291381, 1844 edn., 7:461–2

/[p. 28] CHAMPLAIN (Samuel), first governor of New France or Canada, born in Brouage, distinguished himself in the marines early on and served against the Spanish on the coast of Brittany in the war of 1595.[9] Immediately after the peace agreement, he made a voyage to the West Indies, where he remained for two and a half years. His fortune was likely very modest, for Henry IV, wishing to attach him to his own person, granted him a stipend upon his return, which accorded him the means to maintain himself honourably at the king's side. Commander de Chaste, governor of Dieppe, not long afterwards, obtained from the king a commission to build new establishments in North America and wished to sign a man of Champlain's calibre onto this great enterprise; the latter consented to it very willingly. Henry IV permitted him to make this voyage and charged him with rendering a faithful account of it to him directly. Champlain embarked at Honfleur on the ship of Pont-Gravé, a very experienced sailor from St-Malo, with

9 This date is accurate. The first mention of Champlain in the Brittany army is 1 March 1595.

Voyage de Samuel Champlain, etc. Le commandeur de Chaste était mort pendant son absence, et le privilége qu'on lui avait accordé avait été donné au sieur de Mons, gouverneur de Pons, qui, voulant faire lui-même le voyage de l'Amérique, engagea Champlain à l'accompagner. Jusqu'alors on avait eu le projet de faire des établissements sur les bords du fleuve Saint-Laurent; mais le sieur de Mons, trouvant le cli-/[p. 29]mat trop rigoureux, porta ses vues vers les côtes de l'Acadie.

5 Anonyme, 1837. "Champlain," *Nouveau dictionnaire biographique*

/[p. 127] CHAMPLAIN (Samuel), célèbre navigateur français, fonda la première colonie française au Canada, et bâtit la ville de Québec. Il a donné son nom à un lac de ces contrées. – 17e siècle.

whom he struck up a close friendship and made many other voyages subsequently. Their ship departed on 15 March 1603 and cast anchor in the St. Lawrence River on 24 May. Next, they set out in small vessels and went up the river as far as the St. Louis rapids, where Jacques Cartier had also stopped on his second voyage in 1535.[10] (*See* CARTIER) Having visited the banks of the river, Champlain returned to France and presented the account of his voyage to the king. His narrative was published in Paris in 1603, in 8°, by this title: *Des Sauvages, ou Voyage de Samuel Champlain*, etc. Commander de Chaste had died during his absence and the licence that had been granted to him had been passed to the sieur de Mons, governor of Pons [Saintonge], who, wishing to make the American voyage himself, brought Champlain on board to accompany him. Until then, they had planned to plant some establishments on the banks of the St. Lawrence, but the sieur de Mons, finding the cli-/[p. 29]mate too harsh, cast his sights on the Acadian coast.

5 Anonymous, 1837. "Champlain," *Nouveau dictionnaire biographique*

Nouveau dictionnaire biographique universel et historique des personnages célèbres … Paris: Imprimerie de Mme Huzard, 1837, 127. Toronto: St. Michael's College, Kelly Library [back shelf: CT103, S62 1837 SMRS]

/[p. 127] CHAMPLAIN (Samuel), famous French navigator, founded the first French colony in Canada and built the city of Quebec.[11] He bestowed his name upon a lake in these regions – seventeenth century.

10 Cartier's second voyage was in 1535–36.

11 Champlain built a small establishment (*habitation*) at the present site of Quebec City.

6 Anonyme, 1845. "Champlain," *Encyclopédie du dix-neuvième siècle*

/[p. 106] CHAMPLAIN (Samuel de), né à Brouage (Saintonge), d'armateur[a] qu'il était à Dieppe, partit, en 1608, avec l'assentiment de Henri IV, et devint le fondateur, le gouverneur de Québec, et reconnut une partie du Canada; il établit des relations avec les indigènes et donna à son gouvernement l'aspect d'une véritable colonie. Attaqué par les Anglais en 1627, il fut obligé de capituler et de se retirer; mais le Canada, ayant été restitué à la France en 1629, Champlain reprit son commandement, qu'il conserva jusqu'à sa mort, 1635. On a de lui: *Voyages de la Nouvelle-France*, de 1609 à 1629. – Il a donné son nom à un lac du Canada qui communique au fleuve Saint-Laurent.

a *Armateur*: celui qui, propriétaire ou non, arme, c'est à dire équipe ou fait équiper un ou plusieurs navires ... pour le transport des marchandises ou des passagers (one who, whether the owner or not, arms [that is to say, equips, or causes to be equipped] one or more boats ... for the transportation of merchandise or passengers), *Trésor de la langue française* (TLF), 3:501.

6 Anonymous, 1845. "Champlain," *Encyclopédie du dix-neuvième siècle*

Encyclopédie du dix-neuvième siècle: répertoire universel des sciences, des lettres et des arts avec la biographie des hommes celebres. Paris: Au bureau de l'encyclopédie du XIXe siècle, 1845, 7:105 [Paris: BnF, FRBNF37310456]

EDITION USED: University of Toronto, St. Michael's College, Kelly Library, Centre Sablé, 1872, 9:106

/[p. 106] CHAMPLAIN (Samuel de), born in Brouage (Saintonge), of a ship outfitter,[12] which he was in Dieppe, set off in 1608 with the consent of Henry IV and became the founder and governor of Quebec; and he explored part of Canada; he established relations with the indigenous peoples and conferred upon his government the appearance of a true colony. Attacked by the English in 1627,[13] he was obligated to capitulate and retreat; but when Canada was restored to France in 1629, Champlain recovered his command, which he maintained until his death in 1635. We have by him *Voyages de la Nouvelle-France*, from 1609 to 1629. He bestowed his name upon a Canadian lake that flows into the St. Lawrence River.

12 *Armateur*. Consistent with usage in 1584; see Pardessus, ed., *Collection de lois maritimes, antérieurs au XVIIIe siècle* (Paris, 1831), 4:317: *Lesdicts bourgeois victuailleurs et armateurs de navires*; common expressions: *fils d'armateur, armateur à la pêche*, among others (ibid.).

13 The English capture of Quebec was in 1629 and its restoration to France in 1632.

7 Rainguet, 1851. "Champlain," *Biographie saintongeaise*

/[p. 140] CHAMPLAIN (SAMUEL DE). Ce navigateur célèbre naquit à Brouage, d'une famille de pêcheurs, en 1567. Il se rendit dans le Nouveau-Monde, par ordre du gouvernment d'Henri IV, en l'année 1603, et renonça, à cet effet, à son emploi d'armateur à Dieppe. Il reconnut les différents ports de l'Acadie, qu'il figura exactement dans les planches de ses *Mémoires*. Il explora la côte, où les Anglais n'avaient alors aucun établissement, jusqu'au-delà de Boston. Dans un autre voyage, il remonta le fleuve Saint-Laurent et y conçut le projet d'un grand établissement colonial. Ayant ensuite poussé ses reconnaissances jusqu'au lac qui depuis a porté son nom, et reconnu l'embouchure de la rivière des Utawas, il se concilia l'affection des sauvages Algonquins-Mickamas, et s'en aida avantageusement contre les Iroquois. Champlain jeta les fondements d'une habitation dans l'île de

7 Rainguet, 1851. "Champlain," *Biographie saintongeaise*

Rainguet, Pierre-Damien. "Champlain (Samuel de)," *Biographie saintongeaise, ou Dictionnaire historique de tous les personnages qui se sont illustrés par leurs écrits ou leurs actions dans les anciennes provinces de Saintonge et d'Aunis ...* Saintes, France: Éditions de Saintes, 1851. Reprint, 1971. Geneve: Slatkine, 1971

EDITION USED: University of Toronto, Robarts Library, DC611 S326R2, 140–1

/[p. 140] CHAMPLAIN (Samuel de). This renowned navigator[14] was born into a family of fishermen[15] from Brouage in 1567.[16] In 1603 he went to the New World by order of the government of Henry IV, and to this end gave up his employment as an *armateur*[17] in Dieppe. He reconnoitred the different ports in Acadia, which he represented accurately on the plates in his *Memoires*, and explored the coast where there were no English establishments at the time, as far as the other side of Boston.[18] On another voyage, he went up the St. Lawrence River and there conceived a plan for a great colonial establishment.[19] Having next pushed his exploration up to the lake that has since carried his name and surveyed the mouth of the Ottawa River, he won the affection of the *Algonquins-Mickamas*[20] and used it to his advantage against the Iroquois. Champlain laid the foundations

14 There is no evidence that Champlain ever served as a captain or navigator on a ship, although he did hold the title of *capitaine ordinaire de la marine* by at least 1610 and *capitaine ordinaire pour le Roy en la marine* by 1612.

15 The earliest reference to what may be Champlain's family (1573) states that his father Anthoyne (Anthine) Chappelin (Chappelain) lived in Brouage and was a pilot of ships (*pilotte de navyres*); see Le Blant and Baudry, *Nouveaux documents*, 1:10–11, doc. 5, 23 December 1573. Champlain's marriage contract, issued on 27 December 1610, states that his late father, Anthoine de Champlain, was, during his lifetime, a captain in the navy (*vivant capitaine de la marine*); see Biggar, *The Works*, 2:315.

16 This date cannot be corroborated.

17 An *armateur* is a shipowner or ship outfitter. The word could also mean a pirate. There are no other references to Champlain's occupation between his return from the Caribbean (1601) and his first voyage to New France (1603). As far as is known, neither Champlain nor his father was ever an *armateur*.

18 The southernmost locality reached by Champlain was Nantucket Sound in 1606.

19 The credit for planning and beginning a colony on the St. Lawrence should really go to Pierre Du Gua de Mons (DCB, 1, "Du Gua de Monts, Pierre").

20 Champlain committed France to the Algonquin/Montagnais alliance. The Mi'kmaq, who lived on the east coast of Canada, occasionally fought against the Iroquois but were not part of that alliance.

Montréal. Il remonta plus haut dans le fleuve Saint-Laurent, et défit de nouveau les Iroquois dans leur propre pays. Son courage, son habilité, ses principes d'economie sociale, lui firent, dans l'Amérique du Nord, une réputation fort éten-/[p. 141]due. On peut avec raison le considérer comme le fondateur de la Nouvelle-France. Il fit bâtir, en 1608, la ville de Québec,[a] au bord d'un bassin admirable formé par le grand fleuve Saint-Laurent. En qualité de fondateur de cette colonie, il lui donna des règlements sages et éminemment civilisateurs. Lors de son voyage en France, dans l'année 1611, les Malouins firent révoquer les priviléges de pêche dans le Canada, accordés précédemment au lieutenant-général Dugua de Monts (voy. ce nom). En 1613, le vice-roi de la Nouvelle-France choisit Champlain pour son lieutenant, et lui fit obtenir les fonds nécessaires pour se fortifier à Québec, où il occupa le même poste, sous les gouverneurs de Montmorency, de Ventadour et de Richelieu. Champlain forma une société commerciale qui, s'étendant par tout le Canada, prit le titre de *Compagnie des cent Associés*. Le cardinal de Richelieu voulut être le patron de cette compagnie.

a Rainguet: Sa population actuelle dépasse 30,000 habitants; par le malheureux traité de Paris, de 1763, cette ville d'origine française fut cédée à l'Angleterre (Its current population surpasses 30,000 inhabitants; by the unfortunate Treaty of Paris, of 1763, this city of French origin was ceded to England).

of a dwelling on the island of Montreal.[21] He went farther up the St. Lawrence River and routed the Iroquois again in their own territory.[22] His courage, his skill, and his principles of social economy established an extensive reputation for him in North America. /[p. 141] We can reasonably consider him the founder of New France.[23] He had the city of Quebec built in 1608[24] on the edge of an admirable basin formed by the great St. Lawrence River. In his capacity as founder of this colony, he conferred upon it some wise and eminently civilized regulations. When he travelled to France in 1611, the Malouins had the fishing privileges in Canada revoked, which had previously been granted to lieutenant general Dugua de Monts (see this name).[25] In 1613 the viceroy of New France chose Champlain as his lieutenant[26] and made him procure the necessary funds to fortify himself in Quebec, where he held the same position under governors Montmorency, Ventadour, and Richelieu. Champlain formed a commercial society that covered the whole territory of Canada and assumed the title *Compagnie des cent Associés.* Cardinal Richelieu wanted to be this company's patron.[27]

21 Rainguet is confusing Quebec and Montreal. The settlement on Montreal Island was begun in 1641.

22 In 1609 Champlain, with two volunteers, accompanied an Algonquin/Montagnais war party south on Lake Champlain, where they successfully attacked a temporary Mohawk encampment. In 1615 he accompanied a large Huron war party with 14 Frenchmen on a raid against the Oneida, Cayuga, or Onondaga. From Champlain's point of view, this raid was not successful.

23 Champlain was called *le Père de la Nouvelle France* for the first time by Father Pierre François-Xavier de Charlevoix (Charlevoix, *Histoire et description generale*, 1:306).

24 Champlain supervised the construction of a fortified trading post, not a city.

25 The fur-trading monopoly (not fishing) held by de Mons was revoked in 1608, opening New France to rival traders. It was restored in 1612 when the prince de Condé became viceroy of New France and secured a monopoly under which de Mons and Champlain could operate (*DCB*, 1, "Du Gua de Monts, Pierre").

26 On 15 October 1612, Champlain became lieutenant to Charles de Bourbon, comte de Soissons, who had been appointed lieutenant general of New France seven days earlier. A month later, Champlain became lieutenant to the prince de Condé, who had replaced Soissons as viceroy of the colony (*DCB*, 1, "Champlain, Samuel de").

27 The Compagnie des Cent-Associés (Company of One Hundred Associates) was founded by Cardinal Richelieu in 1627, not by Champlain, who was only a member.

Attaqué par un fort parti d'Anglais, en 1627, Champlain fut obligé de plier;[a] mais, en 1629,[b] le Canada étant revenu à la France, notre compatriote en fut nommé gouverneur-général; il s'y rendit avec quatre vaisseaux.

Champlain nous a laissé les *Voyages de la Nouvelle-France* ou *Canada*, in-4°, Québec, 1632. Son ouvrage, qui a déjà le mérite d'une médaille frappée sur les lieux mêmes de la conquête, embrasse le temps qui s'est écoulé depuis la découverte de Verazani jusqu'à l'an 1631. L'auteur se montre animé d'un grand zèle pour la religion, plein de fermeté, de désintéressement et d'amour de l'ordre. Son style est simple, sans prétention. Champlain mourut regretté à Québec en 1635, après avoir consacré 33 ans de peines et de soucis à fonder une belle colonie peu éloignée de la mère-patrie, mais dont les malheurs des temps nous ont empêché de tirer profit.

8 Trousset, 1885. "Champlain," *Nouveau dictionnaire encyclopédique*

/[p. 789] CHAMPLAIN (Samuel de), explorateur, né vers 1567 à Brouage (Saintonge), d'une famille protestante, mort à Québec en 1635. D'abord marin, il entra

a Rainguet: V. *Encyclopédie du* XIX*e siècle,* V° Champlain. (See extract 6, above.)
b Rainguet: V. Moréry, *Dict. historique*

Attacked by a strong party of English in 1627, Champlain was obliged to bend;[28] but in 1629,[29] with Canada having reverted to France, our compatriot was named its governor general;[30] he went there with four vessels.

Champlain has left us his *Voyages de la Nouvelle-France* ou *Canada*, in 4°, Quebec, 1632.[31] His literary output, which already has the value of a work done on site, like a medal struck in the very region of the conquest, encompasses the period of time from the exploration of Verazani [Verrazano] up to 1631. The author reveals himself to be animated with a great zeal for religion, filled with strength of purpose, selflessness, and a love of order. His style is simple and unpretentious. Champlain died, to the sorrow of all, at Quebec in 1635, having consecrated 33 years of effort and worry to establishing a beautiful colony not far from the motherland, but from which the seasonal hardships have prevented us from deriving any profit.

8 Trousset, 1885. "Champlain," *Nouveau dictionnaire encyclopédique*

Trousset, Jules. *Nouveau dictionnaire encyclopédique universel illustré: répertoire des connaissances humaines*. 7 vols. Paris: Libraire illustrée, 1885

EDITION USED: University of Toronto, St. Michael's College, Kelly Library, Centre Sablé, 1:789

ONLINE: Gallica. Bibliothèque nationale de France, FRBNF31500447

/[p. 789] CHAMPLAIN (Samuel de), explorer, born c. 1567 in Brouage (Saintonge) of a Protestant family; died at Quebec in 1635. A sailor at first, he next entered

28 The attack on the St. Lawrence colony actually began in the summer of 1628 and ended with the surrender of Quebec on 20 July 1629. Here Rainguet has a reference to "*Encyclopédie du XIXe siècle,* Champlain"; see extract no. 6, above.

29 Canada reverted to France under the Treaty of Saint-Germain-en-Laye, 29 March 1632, and was reoccupied by the French in the summer of that year. Here Rainguet has a reference to Moréry, *Dict. Historique*, which we were unable to find.

30 Champlain was never appointed governor of New France. See appendix 2, "Champlain's Signature and Titles." In 1629 the directors of the Company of New France, with the power vested in them by the king and Cardinal Richelieu, gave Champlain the power of a governor but without the title. These were "to command the full extent of the said country, to rule and govern as much the Aboriginals of the place as the French who currently reside there and [those who] will settle there subsequently" (Biggar, *The Works*, 6:152).

31 Champlain, *Les Voyages de la Novvelle France …* (1632), was published simultaneously by three different printers: Claude Collet, Pierre Le-Mur, and Louis Sevestre, not in Quebec but in Paris.

ensuite dans l'armée de terre et devint quartier-maître dans l'armée de Bretagne contre la Ligue. En 1599, il commanda l'une des flottes espagnoles qui firent voile pour le Mexique et qu'il ramena en Espagne deux ans après. En 1603, il quitta Honfleur sur le navire d'un Malouin expérimenté nommé Pontgravé et mouilla, le 24 mai 1603, à l'embouchure du Saguenay, où il abandonna son navire au mouillage, pour remonter en canot le Saint-Laurent. En 1604, il quitta de nouveau la France pour accompagner de Monts et Pontgravé qui voulaient fonder un établissement en Acadie, et pendant les années 1604-5-6, il explora jusq'au cap Cod, la côte américaine dont il dressa soigneusement le plan. Il revint en France en 1607 et repartit l'année suivante pour établir un poste sur le Saint-Laurent: l'accomplissement de cette mission fut la fondation de Québec. En 1609, Champlain associé aux Montaignais contre les Iroquois, visita le beau lac auquel il a laissé son nom. A la tête des Algonquins et des Hurons, il entreprit une longue lutte contre les Iroquois secrètement soutenus par les Hollandais. En 1612, il fut nommé lieutenant gouverneur de la *Nouvelle France* (Canada). Il eut à modérer l'ardeur des jésuites qui nous attiraient la haine des indigènes et semaient la discorde entre les colons catholiques et les réfugiés protestants. En 1628, une flotte anglaise parut devant Québec: Champlain ne capitula, au but d'un an, qu'après avoir essuyé toutes les horreurs de la famine. Le traité de Saint-Germain-en-Laye nous ayant rendu le Canada, l'Acadie et l'Ile du cap Breton ou île Royale, Samuel Champlain, fondateur et père de cette colonie, revint à Québec en 1633, avec le titre de gouverneur de la Nouvelle-France. On lui doit un *Traité de navigation* (1632) et la relation de ses *Voyages*, dont les meilleures éditions sont celle de Paris (1640, in-4°) et celle des abbés Laverdière et Casgrin, de Québec, avec notes, et fac-simile de cartes et d'illustrations (6 vol. in-4°, 1870).

the land army and became a quartermaster in the army of Brittany against the League. In 1599 he commanded one of the Spanish fleets which set sail for Mexico and which he brought back to Spain two years later. In 1603 he left Honfleur on the boat of an experienced man from Saint-Malo named Pontgravé and cast anchor at the mouth of the Saguenay on 24 May 1603, where he abandoned his boat at its moorings in order to go up the St. Lawrence by canoe. In 1604 he left France again in order to accompany De Mont and Pontgravé, who wished to found an establishment in Acadia, and in the years 1604–06 he explored the American coast up to Cape Cod, of which he carefully prepared a map. He came back to France in 1607 and left again the following year to establish a post on the St. Lawrence; the accomplishment of this mission was the foundation of Quebec City. In 1609 Champlain, allied with the Montagnais against the Iroquois, visited the beautiful lake to which he gave his name. At the head of the Algonquins and Hurons, he undertook a long struggle against the Iroquois, secretly supported by the Dutch. In 1612 he was named lieutenant governor of New France (Canada). He had to moderate the zeal of the Jesuits, who drew the hatred of the indigenous people against us and sowed discord between the Catholic colonists and the Protestant refugees. In 1628 an English fleet appeared before Quebec; Champlain capitulated at the end of a year only after having endured the horrors of famine. The treaty of Saint-Germain-en-Laye having returned to us Canada, Acadia, and Cape Breton Island, or Royal Island, Samuel Champlain, the founder and father of this colony, returned to Quebec in 1633 with the title of governor of New France. We are indebted to him for a *Treatise on Navigation* (1632) and the account of his *Voyages*, the best editions of which are those of Paris (1640, in 4°) and of the abbé Laverdière and abbé Casgrain of Quebec, with notes and facsimiles of maps and illustrations (6 vols, 4°, 1870).

DOCUMENT B

Personnel and Pay Records from the Garrisons in Brittany, 1595–1597 / *Paiement dans l'armée royale de Bretagne, 1595–1597*

INTRODUCTION TO THE DOCUMENT

The extracts included in this section are parts of extraordinarily detailed inventories of the thirty-four garrisons stationed in Brittany in 1595, the year Champlain was sent to the Quimper garrison.

The first document, compiled by François Miron,[1] under the command of maréchal Jean d'Aumont,[2] is dated 16 February 1595 and lists the names and ranks of the commanding officers of each garrison, the major administrative personnel, and the number of soldiers and officers by rank, type of military unit, and their wages. At that time, the anticipated total pay for thirty-one of the major garrisons was 38,167 *escus* (écus), 20 *sols* (sous) per month.[3] There were three additional, smaller garrisons, which were maintained at the expense of their governors. The estimate of monthly expenses dated 16 February was collectively forwarded to the Brittany Estates on 20 February[4] by maréchal d'Aumont, maréchal d'Espinay, François Miron, and Claude Cornullier,[5] with a request for 373,000 écus to cover expenses to the end of the year.

A notation in the margin of the entry for the garrison at *Kempercorantin* (Quimper), reproduced here, notes the arrival of an additional fourrier (no name given) at the time that Champlain joined that garrison. His anticipated salary

1 Barthélemy, *Choix de documents inédits*, 179–97; François Miron (Myron) was *trésorier général des finances en Bretagne*

2 Jean d'Aumont (1522–95), named *maréchal* by Henry III and sent to Brittany by Henry IV to oppose the Duke of Mercœur, subdued all of Cornouaille; wounded at the siege of Comper, he died in Rennes on 19 August 1595, aged 73. The Château-de-Comper-en-Brocéliande is 3 km east of the town of Concoret, Brittany.

3 The *escu* (modern "écu") was equal to 3 *livres* or 60 *sols*. One *sol* equalled 12 *deniers*.

4 Barthélemy, *Choix de documents inédits*, 198–204

5 Claude Cornullier, seigneur de la Touche (1568–1645), trésorier de France, général des finances de Bretagne, maire de Nantes in 1605–06

was 20 écus per month, equal to that of a captain, well above the average rate of 4 écus per month for a regular fourrier in the army. The entire pay for the Quimper garrison was 1,875⅓ écus per month.

The second set of documents are the earliest that mention Champlain by name. These manuscripts are in the Archives régionales de Bretagne, Rennes.[6] Although a French transcript was published in 1967,[7] the documents reproduced here have been newly transcribed, with marginal notes that were omitted from the earlier transcript, from digital photographs of the original manuscripts donated to us by the Archives régionales. These records were kept by Gabriel Hus, treasurer of the Brittany Estates.[8] They show that even at this early date Champlain was titled "Sieur de Champlain" by his superiors; that he held the rank of *fourrier* and *ayde* to Sieur Jean Hardy, *maréchal des logis* for the army in Brittany; that he transmitted sensitive documents from the king to maréchal Jean d'Aumont, Henri IV's senior officer in the Brittany army; and that his wages were authorized by d'Aumont and his successor, maréchal d'Espinay Saint-Luc.[9] Champlain's pay was listed in separate accounts under wages and extra pay for special tasks. His wages were 33⅓ écus per month from 1 March to 30 April 1595, and 25 écus per month to the end of the year. Extra pay was ordered on three separate occasions, twice in March and once in July. These records suggest that Champlain was not a member of the fighting forces and that he arrived in Brittany after the major battles with the Spanish forces had ended and shortly after the coronation of Henri IV. As treasurer of the Brittany Estates, Gabriel Hus kept the records for all transactions. The recurring formulas in these documents, including the

6 France, Département d'Ille-et-Vilaine, Archives régionales de Bretagne, Rennes, C2914, ff. 185, 192^{v}, 194^{v}, 195, 229, 523–5, 526^{v}–7

7 Le Blant and Baudry, *Nouveaux documents*, 17–21

8 Gabriel Hus (Hux) (15 ?–1609), sieur de la Bouchetière, trésorier des États de Bretagne, receveur des fouages de Saint-Malo, maire de la ville de Nantes (1599–1601). Hus was a member of the third estate of Brittany. He became *trésorier* on 24 December 1578, succeeding Jean Avril, sieur de Lormaye de la Grée, who held the position from 1545 to 1578 (La Lande de Calan, *Documents inédits*, 2:114). See "Les États de Bretagne," 30.

9 François d'Espinay, seigneur de Saint-Luc, governor of Brouage and Saintonge (1554–97), rallied in support of King Henry IV. He was named lieutenant general in Brittany and sent there in 1594 to reinforce the army of maréchal d'Aumont. When d'Aumont was killed in the summer of 1595, Saint-Luc replaced him as commander-in-chief. In 1596 Saint-Luc was transferred out of Brittany and made grand marshal of artillery and a councillor of state. He was killed at the siege of Amiens on 8 September 1597. Champlain therefore served in Brittany under maréchal d'Aumont, maréchal d'Espinay de Saint-Luc, and maréchal de Cossé-Brissac, brother-in-law of Saint-Luc, who replaced the latter after his death.

marginalia, show that Hus collected both the order and acquittal from the recipients, and submitted them to a financial bureau for final deliberation after letters of confirmation had been received from the maréchal – either Jean d'Aumont or Saint-Luc – when the bureau would issue payment. On 31 March and 8 July, Champlain submitted a claim (*blanc signé en parchemin*) in place of the acquittal when there was no order in the customary way.

There is a third detailed inventory of the Brittany army pay records, which has not been reproduced here. Titled "Extrait du registre des procès-verbaux des États de Bretagne,"[10] it is a lengthy statement of accounts drafted on 3 October 1595, by maréchal François d'Espinay Saint-Luc, addressed to Monsieur de Rennes, president of the Brittany Estates. This statement of accounts is more extensive than the first one of 16 February, including not only the wages of the officers and various regiments, but also those of the provisions officers, officers and men of the artillery and their equipment, and the cavalry units. Together, these costs had risen to 65,621 écus per month. Of this total, the largest single expense was 25,000 écus for 100,000 pounds of gunpowder at 15 sols per pound. The cost of an unspecified number of auxiliary Swiss companies of soldiers, artillery, and cavalry were listed separately. Champlain's superior officer, Jean Hardy, maréchal des logis, is listed here as receiving 66⅔ écus, and two of his fourriers, René du Boys and Paul Hubert, as receiving 33⅓ écus each.[11]

What is clear from these three sets of documents is that the royal army stationed in Brittany was paid from provincial taxes (*fouages*) raised by the Brittany Estates. Unlike most of the French provinces, Brittany was not taxed directly by the crown but maintained local autonomy in taxation. In the case of Champlain, the accounts entry under which his name appears suggests that he was not paid from the royal accounts, although the title to the entry may have been altered.[12] The Estates were the provincial government of Brittany, composed of members from the three estates: clergy (*l'église, premier état*), nobles (*noblesse, deuxième état*), and commoners (*tiers état*). The latter were neither from the clergy nor the aristocracy. Normally the Estates met once a year[13] to discuss upcoming expenses and make estimates of the revenues needed to cover them through the tax system by assessing a tax (*fouage*) on some lands and every hearth (*feu*) in a

10 *Bulletin du Comité de la langue*, 3:480–7

11 Ibid., 481–2

12 See /[f. 192ᵛ], below, and footnote 22.

13 The minutes of the États de Bretagne from 1491 to 1589 have been assembled and published by La Lande de Calan, *Documents inédits*. They give an interesting view into the operations of this system of government.

house, a crude surrogate measure for population. Most of these taxes were collected from the commoners.[14]

Note: Sums written out in letters in the original text are rendered likewise in English. Other sums, which were shortened into Roman numerals in the text, are represented here in Arabic numbers whenever they appear, regardless of current conventions in English.

INTRODUCTION TO THE FRENCH TRANSCRIPTS

Of the French transcripts, only the second set (2, below) has been newly edited from digital photographs of the original manuscripts in the Archives régionales de Bretagne, Rennes. In this case, we have restored the original readings and signalled all the expanded abbreviations in italics. Abbreviations that have not been expanded are the ampersand (&), in accordance with our general policy, and all the compounds of *dit* which could be written led*ict*, lad*icte*, lesd*icts*, etc. We have added the occasional apostrophe, final acute accent on a masculine ending, and grave accent on *a* (i.e., *à*); we have also regularized *i/j*; *u/v*. Capitals and punctuation have been given as written. The original marginalia are in the footnotes for the second set of documents only. The first set here, published in Nantes in 1880, are reproduced from the publication.

Anatole Barthélemy, with help from MM Courcy, Gaultier du Mottay, and l'abbé Chauffier, also maintained the s^{r} abbreviation (found only once with superscript *r* in the Rennes documents) as well as "lad." and other compounds of *dit*. We have only interfered to remove [*sic*] after transcriptions of the surnames s^{r} de Bastenay and variations upon M. de Kermouget, since we are more tolerant of orthographic variants in the twenty-first century. Anatole Barthélemy used both square and round brackets to make editorial interventions.

14 Chateaubriand, *Œuvres complètes*, 247–8. This is a short but clear exposition of the hated *fouage* system.

1 Le 16 février 1595. Garnisons du parti du Roi en Bretagne

/[p. 179] Estat des garnisons par nous Jan d'Aumont, comte de Chasteauroux, mareschal de France, gouverneur de Dauphiné, et lieutenant general pour le Roy en ses pays et armée de Bretaigne, establyes ès villes et places de ced. pays et duché pour la presente année MV^{c} IIII^{xx} quinze, suyvant le pouvoir à nous donné par Sa Majesté et ses lettres-patentes du 30e jour de janvier dernier …

/[p. 193] 26. *Kempercorantin*

A 50 hommes de guerre montez et armez à la legère, commandez par le s^{r} de Bastenay, pour ung moys, – le s^{r} de Bastenay, capitaine, lieutenant, cornette, marchal des logis, et 46 chevaulx legers, 563 éc. ⅔.

A une autre compaignie de 33 hommes de guerre montez et armez à la legère soubz la charge du s^{r} du Clou, au lieu de la compaignie du s^{r} de la Mouche cy devant establye, pour ung moys, – du Clou capitaine, lieutenant, marchal des logis et 30 chevaux legers, 387 éc.

[En marge de l'article précédent se trouve la note suivante]: Ceste garnison a esté augmentée d'ung fourier, par moys, 20 escuz.

1 The King's Garrisons in Brittany, 16 February 1595: Quimper

TRANSCRIPTION: Barthélemy, Anatole de, ed. *Choix de documents inédits sur l'histoire de la Ligue en Bretagne*. Nantes: Société des bibliophiles bretons et de l'histoire de Bretagne, 1880, doc. 19, pp. 179–97

ONLINE: Gallica. Bibliothèque nationale de France, FRBNF30061429

/[p. 179] Statement of accounts of the garrisons by us, Jean d'Aumont, comte de Chasteauroux, maréchal of France, governor of Dauphiné, and lieutenant general for the King in these lands and army in Brittany, established in villages and places of this said land and duchy for the current year, 1595, according to the power given to us by His Majesty and his letters patent of the 30th day of January last.

/[p. 193] [Garrison] 26. *Kempercorantin*[1]

To 50 *hommes de guerre*,[2] mounted and lightly armed, commanded by s^{r} de Bastenay, for one month, – the s^{r} de Bastenay (*capitaine*), *lieutenant*, *cornette*,[3] *maréchal des logis*,[4] and 46 *chevaux légers*,[5] 563⅔ écus.

To another company of 33 hommes de guerre, mounted and lightly armed, under the command of the s^{r} du Clou, instead of the company of the s^{r} de la Mouche formerly established, for one month – capitaine du Clou, lieutenant, maréchal des logis, and 30 chevaux légers, 387 écus.

[The following note is found in the margin of the preceding article]: This garrison was augmented by one *fourrier*, 20 écus per month.[6]

1 More commonly Quimper-Corentin, the ancient name for Quimper. "Corentin" was in honour of St. Corentin, first bishop (sixth century) of the bishopric of Cornouaille, of which Quimper was the seat.

2 *Hommes de guerre*, a generic term for soldiers, whether on foot, horse, or ship; literally "men of war."

3 *Cornette* was third in rank of commissioned officers below *capitaine* and *lieutenant* in a French cavalry unit. A cornet was equivalent to an ensign in the infantry. Both the cornet and ensign carried the regimental flags.

4 The *maréchal des logis* (*marchal des logis*) in a cavalry unit served the same function as a *fourrier* in an infantry unit, which was to provide lodgings for the men and their horses.

5 The *chevaux légers* (*chevaulx legers*), lit. "light horses," were lightly armed cavalry units. Their main function was reconnaisance and quick raiding.

6 In view of the fact that Champlain had just arrived at Quimper, it is likely that this is a reference to him.

/[p. 194] Le régiment du s[r] de Champfleury, de 7 compaignies de gens de pied, servira en lad. garnison avec 30 harquebuziers à cheval capitaine La Croix, avec 5 autres compaignies du régiment du s[r] de Courbouzon jusques à ce que la citadelle soyt bastye et construicte; desquelles 8 (7) compaignies du régiment de Champlfleury, 3 serviront en lad. citadalle et payez des derniers des garnisons, et les autres restantes sur les deniers des 3 escuz par feu.
[Comme à l'ordinaire, chacune de ces compaignies se compose de capitaine, lieutenant, enseigne, sergeant, 2 corporaux, et 44 soldats, soit en tout 50 hommes, dont la solde mensuelle est de 256 éc., – pour les 3 compaignies 768 éc.]
[En marge de l'article relative au régiment de Champfleury se trouve cette note]: Six compaignies de gens de guerre à pied, François, outre le contenu au present estat, ont esté paiez pour 2 mois par ordonnance particullière, qui monte, pour lesd. 2 mois, 3072 éc.

/[p. 194] The regiment[7] of the s^r de Champfleury,[8] [being] with 7 companies of *gens de pied*,[9] capitaine La Croix will serve in the said garrison with 30 *harquebuziers à cheval*,[10] with 5 other companies from the regiment of s^r de Courbouzon,[11] until the citadel has been built and constructed; of the which 7 companies[12] of the regiment of Champfleury, 3 will serve in the said citadel and be paid from the garrison's money, the other [4 companies] remaining with the pay of 3 écus per fire.[13]

[As ordinarily each of these companies is composed of a captain, a lieutenant, an ensign, a sergeant, 2 corporals, and 44 soldiers, to a total of 50 men, whose monthly salary is 256 écus – for the 3 companies, 768 écus.]

[This note is found in the margin of the article relating to the regiment of Champfleury]: Six companies of French foot soldiers, beyond the contents of the present reckoning,[14] were paid for two months by a special order,[15] which amounts, for the said two months, to 3,072 écus.

7 A regiment consisted of 10 companies. Each company had 50 men.

8 Champfleury was still at Quimper when Champlain was there in 1597.

9 *Gens de pied* are foot soldiers, infantry.

10 *Harquebuziers à cheval* were cavalry armed with harquebuses. The harquebus is a type of musket that was standard equipment in European armies during the sixteenth and into the seventeenth century. It was a heavy, hand-held firearm, consisting of a long, smoothbore, metal gun barrel mounted in a wooden stock. A matchlock (serpentine) was mounted on the right side of the stock, holding a slow-burning fuse. Powder and shot were rammed down to the base of the barrel. When the lock was released, the burning fuse hit a pan of powder that ignited powder in a duct leading into the base of the barrel, which touched off the gunpowder and with it one or more lead balls. The heavier harquebus was shot from a forked gun-rest. Beginning in the early seventeenth century, the snaphaunce and flintlock muskets gradually replaced the harquebus (Gooding, *The Canadian Gunsmiths*, 13–29).

11 Jacques de Montgomery, sieur de Corbouzon, was commandant of the garrison at Morlaix. The inventory of Morlaix shows that five companies of gens de pied (250 men), commanded by de Corbouzon, were on loan to Quimper and were paid by Morlaix. Morlaix ordinarily had a garrison of 595 men.

12 The text has 8 companies, which is clearly in error.

13 A "fire" (*feu*) in this sense was a "hearth." The *fouriers* worked on their maps in terms of the number of "fires" to allocate as accommodation for the troops (Communication from David Buisseret, 18 April 2007).

14 *Estat*, reckoning or statement of accounts

15 *Ordonnance*, a special order for payment

[Et un peu au dessous en marge de l'art. relative à la première compaignie du régiment de Champleury cette autre note]: Lad. garnison est aussi augmentée d'une compaignie de 30 harquebuziers à cheval, par mois, par estat particullier, 186 éc.

A 15 hommes de guerre ordonnez à la suite de M. de Kermouget, gouverneur de lad. ville, desquelz y en aura 5 montez et armez à la legère, et 10 harquebuziers à pied, pour ung moys, 90 escuz, sçavoir à chacun chevau-leger 10 éc., et à chacun desd. harquebusiers 4 éc., cy .... 90 éc.

Aud. s[r] de Kermogué gouverneur de lad. ville, 33 éc. ⅓.

/[p. 195] Et au sergent-major de lad. ville, aussy pour ung moys, 33 éc. ⅓.

Somme 1875 éc. ⅓.

2 1595 et 1597. Paiement de diverses sommes à Jean Hardy et Samuel de Champlain, pour leurs gages dans l'armée royale de Bretagne, et frais de voyages[a]

Compte troisième de M. Gabriel Hus, Trésorier des États de Bretaigne de l'extraordinaire de la guerre. Année XV[c] IIII[xx] quinze.

[1 janvier – 30 juin 1595]
/[f. 185] A Jan hardy Mar*ech*al des logis du Roy en cested. armee la somme de six cens esçuz pour ses gaiges tant ordinaires q*u*'extraordinaires des six premiers

a This title, as published by Le Blant and Baudry, omits the particle *de* in "Samuel Champlain" (Le Blant and Baudry, 17), although the documents witness to it (see ff. 194[v], 229[v], 524[v], and 526[v]).

[And a little lower in the margin of the article, relating to the first company of the regiment of Champfleury, this other note]: The said garrison is also increased by one company of 30 harquebusiers à cheval, per month by special reckoning, 186 écus.

To 15 hommes de guerre commanded in the company of M. de Kermouget, governor of the said city, of whom there would be 5 mounted and lightly armed, and 10 harquebuziers on foot, for a month, 90 écus; that is to say to each *chevau-léger*,[16] 10 écus, and to each of the said harquebusiers 4 écus, here 90 écus

To the said s^{r} de Kermogué, governor of the said city, 33⅓ écus.

/[p. 195] And to the sergeant-major of the said city, also for a month, 33⅓ écus.

Total, 1,875⅓ écus.

2 Payment of Various Sums to Jean Hardy and Samuel de Champlain, for Their Wages and Travelling Expenses in the Royal Army of Brittany, March 1595 – April 1597

ORIGINAL DOCUMENTS: Département d'Ille-et-Vilaine, Archives régionales de Bretagne, Rennes, C2914, ff. 185, 192^{v}, 194^{v}, 195, 229, 523–5, 526^{v}–7

TRANSCRIPTION: Le Blant, Robert, and René Baudry, eds. *Nouveaux documents sur Champlain et son époque*. Vol. 1: *1560–1622*. Ottawa: Public Archives of Canada, 1967, doc. 9, pp. 17–21

Third account of Monsieur Gabriel Hus, treasurer of the Brittany Estates, concerning the extraordinary expenses[17] *of the war, year 1595*

[Wages for 1 January – 30 June 1595, receipt dated 6 June]
/[f. 185] To Jean Hardy, maréchal des logis du Roy[18] in this said army, the sum of six hundred écus for his wages, as much regular as supplementary, for the first

16 *Chevau-leger*: light cavalry; i.e., the five "mounted and lightly armed."

17 *l'extraordinaire*: out of the ordinary, irregular, or supplementary expenses. The adjective was used in this way as a noun by 1480 (*TLF*, 8:531).

18 The *Brief discours*, written after Champlain's West Indies voyage (1599–1601), states that he occupied the position of maréchal des logis in the Brittany army (Biggar, *The Works*, 1:3). The above account (1595), lists Champlain as a fourrier in the army and an aide to

Mois de lad. annee IIII[xx] XV savoir LXVI escuz II tie*r* suivant l'estat des officiers de lad. armee et XXXIII escuz I tiers d'app*ointement* su*pplementai*re par ordonnance dud. feu s*ieu*r Mar*ech*al [identity explained in English] & datté du VI[e] de juin IIII[xx] XV et rendu[a] avecq la quitan*ce* dud. s*ieu*r hardy de luy signee en datte desd. jour et an[b]

pour ce cy IIII[c] escuz

/[f. 192[v]] *Autres officiers de lad. armee qui ne sont*[c]
Emploiez en l'estat du Roy

[1 mars – 30 avril 1595]
/[f. 194[v]] A Samuel de Champlain aultre fourier estant en ladicte armee la somme de soixante six escuz deux t*ie*rs po*ur* ses gaiges des mois de mars et avril de lad*icte* ann*ee* IIII[xx] quinze par ordonnan*ce* dud*ict* feu s*ieu*r mar*ech*al du douziesme de

a et rendu] de luy signé *has been crossed out before* et rendu

b Left margin (hereafter LM)] Ceste partie est aud. estat au vray et ont esté cy rend*u* les ordonn*ance* et quictan*ce* [ciuurtir] de quoy et des le*tt*res de vallidaci*o*n aud. rend*u* est et passé par deliberac*io*n du bureau po*ur* *l'apoinctement desd.* IX premiers mois a raison de LXVI ecuz II tier par mois la somme de IIII[c] ecuz et le surplus payé a [recouvrir] sur led. hardy contre lequel ceste [execution de …].

c qui ne sont: *these three words in a cursive hand, distinct from and smaller than the display script of the title, jut out into the right margin. An* in situ *examination of the original document would help to determine if there is any indication of a change of ink, which would confirm that they were not part of the original title.*

six months of the said year of '95, that is to say 66⅔ écus in accordance with the officers' record of the said army and 33⅓ écus of an additional grant by order of the said late sieur maréchal, and dated on the 6th of June '95 and rendered with the acquittal[19] of the said sieur Hardy, signed by him on the date of the said day and year.[20]

For this, 400 écus[21]

/[f. 192ᵛ] *Other officers of the said army who are not employed[22] in the King's household*

/[f. 194ᵛ] [Wages for 1 March – 30 April 1595, acquittal dated 12 May]
To Samuel de Champlain, another fourrier, being in the said army, the sum of sixty-six écus, two-thirds for his wages from the months of March and April of the said year '95, by order of the said late sieur maréchal of the twelfth of May in

Jean Hardy, maréchal des logis de l'armée du Roy. Both the maréchal des logis and the fourriers belonged to the department named the *fourrière* in the royal household. Since the main responsibility of this department was to move the court in its royal progress around the realm, the staff supervised the move, allocated rooms, and issued lodging permits (Knecht, *Renaissance Warrior,* 118; Guyot, *Traité*, 612–17). When the duties of this department were extended to the king's army, it constituted the "quartermaster service" (Bishop, *Champlain,* 7).

19 acquittal (*quictance)*: discharge of duty. The old juridical meaning was to put a debt or payment to rest (*DHLF*, 3:3049). See Nicot, *Thresor*, 532. The *Quittance de finance* had to be submitted before an outstanding debt could be settled with the Royal Treasury (Furetière, *Dictionnaire universel*, 3:Ff).

20 Left margin (hereafter LM)] This part is truly in the said statement of accounts and the order and acquittal have been rendered here [...?] of which and letters of confirmation rendered to the said. The sum of 400 écus is rendered and passed by deliberation of the bureau for the granting of the said nine months by reason of 66⅔ écus per month and the surplus paid to be [recovered] from the said Hardy against which this [execution of ...?].

21 The word *escuz* (écus), for "crowns," does not appear in the original document but is represented by a conventional sign in the form of a shield or crown.

22 who are not employed (*qui ne sont*); these words have been added to the display script of the title in a cursive hand, suggesting that they were not originally intended for this title. See Document B, page 160, note c.

may oud*ict* an cy rendu avec quictan*ce* dud*ict* champlain de luy signee & dattee desd*icts* jo*ur* et an[a]

po*ur* ce cy Lxvj escuz ij tier

[1 mai – 30 novembre 1595]
/[f. 194v] Aud*ict* Champlain la somme de huict vingtz quinze escuz po*ur* ses gaiges de Sept mois de lad*icte* annee commancez le premier jo*ur* de may mil v^c^ IIIIxx quinze et finy le dernier jo*ur* de novembre ens*uivant* aud*ict* an à raison de XXV escuz par ch*ac*un d'iceulx par ordonnan*ce* dud*ict* s*ieu*r de sainct Luc des XXVIme Septembre et XXIIIIme novembre aud. an cy rend*u* avec la quictan*ce* dud*ict* Champlain de luy sign*ee* en datte desd*icts* jo*ur* et an[b]

po*ur* ce cy VIIIxx XV escuz

[28 decembre 1595]
/[f. 195] Aud*ict* Champlain la somme de vingt cinq escuz po*ur* ses gaiges du mois de decembre aud*ict* an, par ordonnance dud*ict* s*ieu*r de sainct Luc du XXVIIIe jo*ur* de decembre oudict an et quictan*ce* en datte du XXVIIIe jour de janvier IIIIxx saize de luy sign*ee*[c]

po*ur* ce cy XXV escus
depport

[15 mars 1595]
/[f. 229v] A Samuel de Champlain ayde du s*ieu*r hardy Mar*ech*al des logis de l'armee du Roy en cedict pais la somme de neuf escuz po*ur* certain voiaige secret qu'il a faict important le service du Roy par ordonnance dud. s*ieu*r Mar*ech*al, contresignee dupré en datte du XVme de mars oudict an IIIIxx quinze & quictan*ce*

a LM] Rend*u* ordonnance et quictan*ce* et en consequence desd. l*ett*res de vallidaci*o*n est la p*ar*tie allouee comme de l'aultre p*ar*t*ie*

b LM] deport / Il a rendu les ordonnan*ce* et quictan*ce*. Est par deliberation du bureau la partie deportee attendant l*ett*res de vallidacion des ordonnan*ce* & action du s*ieur de* S*ainct Luc*

c LM] Deport / [Ainsi so*nt* devant] et ont esté les ordonnan*ce* et quictan*ce* rend*u*

the said year, rendered here with the acquittal of the said Champlain, signed by him and dated on the said day and year.[23] For this 66⅔ écus.[24]

/[f. 194ᵛ] [Wages for 1 May – 30 November 1595, acquittal dated 24 November]
To the said Champlain, the sum of one hundred and seventy five écus, for his wages for seven months of the said year, beginning with the first day of May 1595 and ending on the last day of the following November in the said year, by reason of 25 écus for each of them, by order of the said sieur de Saint-Luc on the 26th of September and the 24th of November in the said year, here rendered with the acquittal of the said Champlain, signed by him on the date of the said day and year.[25]

For this 175 écus

/[f. 195] [Wages for December 1595, receipt dated 28 January 1596]
To the said Champlain, the sum of twenty-five écus for his wages of the month of December in the said year, by order of the said sieur de Saint-Luc, of the twenty-eighth day of December in the said year; and an acquittal signed by him on the 28th day of January '96.[26]

For this 25 écus
Deferred

/[f. 229ᵛ] [Extra pay ordered 15 March, acquittal dated 20 March 1595] (fig. 4)
To Samuel de Champlain, aide to sieur Hardy, *maréchal des logis de l'armée du Roy* in this said country, the sum of nine écus for a certain secret voyage that he made involving the king's service, by order of the said sieur maréchal, countersigned by Dupré on the date of the 15th of March in the said year '95, and an acquittal from

23 LM] Order and acquittal rendered, and as a consequence of the said letters of validation are both parts allocated.

24 The payment would have been delayed somewhat after the death of maréchal d'Aumont, when a control commission was established to verify his accounts before making any decisions to disperse funds. The letters of validation must refer to official confirmation that the new *maréchal* would honour the debt.

25 LM] Deferred. He rendered the order and acquittal. The portion is deferred by deliberation of the bureau awaiting the letters of validation of the order and action from the sieur de Saint-Luc.

26 LM] Deferred. [Thus are owing?] and the order and acquittal have been rendered.

dud*ict* champlain du XXme desd. mois & an de luy signee et derien et goulhere no*tai*res[a]

po*ur* ce cy lad. somme de IX escuz

[f. 523] *Autres deniers paiez suivant les ordonnan*ces *de Monseigneur le Marechal*

[6–11 juillet 1595]
/[f. 524[v]] Au s*ieu*r de Champlain la somme de trente escuz à luy ordonnez par Monsieur de S*ain*t luc sur les III escuz pour feu suivant une l*ett*re missive escritte & adressee par Monsieur le general de la tousche Cornullier et Mr Claude Rouxeau Receveur des fouaiges en l'evesché de Cornouaille dattee à Guingamp du sixiesme de juillet aud. an IIII[xx] quinze qui en auroit faict l'acquict /[f. 525] aud. Champlain comme appert par le blanc signé en parchemin endossé pour servir de quictance de lad*icte* somme de XXX escuz et dattee audict Qempercorentin le XIme jour dudict mois de Juillet audict an lequel Rouxeau auroit depuis baillé audict Loppin pour deniers contans les susd. acquictz & paiement des sommes que luy a fournies & dont il faict recepte soubz son nom cy devant ainsi qu'il appert par la certiffica*ci*on dudict Rouxeau estant au pied de lad. missive dudict s*ieu*r de la tousche dattee le huictiesme dud. Mois de Juillet aud. an[b]

po*ur* ce cy lad. somme de XXX escuz
deport

a LM] En vertu des ordonnan*ce* et quictan*ce* desd. *lettres* de vallida*ci*on
b LM] deport / Comme [aux precendentes] parties

the said Champlain of the 20th of the above-said month and year, signed by him and by the notaries Derien and Goulhere.[27]

For this, the said sum of 9 écus

/[f. 523] *Other funds*[28] *paid in accordance with the orders of Monseigneur le Maréchal*

[Extra pay for fire duties, ordered 6–11 July 1595]
/[f. 524v] To sieur de Champlain, the sum of thirty écus, ordered for him by Monsieur de Sainct-Luc from the three écus per fire [hearth],[29] in accordance with a letter missive, written and addressed by Monsieur the general de La Tousche Cornullier and Mr. Claude Rouxeau, *receveur des fouages* [receiver of duties] in the diocese of Cornouaille,[30] dated at Guingamp[31] on the sixth of July in the said year '95, who would have made out the receipt for it /[f. 525] to the said Champlain, as it appears by the signed blank on parchment, endorsed to serve as a claim[32] for the said sum of 30 écus and dated at the said Quimper-Corentin on the eleventh day of the said month of July in the said year, which Rouxeau would have handed over since then to the said Loppin for cash, counting the above-said receipts and payment of sums which he furnished and for which he makes a receipt under his name here-above, as it appears by the witnessing of the said Rouxeau, being at the foot of the said missive of the said sieur de La Tousche, dated on the eighth of the said month of July in the said year.[33]

For this, the said sum of 30 écus
Deferred

27 LM] By virtue of both the said order and acquittal, of the said letters of validation.
28 The term *deniers* means literally "pennies" and was used in the Middle Ages in the expression *le denier de Saint Pierre*, "Peter's pence," an ecclesiastical tax or levy. It can also be used less specifically as a synecdoche for money or funds.
29 See Document B, no. 1, above, where the pay was also 3 écus per fire or hearth. Champlain must have billed for 10 hearths in order to earn 30 écus. See also n. 13 above.
30 The diocese of Cornouaille is an ancient region in Brittany comprising parts of the current departments of Finistère, Morbihan, and Côtes d'Armor. The administrative centre was Quimper.
31 Guingamp, a small town in north-central Brittany (département Côtes d'Armor).
32 The *blanc signé en parchemin* served as an acquittal of the duty with the words "For the sum of, etc." (Nicot, *Thresor*, 79). In this case, the bill or claim was not completely filled in. "On laisse quelques lignes en *blanc*, que l'on confie à la discretion de quelqu'un pour le remplir" (They leave some lines blank, which they trust to the discretion of someone else to fill in); Furetière, *Dict. universel*, 1:Ee2.
33 LM] Deferred as [with the preceding] parts

[f. 526^{v}] *Fraiz inopinez*

Au s*ieu*r de Champlain la somme de trois escuz pour aller trouver Monsieur le Mar*ech*al /[f. 527] & luy representer quelque chose important le s*er*vice du Roy laquelle somme luy [apa …?] suivant la Rescription du sieur de Lobier commis-s*air*e et intendant aud*ict* Qemp*ercoren*tin de par led*ict* seigneur datté du dernier jo*ur* de Mars aud*ict* an IIII^{xx} quinze par laquelle Il promet d'en fournir l'ordonnan*ce* de Mond. seigneur le Mar*ech*al & blanc signé sur p*a*rchemin dudict Champlain endossé pour servir de quictance de lad*icte* somme de[a] III escuz deport

Champlain, enseigne d'une compagnie à Quimper

[2 avril 1597]
Quimper: Julien du Pou, sieur de Kermouguer, gouverneur pour le roi, assisté des sieurs capitaines Magence, Lespine, Lavallée, lieutenant du sieur Champfleury, Champlain, enseigne du sieur de Milleaubourg, tous capitaines, commandants aux compagnies établies en garnison (2 avril 1597).

[Extrait][b]

a LM] desport / Somme de l'autre part

b Archives d'Ille-et-Vilaine, série B, Parliament, year 1597. From the notes of M. Henri Waquet, in the Archives du Finistère, 76-J, 16. Communication from M. J. Charpy, director of services in the Archives du Finistère. See also the *Mémoires du Chanoine Jean Moreau sur les Guerres de la Ligue en Bretagne* published by Henri Waquet (Quimper, 1960), p. 215n. In spite of careful research in the numerous subseries of the Parliament of Rennes, we have not been able to recover the original document from which this note was taken (Le Blant and Baudry, *Nouveaux documents*, 1:21).

[f. 526^{v}] *Unexpected expenses*

[Extra pay ordered 31 March 1595]
/[ff. 526^{v}] To sieur de Champlain the sum of three écus for going to find Monsieur le maréchal [d'Aumont] /[f. 527] and for representing to him some matter involving the service of the King; the which sum [...?] to him in accordance with the writ returned from the sieur de Lobier, commissioner and intendant in the said Quimper-Corentin on behalf of the said seigneur, dated on the last day of March in the said year '95; by which he promises to provide the order for it from my said seigneur le maréchal and the signed blank on parchment of the said Champlain, endorsed to serve as a claim for the said sum of[34] 3 écus
Deferred

Champlain, enseigne *in a company garrisoned at Quimper*[35]

[2 April 1597]
Quimper: Julien du Pou, sieur de Kermoguer, governor for the King, assisted by the sieurs capitaines Magence, Lespine, Lavallée, lieutenant of sieur Champfleury, Champlain, *enseigne* of the sieur de Milleaubourg,[36] all capitaines and commandants in the companies established in the garrison (2 April 1597).[37]

34 LM] Deferred – sum of the other portion

35 Robert Le Blant and René Baudry were unable to locate the original document in the subseries of the Parliament of Rennes from which this extract is taken. This notice is taken from notes made by Henri Waquet in the Archives of Finistère, 76-J, 16.

36 One sieur de Milleaubourg, a refugee in Brouage after the capture of Soubise, participated in the defence of his adopted town, Brouage, led by François d'Espinay de Saint-Luc against the unsuccessful siege of Condé, the Protestant duke, in 1585. At least one historian has suggested that Champlain would naturally have participated in the same naval defence of his native town, although no evidence survives to support such an assumption (Bishop, *Champlain*, 14).

37 It is not clear from the wording of this sentence whether Champlain was a *capitaine* or *commandant* as well as an *enseigne*. Since an *enseigne* cannot be a *capitaine* at the same time, let alone a *commandant*, we assume that Champlain at this point in time was an *enseigne*.

DOCUMENT C

Gift from Guillermo Elena[1] to Samuel de Champlain of a Vineyard and a Certain Quantity of Money Deposited in San Sebastián, 1601[2] / *Donación de Guillermo Elena a Samuel de Champlain de una heredad de viñas y cierta cantidad de dinero depositada en San Sebastián*[3]

INTRODUCTION TO THE DOCUMENT

Only two previous translations of this document have been published prior to this one: an English one commissioned by Joe Armstrong, in 1987, and a French translation of the same by Normand Paiement, in collaboration with Christiane Lacroix, in 1988.[4] Paiement, working directly from the original Spanish transcription of 1975 and a French version, was able to improve upon the Armstrong one. One such improvement consisted in adding ellipses every time a portion of the document was omitted. Others are more in harmony with our English trans-

1 Guillermo Elena was Champlain's uncle by marriage to his aunt, a sister of Champlain's mother. His name is given most often in French as Guillaume Allene. Variants are Allenne, Allaine, Alayne, and Hellaine. In Spanish, the name is either Elena or Eleno. He was also known as *capitán provenzano* (and in French, *capitaine provençal*).

2 Translation of the title bestowed on the document by the Spanish Archivo Histórico Provincial in Cádiz, when it sent the transcription to the Public Archives of Canada in 1975 (LAC, MG31, D136, vol. 5). That transcription identified the document as found on ff. 256v–259v, but an *in situ* examination discovered its true position ten folios ahead in the bundle.

3 Original Spanish title bestowed on the document by the Spanish Archivo Histórico Provincial.

4 Armstrong, *Champlain*, app. 2, 274–8; Paiement and Lacroix, *Samuel de Champlain*, 335–9. Since Normand Paiement explicitly states that Armstrong had sent him the original Spanish transcription and another French translation of this document, which permitted him to make certain corrections and exclude numerous redundancies (*ses nombreuses redondances*), and to write from a first-person-singular perspective, we attribute this translation to Paiement alone (336 n).

lation below. The most striking examples of absurdities in the Armstrong edition include the opening formula, translated as "Those who are looking at this letter as at me, Guillermo Elena," the juxtaposition of "hostess Ojelde Zalan," clearly a man, and the "purple house" of Antonio de Villa.[5]

Other misinterpretations are more serious with the potential to confuse popular readers and historians alike. One such detail involves the location of the vineyard gifted to Champlain. Whereas the Armstrong edition places it "more or less in the city of La Rochelle," an illegible word before the place name, probably *vecina*, locates it *near* La Rochelle. When the Armstrong edition attributes to this vineyard "income … to be paid to the said Samuel Zamplen," these are actually taxes (*tributos*), which Champlain himself would have to pay when he took possession of the estate. The only significant misinterpretation that Paiement has not corrected lies in Elena's claim to have set aside the will he had made in Quimper (*Queripercorant*) some four years earlier to the benefit of Champlain. If Paiement had acknowledged the true intent of the Spanish, "*que en quanto a los vienes de que yo agora hago esta donascion a el dicho Samoen Zamplen yo revoco el dicho testamento*" (f. 249), signifying a revocation from the previous will only of those goods that he was now giving to Champlain, Armstrong's interpretation of this document as the last will and testament of Guillermo Elena would have been called into question.

Summary of the document, dated 26 June 1601

Champlain's uncle, Guillermo Elena (Guillaume Allene), living in Cádiz, is very ill and possibly dying. As an expression of his gratitude to Champlain for care during his illness, and of love for his nephew through his wife, a sister of Champlain's mother, Elena cancels parts of a previous will to the benefit of Champlain. The *donación* (gift), as Elena calls it, is really a codicil to his previous will. As is proper, Champlain requests a copy for evidence. This codicil bequeaths to Champlain his uncle's estate near La Rochelle in southern France. A second bequest is a sum of money deposited in San Sebastián with Ojelde Zalan, this being the proceeds from the sale of a ship which the two formerly owned together. Elena's claim to this investment had been placed in the hands of his former hostess (innkeeper), Maria Agustín, but had become bogged down in litigation. Elena charges Champlain to settle the suit in place of Maria Augustín. A third bequest transfers to Champlain Elena's investment of one-eighth of the ship *San Julián* and the proceeds from the sale of its cargo. This is the ship on which Champlain

5 Armstrong, *Champlain*, 275, 277

had sailed in his uncle's place for about two and a half years to the West Indies and the Gulf of Mexico. Guillermo Elena had owned one-eighth of the *San Julián* when the ship was hired by Felipe II of Spain in 1598 to transport Spanish troops garrisoned in Blavet, Brittany, home to Cádiz. The *San Julián* set sail for the West Indies in 1599 under the authority of the Spanish king. While Champlain was to wait until his uncle died to inherit his estate near La Rochelle, Elena authorized him to claim the investment on the ship immediately so that the crew could be paid. Another entry in the same document, ascribed to Marcos de Rivera (f. 284), proves that Champlain waited only a week (on 3 July following this date of 26 June) before beginning the process. On 3 July Champlain appointed as his power of attorney the same Bartolomé Garibo who witnessed the will to represent him before Spain's Council of War and Council of Finance and Revenue. Champlain's purpose was to claim the value of an eighth of the *San Julián*, in accordance with his uncle's wishes.

It should be noted that this new translation of the *donación*, with the addition of the *poder* (power of attorney), lends considerable credence to Champlain's role in accompanying the *San Julián* to the West Indies as well as from Blavet to Cádiz. Both the *Brief discours* and the *donación* describe the Spanish use of the ship not only for returning the Spanish soldiers to Spain after the Treaty of Vervins but also for the subsequent voyage. The claim in the *Brief discours* that Champlain's uncle piloted the *San Julián* from Brittany to Spain ("*yo vine en ella* [the *San Julián*] *por capitan*"), when General Pedro de Zubiaur was responsible for the Spanish evacuation, is here confirmed.[6] The date of this gift also agrees with the timing of the voyage to the West Indies from the perspective of the *Brief discours*. The latter claims that the *San Julián* (with Champlain) left in early January 1599 and returned after two years and two months at sea; i.e., sometime in March 1601.[7] This would allow about three months for Champlain to care for his sick uncle and thereby earn his gratitude. Although there is no explicit allusion to Champlain's embarkation to the West Indies on board the *San Julián*, his presence in Cádiz as late as June 1601 can only be explained by it. Furthermore, the *poder* proves that Elena still laid serious claims to investment in this ship, even if it was sold or separated from the rest of the Spanish fleet on the voyage.[8]

6 Biggar, *The Works*, 1:4–5; and *Donascion*, f. 247^{v}, below

7 Biggar, *The Works*, 1:10, 80

8 Laura Giraudo, "Research Report," 96 and notes 28–9, makes these claims from documents in the Archivo General de Indias (AGI), contratación 64, 2956–8, and the Archivo de Simancas, estado K103, B86, n. 115 e.

DESCRIPTION OF THE ORIGINAL DOCUMENT

The original document appears in an unbound manuscript, in very poor condition, as a collection of legal documents recorded by the notary Marcos de Rivera for the year 1601. The collection, in nine large bundles of paper (c. 310 mm x c. 215 mm), consists of 435 numbered folios, of which the first four are only half pages and the last continue up to 453 folios without numbers. Folio 5 displays a list of names. More pages may be missing at the end. The paper has been damaged both by wormholes and by water, mainly along the top and spine; the watermarks taper off towards the bottom. Two white labels on the outside of the brown paper packaging, in which the fragile bundle is wrapped, catalogue the collection as "Notario Marcos de RIVERA Año 1601" (top) and "Cadiz: Archivo Historico Provincial, No 1512" (bottom). There is no colour or decoration, although the marginalia include crammed writing in the wide left margins and the occasional mathematical sums. In the folios where the gift appears, the pages are severely frayed at the bottom. A regular formulaic opening found in this collection is *Sepan quantos esta carta vieren* ... (May as many as see this document know ...).

The diplomatic transcription

The following diplomatic transcription of Cádiz: Archivo Histórico Provincial, No 1512, ff. 246^{v}–249^{v}, depends heavily on the original copy prepared by an archivist in Cádiz for Library and Archives Canada, Ottawa, in 1975, with a few corrections justified by an *in situ* examination of the primary document, 3–4 October 2006, and digital photographs. A fully new transcription is no longer possible, owing to the severe damage inflicted upon the bundle by both water and the hot climate. The *in situ* examination, nevertheless, was sufficient to observe that the transcriber of 1975 silently modernized the text by cancelling the doubled *r* of the seventeenth-century Spanish, introducing capital letters and accents, and expanding the abbreviations. With the exception of the accents and some capitalization near the beginning, these interventions have been maintained, because the entire document can no longer be deciphered on account of the water damage. Nevertheless, every effort has been made to correct the 1975 transcription wherever legible in the original. An original reading has been identified in the notes by its date, 1601, following a square bracket, while a mistranscription bears the date 1975. Corrected readings in the text proper are signalled in the notes in contrast to the 1975 reading. However, despite all our best efforts, our perception of the text still remains, to some extent, imperfect.

Following the norms recommended by the Comisión internacional de diplomática, the original orthography has been favoured wherever possible, with acknowledged expansions placed in italics and the addition of final acute accents to aid comprehension.[9] Since the lack of distinction between such letters as *b/v*, *e/y*, *g/j*, *g/h*, *q/c*, *s/c* (pronounced ç), and *x/j*[10] in such words as *vienes* (for *bienes*), *e* (for *y*), *agora* (for *ahora*), *quales* (for *cuales*), and *dexe* (for *dejé*) might confuse those who would otherwise be able to grasp the meaning in Spanish, the modern equivalents will be given for the first instance of each in the footnotes. Although the *g* in this script looks more like an *h* to a Canadian eye, we have respected the 1975 transcription of it in this regard. Spanish readers should also know that seventeenth-century Spanish allows for such contractions as *della* (*de ella*), *dello* (*de ello*), and *desto* (*de esto*) as well as the normal *del* (*de el*), and adds or drops the initial *h*, apparently at whim, since the Latin letter was no longer pronounced.

Square brackets have been introduced for editorial conjectures or when the complete word is unclear but certain letters can be discerned. Occasionally, the missing letter *g* has been supplied in square brackets; i.e., envar[g]ada and val[g]a, below. Modern equivalents are given after a colon in the footnotes. Since there are no paragraph breaks in the original, these have been supplied wherever logical; i.e., at a change of subject or following the formula *como dicho es* (as stated).

9 Villalba, "Normas españolas," *Serie III, H.ª Medieval* 11 (1998): 285–306

10 César Oudin describes the phonetic similarities of *b/v*, *s/c*, and the Spanish *x* for a French audience of Champlain's contemporaries (Klump, *Grammaire et observations de la langue espagnolle*, 55–6).

Donación de Guillermo Elena a Samuel de Champlain, 1601

Donascion [left margin, top of f. 246v]
Sepan quantos esta carta vieren como yo Guillermo Elena, de nascion marselles, natural que soy de la ciudad de Marsella, estante al presente en esta ciudad de Cadiz, digo que yo tengo mucho amor y voluntad a Samuel Zamplen, franses, natural del bruaze en la provincia que llaman santonze que esta press*en*te por muchas y buenas obras que me a hecho ansí en la enfermedad que he tenido y tengo de muchos[a] dias a esta parte curandome [...?] landome y acudiendo a todas las cosas deque e tenido nesesidad y tambien por el amor que le tengo por aver sido yo casado con una tia del suso dicho hermana de su madre y por otras caussas e justos respectos que a ello me mueven, de la prueva[b] de todas las cuales caussas lo relievo de las quales caussas e respectos de mi voluntad libre y en la más vastante[c] forma que yo puedo otorgo y conozco que hago graciable donascion [lo que ... creyo? ...] y tan vastante como ės nescessario a el dicho samuel zamplen presente para el y para quien el quisiere de una heredad de vinas en que ay[d] diez e siete carties medida de Francia con las cassas y bodega e tierras para sembrar[e] y campo y un za[rno][f] con una guerta[g] que todo esta en la dicha heredad la qual esta como

a muchos] *1601* munchos
b prueva] prueba
c vastante: bastante *et seq.*
d ay: hay *et seq*
e sembrar] *1975* senvrar
f un za[rno]] *1975* un zarna
g guerta: huerta

Gift from Guillermo Elena to Samuel de Champlain of a Vineyard and a Certain Quantity of Money Deposited in San Sebastián

ORIGINAL DOCUMENTS: Cádiz: Archivo Histórico Provincial de Cádiz, no. 1512, ff. 246v–249v; f. 284, Notario Marcos de Rivera, 1601

TRANSCRIPTION: Library and Archives Canada, fonds René Beaudry, MG31, D136, vol. 5

Gift[1] [Cádiz, Spain, 26 June 1601]

/[f. 246v] May as many as see this document know how I, Guillermo Elena, Marseillise by birth, native as I am of the city of Marseilles, being at present in this city of Cádiz, say that I have much love and goodwill towards Samuel Zamplen,[2] Frenchman and native of Bruaze in the province they call Santonze,[3] who is here present, for the many good works that he has done for me in this way during the illness which I have had, and still have, these past many days, caring for me, [...ing] me, and coming to the rescue in everything as soon as I was in need;[4] and also for the love that I bear him, on account of having been married to an aunt of the aforesaid, a sister of his mother; and for other grounds and just considerations that endear him to me; as a token of all the which grounds, I relieve him from the which grounds and considerations of my own free will,[5] and in the most sufficient way that I am able, I authorize and recognize that I am making a gift, easily granted [which ... I believe ...] and as abundant as is necessary, to the said Samuel Zamplen here present, for him and for whomever he may wish, of a vineyard in which there are seventeen plots,[6] a French measurement, with houses and cellar, arable lands and fields, a fenced-in plot with a vegetable garden, all of which is contained in the said estate, situated more or less in the neighbourhood

1 The title appears in the left margin at the top of f. 246v.

2 Samuel de Champlain, written phonetically by the Spanish notary. This document contains other variants in the spelling of his name.

3 Brouage in Saintonge

4 Perhaps Elena suffered from spotted fever (*tabardillo de pintas coloradas*) which nearly killed General Pedro de Zubiaur in the early months of 1597 (Garcias Rivas, "En el IV centenario," 165).

5 Elena seems to be relieving Champlain of providing more evidence to justify the gift, apart from their familial relationship and the care that Champlain has given him. A court might demand more proof, because there are other claimants to the property and Elena's illness might be seen to call his mental ability into question. We are grateful to Juan Luis Suarez for comments on the legal implications here.

6 *Carties* (*quartiers*: mod. Fr.) is a synonym for plots of land (*lopins*).

v[ecina] poco mas o menos de la ciudad de la Rochele que por una parte alinda con heredad del senor de la zarna y por la otra con el senor delus.

Y[a] la qual heredad /[f. 247] le doy a el susodicho y della y de todo ello le hago esta donascion con todas sus entradas y salidas, ussos y costumbres, derechos e servedumbres quantos la dicha heredad, cassas e huertas e lo demas que dicho es tiene e le pertenese y perteneser le pueden de hecho[b] e derecho y de uso y de costumbre y con qualesquier tributos que sobre ello estubieran cargados que todo ha de pagar el dicho Samooel Zamplen y los que la poseyeren adelante; y esta donascion hago a el suso dicho para que lo aya[c] e gose dello desde el dia de mi fallescimiento en adelante, desde agora[d] la otorgo para entonsses porque por los dias de mi vida yo tengo de gosar de ello como cosa mia desde el dia que yo fallesciere en adelante queda la dicha heredad con todo lo demas a ella perteneciente para el dicho Zamoen Samplen como dicho es.

Y porque yo tengo derecho a cierta cantidad de dinero que esta en Vizcaya en la villa de San Sevastian que esta depositado por vienes[e] de Ojelde Zalan viscayno prosedidos de una quenta[f] que yo con el tuve de una nao[g] de ciento y cinquenta toneladas que yo le entregue, el qual dinero yo envar[g]ue[h] en la dicha San Sevastian de que ay proceso el qual esta en la dicha San Sevastian en poder de una muger[i] a quien yo los dexe[j] que se desia Maria Agustin que hera[k] mi guespeda.[l]

a delus. Y] *1601* delus y

b hecho] *1975* fecho (*the latter derivative from the past participle of the Latin verb* facere *was used until the 16th c. in such legal formulas, recognizable as*: de hecho y de derecho (Lorenzo, *El Español del siglo XVI*, 70).

c aya: haya *et seq.* (*would be* tenga *in modern Spanish*)

d agora: ahora *et seq.*

e vienes: bienes *et seq.*

f quenta: cuenta *et seq.*

g nao] *Portuguese term for a ship*

h envar[g]ue] *1975* envarque (*In the* 1975 *reading, the preterite* embarqué (embarked) *would not make sense here, since we are told below that the money is still in San Sebastián. We have therefore chosen to relate this verb to* envar[g]ada *below* (page 182, n. c) as *embargué*? The translation is consequently 'impounded' (*see English n. 11*)

i muger: mujer

j dexe: dejé *et seq.*

k hera: era *et seq.*

l guespeda: huespeda *et seq.*

of the city of La Rochelle, which on one hand borders the estate of Señor de la Zarna and on the other that of Señor Delus.

And the which estate /[f. 247] I give to the aforesaid, and of it and all of it I make this gift with all its entrances and exits, uses and customs, rights and easements, for as much as the said estate, houses and gardens, and the rest that I say it holds and that belong to it and may belong to it by deed and by law,[7] and by use and by custom, and with whatever taxes have been loaded upon it, so that it all has to be paid by the said Samooel Zamplen and those who may possess it thereafter; and this gift I make to the aforesaid so that he may have it and enjoy it henceforth from the day of my demise.[8] From this time forth I authorize it for that time, so that I have the enjoyment of it as mine for all the days of my life. From the day that I die onwards, the said property remains, with all else belonging to it, to the said Zamoen Samplen, as stated.

And because I have the right to a certain amount of money that is in Vizcaya[9] in the city of San Sebastian, which is deposited as the property of Ojelde Zalan[10] from Vizcaya, being the proceeds of an account that I held with him from a ship of 150 tons, which I handed over to him, the which money I [impounded] in the said San Sebastian[11] concerning which there is a lawsuit, which is in the said San Sebastian in the power of a woman named Maria Agustin,[12] to whom I left the

7 Legal formula in Spanish from the Latin *de facto de jure*: *de hecho* [d]*e derecho*, translated by Paiement as *de fait et de droit* (Armstrong, *Champlain*, 337). Warm thanks to Nubia Soda for pointing out Latin etymologies, here and elsewhere.

8 The remainder of this paragraph is omitted by Paiement and compressed into a confusing one-line paragraph in the Armstrong edition (*Champlain*, 276).

9 Vizcaya was a province in the Basque country at that time.

10 This may be a reference to *Ogero de Challa de Byarriz* (Biarritz), living in the French part of the Basque country, to whom Elena sold his share of the 150-tun *hourque* (hooker) *Adventureuse*, which sailed out of La Rochelle (Le Blant and Baudry, *Nouveaux documents*, 1:7). In the document, the word *hourque* seems to be given as an alternate name for *Adventureuse*.

11 The Spanish verb *envarque en* (*embarqué en*) of the 1975 reading is problematic, since Elena is clearly passing to Champlain a claim of goods held in San Sebastián and not goods that were shipped out of it. The Armstrong edition interprets it as "shipped to San Sebastián" (*Champlain*, 276), translated by Paiement as "sent" (*envoyé* … *dans ladite ville*, *Samuel de Champlain*, p. 337). Since Elena later claims to have "impounded" the money (f. 248^{v}), it seems more logical to relate it to the *envargada* (*embargué*) below. See the original Spanish, page 182, note c, and page 186, note b.

12 Maria Augustín, Guillermo Elena's former hostess in San Sebastián, who had power of attorney (*poder*) to handle the debt owed to him by Zalan.

/[f. 247[v]] Digo que tambien desto hago donascion a el dicho Samoel Zamplen para que lo aya para el de la propia suerte, forma e manera que a de aver[a] la dicha heredad con lo que mas es dicho.

Y[b] porque en una nao que se nombrava San Julian que hera de porte de quinientas toneladas yo tenia la otava parte della que lo demas hera del Senor de Landricart Endebanes en Bretania y esta nao vino a Espana e yo vine en ella por capitan e fue envar[g]ada[c] por mandado de Su Magestad y por sus oficiales e vino a esta baia con soldados que trujo del tercio del Cavete en el Armada que trujo el General Pedro de Suviaure y se me deven fletes y aprovechamientos della dicha, que ansi mismo esto hago la dicha donascion a el dicho Samoen Zamplen para

a a de aver: ha de haber
b dicho. Y] *1601* dicho, y
c envar[g]ada] *1601* envarada *(embargada; see English translation, n. 15 for an explanation)*

[proceeds], who was my hostess[13] /[f. 247v]. I say that of this also I make a gift to the said Samoel Zamplen so that he may have it for himself, in his own way, form, and manner that he shall have the said estate with the rest stated.

And because in a ship that was named San Julian, which could carry 500 tons, I had the eighth part of it, since the rest belonged to the Señor de Landricart Endebanes in Bretania.[14] And this ship came to Spain, and I came in it as captain; and it was [seized][15] by order of His Majesty[16] and by his officials. And I came to this bay with soldiers whom he brought from the regiment of Cavete[17] in the fleet of warships of General Pedro de Suviaure.[18] And I am owed freightage and profits from the said ship so that in this very same way I make the said gift to

13 The Spanish noun *huespeda* appears frequently in the first part of *Don Quixote*, chapter 32, which Cervantes published in 1605. A recent translation of this chapter employs "innkeeper's wife" for the term, since the innkeeper (*huesped*: host) is also present at the inn where Don Quixote stops (Grossman, *Miguel de Cervantes*, 266 ff; Flores, *Cervantes*, xvi, 354–5). Elena's hostess seems to be acting independently, however. In contemporary English, the term "hostess" was not necessarily pejorative, since King Duncan of Scotland referred to Lady Macbeth, ironically, as "our honoured hostess"; see Harbage, *The Complete Works*, (*Macbeth*) 1:vi, 10. Written likewise c. 1605 (Harbage, *The Complete Works*, 1107 ff).

14 In French documents, the owner of the *San Julián* is given as Julien (Jullian) de Montigny, sieur de la Hautière (Hottière, Haultière), at that time living in Vannes, Brittany (Mercoeur, *Documents*, 2:159, 177). In 1598–99 the ownership of the vessel became a matter of dispute between France and Spain (Vigneras, "Le voyage," 169–71). We have not been able to identify Landricart Endebanes, evidently a resident of Brittany.

15 If *envarada* is correctly identified as *embargada*, the past participle would be related to the Latin *sequestratus, impeditus* (*Diccionario de autoridades*, 2:382) and *embargo*, which in seventeenth-century Spain meant seizure or confiscation (Alonso, *Enciclopedia del idioma*, 2:1645).

16 His Majesty: the Spanish king, Felipe II (1527–98), great-grandson of Ferdinand and Isabella, who ruled from 1556 until his death. Ardent defender of Catholicism, he married four times: to Mary of Portugal (1543), Mary Tudor of England (1554), Elisabeth de Valois (1560), and the Habsburg Anna of Austria (1570).

17 Blavet, now Port-Louis, is on the east bank, near the mouth of the Blavet River, Brittany, France.

18 General Don Pedro de Zubiaur (c. 1540–1605), a fiery character who was imprisoned by the English and tortured by the Dutch, was sent to Brittany, with the division commanded by D. Juan del Aguila, to support the French Catholic League, 1590–98. During this time, his boats supplied Blavet with provisions and reinforcements. His fame increased when he saved his company from an attack by eleven ships from La Rochelle and Brouage in April 1593, as documented by his detailed correspondence with King Felipe II. As a reward, he was appointed *capitán general* of a squadron of the Armada on

que lo aya para el e gose dello dende[a] el dia de mi fallescimiento en adelante como dicho es.

De todo lo qual hago esta donascion a el suso dicho por las caussas e respectos arriva[b] dichas y porque esta es ansi mi voluntad de la prueva de las quales caussas le[c] relievo sequn esta dicho y porque toda donascion que se hase[d] en mayor numero de quinientos sueldos no vale si no se insinua digo que tantas quantas vesses esta donascion excede la cantidad de los dichos /[f. 248] quinientos sueldos tantas donasciones le hago e se entiende de mi ser hechas[e] en dias y tiempos divididos cada una en el dicho numero, serca[f] de lo qual renuncio las leyes de las ynsinuaciones y de las cantidades en que se puedan haser e las demas de que en este casso me puedo aprovechar que me non valan desde oy[g] para en fin de los dias de mi vida como dicho es.

Me[h] desapodero del derecho, titulo y posecion que tengo a los dichos viene y a los demas que dicho es y en el dicho derecho[i] apodero y entrego a el dicho Samoen Zamplen e desde ahora para en fin de los dias de mi vida le doy la posecion de los dichos vienes en la mas vastante forma que es nescesario e tan bastante como conviene y si es nescesario le otorgo poder bastante y yrrevocable[j] para la poder tomar e aprehender la qual desde ahora para entonses le doy y en senal della esta escriptura para que la tenga por titulo cierto de lo en ella contenido para que aya para el en fin de los dias de mi vida los dichos vienes e haga dellos como de cosa suya que yo desde agora para entonses se los doy.

Y[k] quiero y es mi voluntad que en fin de los dichos dias de mi vida el dicho Samoen Zamplen e quien /[f. 248^{v}] por el fuere puedamos pedir quenta de la dicha heredad a las personas que la han tenido e tienen a su cargo y hazer[l] la quenta con ellos y cobrar y aver[m] a su poder para el todo lo que ubieren[n] rentado y podido

a dende: desde
b arriva: arriba *et seq.*
c *le* (dative): *lo* (accusative). *Such confusion between the dative and the accusative is to be expected for this time period.*
d se hase: se hace *et seq*
e hechas] *1975* fechas
f serca: cerca
g oy: hoy
h es. Me] *1601* es, me
i derecho] *1975* derechi
j yrrevocable] *1975* ynrrevocable
k doy. Y] *1601* doy y
l y hazer] *1975* y e hazer (*hacer*)
m aver: haber
n ubieren: hubieren

the said Samoen Zamplen so that he may have it for himself and enjoy it from the day of my decease onwards, as stated.

Of all of which, I make this gift to the aforesaid man for the reasons and considerations stated above; and because this is thus my will. For the proof of such grounds, I relieve him according to what has been said.[19] And because every gift that is made above 500 maravedis[20] does not count, if it is not properly registered before a public notary, I say that for as many times as this gift exceeds the quantity of the said /[f. 248] 500 maravedis, that is the number of gifts that I make him. And it is understood by me to have been done, divided into days and times, each one to the said number; concerning which, I renounce the laws of public registrations for donations and amounts by which one can act and everything else of which I can avail myself in this case, because they are not worth anything to me from today until the end of the days of my life, as stated.

I dispossess myself of the right, the title, and possession that I hold of the said goods and the rest, as is stated; and in the said right, I empower and hand over to the said Samoen Zamplen, and from now until the end of the days of my life, I give to him the possession of the said goods in the most substantial way that is necessary and as sufficient as is suitable; and if it is necessary, I authorize him sufficient and irrevocable power in order to take it and seize whatever I give him, from now on, for that time; and as a sign of that, I give him this writing so that he may hold it for a sure entitlement to that contained in it so that he may have it for himself, at the end of the days of my life, the said goods, and that he may do with them as with his own possessions; that I, from now on, give them to him for that time.[21]

And I wish, and it is my will, that at the end of the said days of my life, the said Samoen Zamplen and whoever /[f. 248^v] were for him, that we may call for an account of the said estate from the people who have held it, and who still hold it in their charge, and reckon with them, recover and have in his power the whole

3 June 1597. During the summer of 1601, he was in Lisbon, preparing to sail to the aid of the Irish. See Garcias Rivas, "En el IV centenario," 157–71, esp. 162–3, 165–6.

19 Both the Armstrong edition and Paiement omit this sentence from their translations.

20 Old Spanish coins, *sueldos* (here and immediately below). The *maravédi* was once equivalent to the English shilling (one-twentieth of a pound). According to Jarman (*Oxford Spanish Dictionary*), "maravedi" is used in English for a Spanish monetary unit.

21 Paiement omits much of the legal circumlocution in the two paragraphs above, marked by four ellipses. One can make less sense of the Armstrong conflated translation at this point.

rentar y [ambali ...?][a] y valieren los frutos e rentas della de todo el tiempo que la han tenido y tienen a su cargo y ansi mismo puedan cobrar la cantidad de dineros desuso declarada que yo tengo envargada[b] en la dicha villa de San Sebastian todo lo demas que paresciere que me deve y me deviere el dicho Ojelde Zalan y para ello pueda cobrar el proceso que yo deje en poder de la dicha Maria Agustin mi guespeda e pueda pedir todo lo demas que en razon dello me convenga.

Y[c] ansi mismo pueda cobrar de la hacienda real del Rey nuestro senor y de quien mas e mejor convenga toda la cantidad de maravedis que paresciere se me deven y devieren por la dicha nao que tengo dicho que vino con soldados del Cavete y por los fletes y aprovechamientos della y por lo que an de aver[d] los marineros e oficiales que en ella sirvieron a Su Magestad y todo lo arriva dicho lo pueda cobrar de quien y como y por la via que convenga e dar cartas de pago /[f. 249] y contender en juicio e haser [procuraciones] y las demas diligencias necesarias que para ellos lo pongo en mi lugar e le sedo[e] e traspaso mis derechos y aciones reales e personales y executivas[f] y esto a de haser en fin de los dias de mi vida como dicho es porque hasta entonses yo reservo en mi vsar de los dichos vienes como de cosa mia.

Y[g] porque estando yo puede aver como quatro anos poco mas o menos en el lugar de Quiripercorant que es en Bretana yo otorgue mi testamento en que dexe ciertos herederos digo que en quanto a los vienes de que yo agora hago esta donascion a el dicho Samoen Zamplen yo revoco el dicho testamento porque mi voluntad es questa donascion valga e tenga efecto como en ella se contiene y en todo lo demas contenido en el dicho testamento se entiende que queda en su

a ambali ...?] *1975* en valido *does not make sense. Water damage impedes a correct reading of it.*

b envargada: embargada

c convenga. Y] *1975* convenga y

d an de aver: han de haber

e sedo: cedo

f executivas: *1975* executivos

g mia. Y] *1975* mía y

amount that they had leased and had been able to lease and [...?] and which were worth the profits and gains of it for the entire time that they have held and still hold it in their charge. And in this same way, they may recover the quantity of money declared above which I hold impounded in the said city of San Sebastian [and] all the rest which it would appear the said Ojelde Zalan owes me and would owe me. And for this, he may take up the lawsuit which I left in the hands of the said Maria Agustin, my hostess, and he may seek all the rest agreed to be mine because of it.[22]

And in this same way, he may recover from the royal estate of our lord the king[23] and from whom may be more suitable, the whole amount of maravedis that, it would appear, I am owed, currently and in the past, for the said ship that, I told you, came with soldiers from Cavete, and for the freightage and profits of it. And for whatever the sailors and officials who served His Majesty in it must have, and for all the aforesaid he may recover it from whomever and however and by whatever way may be suitable, and give receipts /[f. 249] and legally contest and appoint [proxies] and do all the remaining necessary formalities; so that for them, I place him in my stead and cede and transfer to him my rights and my royal, personal, and executive actions.[24] And this has to be done at the end of the days of my life, as is stated, because until then I reserve for myself the use of the said goods as my own.[25]

And because being myself in Quiripercorant, which is in Bretana,[26] it could have been four years ago, more or less, I authorized my will in which I designated certain heirs, I now say that for as much as the goods of which I now make this gift to the said Samoen Zamplen, I revoke the said will, because it is my wish that this gift may be valid and take effect as is contained in it. And in all the rest contained in the said will, be it understood that it should remain in its force

22 Here Elena (Allene) is charging Champlain to take responsibility for settling the lawsuit against Zalan.

23 This must refer to the same Spanish king, Felipe II, mentioned above (n. 16). Elena's identification with this king is less surprising than his assumption that Champlain would make the same identification! Perhaps the formula is used here (*del Rey nuestro senor*) to meet the expectations of the witnesses and public scribe of the legal *donación*.

24 These actions are legal actions. Executive actions are those to be enforced after the decision or sentence of a judge or tribunal.

25 Most of the above paragraph is omitted from both the Armstrong and the Paiement translations. The last sentence must refer back to the previous paragraph.

26 Quiripercorant (Quimper-Corentin), now Quimper, a town in Brittany at the junction of the Rivers Odet and Stier. Champlain saw service there under the maréchals Jean d'Aumont in 1595, Saint-Luc in 1595–96, and Cossé-Brissac in 1597–98.

fuerca e vigor para se usar del como de testamento mio y me obligo que agora ni en el articulo de la muerte no revocare[a] esta donascion que agora hago a el dicho samoen zamplen aunque diga e alegue que tengo caussas bastantes para la poder revocar que aunque estas [caussas] alegue quiero que esta donascion se cumpla e tenga efecto en todo e por todo /[f. 249v] como en ella se contiene para la qual cumplir segun e dicho obligo mi persona e vienes avidos e por aver[b] y doy poder a las [partes] para ser a ello compelido y quiero que valga como si ansi fuese jusgado y sentenciado por sentencia pasada en cosa juzgada e renuncio e apelaciones y las demas leyes de mi favor que me no valan y la general que dize que general renunciacion no val[g]a.

En testimonio dello ansi lo otorgo en su fecha en la ciudad de Cadiz estando en las cassas morada de Antonio de Villa vesino desta ciudad en veinte e seys dias del mes de junio ano del Senor de mille seiscientos e uno anoss.

Testigos que fueron presentes Bartolome Gariboy y Jacques Aleman vesinos de Cadiz que juraron en forma de derecho que conosen a el dicho Guillermo Elena e que es el mismo contenido en esta escriptura e se nombra ansy. Mas fueron testigos el dicho Antonio de Villa e Pedro Bautista Asirio ginoves[c] e Francisco Sanchez Ahumada vesinos y […?] y e[d] sido presente a todo lo suso dicho. E dicho Samoel Zamplen dixo[e] que aseta la donascion y la quiere en todo y por todo como en ella se contiene y que agradese al susodicho la merced que en ella se le hase e lo pidio por testimonio. Testigos los dichos a lo mesmo los testigos de conosimiento jura-

a revocare: revocaré (*future*)
b avidos e por aver: habidos y por haber
c ginoves: genovés
d e: he
e dixo: dijo

and vigour to be used as my will.[27] And I am bound to it as of now; nor in the last extremity of death[28] will I revoke this gift that I am now making to the said Samoen Zamplen, even though I may say and allege that I have sufficient cause to be able to revoke it; and even though I may allege these [causes], I wish that this gift may be accomplished and take effect in every respect /[f. 249ᵛ], including all that is contained therein; by which to accomplish, according to what has been said, I bind my person and goods, already possessed and those to come. And I give power to the [parties] to be compelled to it and I wish that it may be valid as if it were thus judged and sentenced by a past sentence as business judged. And I renounce both appeals and the other laws on my behalf which are not valid for me, and the general law that says that a general renunciation has no worth.

In testimony of which, I authorize it on its date thus, being in the city of Cadiz, in the houses, the abode[29] of Antonio de Villa, a neighbour of this city, on the 26th day of the month of June in the year of the Lord one thousand six hundred and one.

Witnesses who were present: Bartolome Gariboy and Jacques Aleman, neighbours of Cadiz who swore legally that they know the said Guillermo Elena and that he is the same as the one contained in this writing and who names himself in this way. More witnesses were the said Antonio de Villa and Pedro Bautista Asirio of Genoa and Francesco Sanchez Ahumada, neighbours and [...?] and I was present for all the above said. And the said Samoel Zamplen said that he accepts the gift, and that he wants it completely and for all that is contained therein, and that he is grateful to the aforementioned for the favour that has been done him by it; and he requested it for evidence.[30] The said witnesses, along with the wit-

27 These statements make it clear that Elena (Allene) only revoked those aspects of his previous will that pertain to the gifts he made to Champlain, making the *donación* a codicil to the previous will written about 1597 in Quimper. Beginning with this sentence up to the end of the paragraph, Armstrong omitted this entire section.

28 *articulo de la muerte*: wherein *articulo* is defined in Latin as *punctum, vel instans temporis* "porque es una parte pequena del tiempo" (Real Academia, *Diccionario de autoridades*, 1:425).

29 The "purple house" of the Armstrong translation is not possible, since *las cassas* are plural and the alleged adjective would still be singular (*morada*). While Paiement jumped the latter word with an ellipsis, we have chosen to interpret *cassas, morada* as two nouns in apposition.

30 Champlain requested a copy of this document, which he produced on 29 December 1625, when he made a gift of Elena's estate near La Rochelle to Charles Lebert du Carlo (Leymarie, "Inédit sur le fondateur," 83–4). The Armstrong edition omits this request, while Paiement interprets the Spanish "*lo pidio por testimonio*" as "il a demandé qu'elle soit attestée" (he asked that the gift be witnessed)!

ron que conosen a el dicho Samoel Zamplen e los dichos en testimonio lo firmaron en el registro. Bar. Gariboy. Guillermon Elena. Por testigo Jacques Aleman. Champlain. Passo ante mi Marcos de Rivera, escrivano publico.

Transcription of poder *(power of attorney)*

pod*er* [left margin just below midway down the leaf, f. 284][a]
3. [Sepan quien essta vieren commo?]
A Zamplen frances natural [...]
bruaje en la provincia [que el ...]
[l. 9] puede parecer ...
[l. 11] cien demas ...
[l. 12] memor de pagar el [vna de] la otaua [...]
[l. 13] San Julian [... cassas]
[l. 15] nao [a la propia] que vino [...?]

Summary of the power of attorney

Cádiz, 3 julio de 1601
Samuel Champlain, natural de Bruage (Francia) otorga su poder a Bartolomé Garibo, para que parezca ante los Consejos de Guerra y Hacienda de España en solicitud de que se le pague el valor de la octava parte de la nao San Julián, de porte quinientas toneladas.

Protocolo de Cádiz, pr. 1512 (ante Marcos de Rivera), f. 284[b]

a This feeble excuse for a transcription was done with a digital photograph, since the summary published in Martín, *Catálogo de la exposición* was seen only after the opportunity to examine the document in 2006.

b Martín, *Catálogo*, 41

nesses of acquaintance, swore that they know the said Samoel Zamplen and the said persons signed the registry as a witness. Bar. Gariboy [signature]. Guillermon Elena [signature]. As a witness, Jacques Aleman [signature]. Champlain [signature]. Accomplished before me, Marcos de Rivera, public scribe.

Summary of the Poder

/[f. 284] Power of Attorney[31] [Cádiz, 3 July 1601]
Samuel Champlain, native of Brouage [France], confers his authority upon Bartolomé Garibo to appear before the Councils of War and of Finance and Revenue to request that he [Champlain] be paid the value of the eighth part of the ship San Julián, which carries 500 tons.

31 Title (p*oder*) found in the left margin.

DOCUMENT D

Factum of the Merchants of Saint-Malo against Champlain, [January 1613][1] / *Factum des marchands de Saint-Malo contre Champlain,* [1613]

INTRODUCTION TO THE DOCUMENT

The merchants of Saint-Malo were essentially free traders who considered themselves successors to the discoveries of Jacques Cartier and pioneers of the emerging Canadian fur trade. Under Henri IV, trading monopolies were given on occasion to those of his early supporters in the religious wars who promised to use the profits from the trade to develop settlements. These policies led to constant agitation at court by the Saint-Malo merchants to have the monopolies rescinded and the trade opened to all. Champlain was a steady factor on the side of the monopolists because of his unswerving dedication to colonization and his belief that only monopolies held by well-financed backers could lead to success. Consequently, he was targeted by the merchants in order to discredit him. This is one of the few documents critical of Champlain. The circumstances that developed between 1610 and the end of 1612 leading up to this factum were described by Champlain as follows.[2]

In 1609 de Mons was unable to get his monopoly renewed. Champlain hurried to France, but, unable to reverse the decision, returned the following spring to Canada. In the summer of 1610 he heard the news that Henri IV, his staunchest supporter for colonization, had been assassinated.[3] Fearing the worst for the future of the fledgling colony, Champlain went back to France to persuade de

1 Le Blant and Baudry, *Nouveaux documents*, 1:245, wrote the following in note 1: "The date 1609 which figures at the head of this document in the Bibliothèque Nationale is a later addition, in another and certainly erroneous hand. Indeed, the text of the factum mentions a man from Rouen, who returned from Canada in 1612, and the protections delivered by the king on November 13, 1612. It is at the end of those protections that the merchants of St-Malo made representations to the king. Therefore, we can date this piece from the beginning of January, 1613."

2 Biggar, *The Works*, 2:145–221, 241–7; 4:206–19

3 Henri IV was assassinated on 14 May 1610.

Mons to continue with the settlement at Quebec and to prepare him to meet stiffer competition from the rival traders. He returned in the spring of 1611 and explored Montreal Island with the hope of establishing a second settlement. Because of the competition, the trade was going badly; fearing ruin, Champlain returned to France in the fall, where he met with de Mons, who charged him with placing the Canadian enterprise on a surer footing. Champlain brought his plans to Pierre Jeannin,[4] who took the matter to the governing council. On the advice of Jeannin and other members of the council, Champlain decided to "put myself in the hands of some great man whose authority could repulse envious attacks."[5] The council's suggestion, Charles de Bourbon, comte de Soissons,[6] agreed to become patron of the Canadian enterprise after Champlain approached him with his maps and plans. On 27 September, Soissons was given a twelve-year monopoly on the trade, and on 8 October he was appointed *lieutenant general au pays de la nouvelle France.*[7] Seven days later, Soissons appointed Champlain as *nostre Lieutenant, pour representer nostre personne audit pays de la nouvelle France*, with very broad powers to govern on his behalf.[8] On 1 November, Soissons died and Champlain persuaded Henri II de Bourbon, prince de Condé,[9] to take an interest in the Canadian ventures. On 13 November, Queen Marie de Medici (regent for the young king, Louis XIII) appointed Condé to succeed Soissons,[10] and on 22 November he appointed Champlain as his lieutenant with the same powers that had been given to him by Soissons. The monopoly they acquired this time was extended along the St. Lawrence to include Tadoussac. Champlain proceeded to

4 Pierre Jeannin (1540–1622) was an important jurist and diplomat who served under Charles IX, Henri III, Henri IV, and Louis XIII. He supported Henri IV in 1595 and was appointed by him *conseiller d'État* in 1601 and *intendant des finance* in 1602. After Henri IV's death, Marie de Medici, regent for the young Louis XIII, appointed Jeannin *superintendent des finances.* Champlain mentions him on several occasions as an important supporter of his plans for colonizing Canada. Lescarbot dedicated his 1618 edition of the *Histoire* in part to Jeannin (Grant, *History of New France*, 1:9–11).

5 Biggar, *The Works*, 4:207

6 Soissons, a Catholic, was a first cousin of Henri IV. An early supporter of the Catholic League, Soissons broke with it and became an important officer in Henri's military campaigns against the League.

7 Trudel, *The Beginnings of New France*, 100

8 Biggar, *The Works*, 4:209–16

9 Henri II de Bourbon, prince de Condé (1588–1646), was a nephew of Soissons. As a member of the House of Bourbon, he was a *premier prince du sang* (first-ranking prince of the blood, in line for the throne).

10 Letters patent appointing the prince de Condé (Le Blant and Baudry, *Nouveaux documents*, 1:233–8).

publish and post the commission in the ports of France, an action that immediately drew condemnation, especially from Saint-Malo and Rouen.[11] Champlain, a good organizer, now also showed that he had important supporters close to the new king and had persuaded them to follow the colonizing aims of the late king, Henri IV, for which, he argued, monopoly control of the trade was necessary.

In this factum, Champlain's detractors repeatedly accuse him of making false claims of "discovery" (*decouverture*) following his activities in 1603. On examining *Des Sauvages*, it seems fairly clear that he considered this a journey of exploration, rather than a discovery of territories where others had not been before him. He made a mistake in *Des Sauvages* by writing that Jacques Cartier, a Malouin, had not been west of Rivière Jacques Cartier, a mistake he later corrected. It was probably this error that prompted the invective of the merchants regarding Champlain's so-called "discoveries." His claims of discovery, if any, pertained to some details of geography not previously described, but not the grand features of the geography of the St. Lawrence Valley, which clearly he knew others had seen before him.

This factum was an attempt to discredit Champlain and to neutralize him by calling into question his qualifications, motives, and claims.

11 Biggar, *The Works*, 2:245–7; Le Blant and Baudry, *Nouveaux documents*, 1:239–41. For protests from Saint-Malo and Rouen, see Le Blant and Baudry, *Nouveaux documents*, 1:263–4; Ramé, *Documents inédits* (1867), 36–8.

Factum des Marchands

/[p. 245] Pour faire cognoistre la surprise manifeste du sieur Champlain contenus en sa requeste presentée au Conseil, sur l'advis duquel les deffences du treiziesme novembre ont esté expediées.

Sera pris que depuis l'an mil six cens quatre, pour parvenir au peuplement de la Nouvelle-France, le sieur du Mont et ses assotiez de Normandie, Bretagne et Guyenne, s'obligerent de y mener par chacun an le nombre de soixante personnes, artisans, laboureurs et autres, pour travailler aux mines que on y trouveroit, et tant et si longuement y continuerent qu'il se trouve que laditte compagnye avoit perdu tout le fonds d'icelle montant à près de cinquante mil escuz, laditte perte arrivée à cause en partie de la pauvreté et infertilité desdittes mines, et intemperie dudit pays, grandz froids desdittes habitations, et partie aussi à cause des traverses que les navires estrangers leur firent, traitant et negotiant par force et à main armée avecques les Sauvages dudit pays, et emportans ce que les subjectz de Sa Majesté par le moyen de ses deffenses n'osoient s'approprier, ce qui occasionna la rupture de laditte compagnie et que Saditte Majesté y remist le trafficq libre, qui a duré jusques à ce jour par la revocation de la commission dudit sieur du Mont,

Factum of the Merchants

Le Blant, Robert, and René Baudry, eds. *Nouveaux documents sur Champlain et son époque*. Vol. 1: *1560–1622*. Ottawa: Public Archives of Canada, 1967, doc. 125, pp. 245–7

/[p. 245] In order to make known the apparent surprise of sieur Champlain contained in his request presented to the Council, on the advice of which the injunction[1] of 13 November has been dispatched.

Given that[2] since the year one thousand six hundred and four, in order to succeed in populating New France, the sieur du Mont and his associates from Normandy, Brittany, and Guyenne[3] were obliged each year to take sixty people there, skilled craftsmen, workers, and others to work in the mines which they would find there. And they continued on with this to such an extent that the said company was found to have lost all its funds, up to the sum of almost fifty thousand écus,[4] the said loss having been sustained, in part, on account of the paucity and infertility of the said mines, and the intemperate climate of the said land, the extreme cold of the said settlements, and in part also on account of the interference that foreign boats caused them, trading and negotiating with the *Sauvages* of the said land, by force and with arms, and carrying away that which the subjects of His Majesty did not dare to appropriate by means of his injunction, all of which occasioned the breakup of the said company. Given also that His said Majesty restored free trade there, which has lasted until today, by revoking the commis-

1 *Deffences*: injunction. In legal terms, a plea for justification to support an argument; an interdiction or injunction prohibiting something from taking place. Nicot defines *defenses ou defences* in the plural as "prohibitions" (*Thresor*, 181), but Cotgrave gives *iniunction* before *prohibition* in the singular (*Dictionarie*, A a). In this case, the merchants were trying to prevent Champlain from being appointed to conduct the affairs of New France and bring about a monopolistic company.

2 *Sera pris que: lit.* It will be taken that – Given that

3 *Guyenne* is a former name for the region of Aquitaine. It included the current *départements* south of the Garonne River: Gironde, Dordogne, Lot et Garonne, Landes, and Pyrénées-Atlantiques; see (maps) Sanson, *Atlas*, map 37 and text 36, 154–5.

4 Le Blant and Baudry (*Nouveaux documents*, 1:245) wrote the following in note 2: "According to the *Voyages* of 1632, de Mons would have spent more than 100,000 livres in Acadia alone, from 1604 to 1607. Laverdière, *Œuvres*, 5:52. This sum seemed so considerable that biographers believed it to be a typographical error for 10,000 livres. Bishop, *Champlain*, 125. But the witness of the merchants of St-Malo, hardly to be suspected of any sympathy for de Mons, confirms the figures of the costs assumed by the two companies of de Mons." Note: The *escu* (modern écu), also called "crown," was worth about 3 livres.

que lesdits habitans de Saint-Mallo et autres marchans de la France obteinrent au Conseil de Sa Majesté, auquel negoce ledit Champlain ny ses pretendus assotiez n'ont contribué d'un seul denier, et par consequent du besoing de societé ny d'assotiez, y ayant esté seullement comme passagers, sa profession de peintre le conviant avecques le lucre de veoyr /[p. 246] ledit pays, tousjours aux despans des compagnies qui l'y ont mené, desquelles il a aulcunes fois tiré de grans sallaires, ce qu'il ne peult desnyer.

Qu'il a en cela manifestement surpris Sa Majesté, Messeigneurs du Conseil, et Monseigneur le prince de Condé et Monseigneur l'admiral, leur donnant à entendre que luy et sesditz pretenduz assotiez avoient tout descouvert au-dessus de Quebecq dans la riviere Saint-Laurens, les mettant au rang des descouvertes nouvelles, en consequence de quoy que euls seulz debvoient jouyr du fruit de leurs travaux, qui est l'unicque fondement de sa pretendue commission, laquelle estant appuyée sur un faux donné à entendre ne peult aussi aulcunement subsister.

Pour la verification de ce, chacun peult recognoistre par les histoires que dès l'an mil cinq cens quatre ce pays fut descouvert par les Normans et Bretons faisans leurs pescheries ès environs dudit pays. En l'an mil cinq cens sept Sebastien Cabot au nom du roy d'Angleterre en descouvrit une partie et Jaspar Colteler,

sion of the said sieur du Mont, which the said inhabitants of *Saint-Mallo* and other merchants from France obtained from His Majesty's Council; for the which business, [neither did] the said Champlain nor his alleged associates contribute a single penny, and consequently from the need of the society, not of the associates, having been there simply as passengers, his profession as artist (*peintre*) inviting him, along with the financial gain of seeing /[p. 246] the said country, still at the expense of the companies that took him there, from which he, on occasion, drew a great salary, which he cannot deny.[5]

That by this he undoubtedly surprised His Majesty, the gentlemen of the Council, and My Lord the prince de Condé and My Lord the admiral,[6] by leading them to believe that he and his said alleged associates had discovered everything in the St. Lawrence River above Quebec, placing them on the level of new discoveries, and that, as a result of which, they alone ought to enjoy the fruit of their work, which is the only foundation of the commission that he claims, which, being supported by a false tale, cannot at all stand its ground either.

For the verification of this, everyone can recognize through historical accounts that since the year one thousand five hundred and four, this land has been discovered by the people of Normandy and Brittany making their fishing grounds in the region of the said country. In one thousand five hundred and seven Sebastian Cabot, in the name of the king of England, discovered part of it and *Jasper*

5 Le Blant and Baudry (*Nouveaux documents*, 1:245) wrote in note 3: "This document is the only one which mentions such a 'profession' for Champlain. This statement leads us to believe that, from that time, he enjoyed a certain notoriety as a painter or draftsman, founded undoubtedly upon the illustrations of his book *Des Sauvages* [1603] and perhaps also upon the abundant series of 62 or 72 illustrations in the *Brief Discours*. Champlain's *Les Voyages, 1613* only appeared in the autumn of this year, probably after this *factum* was drawn up." We would like to point out four problems concerning Le Blant and Baudry's note 3: First, the merchants called Champlain a *peintre*, which is an artist or painter, not a draftsman, which is a *dessinateur*. Second, Champlain's *Des Sauvages*, [1603] 1604, did not contain illustrations. Third, in his book *Voyages et Descovvertvres, 1619* (f. 11) and (f. 17^{v}), Champlain mentioned on two occasions that *Les Voyages, 1613*, was not printed until 1614. Although the *privilège* was assigned on 9 January 1613, *Les Voyages* contains the entire *Quatriesme Voyage*, which was written after Champlain returned from Canada on 26 August 1613. Fourth, Champlain's *Brief discours* with the illustrations was first printed in 1859 by the Hakluyt Society. None of these publications could have been seen by the merchants of Saint-Malo.

6 Henri de Bourbon, prince de Condé, was appointed viceroy of New France on 13 November 1612. He appointed Champlain as his lieutenant on 22 November. The admiral of France at the time was Charles de Montmorency, duc de Damville.

Portugois, descouvrit le reste, se promettant trouver passage pour aller à la Chine ou aux Moulucques; mais, comme dict l'histoire intitulée le *Suppleamant de Ptholomée* imprimée à Douay an mil six cens trois, telz effectz et desseings furent en vain, auquel livre se trouve une carte imprimée l'an MV^{e} IIII^{xx} dix-sept qui contient plus de deux cens vingt lieues de descouverture dans laditte riviere de Saint-Laurens et plusieurs veritables particularitez y mentionnez des Sauvages qui y faisoient leurs demeures, depuis lequel temps Jacques Cartier, habitant dudit Saint-Mallo, à ses frais et perilz de sa vie soubz l'autorité et nom du roy de France, entra et penetra fort avant dans laditte riviere, passa Quebecq et le lac, et après luy plusieurs Normans, Biscains, Bretons, et entre autres les sieurs du Pontgravé et Prevert, dudit Saint-Mallo, Fabien de Mescoroua lesquelz, trente-cinq ans ou environ, ont trafficqué dans ledit lac et au-dessus avecques lesdits Sauvages, et y a environ de dix à douze ans seullement [1603] que ledit Champlain fut comme passager mené au premier sault par ledict sieur du Pontgravé, de Saint-Mallo, et en laquelle année se trouva une infinité de personnes de toutes contrées de la France jusques au nombre de neuf ou dix barques qui toutes ensemble negotierent

Colteler,[7] a Portuguese, discovered the rest, promising himself to find a passage by which to go to China or to the Molucca Islands.[8] But as says the historical account, entitled the *Supplement of Ptholomeus*, printed in Douay in one thousand six hundred and three,[9] such effects and plans were in vain, in the which book is found a map, published in 1597,[10] that contains more than two hundred and twenty places of discovery on the said *Saint-Laurens* River and many true details mentioned therein about the *Sauvages* who made their dwellings there. From the which time, Jacques Cartier, an inhabitant of *Saint-Mallo*, at his own expense and at the risk of his life,[11] on the authority and in the name of the king of France, entered and penetrated far inland on the said river, passed Quebec and the lake.[12] And after him, many people from Normandy, the Bay of Biscay area (*Biscains*), Brittany, and, among others, the sieurs du Pontgravé[13] and Prévert of the said *Saint-Mallo*, and Fabien de Mescoroua,[14] all of whom, for thirty-five years or thereabouts, have traded with the said *Sauvages* in the said lake and above. And it is only about ten to twelve years ago [1603] that the said Champlain was brought to the first rapids as a passenger[15] by the said sieur du Pontgravé of *Saint-Mallo*. And in the which year[16] there were found countless people from all parts of France, up to the number of nine or ten barques, which all did business together

7 Read Gaspar Corte-Real for Jasper Colteler. See DCB, 1:234–6, "Corte-Real, Gaspar."

8 *Mouluques:* The Molucca Islands are between New Guinea and Sulawesi (Celebes), north of Australia.

9 The most likely work mentioned here is (maps) Wytfliet, *Descriptionis Ptolemaicae augmentum.* Published in 1597 (Louvain) and 1603 (Douay), this atlas of 19 maps was the first devoted exclusively to the New World (Tooley, *Maps and Map-Makers*, 112).

10 It is likely that this map is (maps) Wytfliet, *Nova Francia, 1597.*

11 He received his answer to this argument in the *Voyages* of 1613 (Biggar, *The Works*, 2:219–21; Laverdière, *Œuvres*, 3:267–8; Le Blant and Baudry, *Nouveaux documents*, 1:246n4).

12 Lac Saint-Pierre

13 Pontgravé, namely Gravé Du Pont.

14 Fabien de Meriscoiena, Basque captain, who shipped a boat to de Mons on 4 April 1605, for whale fishing and the trade in Canada (Le Blant and Baudry, *Nouveaux documents*, 1:246).

15 This is the second time that Champlain is simply called a "passenger," giving the impression that he did nothing of importance on the voyage.

16 The year was 1603.

audit lieu. Pour monstrer que cet endroit estoit dès lors fort frequenté et dès lors /[p. 247] et de par avent descouvert, de tout ce que dessus se trouvera assez de tesmoignage pour revaincre la commission dudit Champlain.

in the said place.[17] In order to show that this place was well frequented from that time on, and since then /[p. 247] and formerly discovered, enough testimony of all that will be found to revoke the commission of the said Champlain.

[This is the first half of this document.]

17 One must remember that in *Des Sauvages* Champlain was not writing a history of the 1603 expedition. He was asked to do a resource survey and comment on the possibilities of exploring westward. On that task he was virtually working on his own. It was Gravé Du Pont who was in charge of trading and would have reported back on such activities. Champlain does not seem to have been aware of prior exploration, not even that of Jacques Cartier, which he does not mention until *Les Voyages, 1613*. Whatever he wrote about that explorer he probably obtained from the men he travelled with. The entire argument presented here by the merchants of Saint-Malo is an attempt to interpret Champlain's actions in the worst light possible.

DOCUMENT E

Decrees and Commissions Preparatory to the Voyage of 1603 / *Arrêts et commissions pour le traffique du Canada, le 13 mars 1603*

INTRODUCTION TO THE DOCUMENTS

Trying to end the squabbles between the "free traders," represented most vociferously by the merchants of Saint-Malo, and those who wanted a trade monopoly in order to finance a colony, Henri IV issued an edict in Paris (document 1, below) on 13 March 1603 permitting three vessels to trade in Canada. This decree was followed the same day by a commission (document 2) outlining its conditions in slightly greater detail. Of the three ships, one was from Saint-Malo captained by Gilles Éberard du Colombier[1] and the two others by Jean Sarcel de Prévert and François Gravé Du Pont.[2]

The groups who represented these three ships were to share all costs equally, and because of the legacy of past enmities, previous monopolies, and current claims, a clause was inserted that Prévert and Gravé were not to interfere with Colombier. Both documents state clearly that part of the mission was "the discovery of lands in Canada and the adjacent countryside." The discovery of lands in Canada was entrusted to Champlain, who was on Gravé Du Pont's ship that

1 Gilles Éberard du Colombier was a merchant-trader from Saint-Malo. Colombier, Prévert, and Gravé had all traded upstream on the St. Lawrence at least as far as the present Trois-Rivières before 1599.

2 Jean Sarcel de Prévert and François Gravé Du Pont were merchant-traders originally from Saint-Malo (*DCB*, 1, "Sarcel de Prévert" and "Gravé du Pont"). Gravé Du Pont (Pontgravé), however, had moved from Saint-Malo to La Rochelle in 1600. Eventually he joined a rival group of traders led by Aymar de Chaste, governor of Dieppe, thus incurring the wrath of his former Malouin colleagues. De Chaste had succeeded to the trading licence of Pierre de Chauvin de Tonnetuit early in 1603 after Chauvin's death. When de Chaste died late in 1603, he was succeeded by Pierre Du Gua de Mons. Prévert, who had a separate licence until 1604, also joined the group headed by de Mons.

headed for the St. Lawrence. Prévert's ship went to Acadia, where he traded and looked for minerals. It is not known where Colombier's ship went.

The French documents have not been newly edited. The previous editors have normalized punctuation, *u*/*v*, *i*/*j*, accents, and the capitalization of proper nouns. We have introduced paragraphs to the second document, in order to facilitate comparisons between the French and the English.

1 Ensuilt la teneur des arrestz et commissions pour le faict du traficq du Canada

Ensuilt la teneur des arrestz et commission pour le faict du traficq du Canada. Extrait des registres du conseil d'Estat.

Sur la requeste presentée par les bourgeoys et habitans de Saint-Malo, tendant à ce qu'il pleust au roy rendre libre en ceste présente année et à l'advenir le traficq du Canada cy davent découvert avecq grande despance par leur prédécesseurs, nonobstant les permissions et defances pretendues par les capitaines Prevert et Pontgravé.

Le Roy, en son conseil, a pour bonnes causes et considérations à ce se mouvans, ordonné et ordonne que le capitaine Coulombier de Saint-Malo, nommé par lesd. habitans dud. Saint-Malo, armera un vaesseau en la presente année, pour, avecq les deux navires desd. Prevert et Pontgravé conjoinctement ou separément, selon que la commodité se fera, aller au traficq et decouverture des terres de Canada et pays adjaczantes, à la charge de contribuer à la tierce partie des loyaulz coustz et fraiz qui se feront en lad. decouverture, faisant Sa Majesté inhibitions et deffances ausd. Prevert et Pont Gravé et à tous autres ses subjectz de quelque qualité et condition qu'ilz soient de le troubler sur les peines qui y escheent.

Fait au conseil d'Estat du roy, tenu à Paris le treziesme jours de mars 1603, ainsi signé Huilliere et scellé.

1 Extract from the Decrees of the Council of State

Michelant, Henri-Victor, and Alfred Ramé. *Relation originale du voyage de Jacques Cartier au Canada en 1534: documents inédits sur Jacques Cartier et Canada*. New series. Paris: Librairie Tross, 1867, 23

ONLINE: Gallica. Bibliothèque nationale de France, FRBNF30201255

MICROFORM: CIHM, 11990

Here follow the terms of the decrees and the commission for the deed of trade in Canada.

Upon the request submitted by the burgesses and inhabitants of Saint-Malo to the effect that it may please the king, in this current year and in future, to liberate the trade in Canada, hitherto discovered by their predecessors at great expense, notwithstanding the permits and prohibitions claimed by the captains Prévert and Pontgravé.

The king in his council, moved to this by sound cause and consideration, has ordered and orders Captain Colombier of Saint-Malo, named by the said inhabitants of the said Saint-Malo, to fit out a vessel in the current year to go with the two ships of the said Prévert and Pontgravé, conjointly or separately at their convenience, for trade and the discovery of lands in Canada and the adjacent countryside, to be paid by contributing to one-third part of the legitimate costs and expenses that will be incurred in the said discovery; with His Majesty inhibiting and forbidding the said Prévert and Pontgravé and all other of his subjects, of whatever quality and condition they may be, to trouble him,[1] upon pain of due punishment.

Executed in the king's Council of State held in Paris on the thirteenth day of March 1603, and thus signed Huillière and sealed.

1 The transcription of the document from the Archives nationales de France (E*5a, f. 248), which reads "*les*" (them) at this point (see http://www.civilization.ca/vmnf/expos/champlain/cont_eng.html), is found to be a misreading for *l'y*, thanks to the verification of Marie-Bethsabée Zarka. The direct object refers to Captain Colombier of Saint-Malo (him) who received the permit, notwithstanding the possible objections of Prévert, Gravé Du Pont, and others.

2 Commission pour le trafficq du Canada

Henry par la grâce de Dieu roy de France et de Navarre, à notre très cher cousin le s[r] de Dampville, amiral de France, ses lieutenans en l'amiraulté, senechal de Saint-Malo ou son lieutenant et alloué de lad. ville et à chacun d'eux en droict soy, salut. Ayant faict veoir en notre conseil la requeste à nous presentée par nos bien amez les bourgeoys et habitans de lad. Ville de Saint-Malo, à ce qu'il nous pleust rendre libre en la présente année et à l'advenir le traficq de Canada, cy davent decouvert avecq grande depance par leurs predecesseurs, nonobstant les permissions et defances pretendue par les capitaines Prevert et Pont Gravé.

Nous, de l'advis de notre conseil, et suyvant l'arrest ce jourd'huy donné et icelluy, dont l'extraict est cy attaché soubz le contrescel de notre chancellerye, avons pour bonnes causes et consideracions à ce nous mouvans, permis et permettons par ces présentes au capitaine Coulombier de Saint-Malo, nommé par lesd. habitans de Saint-Malo, d'armer un vaesseau en la presente année, pour, avecq les deux navires desd. Prevert et Pontgravé, conjoinctement ou separement, selon que la commodité s'ofrira, aller au traficq et decouverture des terres de Canada et païs adjaczans, à la charge de contribuer à la tierce partye des loyaulz coustz et fraiz qui se feront en ladite decouverture, faisant expresses inhibitions et defances ausdits Prevert et Pontgravé et tous aultres nos subjectz de quelque qualité et condition qu'ilz soient de l'y troubler, sur les peines qui y escheent.

2 Commission for Trade in Canada

Michelant, Henri-Victor, and Alfred Ramé. *Relation originale du voyage de Jacques Cartier au Canada en 1534: documents inédits sur Jacques Cartier et Canada.* New series. Paris: Librairie Tross, 1867, 24–5. Also in Beaulieu, Alain, and Réal Ouellet, eds. *Champlain: Des Sauvages.* Montréal: Éditions TYPO, 1993, 224–5
ONLINE: Gallica. Bibliothèque nationale de France, FRBNF30201255
MICROFORM: CIHM, 11990

Henry, by the grace of God, King of France and of Navarre,[2] to our very dear cousin, the sieur de Dampville, admiral of France,[3] his lieutenants in the Admiralty, seneschal[4] of Saint-Malo or his deputy, and [the one] approved by the said city, and to each of them in his own right, greetings. Having had the petition reviewed at our council, which was presented to us by our beloved burgesses and inhabitants of the said city of Saint-Malo, to the effect that it may please us, in this current year and in future, to liberate the trade in Canada, hitherto discovered by their predecessors at great expense,[5] notwithstanding the permits and prohibitions claimed by the captains Prévert and Pontgravé.

Upon the advice of our council, and in accordance with the decree given this day, the extract of which is attached beneath the counter-seal of our chancellery, and moved to this by sound cause and consideration, we have permitted and do permit by these present letters Captain Colombier of Saint-Malo, nominated by the said inhabitants of Saint-Malo, to fit out a vessel in the current year to go with the two ships of the said Prévert and Pontgravé, conjointly or separately at their convenience, for trade and the discovery of lands in Canada and the adjacent countryside, to be paid by contributing to one-third part of the legitimate costs and expenses that will be incurred in the said discovery. This is done expressly to inhibit and forbid the said Prévert and Pontgravé and all other subjects of ours, of whatever quality and condition they may be, from troubling him, upon pain of due punishment.

2 Henry IV, King of France, born 1553, reigned 1594–1610.

3 Charles de Montmorency, duc de Damville, etc. (c. 1537–1612), was *amiral de France* from 1596 to 1612. Champlain dedicated *Des Sauvages* to him in 1603.

4 The sénéchal was the representative of the king for administrative and judicial matters.

5 The "predecessors" to which the Saint-Malo merchants refer are Jacques Cartier and others who followed, especially after 1581.

Sy vous mandons et ordonnons que de notre presente permission et de tout le contenu en les presentes vous faictes soufrir et laisser jouir[a] lesd. habitans de Saint-Malo et led. capitaine Coulombier plainement et paisiblement, cessans et faizans cesser tous troubles et empeschemens au contraire. Mandons au premier notre huissier ou sergent sur ce requis signifier notredit arrest et lesd. presentes ausd. Prevert, Pont Gravé et à tous autres qu'il apartiendra, et leur fere de par nous les deffances y contenues, luy donnant pouvoir de ce fere sans pour ce demander permission, placet, visa, ni pareatis. Car tel est notre plaisir, nonobstant quelconques aultres lettres, mandemens et deffances à ce contraire.

Donné à Paris le XIIIe jour de mars l'an de grace 1603, et de notre règne le quatorziesme. Ainsi signé: Par le Roy en son conseil, Huillere, et scellé du grand seau de cire jaulne.

(*Archives de Saint-Malo*, Reg. 5)

a soufrir et laisser jouir: "Vtenti et fruenti patientiam accommodare" (Nicot, *Thresor*, "jouir").

Here we command you and order that, with our present permission and with all the contents in the present letters, you cause the said inhabitants of Saint-Malo and the said Captain Colombier to apply patience to the using and enjoying, fully and peaceably, ceasing and causing to cease all troubles and impediments to the contrary. We command first our bailiff, or officer required for this, to serve notice of our said decree and the said present letters to the said Prévert, Pontgravé, and all others whom it will concern and to make for them, on our behalf, the prohibitions contained therein, giving him[6] power to do this without, for this purpose, asking for any permission, *placet, visa,* or *pareatis.*[7] For such is our pleasure, notwithstanding any other letters, commands, and prohibitions whatsoever contrary to this.

Given in Paris on the thirteenth day of March in the year of grace 1603 and of our reign the fourteenth. Thus signed by the king in his council, Huillère, and sealed with the great seal of yellow wax.

(Archives de Saint-Malo, reg. 5)

6 Him: the bailiff or officer

7 *placet, visa,* or *pareatis.* Each of these constitutes a form of royal authorization for the commission. The first and last are both letters, while the *visa* could be just a signature or seal. Originating from three Latin verbs (*placet*: it is pleasing; *visa*: things seen; *pareatis*: obey), all three terms had been adopted into the French terminology of the royal chancellery (*TLF*, 13:456, 16:1190; Littré, *Dict.*, 5:1381). The *paréatis* (n.m.) denotes a letter ordering the execution of the act in a jurisdiction foreign to the one in which the act was made. The negation of them all, following the preposition "without" (*sans*), however, indicates that the commission, having received royal assent, required no further authorization or documentation. We are grateful to Bernard Barbiche for clarifying the use of these French terms in the royal chancellery.

DOCUMENT F

Excerpts from Champlain's Works Related to Events before 1604 / *Extraits des Œuvres de Champlain avant 1604*

INTRODUCTION TO THE DOCUMENTS

In his various writings Champlain made some references to his life before he returned from his voyage to Canada in 1603. These are brought together here as a series of excerpts. Each has been newly translated from the original texts. The pagination from the original texts is given as, for example, /[p. 3], and where appropriate the pagination from the Champlain Society edition of Biggar's *Works* is given as {3}. The excerpts cited from *The Works* are reproduced without any editorial intervention with respect to orthography, but apart from the ampersand, abbreviations have been resolved and *i/j*, *u/v* normalized for those cited from the original editions.

1 Blavet à Cádiz. *Brief discours*

Biggar, *The Works* 1:3–6

{3} Ayant esté employé en l'armée du Roy qui estoit[a] en Bretaigne soubz messieurs le mareschal d'Aumont, de St. Luc, et mareschal de Brissac, en qualité de marechal des logis de ladicte armée durant quelques années, et jusques à ce que sa maiesté eust en l'année 1598 reduict en son obeissance ledit[b] païs de Brestaigne, eust licencié[c] son armée, me voyant par ce moyen sans aucune charge ny enploy, je me resolus, pour ne demeurer oysif, de trouuer moyen de faire vng voiage en {4} Espaigne, en y estant pratiquer et acquerir des cognoissances pour par leur faueur [et] entremise faire en sorte de pouuoir m'enbarquer dans quelqu'vn des nauires de la flotte que le Roy d'Espaigne enuoye tous les ans aux Indes occidentalles, affin d'y pouuoir m'y enbarquer des particuliarités qui n'ont[d] peu estre recongneues par aucuns Françoys, à cause qu'ilz n'y ont nul accez libre, pour à mon retour en faire rapport au vray à sa magesté. Pour donc paruenir à mon desseing, je m'en allay à Blauet, où lors il y auoit garnison d'Espaignolz, auquel lieu je trouuay vng mien oncle nommé le Cappitainne Pr[ouençal,] teneu pour vng des bons mariniers de France, et qui en ceste qualité auoict esté entreteneu par le Roy d'Espaigne comme pillotte general en leurs armées de mer. Mondict oncle ayant receu commandement {5} de monsieur le mareschal de Brissac, de conduire les nauires dans

a estoit] *corrected from* estois
b ledit] *corrected from* ledis
c eust licencié] *corrected from* eus licencier
d ont] *corrected from* ons

1 Blavet to Cádiz. *Brief discours*

ORIGINAL MANUSCRIPT: *Brief discours des choses plus remarquables que Sammuel Champlain de Brouage a reconneues aux Indes occidentalles* ... John Carter Brown Library, Providence, RI, Hay Microfilm F6127, 46 leaves, bound; illustrations and maps; 30 cm

TRANSCRIPTION: Biggar, Henry Percival, gen. ed. *The Works of Samuel de Champlain*. 6 vols. Toronto: Champlain Society, 1922–36, 1:3–80

Biggar, *The Works*, 1:3–6

{3} Having been employed for some years in the king's army that was in Brittany under Messieurs the maréchal d'Aumont, de St. Luc, and the maréchal de Brissac,[1] in the capacity of *maréchal des logis* of the said army, and until His Majesty, in the year 1598, had subjugated the said land of Brittany and dismissed his army;[2] and seeing myself thereby without any charge or employment, I resolved, so as not to remain idle, to find a way to make a voyage to Spain and, being there, to {4} acquire and cultivate acquaintances in order, by their favour and mediation, to be able to embark upon one of the ships of the fleet that the King of Spain sends every year to the West Indies and there to launch into details that no Frenchmen have managed to identify, by reason of not having free access to them, in order to make a true report of them to His Majesty upon my return.[3] Therefore, to accomplish my plan, I went along to Blavet,[4] where there was a Spanish garrison at that time, where I found an uncle of mine named Captain Provençal,[5] who was considered to be one of France's first-rate seamen and who, in that capacity, had been commissioned by the King of Spain as *pilotte general* of their sea forces.[6] My said uncle, having received an order {5} from the maréchal de Brissac to command the ships onto which they made the Spaniards from the gar-

1 Maréchal Jean d'Aumont was appointed in 1592 to lead the royalist forces against the duc de Mercœur in Brittany. He was killed in 1595 and was succeeded by maréchal d'Espinay de Saint-Luc until 1596, when the latter was replaced by maréchal de Cossé-Brissac.

2 The army was not "dismissed" as such because there was still resistance and possible rebellion. It is likely that superfluous personnel were dismissed, such as the lodgings officers, because the army was no longer on the move.

3 Here Champlain admits that he took on the role of a spy for Henri IV. We do not know if the king authorized this role or was even aware that Champlain was taking it on.

4 Now Port-Louis on the south coast of Brittany

5 His uncle Provençal was Guillaume Allene, a ship's captain, who was born in Marseille.

6 This title is unlikely. No Frenchman would be eligible to be a pilot-general in the Spanish navy.

lesquelz l'on feist embarquer les Espaignolz de la garnison dudict Blauet, pour les repasser en Espaigne, ainsy qu'il leur auoit[a] esté promis, je m'enbarquay auec luy dans vng grand nauire du port de cinq centz thonneaux, nommé le St Gulian, qui auoit esté pris et arresté pour ledict voiage. En estans partis dudict Blauet au commencement du moys d'aoust …

2 Départ depuis Cádiz. *Brief discours*

Biggar, *The Works*, 1:9–10

{9} Or en mesme temps l'armée du Roy d'Espaigne, qui a accoustumé d'aller tous les ans aux Indes, s'appareilloit audict S[t] Luc où il vins de la part dudict Roy vng seigneur nommé Domp Francisque Colombe, Cheualier de Malte, pour estre general de ladicte armée, lequel voiant nostre vaisseau appareillé et prest à seruir, en sachant par le rapport qu'on luy auoit[b] fais, qu'il estoit fort bon de voille pour son port, il resolut de s'en seruir, et le prendre au fraict ordinaire, qui est vng escu pour thonniau par mois, de sorte que j'eus occasion de me resiouir voiant naistre {10} mon esperance, d'autant mesme que le Cappitaine Prouençal, mon oncle, ayant esté reteneu par le general Soubriago pour seruir ailleurs, et ne pouuant faire le voiage, me commist la charge dudict vaisseau pour auoir esgard à iceluy, que j'acceptay fort vollontiers, et sur ce nous fusmes trouuer ledict sieur general Colombe pour sauoir s'il auroit agreable que je fisses le voiage; ce qu'il me promist librement, auec des tesmoignages d'en estre fort aise, m'ayant promis sa faueur en assistance, qu'il ne m'a depuis desniés aux occasions.

a auoit] *corrected from* auoir
b auoit] *corrected from* auois

rison of the said Blavet embark, in order to cross over to Spain, as had been promised them, I embarked with him in a great ship of five hundred tuns burthen, named the *St Gulian*, which had been taken and seized for the said voyage.[7] And having departed from the said Blavet at the beginning of the month of August …

2 Departure from Cádiz. *Brief discours*

Biggar, *The Works*, 1:9–10

{9} Now at the same time, the King of Spain's army, which was accustomed to go to the Indies every year, was being outfitted in the said *St Luc*,[8] to which a seigneur named *Domp* Francisque Colombe,[9] a knight of Malta, came on behalf of the said king, to be general of the said army; who, seeing our vessel fitted out and ready for service, and knowing by the report that had been made to him that, for its tunnage, it was a very good sailing ship, resolved to make use of it and to take it at the ordinary freight rate, which is one écu per tun per month; so I had occasion to rejoice seeing my hopes rise, the {10} more so that Captain Provençal, my uncle, having been retained by General Soubriago[10] to serve elsewhere and being unable to make the voyage, committed the charge of the said ship to me to watch over it,[11] which I accepted very willingly: and thereupon we went to find the said sieur General Colombe, to know if he would find it agreeable that I should make the voyage. This he freely granted me, with evidence of being very easy with it, having promised me his favour in assistance, which he has not since denied me when the occasion called for it.[12]

7 *San Julián pris et arresté*, taken and seized (intercepted or apprehended). The documentary record states the ship was leased by de Brissac under contract from Julien de Montigny, sieur de la Hautière, at that time living in Vannes, to ship the Spaniards to Cádiz (Carné, *Documents sur la ligue de Bretagne*, 12:159, 177). The ship was seized by Spanish officials once it reached Cádiz.

8 Sanlúcar

9 Admiral Don Francisco Coloma

10 General Don Pedro de Zubiaur

11 *Avoir esgard à iceluy*: "to watch over" or "to supervise"

12 … *desniés aux occasions* … "when the need arose," "on suitable occasions," or "when the occasion called for it"

3 Retour à Cádiz. *Brief discours*

Biggar, *The Works*, 1:80
{80} Ayantz passé lesdictes isles des Essores, nous feusmes recognoistre le cap St Vincent, où nous prismes deux vaisseaux Anglois qui estoient en guerre, que nous menames en la riuiere de Seuille, d'où nous estions partis, et où fust l'acheuement de nostre voiage; auquel je demeuray depuis nostre partement de Seiuille, tant sur mer que sur terre, deux ans deux mois.

4 Henry le Grand, qui me commanda de faire les recherches les plus exactes. *Les Voyages, 1613*

/[f. ãij] AV ROY

...ce petit traicté, que je prens la hardiesse d'adresser à vostre Majesté, intitulé Journalier des voyages & descouvertures que j'ay faites avec le sieur de Mons, vostre Lieutenant, en la nouvelle France: & me voyant poussé d'une juste recognoissance de l'honneur que j'ay reçeu depuis dix ans, des commandements, tant de vostre Majesté, Sire, que du feu Roy, Henry le Grand, d'heureuse memoire, qui me commanda de /[f. ãij[v]] *faire les recherches: & descouvertures les plus exactes qu'il me seroit possible: Ce que j'ay fait auec les augmentations, representées par les cartes, contenues en ce petit liure* ...

3 Return to Cádiz. *Brief discours*

Biggar, *The Works*, 1:80

{80} Having passed the said islands of Azores, we went to explore Cape St. Vincent,[13] where we captured two English ships which were at war, and which we conducted into the river of Seville, from where we had set out, and where the termination of our voyage was; on which voyage, by both land and sea, from the time of our departure from Seville, I spent two years and two months.[14]

4 Henri the Great commanded me to undertake the most exact investigations. *Les Voyages, 1613*

Champlain, Samuel de. *Les Voyages dv Sievr de Champlain Xaintongeois, capitaine ordinaire pour le Roy, en la marine* … Paris: Jean Berjon, 1613, unnumbered page, /[f. ãij]

EDITION USED: Library and Archives Canada, Ottawa, Reserve, FC330, C3 1613

ONLINE: Early Canadiana Online. *Les Voyages du Sieur de Champlain Xaintongeois*

MICROFORM: CIHM, 90024

CHAMPLAIN SOCIETY EDITION: Biggar, *The Works*, 1:207–8

/[f. ãij] TO THE KING[15]

… *this little treatise, which I make bold to dedicate to Your Majesty, entitled a Journal of the Voyages and Discoveries that I made with the Sieur de Mons, your lieutenant, in New France. And seeing myself compelled by a just acknowledgment of the honour that I have received over the past ten years from the commissions as much from Your Majesty, Sire, as from the late king, Henry the Great, of happy memory, who commanded me to undertake the most exact investigations and discoveries possible; which I did with the additions represented on the maps contained in this little book* … [Privilège, 9 January 1613]

13 Cabo de São Vincente, the southwestern tip of Portugal

14 This span of time is widely disputed by historians, but in 1632 Champlain wrote that he returned from the West Indies "nearly two and a half years after the Spanish had left Blavet." See below, excerpt 7.

15 King Louis XIII (27 September 1601 to 14 May 1643)

5 Navigation: C'est cet art qui m'a des mō bas aage attiré à l'aimer. *Les Voyages, 1613*

/[f. ãiij] A LA ROYNE REGENTE
MERE DV ROY

MADAME, Entre tous les arts les plus utiles & excellens, celuy de naviger m'a tousjours semblé tenir le premier lieu: Car d'autant plus qu'il est hazardeux & acco*m*pagné de mille perils & naufrages, d'autant plus aussi est-il estimé & relevé par dessus tous, n'esta*n*t aucunement convenable à ceux qui ma*n*quent de courage & asseurance. Par cet art nous avo*ns* la cognoissance de diverses terres, regions, & Royaumes. Par iceluy nous attirons & apportons en nos terres toutes sortes de richesses, par iceluy l'idolatrie du Paganisme est renversé, & le Christianisme annoncé par tous les endroits de la terre. C'est cet art qui m'a des mo*n* bas aage attiré à l'aimer & qui m'a provoqué à m'exposer presque toute ma vie aux ondes impetueuses de l'Ocea*n*, & qui m'a fait naviger & costoyer une partie des terres de l'Amerique & principalement de la Nouvelle France, où j'ay tousjours en desir d'y faire fleurir le Lys avec l'uni-/[f. ãiij^{v}]que Religion Catholique, Apostolique & Romaine. Ce que je croy à present faire avec l'aide de Dieu, estant assisté de la faveur de vostre Majesté, laquelle je supplie tres-humblement de continuer à nous maintenir, afin que tout rëussisse à l'honneur de Dieu, au bien de la France & splendeur de vostre Regne, pour la grandeur & prosperité duquel, je prieray Dieu, de vous assister tousjours de mille benedictions, & demeureray.

MADAME,
Vostre tres-humble, tres-obeissant
& tres-fidele serviteur & subject.
CHAMPLAIN

5 Navigation: This art has drawn me to love it from a tender age. *Les Voyages, 1613*

Champlain, Samuel de. *Les Voyages dv Sievr de Champlain Xaintongeois, capitaine ordinaire pour le Roy, en la marine* … Paris: Jean Berjon, 1613, unnumbered page, /[f. ãiij]

EDITION USED: Library and Archives Canada, Ottawa, Reserve, FC330, C3 1613

ONLINE: Early Canadiana Online. *Les Voyages du Sieur de Champlain Xaintongeois*

MICROFORM: CIHM, 90024

CHAMPLAIN SOCIETY EDITION: Biggar, *The Works*, 1:209–10

/[f. ãiij] TO THE QUEEN REGENT
MOTHER OF THE KING[16]

MADAM. Among all the most useful and admirable arts, that of navigating has always seemed to me to hold first place; for the more hazardous it is and the more attended by innumerable perils and shipwrecks, so much the more is it esteemed and exalted above all others, being in no way suited to those who lack courage and resolution. Through this art, we gain knowledge of diverse lands, regions, and kingdoms; through it, we attract and bring into our lands all kinds of wealth; through it, the idolatry of paganism is overthrown and Christianity proclaimed in all parts of the earth. It is this art which has drawn me to love it from a tender age and which, for almost my entire life, has stirred me to venture out upon the turbulent waves of the ocean; which has made me sail[17] and coast along part of the shores of America and especially of New France, where I still wish to make the Lily flourish, along with the one /[f. ãiij^v] and only Catholic, Apostolic, and Roman religion. This I now trust to accomplish with God's help, being aided by the favour of Your Majesty, whom I most humbly entreat to continue to maintain us, in order that everything may succeed for the honour of God, the welfare of France, and the splendour of your reign, for the grandeur and prosperity of which I shall pray to God to assist you always with a thousand blessings, and shall remain,

MADAM,

Your most humble, obedient, and
faithful servant and subject,
CHAMPLAIN

16 The queen regent was Marie de Medici, widow of Henri IV and regent for the youthful king Louis XIII, who was 12 years old in 1613.

17 *Naviguer*: "sail" (1516) throughout the sixteenth century, but "navigate" (i.e., directing the course of the ship) by 1678 (HU, *Dictionnaire*; DHLF, *Dictionnaire historique*).

6 Commandement du feu Roy HENRY LE GRAND en l'an 1603. *Les Voyages, 1613*

/[p. 161] Estant de retour en France aprés avoir sejourné trois ans au pays de la nouvelle Fra*n*ce, je fus trouver le sieur de Mons, auquel je recitay les choses les plus singulieres que j'y eusse veues depuis son parteme*n*t, & luy donnay la carte & plan des costes & ports les plus remarquables qui y soient.

Quelque temps aprés ledit sieur de Mons se delibera de continuer ses dessins, & parachever de descouvrir dans les terres par le grand fleuve S. Laurens, où j'avois esté par le commandement du feu Roy HENRY LE /[p. 162] GRAND en l'an 1603. quelque 180. lieues, commençant par la hauteur de 48. degrez deux tiers de latitude, qui est Gaspé entree dudit fleuve jusques au grand saut, qui est sur la hauteur de 45. degrez, & quelques minuttes de latitude, où finist nostre descouverture, & où les batteaux ne pouvoie*n*t passer à nostre jugement pour lors: d'auta*n*t que nous ne l'avions pas bien recogneu comme depuis nous avons fait.

7 Quatriesme entreprise en la Nouvelle France, 1603. *Les Voyages, 1632*

/[p. 38] *Quatriesme entreprise en la Nouvelle France par le Commandeur de Chaste. Le Sieur de Pont Gravé esleu pour le voyage de Tadoussac. L'Autheur se met en voyage.*

6 Command from the Deceased King HENRI THE GREAT in the Year 1603. *Les Voyages, 1613*

Champlain, Samuel de. *Les Voyages dv Sievr de Champlain Xaintongeois, capitaine ordinaire pour le Roy, en la marine* … Paris: Jean Berjon, 1613, 161–2
EDITION USED: Library and Archives Canada, Ottawa, Reserve, FC330, C3 1613
ONLINE: Early Canadiana Online. *Les Voyages du Sieur de Champlain Xaintongeois*
MICROFORM: CIHM, 90024
CHAMPLAIN SOCIETY EDITION: Biggar, *The Works*, 2:3–4

/[p. 161] Having returned to France after a stay of three years in the country of New France,[18] I went to find the sieur de Mons, to whom I related the most striking things that I had seen there since his departure, and I gave him the map and layout of the most remarkable coasts and harbours in those parts.[19]

Some time afterwards, the said sieur de Mons determined to continue his plans and finish discovering the interior along the great river *S. Laurens*, where I had been in the year 1603 by the command of the late King HENRY THE /[p. 162] GREAT; some 180 leagues, beginning 48 and two-thirds degrees of latitude, taking the height of the sun, which is Gaspé, the entrance to the said river, as far as the great rapids at latitude 45 degrees and some minutes, taking the same height, where our discovery ended and where, in our judgment at that time, the boats could not pass, for as much as we had not explored it as thoroughly as we have done since then.

7 The Fourth Undertaking to New France, 1603. *Les Voyages, 1632*

Champlain, Samuel de. *Les Voyages de la Novvelle France Occidentale, dicte Canada* … Paris: Louis Sevestre, 1632, pt. 1: 38–41
EDITION USED: Library and Archives Canada, Ottawa, Reserve, FC330, C5 1632b
ONLINE: Early Canadiana Online. *Les Voyages de la Nouvelle France occidentale*
MICROFORM: CIHM, 90023
CHAMPLAIN SOCIETY EDITION: Biggar, *The Works*, 3:312–18

/[p. 38] *Fourth undertaking to New France by Commander de Chaste. The sieur de Pont Gravé selected for the voyage to Tadoussac. The author sets out on the voyage.*

18 1604–07 in Acadia
19 This map is Champlain's manuscript map: "descr[i]psion des costs p[or]ts rades Illes de la nouuelle france … 1607."

Leur arrivée au Grand sault Sainct Louys. Sa difficulté à le passer. Leur retraite. Mort dudit Commandeur, qui rompt le 6. voyage.

CHAPITRE VII

La quatrième entreprise fut celle du Sieur Commandeur de Chaste, gouverneur de Dieppe, qu estoit homme tres-honorable, bon Catholique, grand serviteur du Roy, qui avoit dignement & fidelement servy sa Majesté en plusieurs occasions signalées. Et bien qu'il eust la teste chargée d'auta*n*t de cheveux gris que d'années, vouloit encores laisser à la posterité par ceste loüable entreprise, une remarque tres charitable en ce dessein, & mesmes s'y porter en personne, pour consommer le reste de ses ans au service de Dieu & de son Roy, en y faisant une demeure arrestée, pour y vivre & mourir glorieusement, comme il esperoit, si Dieu ne l'eust retiré de ce mo*n*de plustost qu'il ne pensoit; & se pouvoit-on bien asseurer que souz sa conduite l'heresie ne se fust jamais plantée aux Indes: car il avoit de tres-chrestiens desseins, dont je pourrois rendre de bons tesmoignages, pour m'avoir fait l'ho*n*neur de m'en communiquer quelque chose.

Donc aprés la mort dudit sieur Chauvin, il obtint nouvelle commission de sa Majesté. Et d'autant que /[p. 39] la despense estoit fort grande, il fit une societé avec plusieurs Gentils hommes, & principaux marchands de Rouen, & d'autres lieux, sur certaines conditions. Ce qu'estant fait, ils font equiper vaisseaux tant pour l'execution de ceste entreprise, que pour descouvrir & peupler le pays. Ledit Pont-Gravé avec commission de sa Majesté (comme personne qui avoit desja fait le voyage, & recognu les defauts du passé) fut éleu pour aller à Tadoussac, & promet d'aller jusques au Sault Sainct Louys, le descouvrir, & passer outre pour en faire son rapport à son retour, & donner ordre à un second embarquement ; & ledit Sieur Commandeur quitter son gouvernement, avec la permission de sa Majesté, qui l'aimoit uniquement, s'en aller au pays de la nouvelle France.

Sur ces entre-faites, je me trouvay en Cour, venu fraischement des Indes Occidentales, où j'avois esté prés de deux ans & demy, aprés que les Espagnols furent partis de Blavet, & la paix faite en Fra*n*ce, où penda*n*t les guerres j'avois servy sadite Majesté souz Messeigneurs le Mareschal d'Aumont, de Sainct Luc, & Mareschal de Brissac. Allant voir de fois à autre ledit Sieur Co*m*mandeur de Chaste, jugeant que je luy pouvois servir en son dessein, il me fit ceste faveur, comme j'ay dit, de m'en communiquer quelque chose, & me demanda si j'aurois agreable de faire le voyage, pour voir ce pays, & ce que les entrepreneurs y feroient. Je luy dis que j'estois son serviteur: que pour me licencier de moy-mesme à entreprendre ce voyage, je ne le pouvois faire sans le Commandement de sadite Majesté, à laquelle j'estois obligé tant de nais-/[p. 40]sance, que d'une

Their arrival at the Great Saint-Louis Rapids. His difficulty to pass it. Their retreat. Death of the said commander, which puts an end to the sixth voyage.

CHAPTER 7

THE fourth undertaking was that of the sieur commandeur de Chaste, governor of Dieppe, who was a very honourable man, a good Catholic, a great servant of the King, and one who had worthily and faithfully served His Majesty on many notable occasions. And although he had a head of hair as grey as his years, he still wished by this commendable undertaking to leave to posterity a very charitable comment on this project, and even to go out in person in order to spend the remainder of his years in service to God and his King by making his final abode there in order to live and die there gloriously, as he had hoped, if God had not withdrawn him from this world sooner than he expected. And one could certainly rest assured that under his administration, heresy would never have taken root in the Indies; for he had very Christian designs, to which I could bear excellent witness, since he did me the honour of communicating something about them to me.

Thus, upon the said sieur Chauvin's death, he obtained a new commission from His Majesty, and for as much as /[p. 39] the expenditure was very great, he created an association with many noblemen and leading merchants of Rouen and other places on certain conditions. That being accomplished, they have ships equipped as much for the execution of this enterprise as to explore and populate the land. The said Pont-Gravé, with His Majesty's commission (as someone who had already made the voyage and realized the defects of the past) was chosen to go to Tadoussac; and he promises to go as far as the Saint-Louis Rapids, explore them, and proceed farther, in order to make his report of it upon his return, and to give order for a second embarkation; and the said sieur commandeur is to leave his government, with the permission of His Majesty, who loved him above all, and to go away to the land of New France.

In the meantime, I found myself at court, newly arrived from the West Indies, where I had been for nearly two and a half years after the Spaniards had left *Blavet* and peace had been made in France, where, during the wars, I had served his said Majesty under my lords the maréchal d'Aumont, de Saint-Luc, and maréchal de Brissac. From time to time, going to see the said sieur commandeur de Chaste in the belief that I could be of service to him in his plans, he did me this honour, as I have said, of communicating something about it to me, and he asked me if I would like to make the voyage in order to see this land, and what the associates would do there. I told him that I was his servant, but for me to give myself leave to undertake this voyage, I could not do so but by the command of his said

pension de laquelle elle m'honoroit, pour avoir moyen de m'entretenir prés d'elle; & que s'il luy en plaisoit parler, & me le commander, que je l'aurois tres-agreable. Ce qu'il me promit, & fit, & receut commandement de sa Majesté pour faire ce voyage, & luy en faire fidel rapport: & pour cet effect Monsieur de Gesvre Secretaire de ses commandemens, m'expedia, avec lettre addressante audit Pont-Gravé, pour me recevoir en son vaisseau, & me faire voir & recognoistre tout ce qui se pourroit en ces lieux, en m'assistant de ce qui luy seroit possible en ceste entreprise.

Me voila expedié, je pars de Paris, & m'embarque dans le vaisseau dudit du Pont l'an 1603 nous faisons heureux voyage jusques à Tadoussac, avec des moye*n*nes barques de 12 à 15. tonneaux, & fusmes jusques à vne lieuë à mont le Grand-Sault Sainct Louis. Le Pont Gravé & moy nous nous mettons dans un petit bateau fort leger, avec cinq matelots, pour n'en pouvoir faire naviger de plus grand, à cause des difficultez. Ayans fait une lieuë avec beaucoup de peine dans une forme de lac, pour le peu d'eau que nous y trouvasmes, & estans parvenus au pied dudit Sault, qui se descharge en ce lac, nous jugeasmes impossible de le passer avec nostre esquif pour estre si furieux, & entre-meslé de rochers, que nous nous trouvasmes contraints de faire presque une lieuë par terre, pour voir le dessus de ce Sault, n'en pouvans voir d'avantage; & tout ce que nous peusmes faire fut de remarquer les difficultez, tout le païs, & le lo*n*g de ladite riviere, avec le rapport des Sauvages de ce qui estoit dedans les ter-/[p. 41]res, des peuples, des lieux, & origines des principales rivieres, & notamment du grand fleuve S. Laurent.

Je fis dés lors un petit discours, avec la carte exacte de tout ce que j'avois veu & recognu, & ainsi nous nous en retournasmes à Tadoussac, sans faire que fort peu de progrés: auquel lieu estoient nos vaisseaux qui faisoient la traitte avec les Sauvages, ce qu'estant fait, nous nous embarquasmes, mettant les voiles au vent, jusques à ce que nous fussions arrivez à Honnefleur, où sceusmes les nouvelles de la mort du Sieur Commandeur de Chaste, qui m'affligea fort, recognoissant que mal-aisément un autre pourroit entreprendre ceste entreprise, qu'il ne fust traversé, si ce n'estoit un Seigneur de qui l'authorité fust capable de repousser l'envie.

Je n'arresté gueres en ce lieu de Honnefleur, que j'allay trouver sa Majesté, à laquelle je fis voir la carte dudit pays, avec le discours fort particulier que je luy en

Majesty, to whom I was under an obligation both by birth /[p. 40] and by a stipend, with which he honoured me as a means to maintain me near him; and that if it pleased him to speak to him about it and command it of me, I would find it very agreeable. This he promised me, and accomplished and received an order from His Majesty [for me] to take this voyage and make a faithful report of it to him. And to this end Monsieur de Gesvre,[20] the secretary of the King's orders, dispatched me with a letter addressed to the said Pont-Gravé to receive me on his ship and to have me see and explore all that could be seen and explored in those parts, assisting me as much as possible in this enterprise.

So I am dispatched. I leave Paris and embark upon the said du Pont's vessel in the year 1603. We have a pleasant voyage as far as Tadoussac, and with medium barques of twelve to fifteen tuns, went as far as a league upstream of the Great Saint-Louis Rapids. Pont Gravé and I get into a very light little boat with five sailors, not being able to make a larger one navigate it on account of the difficulties. Having accomplished a league, in the form of a lake, with a lot of trouble for the shallow water that we found there, and having reached the foot of the said rapids which empty into this lake, we judged it impossible to traverse it with our skiff, on account of it being so torrential and interspersed with rocks that we found ourselves obliged to do almost a league on land in order to view the top of these rapids, being unable to see any farther; and all that we were able to do was to note the difficulties, the whole countryside and along the said river, with the *Sauvages*' report of what /[p. 41] lay inland: the people, places, and sources of the principal rivers, and especially of the great St. Lawrence River.

Since then I have written a short account with a precise map[21] of all that I had seen and explored. And so we turned back from there to Tadoussac, making only very little progress. Our vessels that were doing trade with the *Sauvages* were in this place. When this was done, we embarked, setting the sails to the wind until we arrived in *Honnefleur*, where we learned the news of the death of commander de Chaste, which grieved me greatly, as I recognized that another could not easily undertake this enterprise without being thwarted, unless it was a nobleman whose authority was apt to thrust envy aside.

I scarcely stopped in this town of *Honnefleur*, since I went to find His Majesty, to whom I showed the map of the said land, with the very special account that I

20 Louis Potier, baron de Gesvres, comte de Tresmes (c. 1542–1630), was secretary of state (*secrétaire d'État*) under Henri III and Henri IV.

21 The "short account" is *Des Sauvages*. The map has never been found.

fis, qu'elle eut fort agreable, promettant de ne laisser ce dessein, mais de le faire poursuivre, & favoriser. Voila le cinquiesme voyage rompu par la mort dudit Sieur Commandeur.

En ceste entreprise je n'ay remarqué aucun defaut, pour avoir esté bien commencé: mais je scay qu'aussitost plusieurs marchands de France qui avoient interest en ce negoce, commençoient à faire des plaintes de ce qu'on leur interdisoit le trafic des pellereries, pour le donner à vn seul.

8 Navigation: Apprenant tant par experience que par instruction de bon navigateurs. *Les Voyages, 1632*

/[p. 2] *AV LECTEVR*

AMY Lecteur, Aprés avoir passé trente huict ans de mon aage à faire plusieurs voyages sur mer & couru maints perils & hasards, (desquels Dieu m'a preservé) & ayant tousjours eu desir de voyager és lieux loingtains & estrangers, où je me suis grandement pleû, principalement en ce qui despendoit de la navigation, apprenant tant par experience que par instruction que j'ay receuë de plusieurs bons navigateurs, qu'au singulier plaisir que j'ay eû en la lecture des livres faits sur ce suject: c'est ce qui m'a meû à la fin de mes descouvertures de la nouvelle France Occidentale, pour mon contentement

made of it for him;[22] and with this he was very pleased, promising not to give up this project, but to have it pursued and privileged. Thus the fifth expedition was torn apart by the death of the said commander.

In this enterprise, I did not notice any defect, for it had been well begun; but I know that, immediately, some merchants in France who were interested in this trade began to complain that they were forbidden to traffic in furs by the trade having been given to one alone.

8 Navigation: Learning as much by experience as from the instruction of many good navigators. *Les Voyages, 1632*

Champlain, Samuel de. *Les Voyages de la Novvelle France Occidentale, dicte Canada …* Paris: Louis Sevestre, 1632 (*Traitté de la marine*, 2)
EDITION USED: Library and Archives Canada, Ottawa, Reserve, FC330, C5 1613b
ONLINE: Early Canadiana Online, *Les Voyages de la Nouvelle France occidentale*
MICROFORM: CIHM, 90023
CHAMPLAIN SOCIETY EDITION: Biggar, *The Works*, 6:255

/[p. 2] TO THE READER
Friendly Reader. Having spent thirty-eight years of my life making many sea voyages,[23] and having run many a risk and peril (from which God has preserved me), and having always wished to travel to faraway and foreign places, where I have enjoyed myself greatly, chiefly in that which depended upon navigation, learning as much by experience as by the instruction that I have received from many good navigators, as well as by the singular pleasure that I have derived from the perusal of books written on this subject; I have been moved by all this, at the conclusion of my discoveries of New France in the west, to compose for my

22 It is likely that the "special account" was a verbal account of Champlain's discoveries as they relate to the objectives set by the king. The map, which may be the one mentioned above, has not been found.

23 The *Traitté de la marine* was published as the last part of Champlain's *Les Voyages, 1632*. At the beginning of *Les Voyages, 1632*, he wrote that he had laboured in New France *depuis vingt-sept ans* (Champlain, *Les Voyages, 1632*, 1). Since Champlain first came to Canada in 1603, this statement implies that he began writing his book in 1630 and published it in 1632. To have been at sea for 38 years implies that Champlain first went to sea between 1592 and 1594, most likely the latter. The first record we have of him is as a *fourrier* and aide to Jean Hardy in March 1595, and the earliest record of his being on board a ship, the *San Julián*, is 1598.

faire un petit traitté intelligible, & proffitable à ceux qui s'en voudront servir, pour sçavoir ce qui est necessaire à un bon & parfait navigateur …

9 Champlain à son retour en fit rapport à sa Maiesté.
Les Voyages, 1632

/[p. 292] Auparavant l'an precedent 1603. ledit Champlain par commandement de sa Majesté fit le voyage de la Nouvelle France, en la grande riviere sainct Laurent, & à son retour en fit rapport à sa Majesté, lequel rapport & description il fit imprimer deslors, partit de Hondefleur en Normandie le 15. de Mars audit an, en ce mesme temps le feu sieur Commandeur de Chaste gouverneur de Dieppe, estoit Lieutenant general en ladite Nouvelle France, depuis le 40. degré jusqu'au 52. de latitude.

own satisfaction a short, comprehensive, and useful treatise for those who would like to make use of it to know what is requisite for a good and accomplished navigator …

9 Champlain, on his return, made a report of it to His Majesty. *Les Voyages, 1632*

Champlain, Samuel de. *Les Voyages de la Novvelle France Occidentale, dicte Canada …* Paris: Louis Sevestre, 1632, pt. 2, 292
EDITION USED: Library and Archives Canada, Ottawa, Reserve, FC330, C5 1613b
ONLINE: Early Canadiana Online. *Les Voyages de la Nouvelle France occidentale*
MICROFORM: CIHM, 90023
CHAMPLAIN SOCIETY EDITION: Biggar, *The Works*, 6:189–90

/[p. 292] Earlier in the preceding year,[24] 1603, the said Champlain, by command of His Majesty, made a voyage to New France, to the great river *sainct Laurent*; and on his return, made a report of it to His Majesty, the which report and description he caused to be printed at once.[25] He left *Hondefleur* in Normandy on the fifteenth of March in the said year. At this same time the late Commander de Chaste, governor of Dieppe, was lieutenant general in the said New France from the 40th to the 52nd degree of latitude.

24 That is, the year preceding 1604
25 This is a reference to *Des Sauvages*. The first edition has no printer's date, but it has a *privilège*, granted on 15 November 1603. This entry suggests the book was printed in 1603.

DOCUMENT G

Des Sauvages, or, Voyage of Samuel Champlain, [1603], 1604 / *Des Sauvages, ou, Voyage de Samuel Champlain,* [1603], 1604

INTRODUCTION TO THE DOCUMENT

In this, his first book, Champlain faithfully recorded the information he had gathered according to the orders he had from King Henri IV and Commander Aymar de Chaste: "to make the voyage in order to see this land, and what the associates would do there."[1] What this amounted to was to conduct a survey in order to assess the potential of the banks of the St. Lawrence for settlement and to determine the possibilities for westward exploration.

Champlain left Honfleur on 15 March 1603, reaching Tadoussac on 24 May. By 2 July, he was at the Lachine Rapids, whence he departed for Tadoussac on the fourth. In recounting some of the circumstances of the 1603 expedition for *Les Voyages, 1632*, he mentioned that on leaving the Lachine Rapids and before reaching Tadoussac, he "wrote a short account with a correct map of all I had seen and discovered."[2] It is therefore likely that a substantial part of the material that went into *Des Sauvages* was written by 23 August, when the expedition left Gaspé for France. On 20 September the ships reached *Havre de Grace*,[3] and a few days later, Honfleur.[4] On learning of de Chaste's death, Champlain left immediately to seek out King Henri IV, who was at Saint-Germain-en-Laye near Paris.[5] There he had an audience with the king and showed him "*la carte dudit pays, avec le discours fort particulier que je luys en fis.*"[6] It may be that the *discours*, as the word suggests, was a verbal rendering of Champlain's most important findings. The map has never been found. Having delivered his report to the king, "he

1 Biggar, *The Works*, 3:315

2 Champlain, *Les Voyages, 1632*, pt. 1, 41

3 *Havre de Grace* is on the north shore of the estuary of the Seine River, now the city of Le Havre.

4 Honfleur is southeast of Le Havre on the south shore of the Seine estuary.

5 Cuignet, *L'Itinéraire*, 117

6 Champlain, *Les Voyages, 1632*, pt. 1, 41

caused it to be printed at once," which suggests some haste in the matter.[7] On 15 November, two months after Champlain had returned, he was granted the king's *privilège* (licence), and the book rolled off the press later that year.[8] The rapidity of the process suggests that there was some urgency to get the narrative published. It is likely that the rapid publishing of John Brereton's report of the Gosnold voyage and Edward Hayes's suggestion that the English lay claim to much of North America had something to do with it.[9]

It is evident that Champlain kept a diary of the journey which he rewrote into a more flowing text. Except for a few descriptive or analytical passages, he followed the same pattern in preparing his subsequent volumes. It is unfortunate that none of his field diaries still exist.

Exactly why Champlain chose the title *Des Sauvages* is not clear, because the volume has less to do with the Native inhabitants of Canada than with his resource survey of the St. Lawrence River. It may have been the publisher who added the words *Des Sauvages* to the rest of the title – *Voyage de Samuel Champlain* – the latter possibly being Champlain's choice. Knowing the American Native to be a popular subject, Claude de Monstr'œil may have aimed to attract more sales in this manner. To Champlain and most of his contemporaries, the term *sauvage* had a neutral meaning, not the pejorative meaning that "savages" has to modern readers. For this reason, we have not translated the French word *sauvage*. Throughout the French regime in Canada, the term was mainly used as a synonym for "Indian" or "Native," someone who lived a "natural life." A *sauvage* may have been a thief, a liar, cruel, generous, honourable, and even held up to European readers as a model of behaviour. In the Dutch colonies, the term *Wilden* (wild, uncivilized) and, in the early English colonies, "salvages" (from "sylvan," the forest) were used initially in much the same manner as the French *sauvage*, but in both cases were soon transformed into the pejorative "savages."[10] As early as 1609, Marc Lescarbot wrote: "[I]f we commonly call them Savages, the word is abusive and unmerited, for they are anything but that, as will be proved in the course of this history."[11] Others, such as the Récollet lay brother Gabriel

7 Biggar, *The Works*, 6:189

8 Ibid., 295. There is no date on the title page of the first printing of *Des Sauvages*. Champlain, however, wrote that his first *rapport* was printed in 1603.

9 See part 1, A, "Champlain and His Times to 1604," subsection "Preparations for the Voyage to Canada, 1603."

10 There is an excellent discussion on this entire topic in Jennings, *The Invasion of America*, 43–84. See also Axtell, *The Invasion Within*, and Jaenen, *Friend and Foe*.

11 Grant, *The History of New France*, 1:32–3

Sagard,[12] linked savagery to "the darkness of unbelief," that is, to being pagan, as did Jesuits such as Claude Allouez: "The name 'Savage' gives rise to so very disparaging an idea of those who bear it, that many people in Europe have thought that it is impossible to make true Christians out of them."[13]

On his first journey to Canada in 1603, although sympathetic to the Natives he met, Champlain often condemned their behaviour, even though he had no experience of them and had no cause to be critical. From the linguistic and textual evidence of *Des Sauvages*, there is no hint of the inclusion of any material that would indicate Champlain's views of Natives came from a source outside his own experience or from what some people told him. Occasionally, his descriptions of the Natives are judgmental or sententious. For example, he twice describes them as wicked because of their perceived propensity to lie and to seek revenge for damages.[14] His sententious "*promettent assez & tiennent peu*" ("they promise much and perform little")[15] seems to repeat a standard judgment on the lines of a French proverb: "Toujours promettre et ne jamais tenir, c'est conforter l'extravagant" (Always to promise and never perform is to confirm extravagance).[16] As with the story of the *Gougou* and other stories related to him by fellow Frenchmen, Champlain seems to have accepted the judgmental opinions of his companions at face value. With more experience, he became less reliant on the opinions of others and eventually used the term *Sauvage* in a somewhat broader context, linking it to paganism and the "absence" of civilization: "They are by no means so savage but that in time and through intercourse with a civilized nation, they may be refined."[17] The civilizing influences of the French presence and the Natives' conversion to Christianity is what Champlain thought were necessary to elevate them from "barbarism." Late in life, he thought that complete equality could be achieved through French-Native intermarriage. On at least two occasions, 24 May 1633 and 22 July 1635, Champlain and the Jesuit Superior at Quebec, Father Paul Le Jeune, met with the Huron to propose formal French-Huron intermarriage: "[O]ur young men will marry your daughters, and we will be one people."[18] The understanding was of course that if such an event

12 Wrong, *The Long Journey*
13 *Jesuit Relations* (*JR*), 58:85
14 See, for example, *Des Sauvages*, ch. 3, f. 8, and ch. 12, f. 33.
15 Ibid., ch. 3, f. 8
16 Le Roux de Lincy, *Le livre des proverbes français*, 2:282; "*Toujhour proumëtrë é noun tënë, ës lou fat ëntrëtënë.*" Trinquier, *Proverbes et dictons de la langue d'oc*, 189, no. 1220
17 Biggar, *The Works*, 3:3–4
18 *JR*, 5:211, 8:47–51

were to take place, both partners had to be Catholic. The *sauvage* would cease to exist upon conversion to Christianity.

Over time, from the French point of view – certainly, that of the Jesuits – the *sauvage* was a "barbarian," devoid of religion and possessed of some objectionable cultural habits, but he was "perfectible" through conversion to Christianity. Already in 1603, Champlain himself shared his compatriots' optimism in the perfectability of the Natives, but he was clearly more concerned with settlement than conversion. Twice in *Des Sauvages*, he tells us: "*Je tiens que qui leur monstreroit à vivre & enseigner le labourage des terres, & autres choses, ils l'apprendroient fort bien*" ("I hold that if anyone were to show them how to live and teach them how to work the land and other matters, they would learn it very well"; f. 8); and, again: "*Je … croy que promptement ils seroient reduicts bons Chrestiens si l'on habitoit leurs terres*" ("I believe that they would quickly be brought round to being good Christians, if their lands were colonized"; f. 11). Whether the *sauvage* should also adopt French customs was debated and initially accepted by the Recollect, but it was later rejected by many Jesuits, who said that in order to Christianize them, it would be best to avoid trying to "frenchify" (*franciser*) them.[19] It is important to realize that the term implied cultural differences and not racial ones.

For a critical study of Champlain's French, the printing history of *Des Sauvages*, the manner in which we have prepared the text for translation, and the editorial principles and procedures we have used, the reader is advised to refer to part 1, B, "Textual Introduction to *Des Sauvages*." The original foliation is given thus: /[f. 4ᵛ].

19 Charlevoix, *Histoire et description generale*, 3:144

Des Sauvages, ou, Voyage de Samuel Champlain, [1603], 1604

DES
SAUVAGES,[a]
OU,
VOYAGE DE SAMUEL
CHAMPLAIN, DE BROUAGE,
fait[b] en la France nouvelle,
l'an mil six cens trois:[c]

CONTENANT[d]

Les mœurs, façon de vivre, mariages, guerres, & habitations
des Sauvages de Canadas.

De la descouverte[e] de plus de quatre cens cinquante lieuës dans le païs[f]
des Sauvages.[g] Quels peuples y habitent, des animaux qui s'y trouvent,
des rivieres, lacs, isles & terres, & quels arbres & fruicts elles produisent.

a SAVVAGES,] A2 SAVVAGES;
b fait] A2 faict
c fait … trois:] A2 *faict en la France nouvelle, l'an/ mil six cens trois:*
d CONTENANT] CONTENANT,
e descouuerte] A2 descouuerture
f païs] A2 pays
g Sauuages.] A2 Sauuages,

Des Sauvages, or, Voyage of Samuel Champlain, [1603], 1604

Champlain, Samuel. *Des Savvages, ov, Voyage de Samvel Champlain, de Brovage, fait en la France nouuelle, l'an mil six cens trois* … A Paris, chez Clavde de Monstr'œil, [1603] (fig. 1)

EDITION USED: Bibliothèque nationale de France, Paris, France, RES-LK12 -719, Tolbiac Magazin

ONLINE: Gallica. Bibliothèque nationale de France, FRBNF30219230

Champlain, Samuel. *Des Savvages, ov, Voyage de Samvel Champlain, de Brovage, faict en la France nouuelle, l'an mil six cens trois* … A Paris, chez Clavde de Monstr'œil, 1604 (fig. 2)

EDITION USED: Bibliothèque nationale de France, Paris, France, RES-LK12 -719(A) Tolbiac Magazin

ONLINE: Early Canadiana Online

MICROFORM: CIHM, 90061

DES
SAUVAGES
OR
VOYAGE OF SAMUEL
CHAMPLAIN, OF BROUAGE
made to New France,
in the year one thousand six hundred and three:

CONTAINING

The customs, manner of life, marriages, wars, and
dwellings of the *Sauvages* of *Canadas*.

Concerning the discovery of more than four hundred and fifty leagues
in the territory of the *Sauvages*: which people live there; some animals
that are found there; some rivers, lakes, islands, and lands, and
what trees and fruits they produce.

De la coste d'Arcadie, des terres que l'on y a descouvertes, & de plusieurs mines qui y sont, selon le rapport des Sauvages.

A PARIS,
Chez CLAUDE DE MONSTR'ŒIL, tenant sa boutique en la Cour du Palais, au nom de Jesus.[a]

AVEC PRIVILEGE DU ROY.[b]

Extraict du Privilege.

PAr Privilege du Roy donné à Paris le 15. de Novembre, 1603. signé Brigard. Il est permis au Sieur de Champlain de faire imprimer par tel Imprimeur que bon luy semblera un livre par luy composé, intitulé, *Des Sauvages, ou, Voyage du Sieur de Champlain, fait*[c] *en l'an 1603,*[d] & sont faictes deffences à tous Libraires & Imprimeurs de ce Royaume, de n'imprimer, vendre, & distribuer ledit[e] livre, si ce n'est du consentement de celuy qu'il aura nommé & esleu, à peine[f] de cinquante escus d'amende, de confiscation, & de tous despens, ainsi qu'il est plus amplement contenu audit[g] Privilege.

Ledit[h] Sieur de Champlain, suivant sondit[i] Privilege, a esleu & permis à Claude de Monstr'œil, Libraire en l'Université de Paris, d'imprimer le susdict[j] livre, & luy a cedé & transporté sondit[k] Privilege, sans que nul autre le puisse imprimer ou faire imprimer, vendre & distribuer, durant le temps de cinq annees, sinon du consentement dudit[l] Monstr'œil, sur les peines contenuës audit[m] Privilege. /[f. ãij]

a Iesus.] A2 Iesus. 1604.
b AVEC PRIVILEGE DV ROY.] A2 *Auec Priuilege du Roy.*
c *fait*] A2 *faict*
d *1603,*] A2 1603,
e ledit] A2 ledict
f à peine] A2 a peine
g audit] A2 audict
h Ledit] A2 Ledict
i sondit] A2 sondict
j susdict] A2 sondict
k sondit] A2 sondict
l dudit] A2 dudict
m audit] A2 audict

Concerning the coast of *Arcadie*,[1] of the lands discovered there, and of several mines to be found there according to the report of the *Sauvages.*

PARIS

From the press of CLAUDE DE MONTSTR'ŒIL, keeping shop in the Court of the Palace, at the name of Jesus.[2]

WITH THE KING'S LICENCE

Extract from the Licence

By licence of the King given at Paris, 15 November 1603, signed Brigard. The Sieur de Champlain is permitted to have printed, by such printer as shall seem good to him, a book written by him entitled *Des Sauvages, ou, Voyage du Sieur de Champlain, fait en l'an 1603,* and all booksellers and printers of this kingdom are forbidden to print, sell, and distribute the said book, except with the consent of him whom he shall have appointed and chosen, under penalty of a fine of fifty crowns, of confiscation, and all costs, as is more fully contained in the said licence.

The said Sieur de Champlain, in accordance with his said licence, has chosen and granted to Claude de Monstr'œil, bookseller in the University of Paris, permission to print the above-named book, and has yielded and transferred to him his said licence, so that no other person may print it or have it printed, sold, and distributed over the course of five years, except with the consent of the said Monstr'œil, under penalties contained in the said licence.

1 *Arcadie* (Arcadia), with its classical resonance, is found consistently throughout Champlain's first publication. In 1524 Verrazano is said to have transferred the mythical Greek toponym to the region in or around Nova Scotia (Acadia) because of the beauty of the trees there (*Dictionnaire historique de la langue française* [DHLF], 1:14; Beaulieu and Ouellet, *Champlain*, 85n2). A thorough discussion of the documentary and cartographic evidence linking Verrazano to the naming of Acadia (originally *Archadia, Arcadia*, and *Larcadia*) is in Ganong, *Crucial Maps*, 101–30; Rayburn, *Naming Canada*, 159–62.

2 If the publisher had wished to state merely "at the sign of Jesus," he would have written *à l'enseigne de*, although *au nom de* is currently accepted as a synonym. In this case, we have chosen to allow the ambiguity to stand, when the meaning of *au nom de* would have resonated with other known usages of the time, e.g., "in honour of" (from 1080 CE) (*DHLF*, 2:2384). This seems reasonable, given contemporary religious controversies.

A TRES-NOBLE, HAUT
ET PUISSANT SEIGNEUR,

Messire Charles de Montmorency, Chevalier des Ordres du Roy, Seigneur d'Ampuille, & de Meru, Comte de Sego*n*digny, Vicomte de Meleun, Baron de Chasteau-neuf, & de Gonnort, Admiral de France & de Bretagne.

M*ONSEIGNEUR*,

*Bien que plusieurs aye*nt *escript quelque chose du païs de Canadas, je n'ay voulu pourtant m'arrester à leur dire, & ay expreßément esté sur les lieux pour pouvoir rendre fidelle tesmoignage de la verité, la-*/[f. ãij[v]] *quelle vous verrez (s'il vous plaist) au petit discours que je vous addresse, lequel je vous supplie d'avoir pour aggreable, & ce faisant, je prieray Dieu, Monseigneur, pour vostre grandeur & prosperité, & demeureray toute ma vie,*

Vostre tres-humble[a] &
obeïssant serviteur[b]
S. CHAMPLAIN. /[f. ãiij][c]

LE SIEUR DE LA FRANCHISE
AU DISCOURS DU
Sieur de Champlain.[d]

MUSES si vous chantez vrayment je vous conseille,
Que vous louez Champlain, pour estre courageux,

a tres-humble] A2 trs-humble
b obeïssant seruiteur] A2 obeyssant seruiteur,
c ãiii] A2 ãii
d Layout differs for the 1604 edition; e.g., A2 FRAN-/CHISE, here as commonly found elsewhere in the layout.

TO THE MOST NOBLE, HIGH AND MIGHTY LORD,

Sire Charles de Montmorency,[3] knight of the King's Orders, seigneur of Ampuille and Méru, count of Secondigny, viscount of Melun, baron of Châteauneuf and Gonnort, admiral of France and Brittany

My Lord,
Although many have written something about the land of Canadas,[4] I have however not been disposed to abide by their descriptions, and have been in the places expressly for the purpose of being able to give faithful testimony of the truth, which /[f. ãij^v] *you shall read (if it pleases you) in the brief account that I dedicate to you and that I beg you graciously to approve; and so doing I shall pray to God, My Lord, for your greatness and prosperity, and shall remain all my life,*
Your most humble and obedient servant
S. CHAMPLAIN /[f. ãiij]

SIEUR DE LA FRANCHISE TO THE DISCOURSE OF Sieur de Champlain[5]

Muses, if you really do sing, I advise you
To praise Champlain for being courageous.

3 Charles de Montmorency, duc de Damville, etc. (c. 1537–1612), was admiral of France from 1596 to 1612.

4 Throughout most of the 1603 and 1604 editions, an *s* regularly appears on the end of *Canadas*. It is all the more curious, as Jacques Cartier defines the meaning of the St. Lawrence Iroquoian word for "town" in the singular: "*Ilz appellent une ville Canada*" (Bideaux, *Relations*, 188; Cook, *The Voyages*, 93). There is no doubt that "Canada" is the St. Lawrence Iroquoian word "town or village," and not *la terre et prouvynce de Canada* (country and province) as implied by Cartier (Biggar, *Jacques Cartier*, 119). For a discussion of the word *Canada* and Iroquoian cognates, see Steckley, *A Huron-English / English-Huron Dictionary*, 145; also Lounsbury, "Iroquoian Languages," in Trigger, *Handbook*, 15:340. One might imagine that Champlain meant a collection of towns when he wrote *Canadas* with a plural inflection, but this cannot be the case, since the proper noun in *Des Sauvages* normally follows either *en* or *de*, not *aux* or *des* as required to mark plurality. For a discussion of the country's number and gender, see part 1, B, "Dealing with Champlain's French."

5 This sonnet follows the classic French form of two quatrains, ABBA (*rimes embrassées*) and two triplets (CCD, EED) in hexameter verse with a regular caesura. No attempt has been made to replicate the rhyme and metre.

Sans crainte des hasards[a] *il a veu tant de lieux,*
Que ses relations nous contentent l'oreille:
Il a veu le Perou, Mexicque, & la merveille
Du Vulcan infernal qui vomit.tant de feux,
Et les saults Mocosans, qui offencent les yeux
De ceux qui osent veoir leur cheute nompareille:
Il nous promet encor de passer plus avant
Reduire les Gentils & trouver le Levant,
Par le Nort, ou le Su, pour aller à la Chine.
C'est cheritablement tout pour l'amour de Dieu,
Fy des lasches poltrons qui ne bougent d'un lieu,
Leur vie sans mentir me paroist trop Mesquine.

De la Franchise. /[f. ãiij[v]]

a *hasards*] A2 *hazards*

Without fear of the risks, he saw so many places
That his accounts are pleasing to the ear.
He saw Peru, Mexico, and the marvel
Of infernal Vulcan, who spews up so many fires.
And the rapids of Mocosa,[6] which offend the eyes
Of those who dare to view their incomparable falls.
He promises us to press ever onwards,
To tame the Gentiles[7] and find the Levant,
And proceed to China, by North or by South.
It's all for the love of God, charitably.
Fie on the mean cowards who don't budge from one spot;
Their life, without lying, seems too petty to me.

De la Franchise[8] /[f. ãiij^v]

6 Mocosa: On some sixteenth-century maps, Mocosa was located in the eastern interior of North America, from about Cape Cod to Virginia. Champlain, with the Jesuits, would later consider it an old name for Virginia (1632; Biggar, *The Works*, 6:187). The place name was likely inspired by the Spanish captain Luis Moscoso (or Mocoso) de Alvarado, who explored Florida with Ferdinand de Soto in 1539 and assumed leadership when de Soto died in 1542 (Campeau, *Monumenta Novæ Franciæ*, 1:569n1). In 1562 the Spaniard Diego Gutiérretz (Jr.) first published *Mocosa* on a map. In 1593 Cornelius de Jode compressed the lands to the south of New France so that Mocosa slipped much farther north to below Hochelaga. This map circulated widely and was popularized in Louvain by Cornelius van Wytfliet (1597) and reproduced in France by Franciscus Fabri in 1603 and subsequent French translations (1605, 1607, 1611). It is likely that the unknown poet De la Franchise was thinking of the rapids that prevented Champlain from progressing down the St. Lawrence as belonging to the northernmost region of Mocosa. They would later be named the Lachine Rapids.

7 *Gentils*: Non-Christians in this case, following such usage in the Hebrew and Christian Bibles (*DHLF*, 2:1577).

8 De la Franchise: In early Canada, La Franchise (Frankness), like La Liberté, La Douceur, was a popular nickname for soldiers in the area of Montreal and Quebec; e.g., Jean Beaune, dit Lafranchise (b. 1633), an unnamed soldier Lafranchise (17th century), and Claude Pastourel dit Lafranchise (b. 1639), etc. (Tanguay, *Dictionnaire généalogique*, 1:34, 337, 467). In Middle French, *franchise* could mean "freedom," but when related to human beings c. 1559 it refers to the psychological character of an individual who tells the truth: *caractère d'une personne qui dit la vérité* (*DHLF*, 2:1482). Such a pen name for this anonymous poet was rather appropriate for staking a claim to the veracity of the poem's statements.

TABLE DES CHAPITRES.

a *receptiõ faicte*] A2 *reception faite*
b *Canada,*] A2 *Canadas,*
c *Sainct*] A1 *Saincte*
d *resiouïssance*] A2 *resiouyssance*
e *isles*] A2 *Isles*
f *pointe*] A2 *poincte*
g *Riuieres,*] A2 *riuieres,*

TABLE OF CHAPTERS

FIN.[c] /[f. 1; A]

a *Canadas: Du*] A2 *Canadas, du*
b *Les*] A2 *Lees*
c FIN.] A2 *FIN.*

Chapter 9
Return from the rapids to *Tadousac*, with a comparison of the reportage of several *Sauvages* concerning the length and source of the great *Riviere de Canadas*; of the number of rapids and lakes that it traverses

Chapter 10
Voyage from *Tadousac* to the *Isle perçee*. Description of the *baye des Moluës*, the *isle de bonne-adventure*, the *baye des Chaleurs* and the many rivers, lakes, and regions where several kinds of mines are found

Chapter 11
Return from the *Isle perçee* to *Tadousac*, with the description of the coves, harbours, rivers, islands, rocks, rapids, bays, and shallows that are found along the north coast

Chapter 12
The ceremonies that the *Sauvages* perform before going to war. Of the *Almouchicois Sauvages* and their monstrous shape. Discourse of the *sieur de Prevert de sainct Malo* on the exploration of the coast of *Arcadie*; what mines are there, and of the goodness and fertility of the land

Chapter 13
Of a frightful monster which the *Sauvages* call *Gougou*, and of our short and felicitous return to France

DES SAUVAGES,
OU[a]
VOYAGE DU SIEUR DE Champlain, faict en l'an 1603.

CHAPITRE I.[b]

Bref discours, où est contenu le voyage depuis Honfleur en Normandie, jusques au port de Tadousac en Canadas.

NOUS partismes de Honfleur le 15. jour de Mars 1603. Cedit[c] jour nous relaschasmes à la Rade du Havre de Grace, pour n'avoir le vent favorable. Le Dimanche ensuyvant 16. jour dudict mois, nous mismes à la voille pour faire nostre route. Le 17. ensuivant[d] nous eusmes la veuë d'Orgny & Grenesey, qui sont des isles entre la coste de Normandie & Angleterre. Le 18. dudit mois eusmes la cognoissance de la /[f. 1v] coste de Bretagne.[e] Le 19. nous faisions estat à 7. heures du soir estre le travers de Ouessans. Le 21. à 7. heures[f] du matin nous rencontrasmes 7. vaisseaux Flamans, qui à nostre jugement venoient des Indes. Le jour de Pasques 30. dudit mois fusmes contrariez d'une grande tourmente,* qui paroissoit estre plustost foudre que vent, qui dura l'espace de dixsept jours: mais non si grande qu'elle avoit fait[g] les deux premiers jours: & durant[h] cedit te*m*ps[i] nous eusmes plus d'eschet[j] que d'avancement. Le 16. jour d'Avril le temps commença à s'adoucir, & la mer plus belle qu'elle n'avoit esté[k] avec contentement d'un chacun: de[l] façon que continuans nostredite route[m] jusques au 28. jour dudit mois que rencontrasmes

a OV] A2 OV,
b CHAPITRE I.] A2 CHAP. I.
c Cedit] A2 Cedict
d ensuiuant] A2 ensuyuant
e Bretagne.] A2 Bretaigne.
f 7. heures] A1 A2 17. heures
* LM] *Grande Tourmente.*
g fait] A2 faict
h durant] A2 durãt
i tẽps] A2 temps
j plus d'eschet] A1 plus de d'eschet A2 plus de d'eschet,
k esté] A2 esté,
l de] A2 *De*
m route] A2 routte

DES SAUVAGES
OR
VOYAGE OF THE SIEUR DE CHAMPLAIN
made in the year 1603

Chapter 1

Brief narrative, in which is contained the voyage from Honfleur in Normandy to the port of Tadousac *in Canada*

We set out from Honfleur on the 15th day of March 1603. This same day we put into the harbour of *Havre de Grace* because we did not have a favourable wind.[9] The following Sunday, the 16th day of the said month, we set sail to make our way. The next day, the 17th, we sighted *Orgny* and *Grenesey*,[10] which are islands between the coasts of Normandy and England. On the 18th of the said month, we sighted the /[f. 1ᵛ] coast of Brittany. On the 19th, at seven o'clock in the evening, we reckoned that we were off *Ouessans*.[11] On the 21st, at seven[12] o'clock in the morning, we met seven Flemish ships, which, by our reckoning, were coming from the Indies. On Easter Day, the 30th of the said month, we were delayed by a great storm, which seemed to be more lightning than wind and lasted the space of seventeen days, but was not as severe as it had been on the first two days; and during this said time, we lost more ground[13] than we gained. On the 16th day of April, the weather began to get milder and the sea more beautiful than it had been, to the contentment of all; so that we continued our said course until the 28th day of the said month, when we encountered a very high iceberg. The next

9 Honfleur is a small port near the mouth of the Seine River on the south shore and *Havre de Grace*, now Le Havre, is on the north side. Champlain sailed on *La Bonne-Renommée*, a ship of 120 tuns burthen.

10 The two Channel Islands, Alderney (*Orgny*) and Guernsey (*Grenesey*) off the west coast of the Cherbourg Peninsula

11 Ushant (*Ouessans*) is an island off the western tip of Brittany.

12 Champlain wrote "seventeen o'clock," evidently a misprint for "seven."

13 The 1604 edition repeats a fault here in the 1603 edition: *nous eusmes plus de d'eschet que d'avancement*, which should read *plus d'eschet* (*plus d'échec* in modern French).

une glace fort haute.[a] Le lendemain nous eusmes congnoissance d'un banc de glace* qui duroit plus de 8. lieuës[b] de long, avec une infinité d'autres moindres, qui fut l'occasion que nous ne peusmes passer: & à l'estime du Pilote lesdictes glaces estoient à quelque 100. ou 120. lieues de la terre de Canadas, & estions par les 45. degrez ⅔, & vinsmes trouver passage par les 44.

Le 2. de May nous entrasmes sur le banc à unze heures /[f. 2] du jour par les 44. degrez ⅓. Le 6. dudit[c] mois nous vinsmes si proche de terre que nous oyons la mer battre à la coste: mais nous ne la peusmes recongnoistre pour l'espoisseur de la brume dont cesdites costes sont sujectes,[d] qui fut cause que nous nous mismes à la mer[e] encores quelques lieuës,[f] jusques au lendemain matin, que nous eusmes cognoissance de terre d'un temps assez beau, qui estoit le cap de Saincte Marie.[g]* Le 12. jour ensuivant nous fusmes surprins d'un gra*n*d[h] coup de vent qui dura 2. jours. Le 15. dudit[i] mois nous eusmes cognoissance des isles[j] de sainct Pierre.* Le 17. ensuivant nous rencontrasmes un banc de glace pres du cap de Raie,[k] qui contenoit six lieuës,[l] qui fut occasion que nous ammenasmes toute la nuict, pour eviter le danger où nous pouvions courir. Le lendemain nous mismes à la voille, & eusmes congnoissance du Cap de Raie,* & isles[m] de sainct Paul,* & Cap de sainct Laure*nt*[n]* qui est terre ferme à la bande du Su: & dudit[o] Cap de sainct

a haute.] A2 haulte
* LM] *Rencontre de plusieurs grandes glaces.*
b lieuës] A2 lieues
c dudit] A2 dudict
d sujectes,] A2 subiectes,
e à la mer] A2 a la mer
f lieuës,] A2 lieues,
g le cap de Saincte Marie.] A2 le Cap de Ste Marie.
* RM] *Cap de S. Marie.*
h grād] A2 grand
i dudit] A2 dudict
j isles] A2 Isles
* RM] *Isles de S. Pierre.*
k cap de Raie,] A2 Cap de Raie,
l lieuës,] A2 lieues,
* RM] *Cap de Raye.*
m isles] A2 Isles
* RM] *Isles de S. Paul.*
n sainct Laurēt] A2 S. Laurent
* RM] *Cap. de S. Laurens.*
o dudit] A2 dudict

day, we sighted an ice floe more than eight leagues in length, with an infinite number of other smaller ones, which prevented us from getting past. And by the pilot's reckoning, the said ice was some 100 or 120 leagues from the land of *Canadas*. We were at latitude 45°40' and came upon a passage at latitude 44°.

On 2 May, at eleven o'clock /[f. 2] in the day, we came upon the Bank at latitude 44°20'.[14] On the 6th of the said month, we came so close to land that we heard the sea beating against the shore, but could not identify[15] it because of the thickness of the fog, to which these coasts are subject, which was the reason we put out to sea for a few more leagues until the next morning when, the weather being very clear, we sighted land, which was *cap de Saincte Marie*.[16] On the next day, the 12th, we were surprised by a great gale of wind, which lasted 2 days. On the 15th of the said month, we sighted the islands of *sainct Pierre*. On the following day, the 17th, we encountered an ice floe near *cap de Raie*, which measured six leagues and caused us to strike sail for the whole night to avoid the danger we might incur. The next day we set sail and sighted *cap de Raie* and the *isles de sainct Paul*, and *Cap de sainct Laure*nt, which is the mainland on the south side.

14 The southeastern edge of the Grand Bank now called the "tail of the bank"

15 *recongnoistre* (*reconnaître*): "en général identifier (une chose, une personne)" is attested in 1608 (*DHLF*, 3:3117).

16 Cape St. Mary, Islands of St. Pierre, Cape Ray, and Cape St. Lawrence have retained these names.

Laurens[a] jusques audit Cap de Raie, il y a dixhuict lieuës,[b] qui est la largeur de l'entree de la grande baie[c] de Canadas. Cedit[d] jour sur les dix heures du matin nous rencontras-/[f. 2[v]]mes une autre glace qui co*n*tenoit plus de huict lieuës de long. Le 20. dudict mois nous eusmes congnoissance[e] d'une isle[f] qui a quelque 25. ou 30. lieuës[g] de long, qui s'appelle Anticosty,* qui est l'entree de la riviere de Canadas. Le lendemain eusmes congnoissance de Gachepé,* terre fort haute, & commençasmes à entrer dans ladicte riviere de Canadas, en rengeant la bande du Su jusques à Mantanne,* où il y a dudict Gachepé 65. lieues. Dudict Mantanne, nous vinsmes prendre congnoissance du Pic, où il y a vingt lieuës,[h] qui est à ladicte bande du Su;[i] dudict Pic[j] nous traversames[k] la riviere jusques à Tadousac, où il y a 15. lieuës.[l] Toutes cesdictes terres sont fort hautes eslevees, qui sont sterilles, n'apportant aucune commodité.*

Le 24. dudit[m] mois nous vinsmes mouiller l'ancre[n] devant Tadousac, & le 26. nous entrasmes dans ledict port, qui est faict comme une ance à[o] l'entree[p] de la riviere du Sagenay,* où il y a un courant d'eau & maree fort estra*n*ge, pour sa vistesse & profondité, où quelquesfois il vient des vents impetueux à cause de la froidure qu'ils amenent avec eux. L'on[q] tient que ladicte riviere a[r] quelque 45. ou

a sainct Laurens] A2 sainct Laurent
b lieuës,] A2 lieues,
c baie] A2 baye
d Cedit] A2 Cedict
e congnoissance] A2 cognoissance
f isle] A2 Isle
g lieuës] A1 lieux
* LM] *Anticosty.*
* LM] *Gachepé.*
* LM] *Mantanne.*
h lieuës,] A2 lieues,
i Su;] A2 Su:
j Pic] A2 Pic,
k trauersames] A2 trauerçasmes
l lieuës.] A2 lieues.
* LM] *Terres fort hautes & mauuaises.*
m dudit] A2 dudict
n mouiller l'ancre] A2 moüiller l'Ancre
o à] A1 a
p l'entree] A2 l'entre
* LM] *Port de Tadousac & sa description.*
q L'on] A1 A2 Lon (*no other instance of* l'on *without the apostrophe*)
r a] A2 à

And from the said *Cap de sainct Laurens* to the said *Cap de Raie* there are eighteen leagues, which is the breadth of the entrance to the great *baie de Canadas*.[17] On this said day, about ten o'clock in the morning, we encountered /[f. 2ᵛ] another ice floe, which measured more than eight leagues long. On the 20th of the said month, we came in sight of an island some 25 or 30 leagues in length, which is called *Anticosty*,[18] which is the entrance to the *riviere de Canadas*[19] [map 2]. The next day we sighted *Gachepé*,[20] a very high land, and began to enter the said *riviere de Canadas*, skirting the south coast as far as *Mantanne*,[21] where it is 65 leagues from the said *Gachepé*. From the said *Mantanne*, we came within sight of *Pic*,[22] a distance of twenty leagues, which is on the said south side; from the said *Pic* we crossed the river as far as *Tadousac*, a distance of 15 leagues. All these said lands are elevated very high and are barren, producing nothing.

On the 24th of the said month, we came to drop anchor before *Tadousac*,[23] and on the 26th we entered the said harbour, which is shaped like a cove at the mouth of the *riviere du Sagenay*. Here there is a current and tide, very unusual for its swiftness and depth, where sometimes raging winds blow because of the cold that they bring with them. It is supposed that the distance up the said river is some 45

17 The entrance of the Gulf of St. Lawrence from Cape Ray to Cape St. Lawrence is 110 km.

18 Anticosti Island is 235 km long. Cartier had called it *l'isle de l'Assumption* (Biggar, *Jacques Cartier*, 110).

19 The St. Lawrence River

20 Gaspé. French sailors of the period reckoned that they had entered the St. Lawrence River once they rounded Cap Gaspé into the Détroit d'Honguedo between Anticosti Island and the mainland to the south.

21 Matane is 280 km from Gaspé and 108 km from Bic.

22 Bic (*Pic*) to Tadoussac is 67 km. Bic Mountain (350 m. AMSL [above mean sea level]) and Île du Bic to the west of the harbour served as landmarks for navigators.

23 The voyage from Honfleur to Tadoussac on *La Bonne-Renommée* took from 15 March to 24 May, 70 days.

/[f. 3] 50. lieuës[a] jusques au premier sault, & vie*n*t[b] du costé du Nort noroüest: Ledit port de Tadousac est petit, où il ne pourroit que dix ou douze vaisseaux: mais il y a de l'eau assez à Est à l'abry de ladite[c] riviere de Sagenay le long d'une petite montagne qui est presque coupee de la mer:[d] le reste se sont montagnes hautes eslevees, où il y a peu de terre, sinon rochers & sables remplis de bois de pins, cyprez, sapins, boulles, & quelques manieres d'arbres de peu: il y a un petit estang proche dudict port renfermé de montaignes couvertes de bois. A l'entree dudict port il y a deux pointes, l'une du costé de Ouest contena*n*t une lieuë[e] en mer, qui s'appelle la pointe de sainct Mathieu,[f] & l'autre du costé de Suest[g] contenant un quart de lieuë,[h] qui s'appelle la pointe de tous les Diables: les vents du Su & Su-suest & Su-sorouest, frape*n*t[i] dedans ledit[j] port. Mais de la pointe de sainct Mathieu[k] jusques à ladite pointe[l] de tous les Diables, il y a pres d'une lieuë:[m] l'une & l'autre pointe[n] asseche de basse mer. /[f. 3v]

Chap. II.

Bonne reception faicte aux François par le grand Sagamo des Sauvages de Canada,[o] *leurs festins & danses, la guerre qu'ils ont avec les Irocois, la façon & dequoy sont faicts leurs Canots & Cabanes: Avec la description de la poincte de Sainct Matthieu.*

Le 27. jour nous fusmes trouver les sauvages[p] à la pointe de sainct Mathieu,[q] qui est à une lieuë de Tadousac, avec les deux Sauvages que mena le sieur du Pont

a lieuës] A2 lieues
b viẽt] A2 vient
c ladite] A2 ladicte
d coupee de la mer:] A2 couppee de la Mer:
e lieuë] A2 lieue
f pointe de sainct Mathieu,] A2 poincte de sainct Matthieu,
g Suest] A2 Suest,
h lieuë,] A2 lieue,
i frapẽt] A2 frappent
j ledit] A2 ledict
k pointe de sainct Mathieu] A2 poincte de sainct Matthieu
l ladite pointe] A2 ladicte poincte
m lieuë:] A2 lieue:
n pointe] A2 poincte
o *Canada,*] A2 *Canadas,*
p sauuages] A2 Sauuages
q pointe de sainct Mathieu,] A2 poincte de Sainct Matthieu,

or /[f. 3] 50 leagues to the first waterfall,[24] and that it comes from the north-northwest. The said harbour of *Tadousac* is small, where it could hold only ten to twelve ships, but there is enough water towards the east, in the shelter of the said *riviere de Sagenay* alongside a little mountain, which is almost cut off from the sea[25] [see frontispiece]. The remainder are mountains, rising very steeply, where there is little soil apart from rocks and sand, overgrown with pine, cedar, firs, birch,[26] and some varieties of trees of small value. There is a little pond[27] near the said harbour enclosed by mountains covered with trees. At the entrance of the said harbour are two points, one on the west called *pointe de sainct Mathieu*, running a league out into the sea, and the other on the southeast side called *pointe de tous les Diables*,[28] running out a quarter of a league. The winds from the south, south-southeast, and south-southwest beat into the said harbour. But from *pointe de sainct Mathieu* to the said *pointe de tous les Diables* there is nearly a league; both these points dry up at low tide.

/[f. 3v] Chapter 2

Good reception accorded to the French by the grand Sagamo *of the* Sauvages *of Canada; their feasts and dances; the war they wage with the* Irocois; *the manner of making their canoes and cabins, and the material with which they are made; with a description of the* poincte de Sainct Matthieu

On the 27th, accompanied by the two *Sauvages* whom Monsieur du Pont[29] brought to report on what they had seen in France and on the good reception the

24 The falls at Chicoutimi are 127 km up the Saguenay River from Tadoussac.

25 Champlain included his chart *port de tadoucac* in *Les Voyages, 1613*, 172.

26 *Pin* is the white pine (*Pinus strobus*). Of the two members of the cypress family, *Juniperus* and *Thuja*, only the white cedar (*Thuja occidentalis*) occurs in the lower St. Lawrence area. *Sapin* is probably balsam fir (*Abies balsamea*), although Champlain could have confused the tree for spruce. *Boulle* is white birch (*Betula papyrifera*).

27 The current name for this little pond is Lac de l'anse à l'Eau.

28 *Pointe de sainct Mathieu* is now Pointe aux Alouettes, and *pointe de tous les Diables* is Pointe aux Vaches. The two points are 4.8 km apart.

29 François Gravé Du Pont, frequently mentioned as Pont-Gravé, Pontgravé, Gravé, or Dupont-Gravé, was a merchant of Saint-Malo who had made trading voyages to Canada before 1600. In 1600 he and Chauvin de Tonnetuit established a trading post at Tadoussac, which was abandoned after one winter. He continued, however, to be interested in the country for trade purposes, and the association with Champlain, which began with this voyage of 1603, continued to the end of their lives. Gravé Du Pont was placed

pour faire le rapport de ce qu'ils avoient veu en France, & de la bonne reception que leur avoit fait le Roy. Ayant mis pied à terre nous fusmes à la cabanne de leur grand Sagamo[a] qui s'appelle Anadabijou, où nous le trouvasmes avec quelque 80. ou 100. de ses compagnons qui faisoie*nt*[b] Tabagie (qui veut dire festin) lequel nous reçeut[c] fort bien selon la coustume du pays,* & nous fist assoir aupres de luy, & tous les Sauvages arangez les uns aupres des autres des deux costez de ladite cabanne. L'un des Sauvages que nous /[f. 4] avions amené commença à faire sa harangue, de la bonne reception que leur avoit fait le Roy, & le bon traictement qu'ils avoient receu en France,* & qu'ils s'asseurassent que sadite Majesté[d] leur vouloit du bien, & desiroit peupler leur terre, & faire paix avec leurs ennemis (qui sont les Irocois) ou leur envoyer des forces pour les vaincre: en leur comptant aussi

a Sagamo] A2 Sagamo,

b faisoiēt] A2 faisoient

c reçeut] A2 receut

* LM] *François bien receus par les Sauuages.*

* RM] *Harangue de l'vn des Sauuages que nous auions ramenez.*

d Majesté] A2 Maiesté

King[30] had given them, we went to find the *Sauvages* at *pointe de sainct Mathieu*, which is a league from *Tadousac*. Having set foot on land, we went to the lodge of their *grand Sagamo*,[31] named Anadabijou,[32] where we found him with some 80 or 100 of his companions, who were making *Tabagie*[33] (which means a feast). He received us very well, according to the custom of the country, and made us sit down beside him, with all the *Sauvages* arranged one beside the other on both sides of the lodge. One of the *Sauvages*, whom we /[f. 4] had brought, began to deliver his oration on the good reception that the king had given them and the good treatment they had received in France, and that they might feel assured that his said Majesty wished them well and desired to people their land, and to make peace with their enemies (who are the *Irocois*) or send forces to conquer them.[34]

in charge of Port-Royal during the winter of 1605–06. For Chauvin, see *Dictionary of Canadian Biography* (DCB), 1:209–10, and Gravé Du Pont, *DCB*, 1:345–6.

30 Gravé Du Pont had taken these two Montagnais men back to France the previous year. There they had an audience with Henri IV, king of France (1589–1610), who made promises to them which they were about to relate to their people. See Pierre-Victor Cayet's introduction to Document H, "Of the French Who Have Become Accustomed to Being in Canada."

31 *Sagamo* (sagamore, sachem, and other variants) is a word belonging to the Eastern Algonquian language family, probably Mi'kmaq or Abenaki in origin, meaning someone in authority, a headman. We have italicized *grand* (Great or Big), since this designation by Champlain distinguishes Anadabijou from the other sagamos (ch. 3, f. 4).

32 Anadabijou was an important headman among the Montagnais (*DCB*, 1:61). Under Montagnais customs, Anadabijou was the spokesman for the group not because he had more authority, as Champlain supposed, but probably because he was highly respected and a good orator.

33 *Tabagie* is a word of unknown origin meaning a feast. It was probably derived from *tabac* (tobacco), which was smoked at a *tabagie*. In modern Canadian French it is a smoking-room. Champlain used both the words *petun(m)* and *tabac* for tobacco. He called a pipe a *petunoir*.

34 This was a fateful meeting that helped to shape the future of New France. Judging from Champlain's description of the solemn *tabagie* he witnessed, the Montagnais were certain that they had been offered an alliance with France by Henri IV. Initially, the French probably did not regard these promises very seriously. In 1608, however, after the habitation at Quebec had been built, and again in the following spring, Champlain was requested by the Montagnais to aid them in a war against the Iroquois. Mindful of the promises exchanged in 1602–03 and the vulnerability of Quebec, Champlain felt that he could not refuse his "allies," as he now called them, and accompanied the Montagnais, *Ochateguins* (*Ouendat*, Hurons), and Algonquins against the Iroquois in July 1609. The Hurons and Ottawa Valley Algonquins, who were allies of the Montagnais, had come to the St. Lawrence to enter the Montagnais alliance with the French. The raid on the

les beaux Chasteaux, Palais, maisons[a] & peuples qu'ils avoient veus, & nostre façon de vivre, il fut entendu avec[b] un silence si grand[c] qu'il ne se peut dire de plus. Or apres qu'il eust[d] achevé sa harangue, ledict grand Sagamo Anadabijou, l'ayant attentivement ouy, il commença à prendre du Petum, & en donner audict sieur du Pont Gravé de S. Malo, & à moy, & à quelques autres Sagamos qui estoient aupres de luy: ayant bien petunné, il commença à faire sa harangue à tous, parlant pozement,[e] s'arrestant quelque fois[f] un peu, & puis reprenoit[g] sa parolle, en leur disant,* Que veritablement[h] ils devoient estre fort co*n*tens[i] d'avoir sadicte Majesté[j] pour grand amy. Ils[k] respondirent tous d'une voix, *ho ho ho*,[l] qui est à dire, *ouy ouy*.[m] Luy conti-/[f. 4[v]]nuant tousjours sadicte harangue, dict, Qu'il estoit fort aise que sadicte Majesté[n] peuplast leur terre, & fist la guerre à leurs ennemis, qu'il n'y[o] avoit nation au monde à[p] qu'ils voulussent plus de bien[q] qu'aux François: En fin il leur fit entendre à tous le bien & utilité qu'ils[r] pourroient recevoir de sadicte Majesté.[s]

a maisons] A2 maisons,
b auec] A1 a-/ auec (*error in the first edition because of the line break*)
c grand] A2 grand,
d eust] A2 eut
e pozement,] A2 posément,
f quelque fois] A2 quelquesfois
g reprenoit] A2 reprenant (*printer's attempt to correct the French, but Champlain commonly drops subject pronouns*)
* RM] *Harangue du grand Sagamo.*
h veritablement] A2 veritablemēt
i cōtens] A2 contens
j Majesté] A2 Maiesté
k amy. Ils] A1 A2 amy, ils
l *ho ho ho,*] A2 *ho, ho, ho,*
m *ouy ouy.*] A2 *ouy, ouy.*
n Majesté.] A1 Majesté: A2 Maiesté:
o n'y] A1 A2 ny
p à] A1 a
q bien] A2 bien,
r qu'ils] A1 qui ils
s Majesté] A2 Maiesté

He also told them of the fine castles, palaces, houses, and peoples they had seen, and of our manner of living. He was heard with such silence that no more can be said. Now after he had finished his oration, the said *grand Sagamo* Anadabijou who, having listened to him attentively, began to take some tobacco[35] and to give some of it to the said sieur du Pont-Gravé of St. Malo and to me, and to some other *Sagamos* who were near him. Having had a good smoke,[36] he began to address the whole gathering, speaking deliberately, pausing sometimes a little, and then resuming his speech, saying to them that in truth they ought to be very glad to have his said Majesty for a great friend. They answered all with one voice, "*ho, ho, ho*,"[37] which is to say, "*yes, yes*." Still continuing with /[f. 4^{v}] his said speech, he said that he was quite content[38] that his said Majesty would populate their land and make war on their enemies, and that there was no nation in the world for which they would wish more good than the French. Finally, he gave them all to understand the advantage and profit they might receive from his said Majesty.

Iroquois cemented the alliance and set a precedent for future French-Native relations. Judging from Champlain's writings, he felt he had little choice in the matter.

35 Here, Champlain uses *Petum*, a loan word from a dialect native to Brazil. Thevet, the royal geographer of Henri II, recorded it in 1557. Although the term *tabac* gradually replaced *pétun* in the seventeenth century, its use persisted in western France, notably in Normandy and Upper Brittany (*DHLF*, 2:2693).

36 Champlain may have invented the verb *petunner* himself as a natural derivation from *pétun* (*pétuner*: to consume tobacco), since the year 1603 is listed as the first record of it (ibid.). Although it was gradually replaced by the verb *fumer*, writers continued to use it in such literary texts as Rostand's *Cyrano de Bergerac*. In a series of stylized responses to his oversized nose, the stage character Cyrano improvises in the classical Alexandrine measure of a Racine or Molière (who was a friend of the real Cyrano): "Truculent: Ça, monsieur, lorsque vous pétunez, / La vapeur du tabac vous sort-elle du nez / Sans qu'un voisin ne crie au feu de cheminée?" (Bird, *Cyrano de Bergerac*, 30). This passage was translated by Brian Hooker as "Insolent: sir, when you smoke, the neighbours must suppose / Your chimney is on fire" (Hooker, *Cyrano*, 31). This famous play was first performed on 27 December 1897 but was set during the lifetime of the real Cyrano de Bergerac (1619–55) in the Golden Age of French literature.

37 The second journal attributed to Jacques Cartier (1536) interprets these three cries (*troys criz*) as a sign of joy and alliance (Bideaux, *Relations*, 142; Cook, *Voyages*, 54), although it was at first *chose horrible à houyr* (a terrible thing to hear) (Bideaux, *Relations*, 141). Sagard likewise interpreted *ho, ho, ho* as a joyful greeting (Sagard, *Histoire du Canada*, 232).

38 *aise*: "content or happy" seems to be the primary meaning since 1164 (*DHLF*, 1:74), related to the obsolete verb *aisir*: to make happy (Huguet, *Dictionnaire* [HU], 1:146).

Apres qu'il eust achevé sa harangue, nous sortismes de sa Cabanne, & eux commencerent à faire leur Tabagie, ou festin, qu'ils font avec des chairs d'Orignac, qui est co*m*me boeuf,* d'Ours, de Loumarins & Castors, qui sont les viandes les plus ordinaires qu'ils ont, & du gibier en quantité: ils avoient huict ou dix chaudieres, pleines de viandes, au milieu de ladicte cabanne,[a]* & estoient esloignees les unes des autres quelque six pas, & chacune à[b] son feu. Ils sont assis des deux costez (comme j'ay dit cy-dessus) avec chacun son escuelle d'escorce d'arbre: & lors que la viande est cuitte il y en a un qui fait[c] les partages à chacun dans lesdictes escuëlles,[d] où ils mangent fort sallement:* car quand ils ont les mains grasses, ils les frotent[e] à leurs cheveux, ou bien au poil de leurs chiens, do*n*t[f] ils ont quantité pour la chasse. Premier /[f. 5; B] que leur viande fut cuitte, il y en eust un qui se leva, & print un chien, & s'en alla sauter autour desdictes chaudieres d'un bout

* LM] *Festin des Sauuages.*

a cabanne,] A2 Cabanne,

* LM] *Comme ils font cuire leurs viandes.*

b à] A1 a (*although* A2 *is preferred here, both variants are possible*)

c fait] A2 faict

d escuëlles,] A2 escuelles,

* LM] *Mangent fort salement.*

e frotent] A2 frottent

f dōt] A2 dont

After he had ended his speech, we went out of his lodge, and they began to make their *Tabagie*, or feast, which they make with moose meat – which is like beef – and with bear, seal,[39] and beaver, which are the most ordinary meats that they have, and with great quantities of wild fowl. They had eight or ten cauldrons[40] full of meats in the midst of the said lodge, removed from one another by some six paces, each [cauldron] at its own fire.[41] The men sat on both sides (as I said before), each with his bowl made of the bark of a tree; and when the meat is cooked, one of them apportions to each man his share in the said bowls, out of which they feed very filthily. For when they have greasy hands, they rub them in their hair, or else in the fur of their dogs, of which they have a great number for hunting. Before /[f. 5] their meat was cooked, one of them rose up and took a dog and went leaping about the said cauldrons from one end of the lodge to the

39 *Loumarins* (m. pl.), the generic term *Loup-marins* (lit. sea wolves) for any kind of seal is still used in the maritime region of Quebec (Bideaux, *Relations*, 318n93).

40 Champlain and his contemporaries Lescarbot and Sagard, as well as later French writers familiar with the French-Native trade, used only the terms *chaudière* and *chaudron* to describe the cooking pots they gave in trade, sometimes modified by *grandes*, *moyenes*, or *petites*. Almost all modern writers translate these two terms as "kettles." *Bouilloire*, the modern French word for kettle, did not come into general use until about 1740. In shape, these *chaudières* were flat bottomed to slightly rounded, straight sided, and slightly larger around the rim than the bottom. They were carried and suspended by a bail or handle made of iron, attached to the rim by two lugs. The body of the vessel was made of a single sheet of copper. Later in the seventeenth century these cauldrons were also made of brass, hence terms such as *chaudière rouge* (copper) and *chaudière jaune* (brass). Lids were not traded. These vessels varied greatly in capacity. Judging from those that have survived as Native grave goods, their capacity could range from 5 to over 100 litres, with an average capacity of about 40 litres (Fitzgerald et al., "Basque Banded Copper Kettles," 50). Those copper *chaudrons* brought by Basque traders in the late sixteenth century were heavier and better made than the brass ones sold by French traders after the early seventeenth century (Turgeon, "French Fishers," 600–3; Kenyon, *Grimsby Site*).

41 This is a most improbable Montagnais lodge. With 8 to 10 fires, 6 paces apart down the middle of the *cabanne*, this structure would be over 60 metres (200 feet) in length, rivalling the largest Iroquoian longhouses in size. It would be difficult to know where they would get the bark cladding in the Tadoussac area for such a huge structure. All references to Montagnais lodges state that they did not hold more than about 20 people (Rogers and Leacock, "Montagnais-Naskapi," 175). In 1632, Father Le Jeune described an unusually large lodge at Tadoussac that had "three fires in the middle, distant from each other five or six feet" (*JR*, 5:27). The lodges that Champlain described at the end of this chapter are typical of the Montagnais ones. It may be that what Champlain described was an uncovered, ceremonial structure of bent saplings.

de la cabanne à l'autre:* Estant deva*n*t le grand Sagamo, il jetta son chien à terre de force, & puis tous d'une voix ils s'escrierent *ho, ho, ho*: ce qu'ayant faict, s'en alla asseoir à sa place.[a] En mesme instant, un autre se leva, & feist le semblable, continuant tousjours, jusques à ce que la viande fust cuitte.

Or apres avoir achevé leur Tabagie, ils commencerent à dancer, en prenant les testes de leurs ennemis, qui leur pendoient par derriere: En signe de resjouïssance, il y en a un ou deux qui cha*n*tent en accordant leur voix par la mesure de leurs mains qu'ils frappent sur leurs genoux, puis ils s'arrestent quelques-fois, en s'escriant, *ho, ho, ho*, & recommencent à dancer en soufflant comme un homme qui est hors d'aleine: Ils faisoient ceste resjouïssance pour la victoire par eux obtenüe[b] sur les Irocois,* dont ils en avoient tué quelque cent, ausquels ils couperent les testes, qu'ils avoient avec eux pour leur ceremonie. Ils estoient trois nations quand ils furent à la guerre,[c] les Estechemins, Algoumequins, & Montagnes,*

* RM] *Sauuages dansent autour des chaudieres.*

a place.] A2 place:

b obtenüe] A2 obtenuë

* RM] *Victoire obtenuë sur les Irocois.*

c guerre,] A2 guerre. (*a period would block the nouns in apposition and convert the next sentence into a fragment*)

* RM] *Trois nations de Sauuages, Estechemins, Algoumequins et Montagnes.*

other.* Positioned in front of the *grand Sagamo*, he threw his dog violently upon the ground, and then all with one voice cried out, "*ho, ho, ho*." Having done this, he went and sat down in his place. In the same instant, another rose up and did likewise, and the whole thing went on continually until the meat was cooked.

Now, after having brought their *Tabagie* to an end, they began to dance by taking up the heads of their enemies,[42] which were hanging behind them. As a sign of rejoicing, one or two of them sing, synchronizing their voices by the beat of their hands, which they strike upon their knees; then they pause sometimes, crying out "*ho, ho, ho*," and begin again to dance, panting like a man out of breath. They were making this celebration for the victory that they had won over the *Irocois*, of whom they had killed about a hundred, from whom they had cut the heads that they had with them for their ceremony. There were three nations when they went to war, the *Estechemins*,[43] *Algoumequins*, and *Montagnes*,

* RM] Sauvages *dance around some cauldrons.*

42 Champlain used the word *teste* (*tête*: head), which has been translated as scalp in the Champlain Society edition of *The Works* on the assumption that Natives only scalped those whom they killed in war. The problem with this interpretation is that the Natives also took heads if the occasion permitted. Champlain was usually quite specific, as in *couperent les testes* (cut the heads off), Biggar, *The Works*, 1:103; *escorcherent les testes* (flayed, or skinned the heads), Biggar, *The Works*, 2:102; or *la peau de la teste, qu'ils auoient escorchee* (the skin of the head which they had flayed), Biggar, *The Works*, 2:103. The record of Cartier's second trip to Canada in 1536 also uses the word *peau* (skin) for scalp: "*nous fut par ledit Donnacona monstré les peaulx de cinq testes d'hommes*" ("Donnacona showed us the scalps of five human heads"), Bideaux, *Relations*, 159. Writing in 1632, Gabriel Sagard, a Recollect lay brother and contemporary of Champlain's, made the role of trophy taking clearer: "*ils en emportent la teste; que s'ils en estoient trop chargez, ils se contentent d'en emporter la peau avec sa chevelure, qu'ils appellent* Onontsira, *les passent et les serrent pour en faire des trophées, et mettre en temps de guerre sur les pallissades ou murailles de leur ville, attachées au bout d'une longue perche*" ("they carry away the head; and if they are too much encumbered with these they are content to take the skins with the hair on them, which they call *Onontsira*, tan them, and put them away for trophies, and in time of war set them on the palisades or walls of their town fastened to the end of a long pole"), Wrong, *Long Journey*, 153, 348. Heads were preferred to scalps as trophies because they were evidence of a more complete victory. See, for example, Pierre Radisson's murderous rampage through Neutral/Erie territory in 1653 with nine Iroquois companions. Many women, men, and children were killed. From these they returned home with 5 captives and 22 heads. Scalps were not mentioned (Adams, *The Explorations*, 32–5).

43 According to Champlain, the *Estechemins* (Etechemin) lived along the Atlantic coast from the St. John to the Kennebec Rivers. Since the Maliseet, Passamaquoddy, and Eastern Abenaki occupied this area at that time, it is likely that *Estechemins* is a term for these Eastern Algonquian groups.

au /[f. 5ᵛ] nombre de mille, qui allerent faire la guerre ausdicts Irocois qu'ils rencontrerent à l'entree de la riviere desdits Irocois, & en assommerent une centeine:[a] la guerre qu'ils font, n'est que par surprises, car autrement ils auroient peur, & craignent trop lesdits[b] Irocois, qui sont en plus grand nombre que lesdits Montagnes, Estechemains, & Algoumequins.

Le vingt-huictiesme jour dudit mois, ils se vindre*n*t caba*n*ner audit port de Tadousac, où estoit nostre vaisseau. A la pointe du jour, leurdit[c] grand Sagamo sortit de sa caba*n*ne,[d] allant autour de toutes les autres caba*n*nes, en criant à haute voix, Qu'ils[e] eussent à desloger pour aller à Tadousac, où estoient leurs bons amis:* Tout aussi tost un chacun d'eux deffit sa cabanne, en moins d'un rien, & ledit grand Capitaine le premier commença à prendre son Canot, & le porter à la mer, où il embarqua sa femme & ses enfans, & quantité de fourreures, & se meirent[f] ainsi pres de deux cents[g] Canots, qui vont estrangeme*n*t: Car[h] encore que nostre Chaloupe fut bien armee, si alloient-ils plus viste que nous. Il n'y a que deux personnes qui travaille*n*t à la nage, l'homme & la femme: Leurs /[f. 6] Canos[i] ont quelque huict ou neuf pas de long, & large comme d'un pas, ou pas & demy par le milieu, & vont tousjours en amoindrissant par les deux bouts:* ils sont fort subjects à tourner si on ne les sçait bien gouverner, car ils sont faicts d'escorce d'arbre appellé Bouille, renforcez par le dedans de petits cercles de bois bien & proprement faicts, & sont si legers, qu'un homme en porte un aisément, &

a centeine:] A2 centaine:
b lesdits] A2 lesdicts
c leurdit] A2 leurdict
d cabāne,] A2 cabāno, (*possible confusion with* cano, *as variant of* canot)
e Qu'ils] A2 qu'ils
* LM] *Deslogement des Sauuages de la pointe de S. Math. pour venir à Tadousa* [sic] *voir les François.*
f se meirent] A2 se mirent
g cents] A2 cens
h Car] A2 car
i Canos] A2 Canots
* RM] *Que c'est, & comment sont faicts les Canos des Sauuages.*

to the /[f. 5ᵛ] number of a thousand who went to wage war on the said *Irocois*, whom they encountered at the mouth of the river of the said *Irocois*.[44] And they slaughtered about a hundred of them. The mode of warfare which they practise is altogether by surprises,[45] for otherwise they would be afraid. And they fear the said *Irocois* excessively, who are greater in number than the said *Montagnes*, *Estechemains*, and *Algoumequins*.

On the twenty-eighth day of the said month, they came and encamped at the aforesaid harbour of *Tadousac*, where our ship was. At daybreak their said *grand Sagamo* came out of his lodge and was going around all the other lodges, crying out with a loud voice that they would have to break camp to go to *Tadousac*, where their good friends were. Immediately, each one of them took down his lodge in less than no time, and the said great Captain was the first to begin to take his canoe and carry it to the sea, where he put his wife and children on board and a great number of furs. And thus about two hundred canoes were launched, which move in an astonishing way;[46] for although our *chaloupe*[47] was well manned, yet they went more swiftly than we did.[48] There are only two people who paddle: husband and wife. Their /[f. 6] canoes are about eight or nine paces long and about a pace wide, or a pace and a half at the middle, always tapering down to both ends. They are very liable to capsize if one does not know how to manage them properly. For they are made out of the bark of a tree called a birch,[49] reinforced within by small ribs of wood,[50] well and carefully fashioned, and are so light that a man may carry one of them easily; and each canoe can

44 The mouth of the Rivière Richelieu

45 We now call war by surprise "guerrilla warfare," a borrowing from the Spanish *guerra*. A *guerrilla* in Spanish is a small war, waged by smaller units in ambushes.

46 *Qui vont estrangement*: lit. which go strangely

47 A *chaloupe* (shallop) is a sturdily built, open, shallow-drafted boat with a capacity of up to 25 tuns, used for fishing and for coastal and river traffic. It carried oars for rowing and a removable mast that could be schooner-rigged or fitted with a lug sail (Kemp, *Oxford Companion*, 775).

48 Champlain's description of the birchbark canoe is the first detailed one by a European. A month later he came to the conclusion that Canada could be explored only by means of the canoe. The word *canot* (canoe) used by Champlain is a loan word from Arawak (*canoa*) that Columbus recorded in Hispaniola (Roberts and Shackelton, *The Canoe*, 1). The Algonquian word for a canoe is *tciman* (*tchimân*) and it is *ąhonta* in Huron (Cuoq, *Lexique algonquine*, 390; Potier, *Huron Manuscripts*, 447).

49 Champlain uses the word *boulle* (or *bouille* in one of five occurrences) instead of the more common French word *bouleau* (*Betula papyrifera*), meaning the white or paper birch.

50 Champlain wrote *petits cercles de bois* (little circles of wood). The modern French word for a canoe rib is *nervure*.

chacun Cano[a] peut porter la pesanteur d'une pipe: Quand ils veulent traverser la terre pour aller à quelque riviere où ils ont affaire, ils les portent avec eux.

Leurs cabannes sont basses, faictes comme des te*n*tes couvertes de ladite escorce d'arbre, & laissent tout le haut descouvert comme d'un pied, d'où le jour leur vient,* & font plusieurs feux droit au milieu de leur cabanne, où ils sont quelques fois[b] dix mesnages ense*m*ble. Ils couchent sur des peaux les uns parmy les autres, les chiens avec eux.

Ils estoient au nombre de mille personnes, tant hommes que femmes & enfans. Le lieu de la pointe de S. Matthieu,[c] où ils estoient premierement cabannez, est assez plaisant,* ils estoient au bas d'un petit costau plein d'arbres de sapins & /[f. 6v] cyprez.[d] A ladicte pointe, il y a une petite place unie qui descouvre de fort loin, & au dessus dudit[e] costau est une terre unie, contenant une lieuë de long, demye de large, couverte d'arbres, la terre est fort sablo*n*neuse,[f] où il y a de bo*ns*[g] pasturages; tout le reste ce ne sont que montaignes[h] de rochers fort mauvais: la mer bat autour dudit costau qui asseiche pres d'une gra*n*de[i] demie lieuë de basse eau.

a Cano] A2 Canot

* RM] *Cabannes des Sauuages, dequoy, et comment ils sont faictes.*

b quelques fois] A1 quelques-fos (*printer's correction in 1604*)

c pointe de S. Matthieu,] A2 poincte de S. Mathieu,

* RM] *Description de la poincte de S. Matthieu.*

d cyprez.] A1 cypres. A2 cypres:

e dudit] A2 dudict

f sablōneuse,] A2 sablonneuse,

g bōs] A2 bons

h montaignes] A2 mōtaignes

i grāde] A2 grande

carry the weight of a pipe.[51] When they wish to go overland to get to some river where they have business, they carry them with them.[52]

Their lodges are low, made like tents covered with the aforesaid tree bark, and they leave the whole top part[53] uncovered for about a foot, through which the daylight comes into them. And they make several fires right in the middle of their lodge, where there are sometimes ten households together.[54] They sleep upon skins among each other, and the dogs with them.

They numbered about a thousand people, as many men as women and children.[55] The place at *point de S. Matthieu*, where they were first encamped, is very pleasant. They were at the bottom of a little slope, covered with fir and /[f. 6v] cedar trees. At the said point there is a little level spot, which is visible from very far away, and upon the top of the said slope is level land, measuring a league long by half a league wide and covered with trees. The soil is very sandy, where there is some good pasture. All the rest is nothing but mountains of very barren rocks. The sea beats around the said slope, which dries up for almost a full half-league at low tide.

51 A pipe is a measure of capacity equivalent to half a tun. It contains about 105 imperial gallons (477 litres). One pipe of water weighs about 840 pounds (380 kg).

52 "Their canoes are some eight or nine paces long … they carry with them." This description of the canoe is essentially repeated at the beginning of Champlain's 1608 journal, as published in the second book of *Les Voyages, 1613*, at the end of the first chapter. The most significant changes involve the names for the wood used in the construction. The archaic word *bouille* is replaced by *bouleau* in 1613, and the ribs are carved, more specifically, from *cedre blanc* (white cedar) (Champlain, *Les Voyages, 1613*, 168–9).

53 *Tout le haut*: lit. all the height, the peak or top of the lodge

54 What Champlain probably observed was that up to ten nuclear families could live in one of these lodges. Among the Montagnais, the people who lived together in a lodge were an extended family of closely related kin, such as brothers and their families, working and living together as a household. The individual nuclear families were the component parts of the extended family. These lodge groups travelled, hunted, and worked together throughout the year. Several of these lodge groups, more distantly related to each other, would form winter hunting groups. In the summer many hunting groups would get together at a traditional meeting place such as Tadoussac, where they formed a band, often taking a common designation from the lake or river where they were camped. The classical early description of the Montagnais lodge and hunting groups is by Father Paul Le Jeune, SJ (*JR*, 7:35–209).

55 If this figure is correct, it attests to the importance of the *tabagie* the Montagnais had with the French. Although Montagnais hunting groups would briefly get together into agglomerations of several hundred for a variety of social occasions during the summer, a group of one thousand would have been exceedingly rare.

CHAP. III.

La resjouïssance[a] *que font les Sauvages apres qu'ils ont eu victoire sur leurs ennemis, leurs humeurs, endurent la faim, sont malicieux, leur croyance & faulses opinions, parlent aux diables, leurs habits, & comme ils vont sur les neiges, avec la maniere de leur mariage, & de l'enterrement de leurs morts.*

LE 9. jour de Juin les Sauvages commencerent à se resjouir[b] tous ensemble & faire leur Tabagie, comme j'ay dit cy dessus, & danser, pour ladicte victoire qu'ils avoient obtenuë contre leurs ennemis.* Or apres avoir fait[c] bonne chere, les Algoumequins une des trois nations /[f. 7] sortirent de leurs caba*n*nes, & se retirerent à part dans une place publique, feirent arranger toutes[d] leurs femmes & filles les unes pres des autres, & eux se mirent derriere chantant tous d'une voix comme j'ay dit cy devant: Aussi tost toutes les femmes & filles commencerent à quitter leurs robbes de peaux, & se meirent toutes nuës monstrans leur nature, neantmoins paree de Matachia, qui sont patenostres & cordons entre-lassez faicts de poil de Porc-espic, qu'ils teignent de diverses coulleurs. Apres avoir achevé leurs chants, ils dirent tous d'une voix, *ho ho ho*,[e] à mesme instant, toutes les femmes & filles se couvroient de leurs robbes, car elles sont à leurs pieds, & s'arrestent quelque peu: & puis aussi tost recommençans à chanter ils laissent aller leurs robbes comme auparavant: Ils ne bougent d'un lieu en dansant, & font quelques gestes & mouvemens du corps levans un pied, & puis l'autre, en frappa*n*t[f] contre terre.

a resiouïssance] A2 resiouissance

b se resiouir] A2 se resiouyr

* LM] *Resiouïssance que les Sauuages firent de la victoire qu'ils auoient obtenuë sur leurs ennemis les Irocois.*

c fait] A2 faict

d toutes] A1 toute

e *ho ho ho,*] A2 *ho, ho, ho,*

f frappãt] A2 frapant

Chapter 3

The celebration which the Sauvages *make after they have had victory over their enemies; their disposition; how they endure hunger, are malicious, and talk to devils; their beliefs and false opinions; their clothing, and how they walk on snow; with the manner in which they marry, and bury their dead*

On the 9th day of June the *Sauvages* all began to celebrate together and make their *Tabagie*, as I have said above, and to dance in honour of the said victory which they had obtained over their enemies. Now, after having made good cheer, the *Algoumequins*, one of the three nations[56] /[f. 7], went out of their lodges, and withdrew apart into an open place. They had all their wives and daughters arranged side by side, and placed themselves behind, singing all in unison as I said before. Immediately, all the women and girls began to cast off their robes of skin and stripped themselves stark naked, showing their private parts, yet adorned with *Matachia*,[57] which are beads and braided cords made of porcupine quills, which they dye various colours. After having brought their songs to an end, they said all with one voice, "*ho, ho, ho*." At the same time, all the women and girls were covering themselves with their robes, since they are at their feet, and they pause somewhat; and then all at once beginning to sing again, they let go of their robes as before. While dancing, they do not budge from one spot, but make some gestures and body movements, lifting up one foot and then the other while stamping upon the ground.

56 Champlain, like all the French of the early contact period, used the term *nation* for the various Native groups they encountered. This term should not be interpreted as meaning a political state, but rather a cohesive group of interlinked families who saw themselves distinct from their neighbours. They had a common identity through their language or dialect, a common past, a common decision-making system that was consultative but not coercive, a common belief system, and a shared, loosely defined territory.

57 *Matachias* were personal adornments worn mainly by women but also by men and children. They were strings of beads fashioned into necklaces, bracelets, and earrings, as well as beadwork and dyed porcupine quills, embroidered, along with some painting, on leather garments (Grant, *History of New France*, 3:157–60). The word is apparently of Mi'kmaq origin, or one of the other Eastern Algonquian languages such as Maliseet-Passamaquoddy. It survives in slightly altered form in Acadian French and Canadian French (Poirier, *Le glossaire acadien*, 297–8; Clapin, *Dictionnaire canadien-française*, machicoté – jupon [petticoat]).

Or en faisant ceste danse, le Sagamo des Algoumequins qui s'appelle Besoüat, estoit assis devant lesdictes femmes & filles, au millieu de deux bastons, où estoient les testes de leurs enne-/[f. 7v]mis penduës: quelque fois[a] il se levoit & s'en alloit haranguant* & disant aux Montaignez[b] & Estechemains, voyez comme nous nous resjouïssons[c] de la victoire que nous avons obtenüe[d] sur[e] nos ennemis, il faut que vous en faciez autant, affin que nous soyons contens, puis tous ensemble disoient *ho ho ho.* Retourné qu'il fut en sa place, le Grand[f] Sagamo avec tous ses compagnons despouillerent leurs robbes estans tous nuds hors mis leur nature qui est couverte d'une petite peau, & prindrent chacun ce que bon leur sembla,* comme Matachias, haches, espees, chauderons, graisses, chair d'Orignac,[g] Loup-marin, bref chacun avoit un present qu'ils allerent[h] donner aux Algoumequins. Apres[i] toutes ces ceremonies la danse cessa, & lesdits Algoumequins ho*m*mes & femmes emportere*n*t leurs presens da*ns*[j] leurs cabannes. Ils fire*n*t encor[k] mettre deux hommes de chacune natio*n* des plus dispos qu'ils feirent courir, & celuy qui fut le plus viste à la course eut un present.

Tous ces peuples sont tous d'une humeur* assez joyeuse, ils rient le plus souvent, toutefois ils sont quelque peu Saturniens; Ils parlent fort pozément,[l] com-/[f. 8]me se voullans bien faire entendre, & s'arrestent aussi tost en songeant une grande espace de temps, puis reprennent leur parolle:[m] ils usent bien souvent de ceste façon de faire parmy leurs harangues au co*n*seil, où il n'y a que les plus principaux, qui sont les antiens:[n] Les femmes & enfans n'y assistent point.

a quelque fois] A2 quelque-fois
* LM] *Sagamo des Algoumequins.*
b Montaignez] A1 Montaignes
c resiouïssons] A2 resiouissons
d obtenüe] A2 obtenuë
e sur] A2 de
f Grand] A2 grand
* LM] *Present des Montagnes & Estechemins.*
g chair d'Orignac,] A2 chair, d'Orignac
h allerent] A2 allerēt
i Algoumequins. Apres] A2 Algoumequins, Apres
j dās] A2 à
k firēt encor] A2 firent encores
* LM] *Humeurs des Sauuages.*
l pozément,] A2 posément,
m parolle:] A2 parole,
n antiens: A2 anciens:

Now, while this dance was being performed, the *Sagamo* of the *Algoumequins*, who is named Besouat,[58] was seated before the said women and girls, between two poles, upon which the heads of their enemies /[f. 7ᵛ] were hung. Sometimes he arose and moved away haranguing the *Montaignes* and *Estechemains* and saying, "See how we rejoice for the victory that we have obtained over our enemies; you must do as much so that we may be satisfied." Then they all cried "*ho, ho, ho*" together. As soon as he had returned to his place, the *Grand Sagamo* with all his companions cast off their robes, being stark naked except for their private parts, which are covered with a small piece of skin,[59] and each one of them took what seemed good to him, such as *Matachias*, hatchets, swords, cauldrons, pieces of fat, moose meat, seal; in short, each one had a present, which he proceeded to give to the *Algoumequins*. After all these ceremonies, the dancing ceased, and the said *Algoumequins,* both men and women, carried away their presents into their lodges. Yet they selected two of the most agile men from each nation, whom they made run, and he who was fastest in the race received a present.

All of these people possess a rather joyful disposition. They laugh most often, yet they are somewhat saturnine.[60] They speak very deliberately, as /[f. 8] if wishing to make themselves well understood, and stop suddenly, reflecting for a long space of time, and then begin to speak again. They very often resort to this way of proceeding with their speeches in council, where there are only the foremost leaders, who are the elders; the women and children are not present.

58 Bessouat (Tessouat) was an influential headman of the *Kichesipirini* (Big River People) Algonquins of Morrison Island on the Ottawa River (DCB, 1:638–9).

59 Loincloth

60 *Saturniens*: of a melancholic, saturnine or gloomy temperament

Tous ces peuples patissent tant quelques-fois,* qu'ils sont presque contraints[a] de se manger les uns les autres pour les grandes froidures & neiges: car les animaux & gibier dequoy ils vivent se retirent aux pays plus chauts.[b] Je tiens que qui leur monstreroit à vivre & enseigner le labourage des terres, & autres choses,[c] ils l'apprendroient fort bien; car je vous asseure qu'il s'en trouve assez qui ont bon jugement, & respondent assez bien à propos sur ce que l'on leur pourroit demander.

Ils[d] ont une meschanceté en eux,* qui est, user de vengeance & estre grands menteurs, gens en qui il ne fait[e] pas trop bon s'asseurer, sinon qu'avec raison & la force à la main; promettent assez & tiennent peu: Ce sont la pluspart gens qui n'ont point de loy, selon que j'ay peu voir,[f] & m'informer audit grand Sagamo, lequel me dit, Qu'ils /[f. 8v] croyoient veritablement, qu'il y a un Dieu qui a creé

* RM] *Les Sauuages endurent la faim.*
a contraints] A2 contraincts
b chauts.] A2 chauts:
c choses,] A2 choses
d demander. ¶ Ils] A1 A2 demander: ils
* RM] *Malice des Sauuages.*
e fait] A2 faict
f voir,] A2 voir

All these people suffer so much sometimes that they are almost forced to eat each other because of the great cold and snows, for the animals and fowl upon which they subsist migrate to warmer climates. I hold that if anyone were to show them how to live and teach them how to work the land and other matters, they would learn it very well; for I assure you that there are plenty of them who have good judgment and respond quite appropriately to whatever one could ask of them.[61]

They have a wickedness in them, in that they resort to revenge and are great liars, a people whom it is not too good to trust, except within reason and with force at hand.[62] They promise much and perform little. They are, for the most part, a people who have no law, according to what I have been able to see,[63] and learn from the said *grand Sagamo*,[64] who told me that they /[f. 8v] really believed

61 "All these people … whatever one could ask of them." When this paragraph is repeated in Champlain's 1608 journal as the foundations of the future Quebec City are being laid, the hint of cannibalism here is replaced with the assertion that the Natives are forced to eat shells, leather, and their dogs when nearing starvation in the winter (Champlain, *Les Voyages, 1613*, 192; Biggar, *The Works*, 2:46).

62 When Champlain wrote this, he had had no experience with the Montagnais or any other Native group. The statement that Natives were "great liars" is here totally unsubstantiated and absurd in view of the fact that Champlain was seeking geographical information from them, which he readily accepted. Being a newcomer, he was probably repeating the commonly held opinion of Europeans that all Natives were liars. Comments like these, to some extent repeated in chapter 12, show us an uncritical side to Champlain that is often ignored by his biographers. On a couple of occasions he freely admits that he lied to the Algonquin and Huron in order to get his way.

63 "They have a wickedness in them … according to what I have been able to see" is repeated almost *verbatim* in the fourth chapter of the 1608 journal (Champlain, *Les Voyages, 1613*, 192–3) with a slight improvement in the French. Champlain adds his subject pronouns to *Ils promettent assez, mais ils tiennent peu* for the 1613 publication. It is notable that the alleged conversation of 1603 with the *grand Sagamo* Anadabijou, in which they compare Native legends with Catholic doctrine, is entirely omitted in 1613, except for the brief discussion of prayer at the end of it.

64 Champlain could not speak any of the Native languages, and it is unlikely that Anadabijou could speak much French. Unfortunately, Champlain says nothing about his interpreters. Although the French party had returned with two Natives from France who may have acted as translators, it is doubtful that after less than a year in France they could have translated aspects of Christian dogma. One must therefore exercise caution in accepting the long conversation that Champlain had, or did not have, with the *grand Sagamo* about Christianity. Lescarbot was the first skeptic in this regard (*Histoire de la Nouvelle France*, 430, cited in Beaulieu and Ouellet, *Champlain*, 113–14n2). For a contrast on the same subject, see the conversations in 1634 between Father Paul Le Jeune,

toutes choses.* Et lors je luy dis, Puis qu'ils croyoient[a] à un seul Dieu, Comme*n*t est-ce qu'il les avoit mis au monde, & d'où ils estoient venus? Il[b] me respondit, Apres que Dieu eut fait[c] toutes choses, il print quantité de fleches, & les mit en terre, d'où il sortit hommes & femmes, qui ont multiplié au monde jusques à present, & sont venus de ceste façon. Je luy respondis que ce qu'il disoit estoit faux: mais que veritablement il y avoit un seul Dieu, qui avoit creé toutes choses, en la terre, & aux cieux: Voya*n*t toutes ces choses si parfaites,[d] sans qu'il y eust personne qui gouvernast en ce bas monde, il print du limon de la terre, & en crea Adam nostre premier pere: comme Adam sommeilloit, Dieu print une cotte dudict Adam, & en forma Eue, qu'il luy donna pour compagnie, & que c'estoit la verité qu'eux & nous estio*ns* venus de ceste façon, & non de fleches comme[e] ils croyent. Il ne me dit[f] rien, sinon, Qu'il advoüoit plustost ce que je luy disois, que ce qu'il me disoit.

Je luy demandis aussi, s'il[g] ne croyoit point qu'il y eust[h] autre qu'un seul Dieu:[i] il me dit[j] que leur /[f. 9; C] croyance estoit, Qu'il y avoit un Dieu, un Fils, une Mere, & le Soleil,* qui estoie*nt*[k] quatre; Neantmoins que Dieu estoit par-dessus[l] tous; mais que le Fils estoit bon & le Soleil, à cause du bien qu'ils recevoie*nt*: Mais la Mere[m] ne valloit rien, & les mangeoit, & que le Pere n'estoit pas trop bo*n*.[n] Je luy remonstray son erreur selo*n*[o] nostre foy, enquoy il[p] adjousta quelque peu de creance. Je luy demandis s'ils n'avoient point veu[q] ou ouy[r] dire à leurs ancestres

* LM] *Croyance des Sauuages & leur foy,*
a croyoient] A2 croyent
b Il] A1 A2 il
c fait] A2 faict
d parfaites,] A2 parfaictes,
e comme] A1 commes
f dit] A2 dict
g s'il] A1 s'ils
h eust] A1 eut
i Dieu:] A2 Dieu,
j dit] A2 dit,
* RM] *Croyent vn Dieu, vn fils, vne mere, & le Soleil.*
k estoiẽt] A2 estoient
l par-dessus] A1 par dessus
m Mere] A1 mere
n bõ.] A2 bon.
o selõ] A2 selon
p il] A1 A2 ils
q veu] A2 veu,
r ouy] A1 ouyr

that there is a God who created all things. And then I said to him, since they believed in only one God, "How had He brought them into the world, and where had they come from?" He answered me, "After God had made all things, He took a great quantity of arrows and stuck them in the ground, whence He drew men and women, who have multiplied in the world up to the present time, and they came in this way." I replied to him that what he was saying was false, but that truthfully there was only one God, who had created all things on earth and in the heavens. Seeing all these things so perfect, without anybody to govern this lowly world, He took the clay of the earth[65] and from it created Adam, our first father. While Adam was sleeping, God took a rib from the said Adam and formed Eve out of it, whom He gave to him for companionship. And I said that it was the truth that both they and we had come in this way and not from arrows as they believe. He said nothing to me except that he acknowledged what I was saying to him sooner than what he was saying to me.

I also asked him whether he did not believe that there was something other than one single God. He told me that their /[f. 9] belief was that there was a God, a Son, a Mother, and the Sun, which made four, although God was above all; but that the Son and the Sun were good, because of the benefit they[66] received, but that the mother was not worth anything and used to eat them, and that the Father was not too good. I pointed out his error to him according to our faith, to which he gave some little credence. I asked him whether they had not seen or heard their

SJ, who spoke some Montagnais, and the shaman Carigonan concerning Montagnais religion and Christianity (*JR*, 6:157–227).

65 *Limon*: clay or silt. The French *limon* is the direct descendant of the Latin *limus* (n.m.), found in the relevant passage of the Latin Vulgate: *formavit igitur Dominus Deus hominem de limo terrae* (Gn. 2:7), although modern French translations of this passage use *poussière* (dust), reflected as well in the 1611 King James Version. Champlain has chosen the older account of Creation (J) for comparison with the Algonquin legends, in which Yahweh is portrayed, anthropomorphically, as a potter fashioning Adam from earth watered by a mist.

66 The humans, rather than the son and the sun, the grammatical antecedents of "they"

que Dieu fust venu au monde:[a] il me dit,[b] Qu'il ne l'avoit point veu: mais qu'anciennement il y eust cinq hommes qui s'en allerent vers le Soleil couchant, qui rencontrerent Dieu, qui leur demanda, Où allez vous? Ils[c] dirent, Nous allons chercher nostre vie: Dieu leur respondit, Vous la trouverrez icy.* Ils passerent plus outre, sans faire estat de ce que Dieu leur avoit dit,[d] lequel print une pierre, & en toucha deux, qui furent transmuez en pierre: Et dit[e] derechef aux trois autres, Où allez vous? & ils respondirent comme à la premiere fois,[f] & Dieu leur dit derechef, Ne passez plus outre,[g] vous la trouverrez icy: Et voyant qu'il ne leur venoit rien, ils passerent outre; & Dieu print /[f. 9v] deux bastons, & il en toucha les deux premiers, qui furent transmuez en bastons, & le cinquiesme s'arresta, ne voulant passer plus outre: Et Dieu luy demanda derechef, Où vas-tu ? Je vois chercher ma vie, Demeure, & tu la trouveras: Il demeura sans passer plus outre, & Dieu luy donna de la viande, & en mangea;[h] Apres avoir faict bonne chere, il retourna avec les autres sauvages,[i] & leur racompta tout ce que dessus.[j]

Il me dit[k] aussi, Qu'une autre fois il y avoit un homme qui avoit quantité de Tabac,[l]* (qui est une herbe dequoy ils prennent la fumee) & que Dieu vint à cest homme, & luy demanda où estoit son petunoir, l'homme print son petunoir, & le donna à Dieu, qui petuna beaucoup;[m] apres avoir bien petuné, Dieu ro*m*pit ledict petunoir en plusieurs pieces, & l'homme luy demanda, Pourquoy as-tu rompu mon petunoir, & tu vois bien que je n'en ay point d'autre? Et Dieu en print un qu'il avoit, & le luy donna, luy disant,[n] en voilà[o] un que je te donne, porte le à ton grand Sagamo, qu'il le garde, & s'il le garde bien, il ne manquera point de chose

a monde:] A1 monde,
b dit,] A2 dict,
c Ils] A1 ils
* RM] *De cinq hommes que les Sauuages croyent auoir veu Dieu.*
d dit,] A2 dict,
e dit] A2 dict
f fois,] A2 fois:
g outre,] A1 outre
h mangea;] A2 mangea:
i sauuages,] A2 Sauuages,
j dessus.] A2 dessus:
k dit] A2 dict
l Tabac,] A2 Tabac
* LM] *D'vn autre homme que les Sauuages croyent auoir parlé à Dieu.*
m beaucoup;] A2 beaucoup:
n disant,] A2 disant:
o voilà] A2 voylà

ancestors say that God had come into the world. He told me that he had not seen Him, but that in the olden days there were five men who went away towards the setting sun and met God, who asked them, "Where are you going?" They said, "We are going to seek our food."[67] God answered them, "You shall find it here." They went on farther without registering what God had said to them. God took a stone and touched two of them, who were transformed into stone. And he said yet again to the other three, "Where are you going?" and they replied in the same way as the first time; and God said to them once more, "Go no farther; you shall find it here." And seeing that nothing happened to them, they went on. And God took /[f. 9v] two sticks and touched the first two of them, who were transformed into sticks; and the fifth halted, not wanting to go any farther. And God asked him yet again, "Where are you going?" "I am going to seek my food." "Stay, and you shall find it." He stayed without going any farther, and God gave him some meat, and he ate of it. After having feasted well, he returned [to be] with the other *sauvages* and recounted to them all the above.

He also told me that another time there was a man who had a good supply of tobacco (which is a plant from which they take the smoke), and that God came to this man and asked him where his tobacco pipe was. The man took his tobacco pipe and gave it to God, who smoked a lot. After having had a good smoke, God broke the said pipe into many pieces, and the man asked him, "Why have you broken my pipe? Surely you see that I have no other." And God took one of the ones that He had and presented it to him, saying, "Here is one of them that I am giving to you. Carry it to your *grand Sagamo*. Let him keep it, and if he guards it

67 *Cercher sa vie*: "*Petere cibum, Quaerere victum*" ("To seek one's life: To seek food"); Nicot, *Thresor* (1606), 665. Cotgrave did not record this idiom, but translated *vie* as to be expected: "life" or "living" (*Dictionarie* [1611], 53:ivv).

quelconque, ny tous ses compagnons: ledit[a] homme print le petunoir, /[f. 10] qu'il donna à son grand Sagamo, lequel tandis qu'il l'eut, les Sauvages ne manquerent de rien du monde: Mais que du depuis ledit[b] Sagamo avoit perdu ce petunoir, qui est l'occasion de la grande famine qu'ils ont quelques fois parmy eux. Je luy demandis s'il croioit tout cela, Il me dit qu'ouy, & que c'estoit verité. Or je croy que voilà pourquoy ils disent que Dieu n'est pas trop bon.

Mais je luy repliquay & luy dis, Que Dieu estoit tout bon, & que sans doubte c'estoit le diable[c] qui s'estoit monstré à ces hommes là, & que s'ils croyoient[d] comme nous en Dieu, ils ne ma*n*queroient[e] de ce qu'ils auroient besoing. Que le Soleil qu'ils voyoient, la Lune & les Estoilles avoient esté créez[f] de ce grand Dieu, qui a faict le ciel & la terre, & n'ont nulle puissance que celle que Dieu leur a donnee, Que nous croyo*ns* en ce grand Dieu, qui par sa bo*n*té nous avoit envoyé son cher fils, lequel conceu du S. Esprit, print chair humaine dans le ventre virginal de la vierge Marie,[g] ayant esté trente trois ans en terre, faisant une infinité de miracles, ressuscitant les morts, guerissant les malades, chassant les diables, illuminant les aveugles, enseignant aux /[f. 10v] ho*m*mes la volonté[h] de Dieu son Pere, pour le servir, honnorer,[i] & adorer, a espandu son sang, & souffert mort & passion pour nous & pour nos pechez, & rachepté le genre humain, estant ensevely,[j] est ressuscité, descendu aux enfers, & monté au ciel, où il est assis à la dextre de Dieu son Pere, Que c'estoit là la croyance[k] de tous les Chrestiens;[l] qui croyent au Pere, au Fils, & au S. Esprit, qui ne sont pourtant trois Dieux, ains un mesme, & un seul Dieu, & une Trinité, en laquelle il n'y a point de plus tost ou d'apres, rien de plus grand ne de plus petit. Que la vierge Marie mere du fils de Dieu, & tous les hommes & femmes qui ont vescu en ce mo*n*de, faisant les comma*n*demens de Dieu, & enduré martyre pour son nom, & qui par la permission de Dieu ont fait des miracles, & sont saincts au ciel en son Paradis, prient

a ledit] A2 ledict
b ledit] A2 ledict
c diable] A2 Diable
d croyoient] A1 croioient
e mãqueroient] A2 manqueroiēt
f créez] A1 crees
g vierge Marie,] A2 Vierge Marie,
h volonté] A2 volonte
i honnorer,] A2 honnorer
j enseuely,] A1 A2 enseuely
k là la croyance] A1 la là croyance A2 la la croyance
l Chrestiens;] A2 Chrestiens,

well, he will not lack for anything whatsoever, nor will any of his companions." The said man took the tobacco pipe /[f. 10], which he gave to his *grand Sagamo*. As long as the latter had it, the *Sauvages* lacked for nothing in the world. But since that time, the said *Sagamo* had lost this pipe, which is the cause of the great famine which they sometimes have among them. I asked him whether he believed all that. He said yes, and that it was truth. Now, I think that is why they say that God is not too good.

But I replied to him and said that God was wholly good, and that without a doubt it was the devil who had shown himself to those men, and that if they believed in God, as we do, they would not lack for whatever they would need; that the sun which they beheld, the moon, and the stars had been created by this great God, who had made heaven and earth; and that these have no power but what God has given them. I said that we believe in this great God who, on account of His goodness, had sent us His dear son, who, being conceived by the Holy Spirit, took on human flesh in the virginal womb of the Virgin Mary; who, having been on earth for thirty-three years, working infinite miracles, raising the dead, healing the sick, casting out devils, giving sight to the blind, and teaching /[f. 10v] men the will of God his Father, in order to serve, honour, and adore Him, poured out his blood and suffered his death and passion for us and for our sins, and redeemed humankind; and who, being buried, rose from the dead, descended into hell, and ascended into heaven, where he is seated on the right hand of God his Father. I told him that this was the belief of all Christians,[68] who believe in the Father, the Son, and the Holy Spirit, yet who are not three Gods but one and the same, a single God, and a Trinity, in which there is no before or after, nothing greater or lesser. I also told him that the Virgin Mary, mother of the son of God, and all the men and women who have lived in this world doing the commandments of God and who have suffered martyrdom for His name's sake, and who by God's permission have wrought miracles and are saints in heaven, in his

68 "... the belief of all Christians": Champlain is citing the Apostles' Creed here, which adds the descent into hell to the other articles of faith found in the Nicene Creed: that Jesus was conceived by the power of the Holy Spirit, born of the Virgin Mary, suffered death and was buried, rose again, and ascended to the right hand of God the Father. Protestants in France requested "the legal co-existence of two religions," pleading, "You ought to recognize us, for we are Christians. We accept the Apostles' and the Nicene Creed; the law cannot touch us." It was the Edict of Nantes (1598) that gave legal sanction to this plea, initiated by the Huguenot Gaspard de Coligny and Michel de l'Hospital, admiral and chancellor of France, respectively (Bersier, *Coligny*, vol. 25, esp. 197, 294–5). Even the Reformer Théodore de Bèze maintained faith in the Creed at the Colloque de Poissy in 1561 (Crété, *Coligny*, 179).

tous pour nous ceste grande Majesté divine, de nous pardonner nos fautes & nos pechez que nous faisons contre sa loy & ses commandemens,[a] Et ainsi par les prieres des saincts au ciel, & par nos prieres que nous faisons à sa divine Majesté, il nous donne ce que nous avons besoing, & le diable n'a nulle puissance sur nous: & ne /[f. 11] nous peut faire de mal, Que s'ils avoient ceste croya*n*ce, qu'ils seroient co*m*me nous, que le diable ne leur pourroit plus faire de mal, & ne manqueroient de ce qu'ils auroient besoing. Alors ledict Sagamo me dit, qu'il advoüoit ce que je disois.

Je luy demandis de quelle ceremonie ils usoient à prier leur Dieu: Il me dist, Qu'ils n'usoient point autrement de ceremonies, sinon qu'un chacun prioit en son cœur co*m*me il vouloit: Voilà[b] pourquoy je croy qu'il n'y a aucune loy parmy eux, ne sçave*n*t que c'est d'adorer & prier Dieu, & vivent la plus part comme bestes brutes, & croy que promptement ils seroient reduicts bons Chrestiens si l'on habitoit leurs terres, ce qu'ils desireroient la plus part.

Ils[c] ont parmy eux quelques Sauvages qu'ils appellent Pilotoua,[d] qui parlent au diable visiblement,* & [il] leur dit ce qu'il faut qu'ils facent, tant pour la guerre,

a commandemens,] A2 commandemens:

b Voilà] A2 Voylà

c part. ¶ Ils] A1 A2 part: Ils

d Pilotoua] (*Champlain's second* Voyage *of 1608 repeats this section [Ch. IV] using the French plural form,* Pillotois, *distancing himself from their beliefs by adding* 'whom they believe': Pillotois, qu'ils croient parler au Diable, leur disant ce qu'il faut …).

* RM] *Quels sauuages parlent au diable.*

paradise,[69] all pray to this great divine Majesty for us, to pardon us for our faults and the sins that we commit against his law and commandments. And thus by the prayers of the saints in heaven and by the prayers that we offer to His divine Majesty, He gives us what we need,[70] and the devil has no power over us and can /[f. 11] do us no harm. And I said that if they held this belief, they would be like us, that the devil could no longer do them any harm and they would not lack for whatever they would need. Then the said *Sagamo* told me that he acknowledged what I was saying.[71]

I asked him what ceremony they used in praying to their God.[72] He told me that they did not make use of ceremonies, other than each person praying in his own heart as he wished. You see why I believe that they have no law among them, nor know what it is to worship and pray to God; and they live, for the most part, like brute beasts. And I believe that they would quickly be brought round to being good Christians if their lands were colonized, which they would desire for the most part.[73]

They have among them some *Sauvages* whom they call *Pilotoua*,[74] who speak to the devil face to face, and [he] tells them what they must do, as much in war

69 "Precious in the sight of the Lord is the death of his saints" (Ps. 116:16)

70 It is notable that Champlain is not explicitly advocating prayer to the saints (as the Catholics of his time were accused of doing by the Protestants), but only joining with them in prayer to God directly.

71 Champlain received a far cooler reception to his efforts at catechism than Jacques Cartier did on his second voyage in 1535, possibly because Cartier allegedly offered little choice: be baptized or go to hell. In response to Cartier's threat, Donnacona brought his entire village, Stadacona at the present Quebec, to be baptized (Bideaux, *Relations*, 160; Cook, *Voyages*, 68).

72 "I asked him ..." This paragraph until the end of the chapter, wherein Native customs are described, is entirely repeated in the 1608 account of the founding of Quebec, yet in a less personal way (Champlain, *Les Voyages, 1613*, 193). In the repeated section, Champlain naturally discards the festive context with Anadabijou. Moreover, his conjecture in the above text that the Natives would desire French settlement is more confident in 1608 when the conditional verb *desireroient* becomes the present indicative *desirent* (they desire).

73 Commenting on the perceived facility of converting the Natives was commonplace in early exploration literature, found equally in the works of Christopher Columbus, Verrazano, and Cartier (Bideaux, *Relations*, 113, 333n247). Champlain is unique in making settlement a priority over conversion, probably because it was a priority ordered by Henri IV.

74 According to Father Pierre Biard, *pilotoua* was a Basque word meaning "sorcerer." It was used to describe the shamans (medicine men) of the Mi'kmaq. The Mi'kmaq word

que pour autres choses, & que s'il leur commandoit qu'ils allassent mettre en execution quelque entreprise, ou tuër un François, ou un autre de leur nation, ils obeiroient aussi tost[a] à son commandement. Aussi ils croient que tous les songes qu'ils font sont veritables,* & de fait,[b] il y en a /[f. 11v] beaucoup qui disent avoir veu & songé[c] choses[d] qui adviennent ou adviendront: Mais pour en parler avec verité, se sont visions du Diable, qui les trompe & seduit: Voilà[e] toute la creance que j'ay peu apprendre d'eux[f] qui est bestiale.*

Tous ces peuples ce sont gens bien proportionnez* de leurs[g] corps, sans aucune difformité, ils sont dispos, & les femmes bien formees, remplies & potelees,[h] de couleur basanee[i] pour la quantité de certaine peinture do*n*t ils se frotent, qui les fait[j] devenir olyvastres. Ils sont habillez de peaux, une partie de leur corps est couverte[k] & l'autre partie descouverte:* Mais l'hyver[l] ils remedient à tout, car ils sont habillez de bo*n*nes fourrures, comme d'Orignac, Loutre, Castors, Ours-marins, Cerfs, & Biches, qu'ils ont en quantité. L'hyver quand les neiges sont grandes, ils font une maniere de raquette* qui est grande deux ou trois fois comme celles de France, qu'ils attachent à leurs pieds, & vont ainsi dans les neiges sans enfoncer, car autrement ils ne pourroient chasser ny aller en beaucoup de lieux.

a obeiroient aussi tost] A2 obeiroyent aussi-tost
* RM] *Sauuages croyent fermement aux songes.*
b fait,] A2 faict,
c songé] A1 songè
d choses] A2 chose
e Voilà] A2 Voylà
f d'eux] A2 d'eux,
* LM] *Humeurs des Sauuages.*
g leurs] A2 leur
h potelees,] A1 potelees
i basanee] A1 basanee;
j fait] A2 faict
k couuerte] A1 A2 couuert
* LM] *Habits des Sauuages.*
l l'hyuer] A2 l'Hyuer
* LM] *Inuention qu'ils ont pour aller sur les neiges.*

as in other affairs; and that if he were to command them to go ahead and execute some enterprise, either to kill a Frenchman or another of their nation, they would immediately obey his command. Moreover, they believe that all the dreams they have are true;[75] and indeed there are /[f. 11^{v}] many of them who say that they have seen and dreamed things that are happening or will happen. But to tell the truth of it, these are visions of the devil who deceives and seduces them. There you have all the beliefs that I could learn from them, which are bestial.

All these peoples are well proportioned in their bodies,* without any deformity. They are agile, and the women are well made, filled out and plump, with a tawny complexion on account of the great quantity of a certain pigment with which they rub themselves and which causes them to assume an olive-like hue.[76] They are dressed in skins; one part of their body is covered and the other part uncovered. But in winter they fix everything, for they are dressed in good furs, like those of moose, otter, beavers, bears, seals,[77] stags, and deer, which they have in abundance. In the winter, when the snows are heavy, they make a kind of racquet that they fasten to their feet,[78] which is two or three times bigger than those of France.[79] And thus they advance in the snow without sinking, for otherwise they could neither hunt nor get into a lot of places.

for their shamans was *autmoins* (*aoutmoins*) (*JR*, 3:117–19). Although *pilotoua* appears to be in the plural form here, Champlain corrects it to *Pillotois, qu'ils croient parler au Diable* (Sorcerers, whom they believe speak to the Devil) when he repeats the same passage in his next publication (*Voyages, 1613*, 193). Beaulieu and Ouellet amended the text in *Des Sauvages* by adding an *s* to *Pilotouas* and the personal pronoun *il* to *dit* (*Des Sauvages*, 110), the latter of which we have adopted.

75 Dream interpretation among the Montagnais is briefly mentioned by Father Paul Le Jeune (*JR*, 5:159–61). The practice was an important feature in the cultures of native North America. For the Iroquoian groups who had similar practices, see *JR*, 10:159–73 and 33:189–97. For an analysis, see Wallace, "Dreams and Wishes of the Soul," 234–48. In later years, on several occasions, Champlain pretended to have dreams about specific situations in order to persuade his Native companions to follow his line of action.

* LM] Disposition of the Sauvages. This marginal note is an error, since it repeats a note from f. 7^{v} that is no longer appropriate. In the previous context, towards the beginning of this chapter, the joyful yet saturnine disposition of the Natives was being discussed.

76 olive-like hue: *olyvastres*, lit. olivish (mod. Fr. olivâtre), recorded as a loan word from the Italian *olivastro* (1522). The suffix -âtre (-ish) is commonly applied to colours in French (*DHLF*, 2:2453; *TLF*, 3:806).

77 *Ours-marins*: conflation of *Ours*, *Loups marins*, as corrected in the 1613 edition

78 Snowshoe

79 The earliest comparison between the Native snowshoe and the European tennis racquet appears in the atlas of Guillaume Le Testu (1556); see Lestringant, *Le Brésil d'André*

Ils ont aussi une forme de mariage, qui est, que quand une fille est en l'aage de 14. ou 15. ans,* elle aura plusieurs ser-/[f. 12]viteurs & amys, & aura compagnie avec tous ceux que bon luy semblera, puis au bout de quelque cinq ou six ans, elle pre*n*dra lequel il luy plaira pour son mary, & vivront ainsi ensemble jusques à la fin de leur vie, si ce n'est qu'apres avoir esté quelque te*m*ps ensemble ils n'ont enfans, l'ho*m*me se pourra desmarier & prendre autre femme, disant, que la sienne ne vaut rien, par ainsi les filles sont plus libres que les femmes: Or depuis[a] qu'elles sont mariees, elles sont chastes, & leurs maris sont la plus part jaloux, lesquels donnent des presents au pere ou parens de la fille qu'ils auront espousee. Voilà[b] la ceremonie & façon qu'ils usent en leurs mariages.

Pour ce qui est de leurs enterremens,* quand un homme ou femme meurt, ils font une fosse, où ils mettent tout le bien qu'ils auront, comme chaudrons, fourrures, haches, arcs & fleches, robbes, & autres choses, & puis ils mettent le corps dedans la fosse, & le couvrent de terre[c] où ils mettent quantité de grosses pieces de bois dessus, & un bois debout qu'ils peignent de rouge par le haut. Ils croyent l'immortalité des ames,* & disent qu'ils /[f. 12v] vont se resjoüir[d] en d'autres pays[e] avec leurs[f] parens & amis quand[g] ils sont morts.

* LM] *Mariage des Sauuages.*
a depuis] A1 despuis
b Voilà] A2 Voylà
* RM] *Comme ils enterrent leurs morts.*
c terre] A2 terre,
* RM] *Sauuages croyent l'immortalité.*
d se resioüir] A2 se resiouïr
e pays] A2 païs
f leurs] A1 leur
g quand] A2 quant

They also have a form of marriage, which is that when a girl is 14 or 15 years old, she will have many suitors /[f. 12] and male friends, and will keep company with all those who suit her. Then, at the end of some five or six years, she will take whichever one she pleases for her husband, and they will live together in this way until the end of their lives unless it happens that after having been together some time, they have no children; then the man will be able to dissolve the marriage and take another wife, saying that his own is not worth anything. Thus the girls are more free than the women. Now, as soon as they are married, they are chaste, and their husbands are, for the most part, jealous, who give presents to the father or relatives of the girl whom they have married. You see the ceremony and the way they practise marriage.[80]

As for their burials, when a man or woman dies, they dig a pit, into which they place all the goods they will possess, such as cauldrons, furs, hatchets, bows and arrows, robes, and other things, and then they place the body in the pit and cover it with earth, upon which they place a great many large pieces of wood and one stake standing upright, which they paint red on the upper part.[81] They believe in the immortality of souls and say that they /[f. 12v] are going to rejoice in other lands with their family and friends when they are dead.

Thevet, 398, note to p. 281. The following year, Thevet reported the description of it that was personally related to him by Jacques Cartier, although it does not figure in the journals attributed to the latter: "*Ils vsent d'vne maniere de raquettes tissues de cordes en façon de crible, de deux piés & demy de long & vn pié de large ... specialement quand ils vont chasser aux bestes sauuages, à fin de n'enfoncer point dans les neiges*" ("They use something like racquets, woven with cords like a sieve, two and a half feet long by one foot wide ... especially when they go hunting for wild animals, so as not to sink in the snow"); Thevet, *Singularitez*, 76:147v, and substantially repeated in later works. The French *jeu de paume* (game of the palm) evolved from handball into a racquet sport in the sixteenth century (McClelland, *Body and Mind*, 78). By the time Champlain was born, every royal residence in France had a court (Babelon, *Henri IV*, 119). By 1596, there were "reputedly 250 *jeux de paumes*" (tennis courts) in Paris, and in 1604 Robert Dallington reported in his *View of France* "that there [were] two tennis courts in France for every church" (McClelland, *Body and Mind*, 56, 123).

80 It is hard to say where Champlain got this information. The customs he described do not fit any of the groups he encountered at Tadoussac. For an analysis of seventeenth-century Montagnais marriage, see Leacock, "Montagnais Social Relations," 190–5. In 1615 Champlain described Huron marriage in similar terms (Champlain, *Voyages et Descovvertvres, 1619*, 89–92).

81 What Champlain described are not Montagnais burial practices, but they bear some relation to those of East Coast Algonquians such as the Mi'kmaq. He was not long enough at Tadoussac to observe marriages or burials and is accepting uncritically what other Frenchmen told him.

Chap. IIII.

Riviere du Saguenay[a] *& son origine.*

Le II. jour de Juin je fus à quelque douze ou quinze lieuës[b] dans le Saguenay,* qui est une belle riviere, & a une profondeur incroyable,[c] car je croy, selon que j'ay entendu deviser d'où elle procede, que c'est d'un lieu fort haut, d'où desce*n*d[d] un torent[e] d'eau d'une grande impetuosité;* mais l'eau qui en procede n'est point capable de faire un tel fleuve comme cestuy-là, qui neantmoins ne tient que depuis cedict torre*nt*[f] d'eau, où est le premier sault, jusques au port de Tadousac, qui est l'entree de ladicte riviere du Saguenay, où il y a quelque 45. ou[g] 50. lieuës,[h] & une bonne lieuë & demye[i] de large au plus, & un quart au plus estroit;[j] qui fait[k] qu'il y a grand courant d'eau. Toute la terre que j'ay veu, ce ne sont que mo*n*taignes[l] de rochers* la pluspart couvertes de bois de sapins, cyprez, & boulles, terre fort mal plaisante, où je n'ay point trouvé une /[f. 13; D] lieuë[m] de terre plaine[n] tant d'un costé que d'autre. Il y a quelques montaignes de sable & isles en ladite riviere[o] qui sont hautes eslevees. En fin ce sont de vrais[p] deserts inhabitables d'animaux,[q] & d'oyseaux; car je vous asseure qu'allant chasser par les lieux qui me sembloient les plus plaisants, je ne trouvay rien qui soit, sino*n*[r] de petits oyseaux qui sont comme

a *Saguenay*] A2 *Saguenay,*
b lieuës] A2 lieues
* LM] *Partement de Tadousac pour aller au Saguenay.*
c incroyable,] A2 incroyable:
d d'où desc'd] A1 d'où il desc'd
e torent] A2 torrent
* LM] *Torrent d'eau.*
f cedict torrēt] A2 cedit torrent
g ou] A2 où
h lieuës,] A2 lieues,
i lieuë & demye] A2 lieue & demie
j estroit;] A2 estroict,
k fait] A2 faict
l mōtaignes] A2 mōtaignes,
* LM] *Terres montagnes de rochers mal plaisantes.*
m lieuë] A2 lieue (*only in the catchword*)
n plaine] A2 plaine,
o isles en ladite riuiere] A2 Isles en ladicte riuiere,
p vrais] A2 vrays
q d'animaux,] A2 d'animaux
r sinō] A2 sinon

Chapter 4

The Riviere du Saguenay *and its source*

On the 11th day of June, I went some twelve or fifteen leagues up the *Saguenay*, which is a beautiful river of incredible depth [map 2]; for according to what I have heard about the place from which it proceeds, I think that it comes from a very high place, from where a torrent of water descends with great vehemence. But the water that proceeds from it is not sufficient to make such a river as that one, which, nevertheless, extends only from this said torrent of water, where the first rapids are, to *Tadousac* harbour, which is the mouth of the said *riviere du Saguenay*, a distance of some 45 or 50 leagues.[82] And it is a good league and a half at its widest point and a quarter of a league at its narrowest, which causes there to be a great flow of water. For as much of the region as I saw, there are only rocky mountains, mostly covered with stands of fir trees, cedars, and birch; a most unpleasant land, where I did not find a /[f. 13] league of level ground, either on the one side or on the other. There are some sand beds[83] and islands in the said river, which are highly elevated. In short, these are real deserts, unfit for animals or birds; for I assure you that in going to hunt throughout the places that seemed the most attractive to me, I found nothing at all except small birds, which are like

82 The torrent of water that comes over the first rapids is at Chicoutimi, 127 km from Tadoussac. The Saguenay River below Chicoutimi is not a river but a fjord, ground by glaciers out of an existing rift valley.

83 Sand beds: *montaignes de sable*, lit. mountains of sand

rossignols, & airo*n*delles, lesquels y viennent[a] en Esté;[b] car autrement je croy qu'il n'y[c] en a point, à cause de l'excessif froid qu'il y fait,[d] ceste riviere venant de devers le Norouest.

Ils me feire*n*t rapport,* qu'aya*n*t passé le premier saut, d'où vie*n*t[e] ce torre*n*t d'eau, ils passent huict autres sauts, & puis vont une journee sans en trouver aucun, puis passent autres dix sauts, & viennent deda*ns*[f] un lac, où ils sont deux jours à rapasser, en chasque jour ils peuvent faire à leur aise quelque douze à quinze lieuës;[g] audit bout du lac, il y a des peuples qui sont cabannez, puis on entre dans trois autres rivieres[h] quelques trois ou quatre journees da*ns* chacune, où au bout desdites rivieres, il y a deux ou trois manieres de lacs, d'où pre*n*d[i] la source du Saguenay, de laquelle source jusques audit[j] port de Tadousac, il /[f. 13ᵛ] y a dix journees de leurs Canos. Au bord desdites rivieres, il y a quantité de cabannes, où il vient d'autres nations du costé du Nort, troquer avec lesdits Montagnez des peaux de castor & martre, avec autres marchandises que donnent les vaisseaux François ausdicts Montaignez.[k] Lesdicts Sauvages du Nort disent, qu'ils voyent une mer qui est salee: Je tiens que si cela est, que c'est quelque gouffre de ceste mer

a lesquels y viennent] A1 lesquelles viennent
b Esté;] A2 esté:
c n'y] A2 ny
d fait,] A2 faict,
* RM] *Rapport que l'on m'a faict du commencement de la riuiere du Saguenay.*
e viēt] A2 vient
f viennent dedās] A2 viennēt dedans
g lieuës;] A2 lieuës:
h riuieres] A2 riuieres,
i prēd] A2 prend
j audit] A2 audict
k Montaignez.] A2 Montagnez.

nightingales[84] and swallows, that come there in the summer; for at other times I think there are none, because of the excessive cold of the weather there, this river coming more or less from the northwest.

They[85] reported to me that having passed the first rapids, where this torrent of water comes from, they pass eight other rapids and then go on for a whole day without finding any more of them; then they pass another ten rapids and come into a lake,[86] which they take two days to cross; each day, they can easily make some twelve to fifteen leagues. At the said end of the lake, there are some people encamped.[87] Then one enters into three other rivers, some three or four days' journey in each;[88] where at the end of the said rivers, there are two or three kinds of lakes, whence the Saguenay takes its source;[89] from which source to the said harbour of *Tadousac* there /[f. 13^{v}] are ten days by their canoes. On the banks of the said rivers there are many lodges, where other nations come from the northern side[90] to trade beaver and marten skins with the said *Montagnez* for other merchandise that the French vessels offer to the said *Montaignez*. The said northern *Sauvages* say that they are in sight of a sea that is salty. I hold that, if this is so, it is some gulf of this sea that discharges inland through the northern part, and

84 The European *rossignol*, nightingale, does not occur in Canada. What Champlain, and Cartier before him, identified as a *rossignol* is the song sparrow (*Melospiza melodia*), still called *rossignol* in French-speaking Canada (Ganong, "The Identity of the Animals and Plants," 237).

85 Meaning the Montagnais he was talking to

86 The route described here is not the difficult upper Saguenay, called *La grande décharge*, but the standard canoe route up the less turbulent but rapid-filled Rivière Chicoutimi to Lac Kénogami and from there via the Belle Rivière to Lac Saint-Jean.

87 The people who lived at "the end" of Lac Saint-Jean (*Lac Piouagamik*), were the Montagnais *Kacouchaki* (Porcupine Nation), reached by Father Jean de Quen in 1647 (*JR*, 31:249–53).

88 The three rivers at the western end of Lac Saint-Jean are the Chamouchouane, Mistassini, and Péribonka. Of these, the main early canoe route to Lac Mistassini was over Rivière Chamouchouane and Rivière Nestaocano to Lac Albanel; see (maps), Laure, "Carte du Domaine," 1631.

89 The lakes referred to are probably Lac Albanel and the two divisions of Lac Mistassini. The source of the Saguenay lies in the drainage systems of the three major south flowing rivers, not the Lac Mistassini system, which drains into James Bay through Rivière Rupert.

90 The "other nations" are various Cree bands that came from the "northern side" of the drainage divide to the rivers that flowed south towards Lac Saint-Jean.

qui desgorge par la partie du Nort dans les terres, & de verité il ne peut estre autre chose. Voilà[a] ce que j'ay apprins de la riviere du Saguenay.

Chap. V.

Partement de Tadousac pour aller au Sault, la description des isles[b] *du Lievre, du Coudre, d'Orleans, & de plusieurs autres isles,*[c] *& de nostre arrivee à Quebec.*

Le Mercredy, dixhuictiesme jour de Juin,[d] nous partismes de Tadousac, pour aller au Sault, nous passasmes pres d'une isle[e] qui s'appelle l'isle au Lievre,[f]* qui peut estre à deux lieuës[g] de la terre de la bande du Nort, & à quelques sept lieuës[h] /[f. 14] dudit Tadousac, & à cinq lieues de la terre du Su. De l'isle[i] au Lievre nous rengeasmes la coste du Nort, environ demie[j] lieüe, jusques à une pointe qui advance à la mer, où il faut prendre plus au large: Ladite pointe est à une lieüe[k] d'une isle qui s'appelle l'isle[l] au Coudre,* qui peut tenir environ deux lieües de large, & de ladite isle à la terre du Nort, il y a une lieüe;[m] ladite isle est quelque peu unie,[n] venant en amoindrissant par les deux bouts;[o] au bout de l'Oüest il y a

a Voilà] A2 Voylà
b *isles*] A2 *Isles*
c *isles,*] A2 *Isles,*
d Iuin,] A2 Iuing,
e isle] A2 Isle
f l'isle au Lieure,] A2 l'Isle du Lieure,
* LM] *Isle au Lieure*
g lieuës] A2 lieues
h quelques sept lieuës] A2 quelque sept lieues
i l'isle] A2 l'Isle
j demie] A2 demye
k lieüe] A2 lieuë
l l'isle] A2 l'Isle
* RM] *Isle au Coudre.*
m lieüe;] A2 lieuë,
n vnie,] A1 vnie;
o bouts;] A2 bouts,

in truth it cannot be anything else.[91] This is what I learned about the *riviere du Saguenay.*

Chapter 5

Departure from Tadousac *to go to the rapids; description of the* isles du Lievre, du Coudre, *and* d'Orleans *and a number of other islands; and of our arrival at Quebec*

On Wednesday, the eighteenth day of June, we set out from the said *Tadousac* to go to the rapids[92] [map 3]. We passed near an island that is called the *isle au Lievre*, which may be about two leagues from the land on the northern shore and some seven leagues /[f. 14] from *Tadousac* and five leagues from the south shore.[93] From the *isle au Lievre* we coasted along[94] the north shore for about half a league, as far as a point that juts out into the sea, where one must keep a wider berth. The said point is a league from an island that is called the *isle au Coudre*, which may extend about two leagues in width; and from the said island to the north shore there is a league.[95] The said island is somewhat level, getting smaller at both ends.

91 It is evident that Champlain was being told of the existence of Hudson Bay, whose southern end, James Bay, was in the territory of the Cree. He reasoned correctly that it is an inland "gulf" of the Atlantic Ocean. This was the first solid evidence for the existence of Hudson Bay, "discovered" seven years later by Henry Hudson.

92 The *grand sault*, called *Sault Sainct Louis* by Champlain in 1608 and *rapides de Lachine* after about 1670, stretched along the south side of Montreal Island for about 9 km.

93 Île aux Lièvres (Hare Island) is 7.2 km from the north shore of the St. Lawrence River, 28 km from Tadoussac and 11.2 km from the south shore. The second expedition of Jacques Cartier, in 1535–36, named it for the many hares they found there (*grand nombre de liepvres*) (Bideaux, *Relations*, 182; Cook, *Voyages*, 87). Please note that through the entire journey to and from the Lachine Rapids, no matter in what direction the St. Lawrence happened to flow, Champlain called the left bank (shore) of the river the "north" side and the right bank the "south side." He seems not to have carried a compass.

94 Champlain and Cartier both use *rengeasmes* and *rangames* (1 p. pl. of *ranger*: *longer* in mod. Fr.) to describe a ship's passage hugging the coast, e.g., to coast along or skirt; to sail close to; to sail or coast by.

95 There is some confusion in the previous two sentences. Champlain is saying that there is a point that juts out into the St. Lawrence that is half a league along the north shore from Île aux Lièvres and one league from Île aux Coudres. This makes no sense because Île aux Coudres is 64 km from Île aux Lièvres. The nearest point to Île aux Coudres is Cap aux Oies, 9.1 km from the eastern tip of the island, unless the little cape at Saint-Joseph-de-la-Rive was meant, which is 3 km away. Île aux Coudres is 10.5 km long, 4.6 km wide, and 2.7 km from the north shore. Laverdière and Biggar surmised that "half

des prairies & pointes de rochers qui advancent quelque peu dans la riviere; ladite isle est quelque peu aggreable pour les bois qui l'environnent, il y a force ardoise, & la terre quelque peu graveleuse; au bout de laquelle il y a un rocher qui advance à la mer environ demie lieue.[a] Nous passasmes au Nort de ladite isle, distante de l'isle au Lievre de 12. lieues.[b] Le Jeudy ensuivant nous en partismes, & vinsmes mouiller l'ancre à une anse dangereuse du costé du Nort,* où il y a quelques prairies, & une petite riviere, où les Sauvages cabannent quelque-fois. Cedit jour rengeant tousjours ladicte[c] coste du Nort, jusques à un lieu où nous relachasmes* pour les vents qui nous estoie*n*t /[f. 14[v]] contraires, où il y avoit force rochers & lieux fort dangereux, nous feusmes trois jours en attendant le beau temps: Toute ceste coste n'est que montaignes[d] tant du costé du Su, que du costé du Nort, la plus part ressemblant à celle du Saguenay.

Le Dimanche vingt-deuxiesme jour dudict mois nous en partismes pour aller à l'isle d'Orleans,[e] où il y a qua*n*tité d'isles* à la bande du Su, lesquelles sont basses & couvertes d'arbres, semblans estre fort aggreables, contenans,[f] (selon que j'ay peu juger) les unes deux lieues,[g] & une lieue,[h] & autre demie: Autour de ces isles ce ne sont que rochers & basses, fort dangereux à passer, & sont esloignez

a lieue.] A1 lieue, A2 lieuë,
b lieues.] A2 lieuës.
* RM] *Anse dangereuse.*
c ladicte] A1 ladite
* RM] *Coste dangereuse.*
d montaignes] A2 montagnes
e l'isle d'Orleans,] A2 l'Isle d'Orleans,
* LM] *Isles belles & dangereuses.*
f contenans,] A2 contenans
g lieues,] A2 lieuës,
h lieue,] A2 lieuë,

At the western end there are meadows[96] and rocky points that jut out somewhat into the river. The said island is somewhat pleasant because of the woods that surround it. There is much slate, and the land is somewhat gravelly, at the end of which there is a rock that juts out about half a league into the sea.[97] We passed to the north of the said island at a distance of 12 leagues from the *isle au Lievre*.[98] On the following Thursday, we set out from there and came to drop anchor at a dangerous cove on the north coast, where there are some meadows and a little river where the *Sauvages* sometimes set up camp. On this said day, still skirting the said north shore as far as a place where we put in, on account of winds that were /[f. 14^{v}] against us, where there were many rocks and very dangerous places, we stayed three days waiting for fair weather.[99] This entire coast is nothing but mountains on both the south and the north shores, for the most part resembling that of the Saguenay.

On Sunday, the twenty-second day of the said month, we set out from there to go to the *isle d'Orleans*, where there are many islands along the southern stretch that are low and covered with trees. They seem to be very pleasant, some comprising (as far as I have been able to judge) two leagues, and one league, and another half a league.[100] There are only rocks and shallows around these islands, very

a league" is likely a misprint (Beaulieu and Ouellet, *Champlain*, 127n2), which seems to be a reasonable explanation.

96 The meadows (*prairies*) are probably grasslands in the area called Petit anse l'Ilet.

97 The rocky Pointe du Bout d'en Bas is at the northeastern end of the island. It is 1.1 km long. At the southwestern end is La Butte à Cailla, about 1 km in length.

98 Île aux Coudres (Hazelbush Island) is 64 km from Île aux Lièvres. Cartier's second expedition, 1535–36, named it (*nous la nommasmes*) for the many hazel bushes (*Corylus cornuta*) to be found there (Bideaux, *Relations*, 136, 359n164; Cook, *Voyages*, 49; Biggar, *Jacques Cartier*, 119, 234).

99 The two places where the ship dropped anchor along the rugged and forbidding north shore between Île aux Coudre and Île d'Orléans are impossible to identify from Champlain's meagre description. Judging from sixteenth- and seventeenth-century maps that depict anchorages west of Baie Saint-Paul the most likely places are Ruisseau de la Martine at Maillard, Petite Rivière Saint-François, Havre à la Gribane, and Cap Tourmente; see (maps), Franquelin, "Carte du grande fleuve," 1685; Franquelin, "Partie de L'Amerique," 1699; Deshayes, *La grande riviere*, 1715; Bellin, *Carte du Cours*, 1761.

100 This passage is not clear. Champlain seems to be saying that the islands in the St. Lawrence between Île d'Orléans and the south shore vary in length between one half and two leagues. The westernmost island of the group is Île Madame, and the one farthest east is Île aux Oies. He is correct in saying that dangerous shoals surround them. Cartier first named Île d'Orléans, *l'isle de Bascuz* (Bacchus Island) on account of the abundance of vines he saw there. On his return journey he gave it its present name (Biggar, *Jacques Cartier*, 126, 232).

quelques deux lieues[a] de la gra*n*d' terre du Su: Et de là vinsmes renger à l'isle d'Orleans* du costé du Su: Elle est à une lieue[b] de la terre du Nort, fort plaisante & unie, contena*n*t de long huict lieues:[c] Le costé de la terre du Su est terre basse, quelques deux lieues[d] avant en terre; lesdites terres commencent à estre basses à l'endroit de ladite isle, qui prend estre à deux lieues[e] de la terre du Su: à passer du costé du Nort, il y faict fort dangereux pour les bancs de sable, rochers qui sont entre ladicte isle & la grand' terre, /[f. 15] & asseche presque toute de basse mer, au bout de ladicte isle je vis un torrent d'eau* qui desbordoit de dessus une grande mo*n*taigne de ladicte riviere de Canadas, & dessus ladite[f] montaigne est terre unie & plaisante à voir, bien que dedans lesdites terres l'on voit de hautes montaignes[g] qui peuvent estre à[h] quelques 20. ou 25. lieuës[i] dans les terres,* qui sont proches du premier sault du Saguenay.

Nous[j] vinsmes moüiller l'ancre à[k] Quebec* qui est un destroit de ladicte riviere de Canadas, qui a quelque 300. pas de large: il y a à ce destroit du costé du Nort

a lieues] A2 lieuës
* LM] *Isle d'Orleans.*
b lieue] A2 lieuë
c lieues:] A2 lieuës:
d lieues] A2 lieües
e lieues] A2 lieües
* RM] *Torrent d'eau.*
f ladite] A2 ladicte
g montaignes] A2 montagnes
h à] A2 a
i lieuës] A2 lieües
* RM] *Montaignes que l'on void estre loing.*
j Saguenay. ¶ Nous] A1 A2 Saguenay: Nous
k à] A1 a
* RM] *Description de Quebec.*

dangerous for passing, which are removed by some two leagues from the mainland to the south.[101] And from there we continued skirting the *isle d'Orleans* on the south side. It is very pleasant and level, measuring eight leagues in length, and is situated a league from the land to the north.[102] The southern shore is lowland for some two leagues inland. The said lands begin to be low in the area of the said island, which is situated[103] at two leagues from the land on the south.[104] It makes it very dangerous to cross from the north side on account of the sandbanks and rocks that lie between the said island and the mainland[105] /[f. 15], and at low tide it almost completely dries up. At the end of the said island I saw a torrent of water that was pouring out from the top of a great mountain of the said *riviere de Canadas*,[106] and on the top of the said mountain the ground is level and pleasant to look at, although in the interior of the said lands one sees high mountains, which may be some 20 or 25 leagues inland, that are near the first rapids of the Saguenay.[107]

We went on to cast anchor at Quebec, which is a strait of the said *riviere de Canadas* and has a width of some 300 paces.[108] At this strait, on the north side

101 This group of islands with related shoals is 5 to 6 km from the south shore.

102 Île d'Orléans is 34 km long and 2 to 3 km from the north shore.

103 *prend estre*: is situated. Both the first Champlain Society edition (1922) and that of Beaulieu and Ouellet (1993) assume that *prend* is an error for *peut* before the infinitive *estre*. Yet *estre* could also be a masculine noun signifying state or situation (DHLF, 1:1333; HU, 3:734). In 1606 Jean Nicot translated an example into Latin as *se maintenir en mesme estre: Tenere statum suam* (to hold one's place or position); Nicot, *Thresor*, 266a. The *prend* above still suggests a kind of animistic decision on the part of this island to take up its position.

104 The island is about 3 km from the south shore.

105 Champlain is describing the northern passage along Île d'Orléans.

106 Montmorency Falls was named by Champlain in honour of Charles de Montmorency, admiral of France, to whom he dedicated his book *Des Sauvages* (Champlain, *Les Voyages, 1632*, 132). The falls are about 80 m high and 45 m wide.

107 The mountains are the Laurentides, but Champlain overestimated their distance from the St. Lawrence River. The "first rapids of the Saguenay" are at Chicoutimi.

108 Quebec (*kepek*) is an Algonquian/Montagnais word meaning "where the river narrows," which is why Champlain wrote that Quebec is the name of a "strait" (Lemoine, *Dictionnaire*, 281). According to Bayfield's survey in 1827, the distance across the river at its narrowest part at Quebec was about 1.26 km; see (maps), Bayfield, *Plan of Quebec*, 1829. Since the French *pas* is 2.5 *pieds*, Champlain's distance of 300 *pas* is about 750 *pieds* (228 m), well short of the true distance of about 4,130 *pieds* (1652 *pas*). The bar scale on Champlain's map ("Quebec") of the area around Quebec is hopelessly in error and cannot be used to check his estimated distance across the river (Champlain, *Les Voyages, 1613*, map facing 176).

une montaigne[a] assez haute qui va en abbaissant des deux costez;[b] tout le reste est pays uny & beau, où il y a de bonnes terres pleines d'arbres, comme chesnes, cyprez,[c] boulles, sapins, & trembles, & autres arbres fruictiers, sauvages, & vignes: qui fait[d] qu'à mon opinion, si elles estoient cultivees,[e] elles seroient bonnes comme les nostres. Il y a le long de la coste dudit Quebec des diamans dans des rochers d'ardoise,* qui sont meilleurs que ceux d'Alançon. Dudict Quebec jusques à l'isle au Coudre, il y a 29. lieuës.[f] /[f. 15v]

Chap. VI.

De la pointe Saincte Croix, de la riviere[g] de Batiscan, des Rivieres, rochers, isles, terres, arbres, fruicts, vignes, & beaux pays, qui sont depuis Quebec jusques aux trois Rivieres.

Le Lundy 23. dudict mois nous partismes de Quebec, où la riviere[h] commence à s'eslargir quelques-fois* d'une lieuë,[i] puis de lieüe & demye ou deux lieües au plus: Le pays va de plus en plus en embellissa*n*t, ce sont toutes terres basses, sans rochers, que fort peu: Le costé du Nort est remply de rochers & bancs de sable, il faut prendre celuy du Su, comme d'une demie lieüe[j] de terre. Il y a quelques petites rivieres qui ne sont point navigables, si ce n'est pour les Canos des Sauvages, ausquelles il y a quantité de saults. Nous vinsmes mouiller l'ancre jusques à saincte Croix,* distante de Quebec de 15. lieües,[k] c'est une pointe basse qui va en haussant des deux costez: Le pays est beau & uny, & les terres meilleures qu'en lieu que j'eusse veu, avec qua*n*tité de bois: mais /[f. 16] fort peu de sapins & cyprez: il s'y trouve en quantité, des vignes, poires, noysettes, serizes, groizelles, rouges & vertes, & de certaines petites racines de la grosseur d'une petite nois, ressemblant

a montaigne] A2 montagne
b costez;] A1 A2 costez,
c cyprez,] A2 cypres,
d fait] A2 faict
e cultiuees,] A2 cultiuees
* RM] *Des diamans que l'on trouue à Quebec.*
f lieuës.] A2 lieües.
g *riuiere*] A2 *Riuiere*
h riuiere] A2 Riuiere
* LM] *Du païs qui est entre Quebec & Saincte Croix.*
i lieuë,] A2 lieüe
j lieüe] A2 lieue
* LM] *Pointe de Saincte Croix.*
k lieües,] A2 lieues,

there is a rather high mountain that slopes down on both sides.[109] All the rest is a level and beautiful country, where there are some good lands full of trees, such as oak, cedar, birch, fir, and aspen, and other wild fruit-bearing trees and vines, which make me think that if these lands were cultivated, they would be as good as ours. Along the shore of the said Quebec there are some diamonds in the slate rocks which are better than those of *Alançon*.[110] From the said Quebec to the *isle au Coudre* there are 29 leagues.[111]

/[15v] Chapter 6

Of the pointe Saincte Croix*; of the* riviere de Batiscan*; of the rivers, rocks, islands, soils, trees, fruits, vines, and fine lands that stretch from Quebec up to* trois Rivieres

On Monday, the 23rd of the said month, we set out from Quebec, where the river begins to broaden, sometimes by a league, then by a league and a half or two leagues at the most. The country becomes more and more beautiful. They are all lowlands without rocks, except for a very few. The north side is replete with rocks and sandbanks; it is necessary to take that of the south about half a league from the land. There are some small rivers, in which there are many rapids that are not navigable except with the canoes of the *Sauvages*. We came to cast anchor at *saincte Croix*, at a distance of 15 leagues from Quebec;[112] it is a low point that gradually rises up on both sides. The land is fine and level, and the soil better than in any place that I had seen, with extensive woods, but /[f. 16] very few fir trees and cedars. In these parts are found a great many vines, pears, hazelnuts, cherries, red and green currants, and certain small roots, the size of a small nut, tasting like

109 Now called Cap aux Diamants. See also next footnote.

110 Diamonds of Alençon are quartz crystals from the French village of Herté near Alençon, famous for its lace and quartz jewellery. Rock crystal from the slate deposits near Quebec were first collected by Cartier on his third voyage in 1541 (Beaulieu and Ouellet, *Champlain*, 127n7; Biggar, *Jacques Cartier*, 255). In recounting aspects of Cartier's voyages, André Thevet commented that, on Cartier's return to France, these worthless quartz crystals were given the name "Canada Diamonds" (Schlesinger and Stabler, *André Thevet's North America*, 51).

111 The distance from Quebec to Île aux Coudres is about 92 km.

112 The point Champlain called *Saincte Croix* is now called Pointe au Platon, 57 km from Quebec. On (maps) Franquelin, "Partie De L'Amerique," 1699, the point is named *Platon S. Croix*.

au goust comme treffes, qui sont tres-bonnes roties & bouillies:[a] Toute ceste terre est noire, sans aucuns rochers, sinon qu'il y a grande quantité d'ardoise: elle est fort tendre, & si elle estoit bien cultivee elle seroit de bon rapport: Du costé du Nort il y a une riviere qui s'appelle Batiscan,* qui va fort ava*n*t en terre, par où quelques-fois[b] les Algoumequins viennent: & une autre du mesme costé à trois lieües dudit saincte Croix sur le chemin de Quebec, qui est celle où fut Jacques Quartier au co*m*mencement[c] de la descouverture qu'il en fit, & ne passa point plus outre: Ladite riviere est plaisante, & va assez avant dans les terres. Tout ce costé du Nort est fort uny & agreable.

¶ Le Mercredy 24. jour dudit mois[d] nous partismes dudict saincte Croix, où nous retardasmes une maree & demye, pour le lendemain pouvoir passer de jour, à cause de la grande quantité de rochers qui sont au travers de ladicte riviere,* (chose /[f. 16v] estrange à voir) qui asseche presque toute de basse mer: Mais à demy flot, l'on peut commencer à passer librement, toutesfois il faut y prendre

a roties & bouillies:] A2 rosties & bouillies;
* RM] *Riuiere qui s'appelle Batiscan.*
b quelques-fois] A2 quelques fois
c cõmencement] A2 commencemẽt
d mois] A2 mois,
* RM] *Rochers dangereux.*

truffles,[113] which are very good roasted or boiled.[114] All this soil is black, without any rocks, except that there is a good deal of slate. The soil is very soft and if it were well cultivated would have a good yield. On the north side there is a river called Batiscan[115] that goes far inland, down which the *Algoumequins* sometimes come; and there is another river on the same side at three leagues from the said *saincte Croix* on the way from Quebec, which is the one that *Jacques Quartier*[116] was on at the beginning of the discovery that he made of it, and he did not proceed farther.[117] The said river is attractive and goes rather far inland. This whole north shore is very level and pleasant.

On Wednesday, the 24th day of the said month,[118] we set out from the said *saincte Croix*, where we delayed for a tide and a half in order to be able to cross by daylight the next day, on account of the great number of rocks that lie across the said river,* which (something /[f. 16v] strange to see) dries up almost entirely at low tide. But at half flood, one can begin to cross freely; yet great caution must be

113 Truffles: *treffes*. The term *treffes* to signify mushrooms that grow below ground has been more commonly written *truffes* (*tubera cibarium*) in France since the fourteenth century. The written form *treuffles noir* was also found in Middle French, however (Wartburg, *Französisches Etymologisches Wörtbuch* [FEW], 13, pt. 2, 384). Champlain's contemporary from Saintonge, Agrippa d'Aubigné (1552–1630), consistently wrote *treffles* (with the *e* vowel, but also an *l*) for the same word, as when he wrote of a relic folded into a napkin, *comme on enveloppe les treffles en Xainctonges* (as one wraps truffles in Saintonge); see D'Aubigné, *Faeneste*, 4:11, p. 604; HU, 7:324. Today *trèfle* (*trifolium*) refers to clover in Saintonge (Musset, *Glossaire*, 5:183, 208). There are no truffles, a type of fungus, in Canada. The "small roots, the size of a small nut," that tasted like *treffes* are undoubtedly *Apios americana*, the wild groundnut.

114 The vine observed by Champlain along the St. Lawrence is the common *Vitis labrusca*. There are no wild pears in eastern Canada. It is generally supposed that a species of *Amelanchier* was meant. The wild currants were probably a species of *Ribes*.

115 The Rivière Batiscan is still so called. It is 40 km upstream from Pointe au Platon.

116 The orthography *Quartier* is not peculiar to Champlain but was used in contemporary documents concerning Cartier. Moreover, Ramusio uses it frequently in his Italian edition of Cartier's first two voyages, followed by Thevet and Lescarbot (Lestringant, *Singularitez*, 278 ff.; Bideaux, *Relations*, 291n291).

117 Rivière Jacques Cartier is 7 km downstream from Pointe au Platon on the north side. Champlain was of course mistaken when he wrote that Cartier went no farther. It is evident that he had not read Cartier's accounts at this time. He corrected his error when he returned to the St. Lawrence in 1608 and searched for Cartier's winter quarters (Champlain, *Les Voyages, 1613*, 184–91).

118 24 June 1603 was a Tuesday.

* RM] *Dangerous rocks*

bien garde avec la sonde à la main: La mer y croist pres de 3. brasses & demie: plus nous allions en avant & plus le pays est beau:* nous fusmes à quelques 5. lieues[a] & demye[b] mouiller l'ancre[c] à la bande du Nort.

Le Mercredy ensuyvant[d] nous partismes de cedict lieu, qui est pays plus plat que celuy de devant, plein de grande quantité d'arbres* comme à saincte Croix: Nous passames pres d'une petite isle,[e] qui estoit remplie de vignes, & vinsmes mouiller l'ancre,[f] à la bande du Su, pres d'un petit costau: mais esta*nt*[g] dessus, ce sont terres unies: il y a une autre petite isle[h] à 3. lieües[i] de saincte Croix, proche de la terre du Su. Nous partismes le Jeudy ensuyvant dudict costau, & passasmes pres d'une petite isle,[j]* qui est proche de la bande du Nort, où je fus, à quelques six petites rivieres, dont il y en a deux qui peuve*nt* porter bateaux[k] assez avant,* & une autre qui a quelque 300. pas de large. A son entree[l] il y a quelques isles,[m] elle va fort avant dans terre, est la plus creuse de toutes les autres, lesquelles /[f. 17; E] sont fort plaisantes à voir, les terres esta*nt*[n] pleines d'arbres qui ressemblent à des noyers,* & en ont la mesme odeur,[o] mais je n'y ay point veu de fruict, ce qui me met en doubte;[p] Les Sauvages m'ont dict, qu'ils porte*nt* leurs fruits[q] comme[r] les nostres.

* LM] *Beau pays.*
a 5. lieues] A1 5. lieüe
b demye] A2 demie
c l'ancre] A2 l'Ancre
d ensuyuant] A2 ensuiuant
* LM] *Isle remplie de vignes.*
e isle,] A2 Isle,
f l'ancre,] A2 l'Ancre,
g estãt] A2 estant
h isle] A2 Isle
i lieües] A2 lieues
j isle,] A2 Isle,
* LM] *Autre petite isle.*
k bateaux] A2 batteaux
* LM] *De deux riuieres auec d'autres petites.*
l large. A son entree] A2 large: à son entree
m isles,] A2 Isles,
n estãt] A2 estants
* RM] *Arbres semblants à noyers.*
o odeur,] A2 odeur;
p doubte;] A2 doubte:
q qu'ils portẽt leurs fruits] A1 qu'il porte son fruict
r comme] A2 cõme

taken there, with sounding line in hand.[119] The tide rises almost 3 and half *brasses* here.[120] The farther we proceeded, the more beautiful the country became. We went to cast anchor on the north shore at some 5 and a half leagues.[121]

On the following Wednesday, we left this said place, which is a flatter land than that preceding and as heavily wooded as at *saincte Croix*. We passed close by a little island, which was covered with vines, and came to anchor on the southern stretch near a small hillside, but once above, they are level lands.[122] There is another small island at 3 leagues from *saincte Croix*, near the southern shore.[123] We set out the following Thursday from the said hillside and passed close to a small island near the north shore,[124] where I was in the neighbourhood of some six small rivers, of which there are two that can carry boats far inland, and another that is some 300 paces wide.[125] At its mouth, there are some islands; it extends very far inland and is deeper than all the other rivers, which /[f. 17] are very pleasing to see, the land being covered with trees that resemble walnut trees and have the same smell as they do. But I did not see any fruit, which makes me doubtful. The *Sauvages* told me that they bear fruit like ours.

119 Champlain is commenting on the Rapides Richelieu, which are dangerous at low tide. These rapids begin just above Portneuf, opposite Pointe au Platon, and extend on both sides of the river upstream for about 10 km. Low tide exposes extensive shoals, studded with rocks and boulders, on both sides of the river, the worst part being from Île Richelieu to Portneuf (Canadian Hydrographic Service [CHS], 166–9).

120 The French *brasse* (*toise marine*) is 1.95 m, while the English fathom is 1.83 m. It is the same as the *toise*, used for horizontal measures, which is also 1.95 m. Currently the difference between high and low tide in June at Pointe au Platon is about 4.7 m (3 *brasses*) (CHS, 10).

121 Five and a half leagues would possibly take him near Deschambault on the north shore.

122 Perhaps the small island that used to be between Lotbinière and Vieille-Église on the south shore. The "hillside" could be the shore west of Lotbinière, about 24 m high.

123 Probably Île Richelieu, about 5.5 km southwest of Pointe au Platon

124 The small Île de la Batture on the eastern side of the mouth of the Rivière Sainte-Anne is a better bet than the sizable Île du Large.

125 The two large rivers are Rivière Sainte-Anne and Rivière Batiscan. The former, at its mouth, is about 230 m wide.

Passant plus outre, nous rencontrasmes une isle,[a] qui s'appelle sainct Eloy,* & une autre petite isle,[b] laquelle est tout proche de la terre du Nort; nous[c] passasmes entre ladite isle[d] & ladite terre du Nort, où il y a de l'un à l'autre quelque cent cinquante pas. De[e] ladite isle[f] jusques à la ba*n*de[g] du Su [à] une lieuë & demie passasmes proche d'une riviere, où[h] peuvent aller les Canos.* Toute ceste coste du Nort est assez bonne, l'on y peut aller librement, neantmoins la sonde à la main, pour eviter certaines pointes. Toute ceste coste que nous rengeasmes est sable mouva*n*t,* mais entrant[i] quelque peu dans les bois, la terre est bonne.

Le[j] Vendredy ensuivant nous partismes de ceste isle,[k] costoyant tousjours la bande du Nort tout proche terre, qui est basse, & pleine de tous bons arbres[l] & en quantité jusques aux trois Rivieres,[m] où il comme*n*ce d'y avoir temperature de te*m*ps, quelque peu dissemblable à celuy de sain-/[f. 17v]cte Croix,[n] d'autant que les arbres y sont plus advancez qu'en aucun lieu que j'eusse encores veu. Des trois rivieres* jusques à saincte Croix il y a quinze lieuës.[o] En ceste riviere il y a six isles,[p] trois desquelles sont fort petites, & les autres de quelque cinq à six cens pas de long, fort plaisantes & fertilles, pour le peu qu'elles contiennent. Il y en a une au milieu de ladite riviere* qui regarde le passage de celle de Canadas, & commande aux autres esloignees de la terre, tant d'un costé que d'autre de quatre à cinq cens pas: Elle est eslevee du costé du Su, & va quelque peu en baissant du

a isle,] A2 Isle,
* RM] *Isle sainct Eloy.*
b petite isle,] A1 petit isle, A2 petit Isle,
c Nort; nous] A1 A2 Nort, nous
d isle] A2 Isle
e pas. De] A1 A2 pas, De
f isle] A2 Isle
g bāde] A2 bande
h où] A2 ou
* RM] *D'vne autre petite riuiere.*
* RM] *Coste sablonneuse.*
i mouuāt, mais entrant] A2 mouuant, mais entrāt
j bonne. ¶ Le] A1 A2 bonne: Le
k isle,] A2 Isle,
l arbres] A2 arbres,
m Riuieres,] A2 riuieres,
n sain-/[f. 17v]cte Croix,] A2 S. [f. 17v] Croix,
* LM] *Des trois Riuieres.*
o lieuës.] A2 lieues.
p isles,] A2 Isles,
* LM] *D'vne isle qui est propre à habiter.*

Passing farther on, we encountered an island that is called *sainct Eloy* and another small island that is quite close to the land on the north.[126] We passed between the said island and the said land on the north, where there are, from one island to the other, some one hundred and fifty paces. From the said island to the southern stretch of land [at] a league and a half, we passed near a river where canoes can go.[127*] This entire northern shore is rather good; one can freely proceed along it, but with sounding line in hand in order to avoid certain points. This entire shore that we skirted around is shifting sand; but upon entering somewhat into the woods, the soil is good.

On the following Friday[128] we set out from this island, still coasting the north shore quite near the land, which is low and filled with all sorts of good trees, and in great number as far as *trois Rivieres*, where the temperature of the air begins to be somewhat unlike that of *saincte* /[f. 17v] *Croix*, inasmuch as the trees are more mature there than in any place I had yet seen.[129] From *trois rivieres* up to *saincte Croix* there are fifteen leagues.[130] In this river there are six islands, three of which are very small, and the others of some five to six hundred paces in length, very pleasant and fertile for their little size. There is one of them in the middle of the said river that faces opposite the channel of the *Canadas* river[131] and dominates the others, removed by four or five hundred paces from the land on both sides.

126 Île de Saint-Eloi is about 8 km upstream from the mouth of Rivière Batiscan.

127 The distance from this island to the south shore is about 3 km. The river he passed is probably Rivière Champlain. We have accepted the emendation of Beaulieu and Ouellet here, the addition of *à*, necessary to link the measurement of a league and a half to the rest of the sentence (Beaulieu and Ouellet, *Champlain*, 132).

* RM] *About another little river*

128 Friday, 27 June

129 Champlain, like most people well into the nineteenth century, judged soil quality and climate from the vegetation at a locality.

130 The distance from Sainte-Croix (Pointe au Platon) to Trois-Rivières is about 65 km.

131 The *Canadas* river is the St. Lawrence.

costé du Nort: Ce seroit à mo*n*[a] jugement un lieu propre pour habiter, & pourroit-on le fortifier pro*m*ptement,[b] car sa situatio*n* est forte de soy, & proche d'un grand lac qui n'en est qu'à quelque quatre lieuës,[c] lequel presque joinct la riviere du Saguenay, selon le rapport des Sauvages qui vont pres de cent lieuës[d] au Nort, & passent nombre de saults, puis vont par terre quelque cinq ou six lieues, & entrent dedans un lac, d'où ledit[e] Saguenay prend la meilleure part de sa source, & lesdits Sauvages viennent dudit lac à Tadousac.

Aussi que l'habitation des trois /[f. 18] Rivieres seroit un bien pour la liberté de quelques natio*n*s qui n'osent venir par là,* à cause desdits Irocois, leurs ennemis, qui tiennent toute ladite riviere de Canadas bordee: mais estant habité, on pourroit rendre lesdits Irocois & autres Sauvages amis, où à tout le moins sous la faveur de ladite habituation, lesdits Sauvages viendroient librement sans crainte & da*n*ger:[f] d'autant que ledit lieu des trois rivieres est un passage. Toute la terre que je veis à la terre du Nort est sablonneuse. Nous entrasmes environ une lieue dans ladite riviere, & ne peusmes passer plus outre, à cause du grand courant d'eau:* Avec un esquif nous feusmes pour voir plus ava*n*t, mais nous ne feismes pas plus d'une lieuë,[g] que nous rencontrasmes un sault d'eau fort estroit,* comme de douze

a mō] A2 mon
b prōptement,] A2 promptemēt,
c lieuës,] A2 lieues,
d lieuës] A2 lieues
e ledit] A2 ledict
* RM] *Le bien que pourroit apporter l'habituation des trois Riuieres.*
f dāger:] A2 danger:
* RM] *Grand cours d'eau.*
g lieuë,] A2 lieue,
* RM] *D'vn petit sault d'eau.*

It is high on the south shore, becoming a little lower on the north side.[132] This would be, in my judgment, a place suitable for habitation; and one could fortify it quickly, for its situation is strong in itself and near a great lake that is only some four leagues away.[133] It almost joins the *riviere du Saguenay*, according to the report of the *Sauvages* who travel nearly a hundred leagues to the north and pass by a number of rapids, then go by land some five or six leagues and enter a lake, from where the said Saguenay takes the better part of its source. And the said *Sauvages* come from the said lake to *Tadousac*.[134]

Moreover, the settlement of *trois* /[f. 18] *Rivieres* would be a benefit for the freedom of some nations, who dare not come through there because of the said *Irocois*, their enemies, who keep watch all along the said *riviere de Canadas*; but being inhabited, one could make friends with the said *Irocois* and other *Sauvages*; or, at the very least, under protection of the said settlement, the said *Sauvages* would come freely without fear or danger, forasmuch as the said *trois rivieres* is a place of passage. All the land that I saw on the north shore is sandy. We entered the said river for about a league and could proceed no farther on account of the strong current. With a skiff we went to look farther ahead, but we did not make more than a league before we encountered a very narrow waterfall of about twelve paces, which was the reason why we could not go on. All the land that I saw on

132 Of the six islands seen by Champlain in the mouth of the *Trois Rivières* (now Rivière Saint-Maurice), the three largest are Île Saint-Christophe, Île La Potherie, and Île Saint-Quentin (formerly Île aux Cochons). The river got its name because it was split into three mouths by the two largest islands, La Potherie and Saint-Quentin. The island referred to by Champlain in the middle of the river is Île Saint-Quentin, about 450 m from the mainland. Trois-Rivières was already so called when François Gravé Du Pont and others were there before 1600. Its first appearance on a map was on (maps), Levasseur, ["Carte de l'océan"], 1601. The name given to the river by Cartier was *la ripvière de Fouez* (Biggar, *Jacques Cartier*, 173). The Algonquin name of the Saint-Maurice was *Metaberoutin* (*JR*, 8:19).

133 Champlain ordered his employee, Laviolette, to begin the construction of a post and Jesuit mission at Trois-Rivières in 1634 (*JR*, 6:43; DCB, 1:432. The post was built on the west side of the river mouth, not on Île Saint-Quentin as Champlain originally proposed in 1603. The large lake referred to by Champlain is Lac Saint-Pierre, some 14.5 km distant.

134 Champlain is describing a Native account of the canoe route up Rivière Saint-Maurice to Lac Saint-Jean. Judging from the maps by Father Laure, the main canoe route was up the Saint-Maurice to Rivière Bostonnaise and up that river to Rivière Métabetchouane, which flowed into Lac Saint-Jean. This is a very difficult canoe route because of rapids, long portages, and high waterfalls, particularly on the Métabetchouane. An alternate route was up Rivière Batiscan and the Métabetchouane to Lac Saint-Jean; see (maps), Laure, "Carte du Domaine," 1731.

pas, ce qui fut occasion que nous ne peusmes passer plus outre. Toute la terre que je veis aux bords de ladite riviere va en haussant* de plus en plus, qui est remplie de quantitez de sapins & cyprez, & fort peu d'autres arbres. /[f. 18v]

Chap. VII.

Longueur, largeur, & profondeur d'un lac, & des rivieres qui entrent dedans, des isles qui y sont, quelles terres l'on void dans le païs,[a] *de la riviere des Irocois, & de la forteresse des Sauvages qui leur font la guerre.*

Le Samedy ensuyvant[b] nous partismes des trois Rivieres & vinsmes mouiller l'ancre à un lac où il y a quatre lieuës,[c] tout ce pays depuis les trois rivieres jusques à l'entree dudict lac, est terre à fleur d'eau;* & du costé du Su quelque peu plus haute: Ladicte terre est tres-bonne & la plus plaisante que nous eussions encores veüe, les bois y sont assez clairs, qui fait que l'on y pourroit traverser aisement. Le lendemain 29. de Juin[d] nous entrasmes dans le lac,* qui a quelque 15. lieües de lo*n*g, & quelque 7. ou 8. lieües[e] de large: à son entree du costé du Su environ une lieüe il y a une riviere qui est assez grande, & va dans les terres quelques 60. ou 80. lieües, & continua*n*t du mesme costé il y a une autre petite riviere qui entre environ deux lieües en terre, & sort de dedans un autre petit /[f. 19] lac qui peut contenir quelques[f] trois ou quatre lieües.[g] Du costé du Nort,[h] où la terre y paroist fort haute,* on void jusques à quelques vingt lieües,[i] mais peu à peu les montaignes viennent en diminuant vers l'Oüest[j] comme pays plat: les Sauvages disent que la plus part[k] de ces montagnes sont mauvaises terres: Ledict lac a quelque trois brasses d'eau par où nous passasmes, qui fut presque au milieu, la longueur gist d'Est

* RM] *Terre allant en haussant.*
a *païs,*] A2 pays,
b ensuyuant] A2 ensuiuant
c lieuës,] A2 lieües,
* LM] *Terres basses.*
d Iuin] A2 Iuin,
* LM] *D'vn lac.*
e lieües] A2 lieues
f quelques] A1 quelque
g lieües.] A2 lieues.
h Nort,] A2 Nort
* RM] *Terres qui paroissent fort hautes.*
i lieües,] A2 lieues,
j l'Oüest] A2 l'Ouest
k la plus part] A2 la plus-part

the banks of the said river rises more and more and is replete with a great many firs and cedars, and very few other trees.

/[f. 18v] Chapter 7

Length, breadth, and depth of a lake; and of the rivers that flow into it; of the islands that are in it; what land one sees in the country; of the riviere des Irocois *and of the stronghold of the* Sauvages *who wage war on them*

On the following Saturday[135] we set out from *trois Rivieres* and came to drop anchor at a lake some four leagues distant.[136] All this region from *trois rivieres* to the entrance of the said lake is land at water level, but somewhat higher on the south side. The said land is very good and the most pleasant that we had yet seen. The woods there are quite open, which would make them easy to go through. The next day, the 29th of June, we entered the lake, which is some 15 leagues long and some 7 or 8 leagues wide.[137] About a league from its entrance on the south side, there is a fairly big river that extends some 60 or 80 leagues inland.[138] And continuing along on the same side, there is another small river that penetrates inland about two leagues and issues from within another little /[f. 19] lake, which may measure some three or four leagues.[139] From the north side, where the land appears very high, one sees as far as some twenty leagues; but little by little the mountains fall away towards the west, as if the country were flat. The *Sauvages* say that the majority of these mountains are badlands. The said lake is some three *brasses* deep where we passed, which was nearly in the middle. The length lies east

135 28 June 1603

136 The entrance of Lac Saint-Pierre is 14.5 km from Trois-Rivières. Literally, the French reads: "a lake where there are four leagues" (*un lac où il y a quatre lieuës*). Since Champlain estimates the size of Lac Saint-Pierre at 15 by 7 to 8 leagues, this new measurement must refer to the distance between the lake and Trois-Rivières, as interpreted by the first translators of the Champlain Society (Biggar, *The Works*, 2:138).

137 29 June is St. Peter's Day, which is probably why the lake was so named. Its earlier name, judging from maps based on the Cartier expeditions, was *Lac d'Angoulême*. The lake is about 29 km long to the islands at its western entrance and 12 km wide.

138 Rivière Nicolet

139 Not known which little river is meant. There are none that issue from a lake south of Lac Saint-Pierre.

& Oüest,[a] & de la largeur du Nort au Su;[b] Je croy qu'il ne laisseroit d'y avoir de bons poissons, comme les especes que nous avons pardeçà. Nous le traversasmes ce mesme jour & vinsmes mouiller l'ancre environ deux lieües[c] da*n*s la riviere qui va au hault à l'entree[d] de laquelle il y a trente petites isles;[e] selon ce que j'ay peu voir, les unes sont de deux lieües,[f] d'autres de lieüe[g] & demye & quelques unes moindres, lesquelles sont remplies de quantité de Noyers, qui ne sont gueres differe*n*s[h] des nostres,* & crois que les noix en sont bo*n*nes à leur saiso*n*; j'e*n* veis en quantité sous les arbres, qui estoient de deux façons, les unes petites, & les autres longues, comme d'un pousse, mais elles estoient pourries: Il y a aussi quantité de /[f. 19v] vignes sur le bord desdictes isles; mais quand les eaües[i] sont grandes, la plus part[j] d'icelles sont couvertes d'eau: & ce pays est encores meilleur* qu'aucun autre que j'eusse veu.

Le[k] dernier de Juin nous en partismes, & vinsmes passer à l'entree de la riviere des Irocois, où estoient caba*n*nez & fortiffiez les Sauvages qui leur alloient faire la guerre:* Leur forteresse est faicte de quantité de bastons fort pressez les uns contre les autres, laquelle vient joindre d'un costé sur le bord de la grand' riviere,[l] & l'autre sur le bord de la riviere des Irocois, & leurs Canos arrangez les uns co*n*tre les autres sur le bord[m] pour pouvoir promptement fuir,[n] si d'aventure[o] ils sont surprins des Irocois: car[p] leur forteresse est couverte d'escorce de chesnes, & ne leur sert que pour avoir le temps de s'embarquer. Nous fusmes dans la riviere des

a Oüest,] A2 Ouest,
b Su;] A2 Su:
c lieües] A2 lieues
d l'entree] A2 lentree
e isles;] A2 isles,
f lieües,] A2 lieues,
g lieüe] A2 lieue
h differẽs] A2 differens
* RM] *Isles à la sortie du lac.*
i eaües] A2 eaues
j la plus part] A2 la plus-part
* LM] *Bonnes terres.*
k veu. ¶ Le] A1 A2 veu: Le
* LM] *Sauuages Cabannez, fortifiez à l'entree de la riuiere des Irocois.*
l la grand' riuiere,] A1 la grand riuiere,
m bord] A2 bord,
n promptement fuir,] A2 prõptement fuyr,
o d'auenture] A2 d'avanture
p car] A2 Car

and west, and the width from north to south.[140] I believe it would not fail to have some good fish, like the species that we have on our side. We crossed it this same day and came to drop anchor about two leagues up the river that leads to the height, at the mouth of which there are thirty small islands.[141] According to what I could see, some are two leagues, others a league and a half, and some less, which abound in walnut trees, scarcely different from our own. And I believe that their nuts are good in season. I saw great quantities of them under the trees, which were of two sorts: some small, others about an inch long, but they were rotten.[142] There are also a great many /[f. 19ᵛ] vines on the shores of the said islands; but when the waters are high, the majority of them are covered with water. And this region is still better than any other that I had seen.

On the last day of June we set out from there and came to pass into the mouth of the *riviere des Irocois* where, encamped and fortified, were the *Sauvages* who were going to make war on them.[143] Their stronghold[144] is made of a great many stakes set very close together and comes to an end on the bank of the great river on one side, and on the bank of the *riviere des Irocois* on the other, and their canoes arranged one against the other on the bank so that they may quickly take flight if by chance they are surprised by the *Irocois*; for their stronghold is covered with the bark of the oak and serves only to give them time to embark. We went

140 Lac Saint-Pierre flows from the southwest to the northeast. Originally it was about 3 to 5 m deep along its length, but has been dredged to a depth of about 11 metres. The French *brasse* (fathom) is 1.62 m.

141 Champlain seems to be saying that he traversed the length of Lac Saint-Pierre on 29 June and anchored on the St. Lawrence among the many islands two leagues west from the entrance to the lake.

142 The European walnut is not native to Canada, and the black walnut (*Juglans nigra*) grows south of the area seen by Champlain. What Champlain must have seen is the butternut (*Juglans cinerea*) and perhaps bitternut hickory (*Carya cordiformis*), which is here at its northern limit. A nut one inch (*pouce*) long suggests the butternut. The thumb was established as a measurement in the Latin root of the French *pouce*: *pollix, pollicis*.

143 The mouth of the Rivière Richelieu was a major launching point for expeditions against the Iroquois, specifically the Mohawk. What Champlain described is a typical temporary encampment. In 1610 he fought the Iroquois at that location.

144 Champlain wrote *forteresse*, which suggests something less than a *fort*.

Irocois* quelques cinq ou six lieues, & ne peusmes passer plus outre avec nostre barque, à cause du grand cours d'eau qui dessent,[a] & aussi que l'on ne peut aller par terre & tirer la barque pour la quantité d'arbres qui sont sur le bord: Voyans ne pouvoir advancer d'ava*n*tage, nous prinsmes nostre esquif, pour voir si le courant /[f. 20] estoit plus adoucy, mais allant à quelques deux lieues[b] il estoit encore[c] plus fort, & ne peusmes avancer plus auant: Ne pouvant faire autre chose nous nous en retournasmes en nostre barque: Toute ceste riviere est large de quelque trois à quatre cens[d] pas, fort saine, nous y vinsmes cinq isles,[e]* distantes les unes des autres d'un quart ou de demye lieue,[f] ou d'une lieue[g] au plus: une desquelles contient une lieue,[h] qui est la plus proche;[i] & les autres sont fort petites: Toutes ces terres sont couvertes d'arbres, & terres basses,* comme celles que j'avois veu auparava*n*t, mais il y a plus de sapins & cyprez qu'aux autres lieux: La terre ne laisse d'y estre bo*n*ne, bien qu'elle soit quelque peu sablo*n*neuse. Ceste riviere va co*m*me au Sorouest.

Les Sauvages disent,* qu'à quelque quinze lieues d'où nous avio*ns*[j] esté, il y a un sault, qui vie*nt*[k] de fort hault,[l] où ils portent leurs canos pour le passer environ un quart de lieue, & entre*nt* deda*ns* un lac, où à l'entree il y a trois isles;[m] & esta*nt*

* LM] *Riuiere des Irocois.*
a dessent,] A2 descend,
b lieues] A2 lieues,
c encore] A1 encores
d cens] A2 cent (*quatre cents*, mod. Fr.)
e isles,] A2 Isles,
* RM] *Isles.*
f demye lieue,] A2 demie lieuë,
g lieue] A2 lieuë
h lieue,] A2 lieuë,
i proche;] A2 proche,
* RM] *Terres basses.*
* RM] *Rapport des Sauuages de la riuiere des Irocois.*
j auiōs] A2 auions
k viēt] A2 vient
l hault,] A2 haut,
m isles;] A2 Isles;

some five or six leagues up the *riviere des Irocois* and could pass no farther with our *barque* because of the great current of water that descends there.[145] Moreover, one cannot go on land and pull the *barque* because of the great number of trees that are on the shore. Seeing it was impossible to advance any farther, we took our *esquif* to see if the current /[f. 20] became gentler; but proceeding for some two leagues, it became yet stronger, and we could not advance any farther.[146] Being unable to do anything else, we returned from there in our boat.[147] This entire river is some three to four hundred paces wide and very clear.[148] We saw five islands in it, at a distance one from the other of a quarter or half a league, or one league at the most; one of which, the nearest, measures a league; and the others are very small.[149] All these lands are covered in trees and are lowlands like those that I had seen previously, but there are more fir trees and cedars than in the other places. The soil does not fail to be good there, although it may be somewhat sandy. This river runs approximately southwest.[150]

The *Sauvages* say that at some fifteen leagues from where we had been, there is a rapids that descends from very high up, where they carry their canoes for about a quarter of a league in order to pass it by and enter a lake, where there are three

145 Champlain and his party had reached the rapids at Saint-Ours, about 21 km from the mouth of the river, in what he called a *barque*. Unfortunately, Champlain was consistently vague about describing the vessels on which he was sailing. Technically, a barque is a small three-masted sailing vessel, square rigged on the foremast and mainmast, and fore and aft rigged on the mizzen. Sometimes the mizzen was rigged for a lateen sail (Kemp, *Oxford Companion*, 61–2). Most of the early barques in Canadian waters had a capacity of 25 to 75 tuns. It is doubtful if anyone would get a barque up the Rivière Richelieu as far as Saint-Ours, let alone contemplate dragging it overland around the rapids, as his next sentence suggests. It is likely that Champlain used the term *barque* in a general sense, such as "ship" or "boat." In a later account of this trip, Champlain wrote that they had travelled in *barques de 12 à 15 tonneaux* (Champlain, *Les Voyages, 1632*, 40). This type of boat may have been a two-masted pinnace, often used for coastal exploration, although Champlain never mentioned a pinnace on any of his journeys.

146 An *esquif* (skiff) is a rowboat equipped with one or two pairs of oars. This one was probably flat bottomed because it was designed to cross the Lachine Rapids. See below.

147 Boat: *barque*. In this case, we have translated *barque* as the generic word for boat, because Champlain was in the skiff at this point, rather than the larger *barque*.

148 The width of the river downstream from Saint-Ours is about 400 m. The French *pas* was 2.5 *pieds* (0.8 m), making Champlain's estimate correct.

149 The nearest and largest of the islands downstream of Saint-Ours is Île Deschaillons. It is about 2 km long.

150 Rivière Richelieu runs almost exactly from south to north.

dedans, ils en re*n*contrent[a] encores quelques-unes. Il[b] peut contenir quelque quara*n*te ou cinqua*n*te lieues de long, & de large quelque vingt cinq lieues, dans lequel descendent quantité /[f. 20v] de rivieres, jusques au nombre de dix, lesquelles portent canos assez avant: Puis vena*n*t à la fin dudit lac, il y a un autre sault,[c] & rentrent dedans un autre lac, qui est de la grandeur dudit[d] premier, au bout duquel sont cabannez les Irocois. Ils disent aussi qu'il y a une riviere qui va rendre à la coste de la Floride, d'où il y peut avoir dudit dernier lac, quelque cent ou cent quarante lieues:[e] tout le païs[f] des Irocois* est quelque peu montagneux, neantmoins païs tresbon,[g] temperé, sans beaucoup d'hyver, que fort peu.

Chap. VIII.[h]

Arrivee au sault, sa description, & ce qui s'y void de remarquable, avec le rapport des Sauvages de la fin de la grande riviere.

PArtant de la riviere des Irocois, Nous fusmes mouiller l'ancre[i] à trois lieues[j] de là, à la bande du Nort, tout ce pays[k] est une terre basse,* remplie[l] de toutes[m] les sortes d'arbres que j'ay dit[n] cy dessus. Le premier jour de Juillet nous costoyasmes la bande du Nort où[o] le bois y est fort clair, plus qu'en aucun lieu que nous eussions /[f. 21; F] encores veu auparavant, & toute bonne terre pour cultiver: Je me meis dans un canot à la bande du Su, où je veis quantité d'isles, lesquelles sont fort fertilles* en fruicts, comme vignes, noix, noizettes, & une maniere de

a rēcontrent] A2 rencōtrent
b quelques-vnes. Il] A1 quelques-vnes, il A2 quelques vnes, il
c sault,] A2 saut,
d dudit] A2 dudict
e lieues:] A2 lieuës:
f païs] A2 pays
* LM] *Quel est le païs des Irocois.*
g païs tresbon,] A2 pays tres-bon,
h VIII.] A1 A2 VII.
i l'ancre] A2 l'Ancre
j lieues] A2 lieuës
k pays] A1 A2 pays,
* LM] *Terres basses.*
l remplie] A2 remplies
n toutes] A2 touste
n dit] A2 dict
o où] A2 ou
* RM] *Isles en quantité fertiles.*

islands at the mouth.[151] And once in the lake, they still encounter some more of them. The lake may measure some forty or fifty leagues in length and about twenty-five leagues in width,[152] into which a number /[f. 20^{v}] of rivers descend – as many as ten – which carry canoes fairly far inland. Then, coming to the end of the said lake, there is another rapids and they enter another lake again, which is the same size as the said first, at the end of which the *Irocois* are encamped.[153] They say, moreover, that there is a river that goes to the coast of Florida, which may be some hundred or hundred and forty leagues from the said lake last mentioned.[154] The entire land of the *Irocois* is somewhat mountainous, nevertheless a very good region, temperate, and without much of a winter, only very little.

Chapter 8[155]

Arrival at the rapids, their description, and what is remarkable to be seen there, with the account given by the Sauvages *of the end of the great river*

Setting off from the *riviere des Irocois*, we went to drop anchor three leagues from there on the north shore. This whole region is lowland, replete with all the kinds of trees which I have mentioned above. On the first day of July we coasted along the north shore, where the woods are very open, more so than in any place that we had seen /[f. 21] up to that point, and all good land to cultivate. I set out in a canoe to the south shore,[156] where I saw a large number of islands that are very fertile with fruits like vines, walnuts, hazelnuts, and a kind of fruit that resembles

151 Champlain is describing a Native account of the route, amid river and lakes, up the Rivière Richelieu to the Atlantic coast. The next rapids, 50 km south of Saint-Ours, are the much larger Chambly Rapids, with a lengthy portage. The lake mentioned next is Lake Champlain, and the islands, Île la Motte and perhaps North and South Hero Island.

152 Lake Champlain is about 145 km long and of variable width.

153 The rapids at the south end of Lake Champlain are at Ticonderoga, and the next lake to the south is Lake George. At that time, the Mohawk villages were on the south side of the Mohawk River between the present Fultonville and Fort Plain, Montgomery County, New York State (Gehring and Starna, *A Journey*, 2).

154 The south river leading to the coast is the Hudson River. The distance from the southern end of Lake St. George, down the Hudson, is about 600 km.

155 The chapter heading in the original editions repeats chapter 7, when it should be 8.

156 As far as is known, this is the first time a European had ventured out in a canoe.

fruict qui semble à des chastaignes, serises, chesnes, trembles,[a] pible, houblon, fresne, erable, hestre, cyprez, fort peu de pins & sapins:[b] il y a aussi d'autres arbres que je ne cognois point, lesquels sont fort aggreables,[c] il s'y trouve quantité de fraises, fra*m*boises,[d] groizelles, rouges, vertes & bleues,[e] avec force petits fruicts qui y croissent parmy gra*n*de[f] qua*n*tité d'herbages: Il y a aussi plusieurs bestes sauvages,* comme orignas, cerfs, biches, dains, ours, porc-epics, lapins, regnards, castors, loutres, rats musquets,[g] & quelques autres sortes d'animaux que je ne cognois point, lesquels sont bons à manger, & dequoy vivent[h] les Sauvages.

Nous passasmes contre une isle[i] qui est fort aggreable,* & contient quelque quatre lieues de long, & environ demie de large. Je veis à la bande du Su deux hautes montaignes, qui paroissoient* comme à quelque vingt[j] lieues dans les

a trembles,] A2 tremble,
b sapins:] A1 sapins,
c aggreables,] A2 agreables,
d fraises, frāboises,] A2 fraizes, framboises,
e bleues,] A2 bleuës,
f grāde] A2 grande
* RM] *Des bestes Sauuages.*
g rats musquets,] A1 A2 rats, musquets, (*error copied in 1604 re-composition*)
h viuent] A1 vinent
i isle] A2 Isle
* RM] *Isle aggreable.*
* RM] *Montaignes qui paroissent dans les terres.*
j vingt] A1 vingts (*vingt normally invar., although it can take an s when multiplied*)

the chestnut, cherries, oaks, aspen, poplar,[157] hops, ash, maple, beech, cedar, and very few pines and firs.[158] There are also other trees with which I am not familiar but which are very pleasant. One finds there large quantities of strawberries, raspberries, red, green, and blue currants,[159] together with a great number of small fruits that grow in the thick grasslands there. There are also many wild animals, such as moose, elk, hinds, deer,[160] bears, porcupines, rabbits, foxes, beavers, otters, muskrats,[161] and some other kinds of animals with which I am unacquainted but which are good to eat and on which the *Sauvages* subsist.

We passed opposite an island that is very pleasant and measures some four leagues in length and about half a league in width.[162] I saw two high mountains on the south shore that appeared to be about twenty leagues inland.[163] The *Sau-*

157 poplar: *pible* (*Populus nigra*). The *pible* (mod. Fr. *peuplier*) is a regionalism from Champlain's home province of Saintonge. Celtic in origin, the occitan form *piboul* is the likeliest influence on the Saintonge (Éveillé, *Glossaire saintongeais*, 293; Musset, *Glossaire des patois*, 4:169; Jonain, *Dictionnaire*, 309). *Populus nigra* is not native to North America, but the leaves resemble *P. deltoides* and *P. tremuloides*, which are native to the areas visited by Champlain.

158 Of this list of fruits, the following need an explanation. The "vines" (grapes) were the common *Vitis labrusca*, which impressed the Recollects enough in 1623–24 to make some into sacramental wine (Wrong, *Long Journey*, 83). The chestnut is probably the American variety, *Castanea dentata*, and hops, the wild *Humulus lupulus*.

159 The three varieties of currant are difficult to identify. They may be "red" (*Ribes triste*), "green" (*Ribes oxyacanthoides*), and "blue" (*Ribes lacustre*). Champlain's use of *groizelles* for the standard French *groseilles* (currants) is another regionalism. Now written *groeséle* in the local usage, where *oe* reflects a pronunciation of [wé] or [oï], it refers to the native gooseberry (*Ribes uva crispa*) of Saintonge (Pivetea, *Dictionnaire*, 17, 165; Musset, *Glossaire*, 3:221).

160 The *orignas* is the moose (*Alces alces americana*); *cerf* is the elk (wapiti) (*Cervus elaphus*); *biches* are probably female wapiti, called "hinds" in Europe. The *dain* is the European fallow deer. Its nearest relative in eastern North America is the common white-tailed deer (*Odocoileus viriginianum*). The other animals present no difficulties to identify.

161 Champlain wrote *rats, musquets*, as if there were two animals. The *rat musqué* is the muskrat (*Ondatra zibethicus*). The Norway rat, which Champlain would have been familiar with, was introduced to the Americas with European settlement (Banfield, *Mammals of Canada*, 221–3).

162 The interconnected group of islands collectively called Îles de Verchères (Bouchard, Ronde, Marie, Demarais, and Beauregard) meets this description. They are some 15 km in length and 1.5 km at their widest point.

163 These are Mont Saint-Bruno (202 m, AMSL) and Mont Saint-Hilaire (408 m, AMSL), about 28 km southeast (not south) of the Îles Verchères.

terres: les Sauvages me dirent, que c'estoit le pre-/[f. 21v]mier sault de ladite riviere des Irocois. Le Mecredy ensuivant nous partismes de ce lieu, & feismes quelques cinq ou six lieues. Nous veismes quantité d'isles,* la terre y est fort basse, & sont couvertes de bois, ainsi que celles de la riviere des Irocois: le jour ensuiva*n*t[a] nous feismes quelques lieues, & passasmes aussi par quantité d'autres isles qui sont tres-bonnes & plaisantes, pour la quantité des prairies qu'il y a, tant du costé de terre ferme, que des autres isles: & tous les bois y sont fort petits,[b]* au regard de ceux que nous avions passé. En fin nous arrivasmes cedit jour à l'entree du sault,* avec vent en poupe, & rencontrasmes une isle qui est presque au milieu de ladite entree, laquelle contie*n*t un quart de lieue de long, & passasmes à la bande[c] du Su de ladite isle, où il n'y avoit que de 3. à quatre ou cinq pieds d'eau, & aucunes-fois une brasse[d] ou deux, & puis tout à un coup n'en trouvio*n*s que trois ou quatre pieds: Il y a force rochers, & petites isles,* où il n'y a point de bois, & sont à fleur d'eau. Du comme*n*cement de la susdite isle, qui est au milieu de ladite entree, l'eau commence à venir de grande force:* bien que nous eussions le vent fort bon, si /[f. 22] ne peusmes nous en toute nostre puissa*n*ce beaucoup ava*n*cer, toutesfois nous passasmes ladite isle qui est à l'entree dudit sault. Voyant que nous ne pouvions avancer, nous vinsmes mouiller l'ancre à la bande du Nort, contre une petite isle* qui est fertile en la plus part[e] des fruicts que j'ay dit cy dessus.

* LM] *Isles en quantité*
a ensuiuãt] A2 ensuiuant
b petits,] A2 petis,
* LM] *Bois fort petit.*
* LM] *Entree du saut.*
c bande] A2 bãde
d brasse] A1 brase
* LM] *Isles.*
* LM] *Grand courant d'eau.*
* RM] *Isle où nous mouillasmes l'ancre.*
e plus part] A2 plus-part

vages told me that the first /[f. 21v] rapids of the said *riviere des Irocois* were there.[164] On the following Wednesday,[165] we set out from this place and made some five or six leagues. We saw a great many islands. The land there is very low, and these islands are covered with trees like those of the *riviere des Irocois*. The following day we made some leagues and passed by many other islands as well, which are very good and pleasant because of the many meadows there are, both on the mainland and on the other islands. And all the woods there are very small in comparison with those that we had passed. At length on this said day[166] we arrived at the entrance of the rapids with the wind in the stern and encountered an island that is almost in the middle of the said entrance, a quarter of a league in length, and we passed along the southern stretch of the said island, where there were only from 3 to four or five feet of water; and sometimes a *brasse* or two; and then suddenly we were finding only three or four feet of it.[167] There are a great many rocks, and small islands at water level where there are no woods at all. From the beginning of the above-mentioned island, which is in the middle of the said entrance, the water begins to come with great force. Although we had a very favourable wind /[f. 22], yet we could not, under all our power, make much headway. Nevertheless, we passed the said island that is at the entrance of the said rapids. Seeing that we could not advance, we came to cast anchor on the north shore across from a small island abounding in most of those fruits that I mentioned above.[168]

164 The "first rapids" on the river are at Chambly, some 10 km south of the two mountains.

165 Wednesday 2 July 1603

166 Wednesday, 2 July

167 The Courant Sainte-Marie is a strong current with associated rapids and shallows along the shore of Montreal Island south to Île Sainte-Hélène, the island in the "middle of the entrance." This island is about 2.8 km long and 0.5 km wide. Champlain passed along the east and then the south shore of Île Sainte-Hélène, probably named by him in honour of his wife Hélène Boullé in 1611, the year after they were married. The small island to the north of Île Sainte-Hélène was Île Ronde, and the one to the south, Île Saint-Pierre (also Île Verte). These three islands are now one.

168 When Champlain's vessel rounded the southern end of Île Sainte-Hélène, it was moving directly into the powerful currents coming from the Lachine Rapids. Unable to make much headway south of the island, the crew steered west (not north) towards Montreal Island and anchored opposite the small Market Gate Island near the mouth of the Rivière Saint-Pierre. This little island no longer exists, and the Rivière Saint-Pierre is now part of the Lachine Canal. Montreal was founded in 1641 on the north bank of the Rivière Saint-Pierre, opposite the "little island," almost where Champlain landed in 1603. It should be noted again that Champlain is still calling the left bank of the St. Lawrence the "north shore" and the right bank the "south shore." His directions in the Montreal

Nous[a] appareillasmes aussi tost[b] nostre esquif, que l'on avoit fait faire expres pour passer ledit[c] sault: dans lequel nous entrasmes ledit sieur du Po*n*t & moy, avec quelques autres Sauvages que nous avions menez pour nous mo*n*strer le chemin: partant de nostre barque, nous ne feusmes pas à trois ce*n*ts[d] pas, qu'il nous fallut descendre,* & quelques matelots[e] se mettre à l'eau pour passer nostre esquif: le canot des Sauvages passoit aisément:[f] nous rencontrasmes une infinité de petits rochers* qui estoient à fleur d'eau, où nous touchions souventes fois.

Il y a deux grandes isles,* une du costé du Nort,[g] laquelle contient quelque quinze lieues de long, & presque autant de large, commence à quelques douze lieues dans la riviere de Canada,[h] allant vers la riviere des Irocois, & vie*n*t tomber[i] par delà le sault. L'isle qui est à la ba*n*de du Su, a quelque quatre lieues /[f. 22v] de long, & demie de large: Il y a encores une autre isle qui est proche de celle du Nort, laquelle peut tenir quelque demie lieue de long, & un quart de large:[j] & une autre petite isle qui est entre celle du Nort,[k] & l'autre plus proche du Su, par où nous passasmes l'entree du sault: esta*n*t passee, il y a une maniere de lac,* où sont toutes ces isles, lequel peut co*n*tenir quelque cinq lieues de long, & presque autant de large, où il y a qua*n*tité de petites isles qui sont rochers: il y a proche dudit sault une montagne* qui descouvre assez loing dans lesdites terres, & une petite riviere qui vient de ladicte montaigne tomber dans le lac. L'on void du

a dessus. ¶ Nous] A1 A2 dessus; Nous
b aussi tost] A2 aussi-tost
c ledit] A2 ledict
d cẽts] A2 cens
* RM] *Passage mauuais.*
e matelots] A2 Matelots
f aisément:] A2 aisement:
* RM] *Rochers.*
* RM] *Deux grandes isles.*
g Nort,] A2 Nort:
h la riuiere de Canada] (*This is the sole witness to* Canada *without an* s *in the second edition of 1604)*
i viẽt tomber] A2 vient tõber
j large:] A2 large;
k laquelle peut tenir quelque demie lieue … celle du Nort,] A1 A2 *(rpt. twice)* A2 … lieuë; rpt. A1: laquelle peut tenir quelq̃ demie lieue A2 … quelq̃ demye lieue
* LM] *Maniere de lac.*
* LM] *Montagne proche du sault.*

We at once made ready our skiff, which had been constructed precisely for the purpose of crossing the said rapids; into which the said sieur du Pont and I entered with some other *Sauvages* whom we had brought along to show us the way. Leaving our *barque*, we had not yet gone as far as three hundred paces when we were forced to get out,* and some sailors had to get into the water to get our skiff across. The *Sauvages'* canoe crossed easily. We encountered an infinite number of small rocks, which were at water level and which we frequently touched upon.[169]

There are two large islands: one on the north side that extends some fifteen leagues in length and almost as many in width. It begins at some twelve leagues down the *riviere de Canada*, heading towards the *riviere des Irocois*, and extends beyond the rapids.[170] The island that is on the southern stretch is some four leagues /[f. 22ᵛ] long and half a league wide. There is yet another island that is near the northern one, which may extend about a half league in length and a quarter league in width;[171] and another little island that is between that of the north and the other nearer one of the south, by which we passed the rapids' entrance. Having gone by this entrance, one finds a kind of lake where all these islands are, the lake measuring possibly some five leagues in length and almost as many in width, and where there are a great number of little islands that are rocks.[172] There is a mountain near the said rapids that is sighted from very far within the said lands, and a little river that descends from the said mountain to

area are therefore confused. What he calls the "south side" is really the east side of the river, and the "north side" is the west side of the river.

* RM] *Bad passage*

169 Gravé Du Pont, Champlain, and some Natives were trying to get the skiff across the shallows towards the Lachine Rapids. Just like the attempt to cross the Saint-Ours Rapids on the Rivière Richelieu, the skiff proved to be inferior to the canoes, even though it was constructed for the purpose, a fact noted by Champlain.

170 The largest of the confusing array of islands mentioned by Champlain is Montreal Island, about 49 km long, on the west, not north side of the river. Although the verb "extends" here best suits the context "beyond the rapids," the French reads literally *vient tomber* (comes to fall), which is more suited to a river than an island. To make matters worse, Champlain does not state his subject before "begins," which we assume to be the "one island" that he has just introduced as his topic.

171 The "island on the southern stretch," really east side, since the St. Lawrence at this point flows almost north-south, may be the Îles de Boucherville, which are some 7.5 km long. The other "island near the northern one" (western shore) could be Île des Sœurs (at one time Île Saint-Paul) which is 3.7 km long.

172 The "lake" mentioned by Champlain seems to be the widening of the St. Lawrence River between the Lachine Rapids and Île Sainte-Hélène, now Bassin de la Prairie.

costé du Su quelques trois ou quatre mo*n*taignes qui paroissent comme à quelque quinze ou seize lieues[a] dans les terres. Il y a aussi deux rivieres, l'une qui va au premier lac de la riviere des Irocois,* par où[b] quelques-fois les Algoumequins leur vont faire la guerre, & l'autre qui est proche du sault qui va quelque peu dans les terres. Venans à approcher dudit[c] sault avec nostre petit esquif,* & le canot, je vous /[f. 23] asseure que jamais je ne veis un torrent d'eau* desborder avec une telle impetuosité comme il faict, bien qu'il ne soit pas beaucoup haut, n'estant en d'aucuns lieux que d'une brasse ou de deux[d]* & au plus de trois: il dessend[e] comme de degré en degré, & en chasque lieu où il y a quelque peu de hauteur,[f] il s'y faict un esbouillonnement estrange de la force & roideur que va l'eau en traversant ledit sault* qui peut contenir une lieue: Il[g] y a force rochers de large, & environ le milieu, il y a des isles* qui sont fort estroites[h] & fort longues, où il y a sault tant du costé desdictes isles qui sont au Su, co*m*me du costé du Nort, où il fait si da*n*gereux, qu'il est hors de la puissance d'homme d'y passer un bateau, pour petit qu'il soit.*

Nous fusmes[i] par terre dans les bois pour en veoir la fin, où il y a une lieue, & où l'o*n* ne voit[j] plus de rochers ny de saults, mais l'eau y va si viste qu'il est impossible de plus; & ce courant contient quelque trois ou quatres lieues:[k]* de façon que c'est en vain de s'imaginer que l'on peust faire passer aucuns bateaux par lesdicts

a lieues] A2 lieuës
* LM] *Riuiere dedans le lac qui va aux Irocois.*
b où] A1 ou
c dudit] A2 dudict
* LM] *Arriuee au sault auec l'esquif.*
* RM] *Torrent d'eau au sault.*
d deux] A2 deux,
* RM] *Hauteur du sault.*
e dessend] A2 descend
f hauteur,] A2 hauteur
* RM] *Rochers dans le sault.*
g Il] A2 il
* RM] *Isles.*
h estroites] A2 estroictes
* RM] *Impossible de passer le sault par basteau.*
i fusmes] A2 feusmes
j voit] A2 veoit
* RM] *Trauerses que nous fismes par terre pour veoir la fin du sault.*
k lieues:] A2 lieues;

fall into the lake.[173] One sees on the south side some three or four mountains, which seem to be about fifteen or sixteen leagues inland.[174] There are also two rivers: one that flows to the first lake of the *riviere des Irocois*, by which the *Algoumequins* sometimes go to make war upon them; and the other that is near the rapids and extends somewhat inland.[175] Approaching the said rapids with our little skiff and the canoe, I /[f. 23] assure you that I have never seen a torrent of water pour over with such a rage as this does, although it is not very high, not being in any place more than a *brasse* or two, or three at the most.[176] It descends as if stepwise, and in each place where there is somewhat of a height, it makes a strange bubbling forth[177] from the force and steepness with which the water flows when crossing the said rapids, which may measure a league.[178] There are many rocks in the open, and around the middle there are some islands that are very long and narrow,[179] where there are rapids as much on the side of the said islands that are to the south as on the north shore, where it is so dangerous that it is beyond the power of a man to get a boat past it, however small it may be.

We went by land into the woods to see the end of it, a league away, and where one no longer sees any rocks or rapids, but the water runs so fast there that any more speed is impossible. And this current extends for some three or four leagues, in such a way that it is vain to fancy anyone being able to make any boats cross

173 The mountain is Mount Royal, which rises 234 m (AMSL). The "little river" is Rivière Saint-Pierre, now replaced by the Lachine Canal over most of its original length.

174 The Monteregian Hills lie east (not south) of Montreal. Those that can be seen from Mount Royal are, from west to east: Saint-Bruno (202 m AMSL), Saint-Hilaire (408 m), Rougemont (364 m), Yamaska (418 m), and Saint-Gregoire (252 m). The latter lies a little south of the other four. The nearest, Saint-Bruno, is 22 km from Mount Royal, and the farthest, Yamaska, about 57 km. It is doubtful that the rest of the Monteregian Hills, far to the east, could have been seen by Champlain.

175 The first of these rivers is the Rivière Saint-Lambert, a well-known canoe route that connects to Rivière L'Acadie by means of a portage. The Rivière L'Acadie in turn flows into the Chambly basin, the "first lake of the *riviere des Irocois*." The second river is probably Rivière La Tortue.

176 This makes the foot of the Lachine Rapids about 2 to 6 m high.

177 bubbling forth: *esbouillonnement*, pronounced today in the Saintonge linguistic region as *éboujhement* and defined as a masculine noun meaning "action de se mettre en mouvement" (a setting in motion) (Pivetea, *Dictionnaire*, 119). The modern French equivalent would be *bouillonnement* (n.m.) meaning bubbling or boiling.

178 The eastern end of the Lachine Rapids is 3 km wide, including the width of Île au Héron.

179 The largest of these islands is Île au Héron.

saults.[a]* Mais qui les voudroit passer,[b] il se faudroit accommoder des Canos[c] des Sauvages,* /[f. 23v] qu'un homme peut porter aisement: car de porter bateaux, c'est chose laquelle ne se peut faire en si bref temps comme il le faudroit pour pouvoir s'en retourner en Fra*n*ce, si l'on n'y hyvernoit: Et outre ce sault premier, il y en a dix autres, la plus part[d] difficilles à passer: de façon que ce seroit de gra*n*des peines & travaux pour pouvoir voir,[e] & faire ce que l'on pourroit se promettre par basteau, si ce n'estoit à grands frais & despens, & encores en da*n*ger de travailler en vain: mais avec les canots des Sauvages l'on peut aller librement & promptement en toutes les terres, tant aux petites Rivieres comme aux grandes: Si bien qu'en se gouvernant par le moyen desdits Sauvages & de leurs canots, l'on pourra voir tout ce qui se peut, bon & mauvais, dans un an ou deux. Tout ce peu de pays du costé dudict[f] sault que nous traversames[g] par terre, est bois fort clair,* où[h] l'on peut aller aiseme*n*t, avecques armes, sans beaucoup de peine; l'air[i] y est plus doux & temperé, & de meilleure terre qu'en lieu que j'eusse veu, où[j] il y a quantité de bois & fruicts, comme en tous[k] les autres lieux cy dessus, & est par les 45. degrez & quelques minu*-/[f. 24]tes.

Voyans que nous ne pouvions faire d'avantage, nous en retournasmes en nostre barque, où nous interrogeasmes les Sauvages que nous avions, de la fin de la riviere,* que je leur fis figurer de la[l] main, & de quelle partie procedoit sa source. Ils nous dirent que passé le premier sault que nous avions veu, ils faisoient quelques dix ou quinze lieues avec leurs canots dedans la riviere, où il y a une riviere qui

* RM] *Cours d'eau au dessus du sault.*
a saults.] A1 saultr.
b passer,] A2 passer
c Canos] A2 Canots
* RM] *Moyen de passer le sault.*
d plus part] A2 plus-part
e voir,] A2 veoir,
f dudict] A2 dudit
g trauersames] A2 trauersasmes
* LM] *Bonnes terres & bois fort clair.*
h où] A1 ou
i l'air] A2 l'ayr
j où] A1 ou
k tous] A1 tout
* LM] *Ledit sault est par les 45. degrez, & quelques minutes.*
* RM] *Sauuages que nous interrogeasmes, où est la fin de la grande Riuiere.*
l la] A1 leur

over the said rapids.[180] But whoever would wish to cross them would have to make the most of the canoes of the *Sauvages* /[f. 23v], which a man can easily carry. For carrying boats[181] is a thing that cannot be done in such a short time as would be necessary to enable one to return from there to France if one did not winter there. And beyond these first rapids, there are ten more of them, for the most part difficult to pass; so that it would be a matter of great difficulty and labour to be able to see and do what one could commit oneself to do by boat, except at great cost and expense, and still at risk of working in vain. But with the canoes of the *Sauvages*, one may travel freely and quickly throughout all the lands, as much in the little rivers as in the big ones, so well that by directing one's course with the help of the said *Sauvages* and their canoes, one would be able to see all that can be seen, good and bad, within the space of a year or two.[182] All this bit of country on the side of the said rapids that we crossed overland is a very open wood where one can travel easily with weapons and without much trouble. The air there is milder and temperate. And there is better soil than in any place that I had seen, with trees and fruits in great quantity, as in all the other above-mentioned places; and it is at 45 degrees and some min-/[f. 24]utes.[183]

Seeing that we could do no more, we returned from there in our *barque*, when we questioned the *Sauvages* whom we had with us about the end of the river, which I made them draw by hand, and about where its source came from[184] [map 4, fig. 8, top]. They told us that having crossed the first rapids that we had seen,[185] they did some ten or fifteen leagues with their canoes in the river, to where there

180 Champlain walked westward along the portage around the rapids as far as what is now the western entrance of the Lachine Canal. The length of the most turbulent part of the Lachine Rapids is about 4.8 km, but the swift water and associated rapids and shallows stretch westward for another 4 km, as Champlain described.

181 A portage

182 This sentence presents the most acute and far-reaching observations in *Des Sauvages*. Champlain had deduced what no one had before him – that in order to explore the interior of Canada, the "help of the *Sauvages*" had to be secured and European boats had to be replaced by canoes.

183 The latitude at the beginning of the Lachine Rapids is 45° 25'N.

184 This is the first of three accounts gathered by Champlain of the geography of the St. Lawrence River and lakes west of Montreal Island. Unfortunately we know nothing about his interpreters. Note that Champlain requested that maps be drawn.

185 The Lachine Rapids, then called *Saut St-Louis*. The seventeenth-century geography of the St. Lawrence River from the Lachine Rapids to Lake Ontario was changed substantially through the construction of the St. Lawrence Seaway, opened in 1959. The rapids and lakes mentioned in the three Algonquin accounts recorded by Champlain are identified here and are named according to early-eighteenth-century maps, espe-

va en la demeure des Algoumequins, qui sont à quelque soixante lieues esloignez de la grande riviere, & puis ils venoient à passer cinq saults[a] lesquels peuvent contenir du premier au dernier huict lieues, desquels il y en a deux où ils portent leurs canots pour les passer: chasque sault peut tenir quelque demy quart de lieuë,[b] où un quart au plus: Et puis ils viennent dedans un lac, qui peut tenir quelque quinze ou seize lieües[c] de long. De là ils rentrent dedans une riviere, qui peut contenir une lieue de large, & font quelques[d] deux lieues dedans, & puis rentrent dans un autre lac de quelque quatre ou cinq lieues de long; venant au bout duquel ils passent cinq autres saults, distant du premier au dernier quelque /[f. 24v] vingt-cinq[e] ou trente lieues, dont il y en a trois où ils portent leurs canots pour les passer;[f] & les autres deux ils ne les font que trainer dedans l'eau, d'autant que le cours n'y[g] est si fort ne mauvais co*m*me aux autres: De tous ces saults aucun n'est si dificille à passer comme celuy que nous avons veu: Et puis ils vie*n*nent[h] dedans un lac qui peut tenir quelques[i] 80. lieues de long, où il y a quantité d'isles, & qu'au bout[j] d'iceluy l'eau y est salubre, & l'hyver doux.

a saults] A2 saults,
b lieuë,] A2 lieue,
c lieües] A2 lieues
d quelques] A2 quelque
e vingt-cinq] A2 vingtcinq
f passer;] A2 passer,
g n'y] A2 ny
h viēnent] A2 viennent
i quelques] A2 quelque
j qu'au bout] A1 que au bout

is a river that enters the home of the *Algoumequins*, who are removed from the great river by some sixty leagues;[186] and then they used to go past five rapids, which may extend eight leagues from first to last, of which there are two of them where they carry their canoes to get past them. Each rapids may extend an eighth of a league or a quarter at the most.[187] And then they come into a lake, which may extend some fifteen or sixteen leagues in length.[188] From there, they enter a river again, which may measure a league in width, and accomplish some two leagues in it; and then they enter another lake again, of some four or five leagues in length,[189] coming to the end of which, they pass another five rapids, distant from first to last some /[f. 24^v] twenty-five or thirty leagues, of which there are three where they carry their canoes in order to cross them. And for the other two, all they do is drag them in the water, for as much as the current there is neither as strong nor as bad as in the others.[190] Of all these rapids, none is so difficult to cross as the one that we saw.[191] And then they come into a lake, which may extend some 80 leagues in length, where there are many islands. And they told us that at the end of that lake the water is wholesome and the winter mild.[192]

cially Deshayes, *La Grande Riviere* [1715]; Chaussegros, "Carte du Lac Ontario," 1726; and Chaussegros and D'Anville, *Le Fleuve Saint-Laurent* [1755].

186 The Ottawa River. It is probable that Champlain was at the Lachine Rapids with Bessouat (Tessouat), the influential headman of the *Kichesipirini* (Big River People) Algonquins whom he had met at Tadoussac. These people lived at and about Morrison Island, some 130 km upstream from the mouth of the Ottawa River where it enters the Lac des Deux Montagnes.

187 This is the first set of rapids west of those at Lachine (*Saut St-Louis*), between Lac Saint-Louis and Lac Saint-François. During the French regime they were named, from east to west: *Les Cascades, Saut du Trou, Saut du Buisson, Rapide du Couteau des Cédres*, and *Rapide du Couteau St-François*. The distance between the first and last is about 19.5 km.

188 Lac Saint-François stretches some 34 km west to Thompson Island and another 19 km to Cornwall Island, where the next set of rapids begin.

189 There is no "lake" west of Lac Saint-François. It may be the open stretch of the river from Thompson to Cornwall Island that was meant.

190 The second set of rapids begins just east of Cornwall Island with the *Rapide de la Pointe Maligne* (also *Mouilliée*), followed by the *Long Saut* around Barnhart and Long Sault Islands; then the *Rapide Plat* at Ogden Island, and finally *Le Galop*, beginning with a set of rapids near the present town of Iroquois, a more open stretch as far as the town of Cardinal, and then the main rapids around Galop, Spencer, and Drummond Islands to Chimney Island, marking *Fin des Rapides*. This entire stretch of fast water, rapids, and portages extended for about 65 km.

191 The Lachine Rapids

192 The western end of Wolfe Island marks the beginning of Lake Ontario, some 110 km from the last rapids at Galop Island. By canoe, along the north shore from Wolfe Island to the Niagara River, the lake is about 370 km long.

A la fin dudict lac ils passent un sault, qui est quelque peu eslevé, où il y a peu d'eau laquelle dessend:[a] là ils portent leurs canots[b] par terre environ un quart de lieüe[c] pour passer ce sault: De là entrent dans un autre lac qui peut tenir quelque soixante lieuës[d] de long, & que l'eau en est fort salubre: estant à la fin ils viennent à un destroit[e] qui contient deux lieues de large & va assez ava*n*t dans les terres: qu'ils n'avoient point passé plus outre, & n'avoient veu la fin d'un lac qui est à[f] quelque quinze ou seize lieues d'où ils ont esté, ny que ceux qui le leur avoient dit[g] eussent veu homme qui l'eust veu,[h] d'autant qu'il est si grand, qu'ils ne se hazarderont pas de se mettre au large, de peur que /[f. 25; G] quelque tourmente ou coup de vent ne les surprint: disent qu'en esté le Soleil se couche au Nort dudict[i] lac, & en l'hiver il se couche comme au millieu;[j] que l'eau y est tres-mauvaise, comme celle de ceste mer. Je leur demandis, si depuis cedict lac dernier qu'ils avoient veu, si l'eau descendoit[k] tousjours dans la Riviere[l] venant à Gaschepay:[m] ils me dirent[n] que non, que depuis le troisiesme lac, elle descendoit seulement vena*n*t audit Gaschepay, mais que depuis le dernier sault, qui est quelque peu haut, comme j'ay dit,[o] que l'eau estoit presque pacifique, & que ledict lac pouvoit prendre cours par autres rivieres, lesquelles vont dedans les terres, soit au Su, ou au Nort, dont il y en a quantité qui y reffluent,[p] & dont[q] ils ne voyent point la fin. Or à mon jugement, il faudroit q*ue* si ta*n*t de rivieres desbordent dedans ce lac, n'aya*n*t que si peu de cours audit[r] sault, qu'il faut par necessité, qu'il reffluë dedans quelque grandissime riviere: Mais ce qui me faict croire qu'il n'y a point de

a dessend:] A2 descend:
b canots] A2 canos
c lieüe] A2 lieue
d lieuës] A2 lieues
e destroit] A2 destroict
f à] A1 a
g qui le leur auoient dit] A1 A2 qui leur auoient dict
h qui l'eust veu,] A1 A2 qui le l'eust veu, (*displacement of* le *here from where it should be in corrected n.* g *above*)
i dudict] A2 dudit
j millieu;] A2 milieu:
k descendoit] A2 descēdoit
l Riuiere] A2 riuiere
m Gaschepay:] A1 A2 Gaschepay,
n dirent] A2 dirēt
o dit,] A2 dit
p reffluent,] A2 reffluēt,
q & dont] A1 dont
r audit] A2 audict

At the end of the said lake they pass a rapid that is somewhat elevated, and where there is a bit of water that flows down. There they carry their canoes overland for about a quarter of a league in order to pass this rapid.[193] From there, they enter into another lake, which may extend some sixty leagues in length.[194] And they told us that the water in it is very wholesome. Being at the end, they come to a strait that measures two leagues in width and leads rather far inland.[195] They told us that they had not gone any farther and had not seen the end of a lake that is located at some fifteen or sixteen leagues from where they were. Neither had those who had told them about it seen anyone who had seen it, forasmuch as it is so big that they will not risk setting out into the open, for fear that /[f. 25] some storm or gale might surprise them.[196] They say that in summer the sun sets to the north of the said lake, and in winter it sets as if in the middle; that the water there is very bad, like that of this sea. I asked them whether, from this said last lake which they had seen,[197] the water still flowed down the river towards *Gaschepay*.[198] They told me that no, it descended only from the third lake[199] to the said *Gaschepay*, but that after the last rapids,[200] which are somewhat high as I said, the water was almost still, and that the said lake[201] could find an outlet by other rivers, which go inland either to the south or to the north, of which there are many that flow out there and whose end they do not see. Now, in my judgment, if so many rivers pour out into this lake,[202] having only such a small discharge at the said rapids,[203] it would require it, by necessity, to flow out into some very great river.[204] But what makes me believe that there is no river by which this said

193 Niagara Falls. The portage around the falls is about 2 km.

194 Lake Erie is about 400 km long from the Niagara River to the St. Clair River.

195 The St. Clair River

196 This huge lake is Lake Huron, which begins 127 km from the southern end of the St. Clair River.

197 Lake Erie

198 The Gaspé, where the St. Lawrence River flows into the Atlantic

199 The "third lake" is Lake Ontario.

200 Niagara Falls

201 Lake Erie

202 Lake Erie

203 Niagara Falls

204 *grandissime riuiere*: very great river. The adjective *grandissime* is a loan word from the Italian *grandissimo*, found very rarely in the Saintonge region as *grandessime* (Gautier, *Grammaire*, 49; *TLF*, 9:422). Champlain uses it three times in chapters 8 and 9 to describe a "very great" river, lake or current (ff. 25, 27ᵛ).

riviere[a] par où cedit lac reffluë,[b] veu le nombre de toutes les autres rivieres qui refflue*nt* dedans, c'est que les Sauvages n'ont veu aucune riviere qui print son cours par dedans les terres, /[f. 25^{v}] qu'au lieu où ils ont esté: Ce qui me faict croire que c'est la mer du Su, estant salee comme ils disent, toutesfois[c] il n'y faut tant adjouster de foy,[d] que ce ne soit avec raisons apparentes, bien qu'il y en aye quelque peu: Voilà au certain tout ce que j'ay veu cy dessus, & ouy dire aux Sauvages[e] sur ce que nous les avons interrogez.

Chap. IX.

Retour du Sault à Tadousac, avec la confrontation du rapport de plusieurs Sauvages, touchant la longueur, & commencement de la grande Riviere de Canadas: Du[f] *nombre des saults & lacs qu'elle traverse.*

NOus partismes dudict Sault[g] le Vendredy quatriesme jour de Juin, & revinsmes cedit jour à la riviere des Irocois.[h] Le Dimanche sixiesme jour de Juin nous en partismes, & vinsmes mouiller l'ancre au lac. Le Lundy ensuivant nous feusmes mouiller l'ancre aux trois Rivieres. Cedit jour nous feismes quelques quatre lieues par delà lesdictes trois Rivieres. Le Mardy ensuivant nous vinsmes à Quebec, & le /[f. 26] lendemain nous feusmes au bout de l'isle d'Orleans, où les Sauvages vindrent à nous, qui estoient cabannez à la gra*n*d' terre[i] du Nort. Nous interrogeasmes deux ou trois Algoumequins,* pour sçavoir s'ils se conformeroient avec ceux que nous avions interrogez touchant la fin & le commencement de ladite[j] riviere de Canadas: Ils dirent, comme ils l'ont figuré, que passé le sault que nous avions veu, environ deux ou trois lieues, il va une riviere en leur demeure, qui est en[k] la bande du Nort, continuant le chemin dans ladicte grand' riviere,[l] ils passent un sault, où ils portent leurs canots, & viennent à passer cinq autres saults,

a riuiere] A2 Riuiere
b reffluë,] A2 refflue,
c toutesfois] A2 toutes-fois
d foy,] A2 foy;
e Sauuages] A2 Sauuages,
f *Canadas: Du*] A2 *Canadas, du*
g Sault] A2 sault
h Irocois.] A2 Irocois, (*introducing a clear error*)
i grād' terre] A2 grand' terre
* RM] *Autre rapport des Sauuages Algoumequins.*
j ladite] A2 ladicte
k en] A2 à
l grand' riuiere,] A1 grand riuiere,

lake flows out, given the number of all the other rivers that flow into it, is the fact that the *Sauvages* have not seen any river taking its course inland, /[f. 25v] except in the place where they were; which makes me believe that it is the South Sea, being salty as they say. Nevertheless, we must not give too much credence to this view, except with manifest reasons, although there may be some small grounds for it. This is unquestionably everything that I have seen in the matter or heard the *Sauvages* say in response to our questions.

Chapter 9

Return from the rapids to Tadousac, *with a comparison of the reportage of several* Sauvages *concerning the length and source of the great* Riviere de Canadas*; of the number of rapids and lakes that it traverses*

We set out from the said rapids on Friday, the fourth day of July,[205] and returned this same day to the *riviere des Irocois*. On Sunday, the sixth day of July,[206] we set out from there and came to cast anchor in the lake.[207] On the following Monday we went to cast anchor at *trois Rivieres*. This said day we accomplished some four leagues beyond the said *trois Rivieres*. On the following Tuesday we came to Quebec; and the /[f. 26] next day went to the end of *isle d'Orleans*, where the *Sauvages* came to us, who were encamped on the mainland to the north. We questioned two or three *Algoumequins* in order to know whether they would agree with those whom we had examined concerning the end and source of the said *riviere de Canadas*[208] [fig. 8, centre]. They said, as they drew it, that about two or three leagues past the rapids that we had seen,[209] a river leads into their home ground, which is on the north shore.[210] Continuing along the said great river, they pass a rapids where they carry their canoes, and come by five other rapids,

205 The French text reads *Juin* (June) for both 4 July and 6 July, but Champlain was near *trois Rivieres* towards the end of June (chapter 7) and "the first day of July" is introduced at the beginning of chapter 8.

206 *Juin* (June) in the original editions is clearly an error for July.

207 Lac Saint-Pierre

208 St. Lawrence River

209 Here begins the second of the three Algonquin accounts of the route west to the Great Lakes. The Natives are drawing a map beginning at the Lachine Rapids.

210 From the head of the Lachine Rapids to Lac des Deux Montagnes, where canoes turn north towards the Ottawa River, is 27 km.

lesquels peuvent contenir du premier au dernier quelque neuf ou dix lieues, & que lesdits[a] saults ne sont point difficiles à passer, & ne font que trainer leurs canots en la plus part desdits saults, horsmis à deux où ils les portent, de là viennent à entrer dedans une riviere, qui est comme une maniere de lac, laquelle peut contenir[b] quelques six ou sept lieues;[c] & puis passent cinq autres saults, où ils trainent[d] leurs canots co*mm*e[e] ausdits premiers, horsmis à deux, où[f] ils les portent comme aux premiers;[g] & que du pre-[f. 26v]mier au dernier il y a quelque vingt ou 25. lieues:[h] puis viennent dedans un lac qui contient quelque cent cinquante lieues de long, & quelque quatre ou cinq lieues à l'entree dudit lac, il y a une riviere qui va aux Algoumequins vers le nort:[i] Et une autre qui va aux Irocois, par où lesdicts[j] Algoumequins & Irocois se font la guerre: Et un peu plus haut à la bande du Su dudit[k] lac, il y a une autre riviere qui va aux Irocois: puis venant à la fin dudit lac, ils rencontrent un autre sault, où ils portent leurs canots: de là ils entrent dedans un autre tres-grand lac qui peut contenir autant comme le premier: ils n'y[l] ont esté que fort peu dans ce dernier, & ont ouy dire qu'à la fin dudit[m] lac, il y a une

a lesdits] A2 lesdicts
b contenir] A2 cōtenir
c quelques six ou sept lieues;] A1 quelque six ou sept lieues;
d trainent] A2 trainēt
e cōme] A2 comme
f où] A1 ou
g premiers;] A2 premiers,
h quelque vingt ou 25. lieues:] A2 quelque 20. ou 25. lieuës:
i nort:] A2 Nort:
j lesdicts] A2 lesdits
k dudit] A2 dudict
l n'y] A2 ny
m dudit] A2 dudict

which may measure some nine or ten leagues from first to last. They said that the said rapids are not hard to cross and that they only drag their canoes in most of the said rapids, excluding two where they carry them.[211] From there, they come to enter a river that is like a kind of lake and may measure some six or seven leagues;[212] and then they pass five other rapids, where they drag their canoes as at the said first ones, excluding two, where they carry them as at the first ones. And they said that from first /[f. 26^{v}] to last, there are some twenty or 25 leagues.[213] Then they come into a lake that measures some hundred and fifty leagues in length and some four or five leagues at the entrance of the said lake.[214] Here there is a river that flows northward to the *Algoumequins*, and another that goes to the *Irocois*, by way of which the said *Algoumequins* and *Irocois* make war upon each other.[215] And a little higher up on the southern shore of the said lake, there is another river that goes to the *Irocois*.[216] Then coming to the end of the said lake, they meet with another waterfall, where they carry their canoes.[217] From there, they enter into another very big lake, which may be as big as the former.[218] They have only very rarely been in this last lake and have heard it said that at the end of

211 The first set of rapids between Lac Saint-Louis and Lac Saint-François spans about 19.5 km.

212 Lac Saint-François, from the end of the first set of rapids to the eastern tip of Cornwall Island, is 53 km.

213 The second set of rapids from the *Rapide de la Pointe Maligne* to *Fin des Rapides* at *Le Galop* is 65 km.

214 Lake Ontario is about 370 km long, along the north shore from Wolfe Island to the Niagara River. The width of the lake at the western end of Wolfe Island is about 17 km. The length of the lake given on Champlain's map *Carte Geographiqve, 1612*, is *15 Iournees des canaux des sauuages*. Since the Algonquins had no concept of a league, this suggests that Champlain converted *15 Iournees* to be equivalent to 150 leagues, making one *journée* (a day's trip) to be equivalent to 10 leagues. The Lake Ontario portion of this map was undoubtedly based on the Native sketch map obtained by Champlain during this interview.

215 The river route north to the Algonquin is the Cataraqui-Rideau system and not the Trent River, as given in Biggar, *The Works*, 1:160. The Trent leads northwest to the Huron, not the Algonquin. The first river to the Iroquois is the Oswego River, which flows from Lake Oneida. Both the river and lake (*lac des Irocois*) are on Champlain's *Carte Geographiqve, 1612*. The three Iroquois groups that can be reached by this route are the Oneida, Onondaga, and Cayuga.

216 The second river to the Iroquois (the Seneca) is the Genesee River, also marked on the 1612 map.

217 Niagara Falls is given as *sault de au* (probably *sault d'eau*: waterfall) on the 1612 map.

218 On the 1612 map, Lake Erie is given as *grand lac contenant 300. lieux de long.*

mer, do*n*t ils n'ont[a] veu la fin, ne ouy dire qu'aucun l'aye veue:[b] Mais que là où ils ont esté,[c] l'eau n'est point mauvaise, d'autant qu'ils n'ont point advancé plus haut:[d] & que le cours de l'eau vient du costé du Soleil couchant venant à l'Orient, & ne sçavent si passé ledit[e] lac qu'ils ont veu, il y a autre cours d'eau qui aille du costé de l'Occident: que le Soleil se couche à main droite dudit lac, qui est selon mon jugement au Norouest, peu plus ou moins, & qu'au /[f. 27[f]] premier lac l'eau ne gelle point, ce qui faict juger que le temps y est temperé, & que toutes les terres des Algoumequins est terre basse, re*m*plie[g] de fort peu de bois, & du costé des Irocois est terre montaigneuse, neantmoins elles sont tresbonnes[h] & fertilles, & meilleures qu'en aucun endroict qu'ils ayent veu: lesdits[i] Irocois se tiennent à quelque cinquante ou soixante lieues[j] dudit grand lac. Voilà au certain ce qu'ils m'ont dit[k] avoir veu, qui ne differe que bien peu au rapport des premiers.

Cedict jour nous feusmes proche de l'isle au Coudre,[l] co*m*me environ[m] trois lieues.[n] Le Jeudy 10. dudit mois, nous vinsmes à quelque lieue[o] & demie de l'isle au Lievre, du costé du Nort, où il vint d'autres Sauvages en nostre barque, entre lesquels il y avoit un jeune ho*m*me Algoumequin,* qui avoit fort voyagé dedans ledit[p] grand lac: nous l'interrogeasmes fort particulierement comme nous avions faict les autres Sauvages: il nous dit,[q] Que passé ledict sault que nous avions veu, qu'à quelque deux ou trois lieuës,[r] il y a une riviere qui va ausdicts Algoumequins,

a dōt ils n'ont] A2 dont ils n'ōt
b veue:] A2 veuë:
c esté,] A2 esté
d haut:] A2 haut,
e ledit] A2 ledict
f 27] A1 29
g rēplie] A2 remplie
h tresbonnes] A2 tres-bonnes
i lesdits] A2 lesdicts
j lieues] A2 lieuës
k dit] A2 dict
l au Coudre,] A1 aux Coudres,
m cōme enuiron] A2 comme enuirō
n lieues.] A2 lieuës.
o lieue] A2 lieuë
* RM] *Rapport d'vn ieune homme Sauuage Algoumequin.*
p ledit] A2 ledict
q dit,] A2 dict,
r lieuës,] A2 lieues,

the said lake there is a sea,[219] whose end they have not seen, nor have they heard that anyone else may have seen it; but they say that there, where they have been, the water is not bad, forasmuch as they have not penetrated farther upstream. They also say that the water flows from the direction of the setting sun towards the orient;[220] and they do not know whether, past the said lake that they have seen, there is another watercourse that might flow from the direction of the occident.[221] They say that the sun sets on the right hand of the said lake, which, in my judgment, is more or less to the northwest; and that /[f. 27] in the first lake, the water does not freeze, which makes me think that the climate there is moderate and that all the territory of the *Algoumequins* is lowland, covered with very few woods. And the land is mountainous on the *Irocois* side, yet very good and fertile, and better than in any other place they may have seen. The said *Irocois* keep a distance of some fifty or sixty leagues from the said great lake.[222] Here you have, for certain, what they told me they had seen, which differs only very little from the former report.[223]

This said day we went near the *isle au Coudre*, like about three leagues. On Thursday, the 10th of the said month,[224] we came within about a league and a half of the *isle au Lievre* on the north side, where other *Sauvages* came into our *barque*, among whom there was a young male *Algoumequin* who had travelled a lot in the said great lake.[225] We questioned him very particularly, as we had done with the other *Sauvages*. He told us that past the said rapids that we had seen[226] [fig. 8, bottom], that at some two or three leagues, there is a river that flows towards the

219 Lake Huron
220 He is saying that the water flows from the west ("setting sun") to the east (*orient*).
221 Champlain is asking them about rivers flowing from the west (*occident*).
222 The main Oneida villages were about 75 km from Lake Ontario; the Onondaga 65 km, the Cayuga 45 km, and the Seneca villages about 30 km.
223 Champlain did in fact get two similar reports reflecting well on his informants and the skill of his interpreters.
224 July
225 Lake Ontario
226 Lachine Rapids. Here begins the third of the three Algonquin accounts of the route westward.

où[a] ils sont cabannez, & qu'allant en ladicte grand' riviere[b] il y a cinq saults, qui peuve*n*t /[f. 27v] contenir du premier au dernier quelque huict ou neuf lieües,[c] dont il y en a trois où ils portent leurs canots, & deux autres où ils les trainent: que chacun desdicts saults[d] peut tenir un quart de lieuë de lo*n*g, puis viennent[e] dedans un lac qui peut co*n*tenir[f] quelques quinze lieues. Puis ils passent cinq autres saults, qui peuvent contenir du premier au dernier quelques vingt à[g] vingt cinq lieues, où il n'y a que 2. desdicts saults qu'ils[h] passent avec leurs canots;[i] aux autres trois ils ne les font que trainer. De là ils entrent dedans un grandissime lac, qui peut contenir quelques[j] trois cents lieues de long: Advançant quelques cent lieues dedans ledict lac, ils rencontrent une isle qui est fort grande, où au delà de ladicte isle, l'eau est salubre;[k] mais que passant quelque cent lieues plus avant, l'eau est encore plus mauvaise arrivant à la fin dudict lac, l'eau est du tout salee: Qu'il y a un sault[l] qui peut contenir une lieue[m] de large, d'où il dessend[n] un grandissime

a où] A1 ou
b grand' riuiere] A1 grand riuiere
c lieües,] A2 lieues,
d saults] A2 sauts
e lōg, puis viennent] A2 long, puis viennēt
f cōtenir] A2 contenir
g à] A1 a
h qu'ils] A2 qu,ils
i canots;] A1 A2 canots,
j quelques] A2 quelque
k salubre;] A2 salubre,
l sault] A2 saut
m lieue] A2 lieuë
n dessend] A2 descend

said *Algoumequins*,[227] where they are encamped, and that, going along the said great river, there are five rapids that may /[f. 27v] measure, from first to last, some eight or nine leagues; of which there are three where they carry their canoes and two others where they drag them.[228] He told us that each of the said rapids may extend for a quarter of a league in length. Then they come into a lake that may measure some fifteen leagues.[229] Then they pass five other rapids that may cover, from first to last, some twenty to twenty-five leagues, where there are only 2 of the said rapids that they cross with their canoes; at the other three, they only drag them.[230] From there, they enter a very large lake, which may measure some three hundred leagues in length.[231] Proceeding some hundred leagues into the said lake, they encounter an island that is very big, where beyond the said island the water is drinkable.[232] But he told us that, continuing on some hundred leagues farther, the water is still worse. Arriving at the end of the said lake, the water is totally salty.[233] He told us that there is a waterfall that may measure a league in width, from where an exceedingly great current of water descends into the said

227 It is 27 km to where the route to the Ottawa River turns north.

228 These are the first sets of rapids west of Lachine, between Lac Saint-Louis and Lac Saint-François. They are 19.5 km apart.

229 Lac Saint-François, some 53 km to the second set of rapids

230 The second sets of rapids from *Rapide de la Pointe Maligne* to *Fin des Rapides* at *Le Galop* are 65 km apart.

231 Lake Ontario is about 370 km long, along the north shore from Wolfe Island to the Niagara River.

232 This island is probably the complex, highly indented peninsula comprising Prince Edward County, Ontario, but it is only 37 km from Wolfe Island. On Champlain's map *Carte Geographiqve, 1612*, it is rendered tolerably well. We have translated *salubre* (wholesome) as "drinkable" from the survival perspective of the Natives. This option also allows a contrast with the "worse" water and "totally salty" water that follow, in a way that does not betray connotations appropriate to *salubre*, unlike the translation published by Purchas in 1625 and unfortunately repeated by the first Champlain Society edition; i.e., "brackish."

233 This statement is of course nonsense. Not only did the two earlier Algonquin accounts declare Lake Ontario to have wholesome (*salubre*) water, but it is physically impossible for a lake that is fed from the west and discharges to the east into the St. Lawrence to be salty at its western end and fresh or wholesome (*salubre*) at its eastern end. It is difficult to believe that an Algonquin who had been as far as Niagara Falls could concoct such a story, unless he was prompted. Did Champlain want to be told of a passage west to salt water? Did Champlain invent all this? Or did he get this information from European verbal or printed sources, such as Edward Hayes? (appendix 5). Why would he place his confidence in a "young male Algonquin" when two other reports by adults, which were similar, contradicted this story?

courant d'eau dans ledit[a] lac. Que passé ce sault,[b] on ne voit pl*us*[c] de terre, ny d'un costé ne d'autre, sinon une mer si grande, qu'ils n'en ont point veu la fin, ny ouy dire qu'aucun l'aye veue: Que le So-/[f. 28]leil[d] se couche à main droite dudict lac, & qu'à son entree[e] il y a une Riviere qui va aux Algoumequins & l'autre aux Irocois, par où ils se font la guerre. Que la terre des Irocois est quelque peu montaigneuse, neantmoins fort fertille, où il y a qua*n*tité de bled d'Inde, & autres fruicts qu'ils n'ont point en leur terre: Que la terre des Algoumequins est basse & fertille. Je leur demandis[f] s'ils n'avoient point congnoissance[g] de quelques mines. Ils[h] nous dirent, Qu'il y a une nation,[i] qu'on appelle les bons Irocois, qui viennent pour troquer des marchandises, que les vaisseaux François do*n*nent aux Algoumequins, lesquels disent qu'il y a à la partie du Nort une mine de franc cuivre, dont[j] ils nous en ont monstré quelques brasselets qu'ils avoie*n*t eu desdicts bons Irocois: Que si l'on y voulloit aller, ils y meneroient ceux qui seroient depputez[k] pour cest effect. Voylà tout ce que j'ay peu apprendre des uns & des autres, ne se differant que bien peu,* sinon que les seconds qui furent interrogez, dirent n'avoir point beu de l'eau salee, aussi ils n'ont pas esté si loing dans ledict lac comme les autres, & different quelque peu du chemin, les uns le faisant /[f. 28v] plus court, & les autres plus long: De façon que selon leur rapport, du sault où nous avons esté, il y a jusques à la mer salee, qui peut estre celle du Su, quelque 400. lieües:[l] sans doute, selon leur rapport, ce ne doit[m] estre autre chose que la mer du Su, le Soleil se couchant où ils disent. Le Vendredy dixiesme dudict mois nous fusmes de retour à Tadousac* où estoit nostre vaisseau.

a ledit] A2 ledict
b sault,] A2 saut,
c plus] A1 pl' (*abbreviation mark)*
d f. 28] A1 f. 27
e qu'à son entree] A1 qu'a son entree
f demandis] (*Literary past for a verb ending in -ir, when* demander *would normally be* demandai *in the first person.*)
g congnoissance] A2 cognoissance
h mines. Ils] A1 A2 mines, ils
i nation,] A2 Nation,
j dont] A2 dōt
k depputez] A2 deputez
* RM] *Peu de difference entre le rapport des Sauuages.*
l lieües:] A2 lieues:
m doit] A2 doibt
* LM] *Retour à Tadousac.*

lake.[234] Past this waterfall, one no longer sees any land, either on the one side or on the other, but a sea so great that they have not seen the end of it or heard tell of anyone who may have seen it.[235] He told us that the sun /[f. 28] sets on the right-hand side of the said lake[236] and that at its entrance there is a river that flows to the *Algoumequins* and another to the *Irocois*, by way of which they make war upon each other;[237] that the land of the *Irocois* is somewhat mountainous yet very fertile, where there is a great deal of Indian corn and other produce that they do not have in their own territory; and that the land of the *Algoumequins* is low and fertile. I asked them whether they had knowledge of any mines. They told us that there is a nation whom they call the good *Irocois*, who come to barter for merchandise which the French vessels furnish to the *Algoumequins*[238] and who say that there is, in the northern parts, a mine of pure copper, from which they showed us some bracelets that they had obtained from the said good *Irocois*.[239] They told us that if someone was wanting to go there, they would guide there those who would be appointed to this effect. This is all that I have been able to learn from both parties, only differing very little from each other, except that the second who were questioned said they had not drunk from the salt water; moreover, they have not been so far into the said lake as the others, and they differ somewhat concerning the path, some making /[f. 28/27^{v}] it shorter and others longer, in such a way that, according to their report, from the rapids where we were as far as the salt sea – which may be that of the South – there are some 400 leagues. Without a doubt, according to their report, this must be nothing other than the South Sea, the sun setting where they say it does.[240] On Friday, the tenth of the said month,[241] we were back in *Tadousac* where our vessel was.

234 The present Niagara Falls (American and Canadian) are about 1.12 km in width.

235 This sea (*mer*) is Lake Erie.

236 Lake Ontario

237 The Oswego River leads to the Iroquois and the Cataraqui-Rideau system to the Algonquins.

238 The "good Iroquois" were the Iroquoian-speaking *Ouendat* (Huron), who had not yet contacted French traders and were receiving French wares through Algonquin traders.

239 The *Ouendat* had trade connections north into Lake Superior, where the copper probably came from (Harris and Matthews, *Historical Atlas*, plates 33 and 35).

240 The *mer du Su* is the Pacific Ocean (*JR*, 45:221–3).

241 Friday was the 11th July.

Chap. X.

Voyage de Tadousac en l'Isle perçee, description de la baye des Moluës, de l'isle de bonne-adventure, de la baye des[a] *Chaleurs, de plusieurs rivieres, lacs, & pays où se trouve plusieurs sortes de mines.*

AUssi tost que nous fusmes arrivez à Tadousac, nous nous rembarquasmes pour aller à Gachepay,* qui est distant dudict Tadousac environ cent lieües.[b] Le treiziesme jour dudict mois, nous rencontrasmes une troupe[c] de Sauvages* qui estoient cabannez du costé[d] du Su, presque au millieu[e] du chemin de Tadousac à Gachepay, leur Sagamo qui les menoit s'ap-/[f. 29; H]pelle Armouchides, qui est tenu pour l'un des plus advisez & hardis qui soit entre les Sauvages: il s'en alloit à Tadousac pour troquer des flesches, & chairs d'Orignac, qu'ils ont pour des Castors & Martres des autres Sauvages Montaignez,[f] Estechemains & Algoumequins.

Le 15. jour dudict mois nous arrivasmes à Gachepay,* qui est dans une baye, comme à une lieüe & demye[g] du costé du Nort: ladicte[h] baye contient quelque sept ou huit lieües[i] de long, & à son entree quatre lieües[j] de large: il y a une Riviere qui va quelque trente lieües[k] dans les terres, puis nous vismes une autre baye que l'on appelle la baye des Molües,[l]* laquelle peut tenir quelques trois lieües[m] de long, autant de large à son entree: De là l'on vient à l'isle[n] perçee,* qui est comme un rocher fort haut, eslevé des deux costez, où il y a un trou par où les chalouppes

a des] A2 de
* LM] *Partement de Tadousac pour aller à Gachepay.*
b lieuës.] A2 lieues.
c troupe] A2 trouppe
* LM] *Rencontre de Sauuages.*
d costé] A2 coste
e millieu] A2 milieu
f Mõtaignez,] A1 Mõtaignez A2 Montaignz
* RM] *Nostre arriuee à Gachepay.*
g lieüe et demye] A2 lieuë et demie
h ladicte] A2 ladite
i huit lieües] A2 huict lieuës
j lieües] A2 lieues
k lieües] A2 lieues
l baye des Molües,] A2 Baye des Mollues,
* RM] *De la baye des Moluës.*
m lieües] A2 lieues
n l'isle] A2 l'Isle
* RM] *L'isle percee.*

Chapter 10

Voyage from Tadousac *to the* Isle perçee; *description of the* baye des Moluës, *the* isle de bonne-adventure, *the* baye des Chaleurs, *and the many rivers, lakes and regions where several kinds of mines are found*

As soon as we had arrived in *Tadousac*, we embarked again for *Gachepay*, which is distant from the said *Tadousac* by about a hundred leagues[242] [map 2]. On the thirteenth day of the said month, we encountered a band of *Sauvages* who were encamped on the south shore, almost midway between *Tadousac* and *Gachepay*. Their *Sagamo* who led them is called /[f. 29] Armouchides and is held to be one of the most informed and daring there could be among the *Sauvages*.[243] He was on his way to *Tadousac* to trade some arrows and moose meat, which they have, for beaver and marten from the other *Sauvages*: the *Montaignez, Estechemains*, and *Algoumequins*.

On the 15th of the said month, we arrived in *Gachepay*, which is in a bay about a league and a half from the north side.[244] The said bay measures some seven or eight leagues in length and, at its entrance, four leagues in width.[245] There is a river that flows some thirty leagues inland.[246] Then we saw another bay which they call the *baye des Molües*, which may extend some three leagues in length and as much in width at its entrance.[247] From there, one comes to the *isle perçee*, which is like a very high rock, lofty from both sides, where there is a hole through

242 The distance from Tadoussac to Gaspé is about 480 km.

243 The name of the headman Armouchides implies that he may have been a member of the *Armouchiquois* (*Almouchiquois*), a group of culturally and linguistically related Algonquian speakers, who lived along the southern New England coast from about the present Kennebunkport, Maine, to at least southern Cape Cod. The main component groups were the Pawtucket, Massachusetts, and Pokanoket. To the north were the Eastern Abenaki and beyond them farther north the Etechemin comprising the Maliseet and Passamaquoddy.

244 What is meant here is Gaspé, the small fishing station on the south side of the Baie de Gaspé, on the eastern tip of the peninsula separating the York and Dartmouth Rivers. The Gaspé fishing station is 4 km from the north side of the bay.

245 Baie de Gaspé is 33 km deep to the mouth of the Dartmouth River and 13 km wide at its mouth.

246 Probably the Rivière York which flows into the basin adjacent to the fishing station, rather than the Rivière Dartmouth which flows into the basin north of the station.

247 *Baye des Molües* is now Baie de Malbaie. It is 8 km deep and 10 km across at its mouth.

& bateaux peuvent passer de haute mer: & de basse mer, l'on peut aller de la gra*n*d terre à ladite isle, qui n'en est qu'à quelque quatre ou cinq cens pas. Plus il y a une autre isle co*m*me au Suest de l'isle percee,[a] environ une lieuë,[b] qui s'appelle l'isle de Bonne adventure,* & peut tenir de long demie lieuë.[c] /[f. 29v] Tous cesdits lieux de Gachepay, baye des Molües,[d] & Isle percee,[e] sont les lieux où il se faict la pesche du poisson sec & verd.

Passant l'isle percee,[f] il y a une baye qui s'appelle la baye[g] de Chaleurs,* qui va comme à[h] l'Ouest Sorouest, quelques[i] quatre-vingts lieues dedans les terres, contenant de large en son entree quelques quinze lieues: Les Sauvages Canadiens disent, qu'à la gra*n*de riviere de Canadas, environ quelques soixante lieues, rengeant la coste du Su, il y a une petite riviere qui s'appelle Mantanne, laquelle va quelques dixhuict lieues dans les terres, & estans au bout d'icelle ils portent leurs canots[j] environ une lieue par terre, & se viennent rendre à ladite baye de Chaleurs, par où ils vont quelquesfois à l'isle percee:[k] Aussi ils vont de ladicte baye[l] à Tregate & Misamichy.* Continuant ladite coste,[m] on renge quantité de rivieres,[n] & vient-on à un lieu où il y a une riviere qui s'appelle Souricoua, où le Sieur Prevert a esté* pour descouvrir une mine de cuivre. Ils vont avec leurs canots[o] dans ceste riviere deux ou trois jours,[p] puis ils traversent quelques deux ou trois lieues de terre, jusques à[q] ladite mine, qui est sur le bord de la mer du /[f. 30][r] costé du

a percee,] A2 perçee,
b lieuë,] A2 lieue
* RM] *L'Isle de Bonne aduenture.*
c demie lieuë.] A2 demye lieue.
d Molües,] A2 Molues,
e percee,] A2 perçee,
f percee,] A2 perçee,
g baye] A2 baie
* LM] *De la baye de Chaleurs.*
h à] A1 A2 a
i quelques] A2 quelque
j canots] A2 Canots
k l'isle percee:] A2 l'Isle perçee.
l baye] A2 Baye
* LM] *De Tregate & Misamichy.*
m coste,] A2 coste
n riuieres,] A2 Riuieres,
* LM] *Riuiere où a esté le Sieur Preuert.*
o canots] A2 Canots
p iours,] A2 jours
q à] A2 a
r f. 30] A2 f. 29

which shallops and boats can pass at high tide; and at low tide, one can walk from the mainland to the said island, which is only some four or five hundred paces away.[248] Moreover, there is another island about a league from the *isle percee* towards the southeast, which is called the *isle de Bonne adventure* and may extend half a league in length[249] /[f. 29v]. All these said places, *Gachepay*, *baye des Molües*, and *Isle percee*, are places where fishing for dry and green fish is done.[250]

Crossing the *isle percee*, there is a bay that is called the *baye de Chaleurs*, which goes about west-southwest some eighty leagues inland, measuring some fifteen leagues in width at its entrance.[251] The *Sauvages Canadiens*[252] say that when they skirt the south shore for about sixty leagues into the great *riviere de Canadas*, there is a little river called *Mantanne*, which flows some eighteen leagues inland; and when they are at the end of it, they carry their canoes overland for about a league and come to the said *baye de Chaleurs*, by which they sometimes proceed to the *isle percee*.[253] They also go from the said bay to *Tregate* and *Misamichy*.[254] Continuing along the said coast, one skirts a great number of rivers and comes to a place where there is a river called *Souricoua*, where the *Sieur Prevert* had gone to discover a copper mine. They go up this river with their canoes for two or three days; then they cross overland for about two or three leagues of land as far as the said mine, which lies beside the sea on /[f. 30] the south shore. At the mouth of

248 The walking distance to Île Percée is about 250 m, which is correct at 400 to 500 paces but can only be reached at low tide. Judging from his narrative and the 1612 map, this is as far along the coast that Champlain got in 1603. The remainder of the geography eastward is from hearsay.

249 The island is still so called and is about 2.8 km long.

250 For the seventeenth-century green and dry fisheries on the coast of Acadia, see Ganong, *Denys: Description*, 257–348.

251 The Baie des Chaleurs was named by Jacques Cartier in 1534. Its entrance at Miscou Island is about 30 km and its length about 130 km.

252 This seems to be an early name for the *Gaspesiens*, the western branch of the Mi'kmaq. Champlain marked them on the north shore of Baie des Chaleurs; see (maps), Champlain, *Carte Geographiqve, 1612*. In the *Relation* of 1634, Father Le Jeune wrote that *les Canadiens ... habitant vers Gaspé* (*JR*, 6:184). Lescarbot wrote that the people of Gaspé and Baie des Chaleurs called themselves *Canadoquoa*, which the French pronounced *Canadaquois* (Lescarbot, *Histoire*, 250). In the 1612 edition of the *Histoire* (230), Lescarbot spelled them *Canadaquoa*.

253 Champlain was being told of the canoe route partway up the Rivière Matane, then by a portage to the Rivière Matapédia, and down to the Rivière Ristigouche, which empties into the Baie des Chaleurs.

254 It may be that the canoe route to Tracadie went via the Big Tracadie River and the one to Miramichi by way of the Nipisiguit River and a portage to the northwest Miramichi River.

Su: A l'entree de ladite riviere,[a] on trouve une isle environ une lieue dans la mer, de ladicte isle jusques à l'isle percee, il y a quelques[b] soixante ou septante lieues, puis continuant ladite coste qui va devers l'Est on rencontre un destroit[c]* qui peut tenir deux lieues de large, & vingt-cinq[d] de long. Du costé de l'Est est une isle qui s'appelle sainct Laurens,[e] où est le cap Breton, & où une nation de Sauvages appellez les Souricois hivernent.* Passant le destroit[f] de l'isle de sainct Laurens,[g] costoyant la coste d'Arcadie, on vient deda*ns* une baye qui vient joindre ladicte mine de cuivre.* Allant plus outre, on trouve une riviere qui va quelques soixante ou quatre-vingts[h] lieues dedans[i] les terres, laquelle va proche du lac des Irocois,* par où lesdicts Sauvages de la coste d'Arcadie leur vont faire la guerre.

a ladicte riuiere,] A2 ladite riuiere.
b quelques] A2 quelque
c destroit] A2 destroict
* RM] *Destroit entre la grande terre & vne isle.*
d vingt-cinq] A2 vingt cinq
e Laurens,] A2 Laurent,
* RM] *Souricois, & où ils hiuernent.*
f destroit] A2 destroict
g Laurens,] A2 Laurent,
* RM] *De la mine de cuiure.*
h quatre-vingts] A2 quatre-vingt
i dedans] A2 dans
* RM] *Riuiere à la coste d'Arcadie, allant proche du lac des Irocois.*

the said river, one finds an island about a league out to sea.[255] From the said island up to the *isle percee*, there are about sixty or seventy leagues.[256] Then continuing along the said coast that goes towards the east, one encounters a strait that may extend two leagues in width and twenty-five in length.[257] On the east side is an island that is called *sainct Laurens*, where the *cap Breton* is and where a nation of *Sauvages* called the *Souricois* winters.[258] Passing the strait of the *isle de sainct Laurens*, coasting along the shore of *Arcadie*,[259] one enters a bay that goes as far as the said copper mine.[260] Going farther, one finds a river that goes some sixty or eighty leagues inland and approaches the *lac des Irocois*,[261] by way of which the said *Sauvages* of the coast of *Arcadie* go to make war upon them.[262]

255 The identity of the *Rivière Souricoua* is conjectural. There is no river by this name on the seventeenth-century maps. The word *Souricois* was an early name for the Mi'kmaq who inhabited what is now Nova Scotia. What is being described here is a short route from Northumberland Strait south to either Chignecto Bay or the Minas Basin, the north-eastern and northwestern arms of the Bay of Fundy. It may be that this route ran from Baie Verte across the Chignecto Isthmus by means of Mill Brook, Portage Lakes, Patten Lake, and the Missaguash River to Cumberland Basin, which forms the northeastern part of Chignecto Bay, a distance of about 30 km. This is the only consistent canoe route between Northumberland Sound and any part of the Bay of Fundy marked on the early maps. This river, swamps, and lakes system now forms part of the boundary between New Brunswick and Nova Scotia; see (maps), [Franquet], "Carte Particuliere," [1751]; Wyld, *A Map of the Provinces*, 1825. According to Gagnon (Biggar, *The Works*, 6:233), there was a route from Tatamagouche Bay, up the French River across to the Chiganois River, and down to the Minas Basin, a distance of at least 50 km. Although much longer and more difficult than the Missaguash route, this route seems to correspond better to the river marked as 29: *Riuiere par où l'on va à la Baye Françoise* on Champlain's map *Carte de la nouuelle france, 1632*.

256 The distance from Île Percé to Baie Verte is about 350 km and to Tatamagouche Bay about 430 km.

257 The Strait of Canso is 27 km long and from 1 to 2 km wide.

258 *Isle Sainct Laurens* is an early name for Cape Breton Island. Cap Breton is the most eastern cape on the island at 45°56.5'N.

259 *Arcadie*, see above, footnote 1.

260 The Bay of Fundy. In 1604 Champlain found some copper at Advocate Harbour, Nova Scotia.

261 The only river of this length flowing into the Bay of Fundy is the Saint John, but it comes from nowhere close to Lake Champlain, unless the Mi'kmaq first travelled to the St. Lawrence and then approached the Iroquois up the Rivière Richelieu; see (maps), Aubry, "Carte Pour les hauteurs," 1715, which shows canoe routes connecting the Saint John River to Rivières du Loup, Ouelle, and La Chaudière.

262 them: the Iroquois. The antecedent of "them" arises in the lake of the Iroquois (*lac des Iroquois*), which Champlain did not necessarily consider a place name.

Ce[a] seroit un grand bien qui pourroit trouver à la coste de la Floride quelque passage qui allast donner proche du susdict gra*n*d lac,[b] où l'eau est sallee, tant pour la navigation des vaisseaux, lesquels ne seroient subjects[c] à tant de perils comme ils sont en Canadas, que pour l'accourcissement du chemin de plus de trois cens[d] lieues. Et est tres-cer-/[f. 30v]tain qu'il y a des rivieres en la coste de la Floride que l'on n'a point encores descouvertes, lesquelles vo*n*t[e] dans les terres, où le pays y est tres-bon & fertille, & de fort bo*ns*[f] ports. Le pays & coste de la Floride peut avoir un autre temperature de temps, plus fertille en quantité de fruicts, & autres choses que celuy que j'ay veu: mais il ne peut y avoir des terres plus unies ny meilleures que celles que nous avons veuës. Les Sauvages disent qu'en ladite[g] grand' baye[h] de Chaleurs il y a une riviere qui va quelques[i] vingt lieues dans les terres, où au bout* est un lac[j] qui peut contenir quelques[k] vingt lieues, auquel il y a fort peu d'eau, qu'en Esté il asseiche, auquel ils trouvent dans la terre, environ un pied ou pied & demy,[l] une maniere de metail qui ressemble à de l'argent que je leur avois monstré, & qu'en un autre lieu proche dudit[m] lac il y a une mine de cuivre. Voilà ce que j'ay appris[n] desdicts Sauvages. /[f. 31]

Chap. XI.

Retour de l'Isle percee[o] *à Tadousac, avec la description des ances, ports, rivieres, isles, rochers, saults,*[p] *bayes, & basses, qui sont le long de la coste du Nort.*

a guerre. ¶ Ce] A1 A2 guerre: Ce
b susdict grãd lac,] A2 susdit grand lac,
c subiects] A2 subiect
d cens] A2 cent
e võt] A2 vont
f bõs] A2 bons
g ladite] A2 ladicte
h grand' baye] A1 grand baye
i quelques] A2 quelque
* LM] *Rapport fait des Sauuages d'vne riuiere qui va dans les terres, ou au bout de laquelle il se trouue vne maniere de metail.*
j lac] A2 lac,
k quelques] A2 quelque
l demy,] A1 demy
m dudit] A2 dudict
n appris] A2 apprins
o *percee*] A2 *perçee*
p saults] A1 ponts (*well corrected, 1604*)

It would be a great asset if someone could find some passage on the coast of Florida that would come out near the above-said great lake where the water is salty,[263] both for the navigation of the vessels that would not be subjected to so many perils as they are in *Canadas*, and for the shortening of the route by more than three hundred leagues. And it is quite certain /[f. 30ᵛ] that there are rivers on the coast of Florida that have not been explored yet, and that go inland where the land is very good and fertile, with very good harbours. The land and coast of Florida may have another seasonal temperature and a greater abundance of produce and other things than that which I have seen, but there cannot be lands more level or of better quality than those that we have seen. The *Sauvages* say that in the said great *baye de Chaleurs*, there is a river that flows some twenty leagues inland, at the end of which is a lake[264] that may measure some twenty leagues, in which there is very little water – they say that it dries up in the summer – and in which they find a kind of metal that resembles the silver I had shown them, about a foot or a foot and a half below the ground. And they say that in another place near the said lake, there is a copper mine. This is what I learned from the said *Sauvages*.

/[f. 31] Chapter 11

Return from the Isle perçee *to* Tadousac, *with the description of the coves, harbours, rivers, islands, rocks, rapids,*[265] *bays, and shallows which are found along the north coast*

263 Champlain probably means the west end of Lake Ontario and Lake Erie, the lakes described as being salty by the "young male Algonquin" in the third account he obtained of the lakes west of Montreal. It is likely that in this passage Champlain is using conjectures from Edward Hayes's *A Treatise*. See appendix 5.

264 This could be the Rivière Matapédia, which flows from Lac Matapédia into the Ristigouche River and from there into the Baie des Chaleurs.

265 rapids: *saults* represents the 1604 reading, which has been preferred over the *ponts* (bridges) of the earlier edition, not only because it makes more sense, but also because the equivalent chapter heading in the "Table of Chapters" at the beginning uses *saults*. Since the chapter headings are, in all other cases, consistently duplicated between their placement in the table and in the text, *ponts* must be an error.

NOus partismes de l'isle perçee* le dix-neufiesme dudict mois pour retourner à[a] Tadousac: Comme nous fusmes à quelque trois lieuës[b] du Cap l'Evesque,[c] nous fusmes contrariez d'une tourmente* laquelle dura deux jours, qui nous fit relascher dedans une grande anse[d] en attendant le beau temps. Le lendemain nous en partismes & fusmes encores contrariez d'une autre tourmente:[e]* Ne voullant relascher, & pensant gaigner chemin nous fusmes à la coste du Nort le 28. jour de Juillet mouiller l'ancre* à une ance qui est fort mauvaise, à cause des bancs de Rochers qu'il y a, ceste ance est par les 51.[f] degrés[g] & quelques minutes.[h] Le lendemain nous vinsmes mouiller l'ancre proche d'une riviere qui s'appelle saincte Marguerite,* où il y a de plaine mer quelque trois brasses d'eau, & brasse & demye[i] de /[f. 31v] basse mer; elle va assez avant. A ce que j'ay veu dans terre du costé de l'Est, il y a un sault d'eau qui entre dans ladicte Riviere, & vient de quelque cinquante ou soixante brasses de haut,[j] d'où procede la plus grand part de l'eau qui dessend[k] dedans: A son entree il y a un banc de sable, où il peut avoir de basse eau demy brasse.

Toute[l] la coste du costé de l'Est est sable mouvant,* où il y a une poincte à quelque demye[m] lieuë de ladicte Riviere qui advance une demye[n] lieuë en la mer: & du costé de l'Ouest, il y a une petite isle,[o] cedict lieu est par les 50. degrez.

* RM] *Partement de l'isle perçee.*
a à] A1 a
b lieuës] A2 lieues
c Cap l'Euesque,] A1 Cap l'Euesque
* RM] *Tourmente.*
d grande anse] A2 grand' ance
e tourmente:] A2 tourmente;
* RM] *Autre tourmente.*
* RM] *Coste du Nort où nous relaschasmes.*
f 51.] A2 15.
g degrés] A1 A2 degré
h minutes.] A2 minuttes.
* RM] *De la riuiere Saincte Marguerite.*
i demye] A2 demie
j haut,] A2 hault,
k dessend] A2 descend
l brasse. ¶ Toute] A1 A2 brasse: Toute
* LM] *Coste sablonneuse.*
m demye] A2 demie
n demye] A2 demie
o isle,] A2 Isle,

We set out from the *isle percee* on the nineteenth of the said month[266] to return to *Tadousac*. When we were some three leagues from *Cap l'Evesque*,[267] we were delayed by a storm, which lasted two days and forced us to put into a great inlet while waiting for fair weather.[268] The following day we set out from there and were again delayed by another storm. Not wishing to take shelter and thinking to make headway, we went to the north shore on the 28th day of July to cast anchor in a cove that is a very bad place because of the rocky banks there. This cove is at 51 degrees and some minutes.[269] The following day, we came to cast anchor near a river that is called *saincte Marguerite*, where there are some three *brasses* of water at high tide and a *brasse* and a half at /[f. 31v] low tide; it goes fairly far inland. By what I saw of the east shore inland, there is a fall of water that enters the said river and falls from a height of some fifty or sixty *brasses*, from where the greatest part of the water proceeds that descends into it. At its mouth there is a sandbank, where there may be half a *brasse* at low tide.[270]

The entire coast on the east side is shifting sand, where there is a point, at about half a league from the said river, which projects half a league out into the sea.[271] And on the west side, there is a small island.[272] This said place is at 50 degrees.[273]

266 19 July 1603

267 On his map of 1612, Champlain marked *C. aleuesque* on a point adjacent to an inlet about halfway between *Gaspay* and *C. de chate*; see (maps), Champlain, *Carte Geographiqve, 1612*. This point seems to be Cap de la Madeleine at the entrance to the mouth of the Rivière Madeleine, which is a prominently marked refuge on many seventeenth-century charts. It is halfway (110 km) from Cap de Gaspé to Cap Chat.

268 The mouth of the Rivière Madeleine

269 Champlain is now on the north shore of the St. Lawrence River. The location of this "cove" is too vague. It is possible he is near Baie de Moisie, marked on his 1612 map as having shoals. The latitude notation of "51° and some minutes" is a full degree too far north for any section of the north side of the St. Lawrence west of Anticosti Island.

270 Still called Rivière Sainte-Marguerite. The falls are 7 km upstream. The sandbank (Plage Sainte-Marguerite) is still present.

271 This is probably Pointe à la Chasse separating Baie des Sept Îles from Rivière Sainte-Marguerite.

272 There is still a small island to the west of the mouth of the river.

273 The mouth of Rivière Sainte-Marguerite is at 50°08.5'N.

Toutes ces terres sont tres-mauvaises,[a]* re*m*plies[b] de sapins: la terre y est quelque peu haute,[c] mais non tant que celle du Su. A quelques trois lieües[d] de là nous passames[e] proche d'une autre riviere[f]* laquelle sembloit estre fort grande, barree neantmoins la pluspart de rochers: A quelque 8. lieuës[g] de là il y a une poincte qui advance une lieue & demye[h] à la mer,* où il n'y a que brasse & demye[i] d'eau: Passé ceste pointe[j] il s'en trouve une autre* à quelque 4. lieues où[k] il y a assez d'eau: Toute ceste coste est terre basse & sablonneuse. A quelque 4. lieues de là il y a une ance* où entre une /[f. 32] Riviere; il[l] y peut aller beaucoup de vaisseaux du costé de l'Ouest; c'est[m] une pointe basse qui advance environ d'une lieue en la mer, [et] il faut renger la terre de l'Est comme de trois cens pas pour pouvoir entrer dedans: Voylà le meilleur port qui est en toute la coste du Nort, mais il y fait[n] fort dangereux y aller pour les basses, & ba*n*cs de sable qu'il y a en la pluspart de la coste pres de deux lieues[o] à la mer. On trouve à quelques[p] six lieues de là une baye,* où il y a une isle de sable; toute ladite baye est fort baturiere, si ce n'est du costé de l'Est, où il peut avoir quelques 4. brasses[q] d'eau.

a tres-mauuaises,] A1 tres-mauuaises A2 tresmauuaises,
* LM] *Terres mauuaises.*
b rẽplies] A2 remplies
c haute,] A2 haulte,
d quelques trois lieües] A2 quelque trois lieues
e passames] A2 passasmes
f riuiere] A2 riuiere,
* LM] *Riuiere.*
g lieuës] A2 lieues
h demye] A2 demie
* LM] *D'vne pointe qui aduance à la mer.*
i demye] A2 demie
j pointe] A2 poincte
* LM] *D'vne autre pointe.*
k où] A1 A2 ou
* LM] *D'vne bonne ance où il peut quantité de vaisseaux.*
l Riuiere; il] A1 A2 Riuiere, il
m l'Ouest; c'est] A1 A2 l'Ouest, c'est
n fait] A2 faict
o de deux lieues] A2 deldeux lieuës
p quelques] A1 quelque
* RM] *Baye.*
q quelques 4. brasses] A1 quelque 4. brasse

All these lands are very poor and filled with fir trees. The land there is somewhat high, but not so much as that to the south. At some three leagues from there, we passed near another river that seemed to be very large, yet blocked off by rocks for the most part.[274] At some 8 leagues from there, there is a point that projects a league and a half into the sea, where there is only a *brasse* and a half of water.[275] Past this point is found another, some four leagues away, where there is enough water.[276] This whole coast is lowland and sandy. At some 4 leagues from there is a bay where a /[f. 32] river enters.[277] Many vessels may go there from the west side. It is a low point that runs out about a league into the sea. One must skirt the land on the east for about three hundred paces to be able to make an entrance. This is the best harbour along the whole of the north coast; but the shallows and sandbanks, which are there for nearly two leagues into the sea along most of the coast, make it very dangerous to go there. At some six leagues from there one finds a bay where there is a sandy island.[278] All the said bay is chock full of rocky tidal flats and reefs,[279] except on the east side, where it is possible to have some 4 *brasses* of water.

274 This is a good description of the Rivière aux Rochers, some 27 km from Rivière Sainte-Marguerite.

275 Pointe des Monts, about 88 km from Rivière aux Rochers

276 The most prominent point is Pointe à la Croix, 35 km from Pointe des Monts.

277 The mouth of Rivière Manicouagan is 30 km from Pointe à la Croix.

278 Baie aux Outardes, 33 km from the Rivière Manicouagan

279 chock full of tidal flats and reefs: *fort baturiere*. The adjective *baturiere* derives from *batture* (s.f.), used in 1529 to designate rocks situated below the surface of the water (i.e., shoals), but then later in Canada, the zone that is uncovered at low tide (*DHLF*, 1:355). Since we translate *basses* as "shoals," which are not uncovered at low tide, we have attempted to distinguish this variation more emphatically; see Beaulieu and Ouellet, *Champlain*, 239.

Dans[a] le canal qui entre dans ladite baye à[b] quelque 4. lieues de là, il y a une belle ance* où[c] entre une Riviere: Toute ceste coste est basse & sablonneuse,* il y dessend[d] un sault d'eau qui est grand. A quelque cinq lieues de là, il y a une pointe qui advance environ demye lieue en la mer où[e] il y a une ance;[f] & d'une poincte à l'autre y a trois lieues; mais ce n'est que battures où[g] il y a peu d'eau. A quelques deux lieues il y a une plage où il y a un bon port, & une petite Riviere, où il y a trois isles, & où des vaisseaux se pourroient mettre à[h] l'abry. A quelque trois lieues de là il y a* /[f. 32v] une pointe de sable qui advance environ une lieue, où au bout il y a un petit islet. Puis allant à Lesquemin vous rencontrez 2. petites isles basses,* & un petit rocher à terre. Cesdictes isles sont environ à demie lieue de Lesquemin, qui est un fort mauvais port,* entourné de rochers, & [qui] asseche de basse mer, & [il] faut[i] variser pour entrer dedans au derriere d'une petite poincte de rocher, où il n'y peut qu'un vaisseau: Un peu plus haut, il y a une Riviere* qui va quelque peu dans les terres: c'est le lieu où les Basques font la pesche des ballaines.[j] Pour dire verité le port ne vaut du tout rien. Nous vinsmes de là audict port de Tadousac* le troisiesme d'Aoust. Toutes cesdictes[k] terres cy dessus sont basses à la coste, & dans les terres fort hautes. Ils ne sont si plaisantes ny fertilles que celles du Su, bien qu'elles soient plus basses. Voilà au certain tout ce que j'ay veu de ceste-dicte coste du Nort. /[f. 33; 1]

a d'eau. ¶ Dans] A1 A2 d'eau: dans
b à] A1 a
* RM] *Ance.*
c où] A1 ou
* RM] *Coste sablonneuse.*
d dessend] A2 descend
e où] A1 ou
f ance;] A2 ance,
g où] A1 ou
h à] A1 a
* RM] *D'vne pointe qui aduance à la mer.*
* LM] *De deux isles.*
* LM] *Port de Lesquemain.*
i faut] A2 fault
* LM] *Riuiere.*
j ballaines.] A2 Ballaines.
* LM] *Arriuee à Tadousac.*
k cesdictes] A2 cesdites

Within the channel that enters the said bay at some four leagues from there, there is a fine cove into which a river empties. This entire coast is low and sandy. A great waterfall descends there.[280] At some five leagues from there, there is a point that projects about half a league into the sea,[281] where there is a cove, and from one point to the other there are three leagues, but where there is little water, it is nothing but rocks.[282] At about two leagues, there is a beach, where there is a good harbour and a small river, in which there are three islands, and where vessels could be sheltered.[283] At some three leagues from there, there is a /[f. 32ᵛ] sandy point that projects out for about a league, where there is a little islet at the end.[284] Then, going on towards *Lesquemin*, you encounter two low little islands and a small rock on land. These said islands are about half a league from *Lesquemin*,[285] which is a very bad harbour, surrounded by rocks, and which dries up at low tide. And it is necessary to tack in order to make an entrance behind a little rocky point, where only one vessel can fit. A little higher up, there is a river that goes a little way inland. This is the place where the Basques go whaling.[286] To tell the truth, the harbour is not worth anything at all. We came from there to the said port of *Tadousac* on the third of August. All these aforesaid lands are low at the shore and very high inland. They are neither as pleasant nor as fertile as those of the south, although they may be lower. This is unquestionably everything that I saw of this said north shore.

280 Champlain is describing the entrance of the Rivière aux Outardes, with its sandy shoals. The "great waterfall" is the Chute aux Outardes near the town of the same name, 11 km from the mouth of the river.

281 Pointe à Michel at the southern entrance to the mouth of the Rivière Betsiamites, some 23 km from Baie aux Outardes

282 rocks: *battures*. See note 279.

283 The Îlets Jérémie. The beaches are still present.

284 This is a good description of Cap Colombier, 9 km from Îlets Jérémie.

285 The Îles Escoumins stretch for about 2.5 km along the coast 3 km north of the mouth of Rivière des Escoumins.

286 Basque activity sites in this area of the St. Lawrence have been located at Les Escoumins and 11 km to the south at Bon-Désir, as well as on Île aux Basques, 26 km across the St. Lawrence River to the southeast from the previous places. The description of the harbour of Les Escoumins is good.

CHAP. XII.

Les[a] *ceremonies que font les Sauvages devant que d'aller*[b] *à la guerre: Des Sauvages Almouchicois, & de leur monstrueuse forme. Discours du sieur de Prevert de sainct Malo, sur la descouverture de la coste d'Arcadie, quelles mines il y a, & de la bonté & fertilité du pays.*

ARrivant à Tadousac nous trouvasmes[c] les Sauvages que nous avions rencontrez en la riviere des Irocois,* qui avoient faict rencontre au premier lac[d] de trois Canots Irocois, lesquels se battire*nt*[e] contre dix autres de Montaignez, & apporterent les testes des Irocois à Tadousac, & n'y eust[f] qu'un Montaignez blessé au bras d'un coup de fleche,[g] lequel songeant quelque chose, il falloit que tous les 10. autres le meissent en executio*n* pour le rendre content, croyant aussi que sa playe s'en doit[h] mieux porter.* Si cedict Sauvage meurt, ses parens vengeront sa mort, soit sur leur nation, ou sur d'autres, ou bien il faut que les Capitaines facent des presens aux parens du deffunct, à fin qu'ils soient contens, ou autrement, comme j'ay dit, ils useroient de vengeance,[i] qui est une grande meschanceté entre eux.

Premier que lesdits[j] Mo*n*taignez par-/[f. 33v]tissent pour aller à la guerre, ils s'assemblerent tous, avec leurs plus riches habits de fourreures, castors, & autres peaux, parez de Patenostres & cordons de diverses couleurs, & s'assemblerent

a *Les*] A2 *Lees*

b *deuant que d'aller*] A2 *deuant que, d'aller* (*mod. Fr.* avant d'aller; *confusion here between prep. and conj.*)

c trouuasmes] A2 trouuasnes

* RM] *Sauuages que nous trouuasmes reuenenans de la guerre, lesquels nous auions rencontré à la riuiere des Irocois.*

d lac] A1 lac,

e battirẽt] A2 battirent

f n'y eust] A2 ny eut

g fleche,] A2 flesche,

h doit] A2 doibt

* RM] *Sauuages couppent la teste à leurs ennemis.*

i vengeance,] A2 vengeance:

j lesdits] A2 lesdicts

/[f. 33] Chapter 12

The ceremonies that the Sauvages *perform before going to war. Of the* Almouchicois Sauvages *and their monstrous shape. Discourse of the* sieur de Prevert de sainct Malo *on the exploration of the coast of* Arcadie; *what mines are there, and of the goodness and fertility of the land*

Arriving at *Tadousac*, we found the *Sauvages* we had met in the *riviere des Irocois,*[287] who, at the first lake, had encountered three *Irocois* canoes that fought against ten others of the *Montaignez*, and who brought the heads of the *Irocois* to *Tadousac*. And there was only one *Montaignez* wounded in the arm by an arrow shot, who on dreaming something, made it necessary[288] for all the other 10 to execute it to make him happy, also believing that his wound ought to be the better for it.[289*] If this said *Sauvage* dies, his relatives will avenge his death, either upon their nation or upon others, or else the captains must give presents to the relatives of the dead man in order that they may be happy. Or otherwise, as I said, they would take revenge, which is a great wickedness among them.[290]

Before the said *Montaignez* would set out /[f. 33v] to go to war, they all assembled in a great public place, with their richest costumes of fur, beaver, and other pelts, adorned with beads[291] and cords of various colours, where there was a

287 Although Champlain was not specific about the identity of the Natives he had previously met at the Rivière Richelieu, a number of references suggest they were Algonquin and Montagnais.

288 Made it necessary: *il falloit que*. The causative has been used to maintain the one Montagnais as subject of the rest of the sentence, although *il falloit que* is an impersonal expression: lit. it was necessary.

289 This is the second time on the 1603 voyage that Champlain mentioned the importance of dream interpretation in native religion and behaviour.

* RM] Sauvages *cut the heads off their enemies.*

290 It was important that a dream be interpreted correctly and then acted out. Failure to do so might result in the death of the afflicted. If that was the case, the relatives of the dead person could avenge themselves not only on those who caused the primary physical injury (in this case the Iroquois) but also those who did not carry out the instructions dictated in the dream that might have cured him. In most cases where neighbours or villagers were unsuccessful in resolving the dream, instead of being subject to vengeance by the relatives of the deceased, they were required to give gifts to assuage their grief.

291 Beads: *Patenostres* (patenôtre) were small French glass beads about the size and shape of those in a rosary (Kenyon, "Sagard's 'Rassade Rouge,'" 6–7). It is probable that the word *patenostres* was a general term for all beads in the sixteenth and early seventeenth centuries (Turgeon, "French Beads," 72).

dedans une grand' place publicque, où il y avoit au deva*n*t d'eux un Sagamo qui s'appelloit Begourat qui les menoit à la guerre, & estoient les uns derriere les autres, avec leurs arcs & flesches, massuës, & rondelles dequoy ils se parent pour se battre:[a] & alloient sautant les uns apres les autres, en faisant plusieurs gestes de leurs corps ils faisoient maints tours de limaçon: apres ils commencerent à danser à la façon accoustumee, comme j'ay dit cy dessus, puis ils feirent leur Tabagie, & apres l'avoir faict, les femmes se despouillerent toutes nues,[b] parees de leurs plus beaux Matachias, & se meirent dedans leurs canots[c] ainsi nues[d] en dansant, & puis elles se vindrent mettre à l'eau en se battant à coups de leurs avirons, se jettans quantité d'eau les unes sur les autres: toutesfois elles ne se faisoient point de mal, car elles se paroient des coups qu'elles s'entre-ruoient:[e] Apres avoir faict toutes ces ceremonies, elles se retirerent en leurs cabannes,[f] & les Sauvages s'en allerent à la guerre con-/[f. 34]tre les Irocois.

Le seiziesme jour d'Aoust, nous partismes de Tadousac,* & le 18. dudit mois arrivasmes à l'isle percee,[g] où nous trouvasmes le sieur Prevert de sainct Malo, qui venoit de la mine où il avoit esté avec beaucoup de peine pour la crainte[h] que les Sauvages avoient de faire rencontre de leurs ennemis[i] qui sont les Armouchicois,* lesquels sont hommes sauvages[j] du tout monstrueux[k] pour la forme qu'ils ont: car leur teste est petite, & le corps court, les bras menus comme d'un schelet, & les cuisses semblablement: les jambes[l] grosses & longues, qui sont toutes[m] d'une venue,[n] & qua*n*d ils sont assis sur leurs talons, les genoux leur passent plus d'un demy pied par dessus la teste, qui est chose estrange, & semblent estre hors de nature: Ils sont neantmoins fort dispos, & determinez: & sont aux meilleures terres de toute la coste d'Arcadie: Aussi les Souricois les craignent fort.

a se battre:] A1 se batre:
b nues,] A2 nuës,
c canots] A2 Canots
d nues] A2 nuës
e s'entre-ruoient:] A2 s'entre-ruoient.
f cabannes,] A2 Cabannes,
* RM] *Partement de Tadousac.*
g percee,] A2 perçee,
h crainte] A2 craincte
i ennemis] A2 ennemis,
* RM] *Sauuages Armouchicois.*
j sauuages] A1 Sauuages
k monstrueux] A2 monstrueux,
l iambes] A2 jambes
m toutes] A2 tontes
n venue,] A2 venuë,

Sagamo named Begourat before them, who was leading them to war.[292] And they were in a row with their bows and arrows, clubs and round shields, with which they deck themselves out for fighting. And they went leaping one after the other, while making many gestures with their bodies and executing many winding manoeuvres.[293] Afterwards they began to dance in the accustomed manner, as I said above. Then they made their *Tabagie*, and after having made it, the women stripped themselves stark naked, being adorned with their finest *Matachias*; and they put themselves into their canoes while dancing naked in this way; and then they put out upon the water, fighting among themselves with strokes of their paddles, splashing a great amount of water at one another. Yet they did themselves no harm, for they warded off the strokes which they flung up at one another. After having performed all these ceremonies, they withdrew into their lodges, and the *Sauvages* went off to war against /[f. 34] the *Irocois*.[294]

On the sixteenth day of August we set out from *Tadousac* [map 2], and on the 18th of the said month we arrived at the *isle percee*, where we found the *sieur Prevert de sainct Malo*,[295] who was coming from the mine where he had been with much trouble, on account of the fear which the *Sauvages* had of encountering their enemies, the *Armouchicois*,[296] who are wild men, totally monstrous because of the shape they have. For their heads are small and their bodies short, their arms slender like a skeleton, as well as their thighs in a similar way. Their legs are thick and long, and straight up and down. And when they are seated upon their heels, their knees go more than half a foot above their heads, which is a strange thing. And they seem to be unnatural. Nevertheless, they are very agile and determined, and inhabit the best lands on the entire coast of *Arcadie*.[297] Moreover, the *Souricois* fear them greatly.[298]

292 Begourat (Bechourat) was a headman of the Montagnais (DCB, 1:86).

293 Winding manoeuvres: *tours de limaçon*; lit. twists of a snail.

294 This is a good description of the ceremonies held by most northeastern native groups before setting out on a campaign against an enemy.

295 For a biography of Jean Sarcel de Prévert, see DCB, 1:601–2.

296 The *Armouchiquois* (*Almouchiquois*) are culturally and linguistically related Algonquian groups who lived along the southern New England coast from about the Saco River in Maine to southern Cape Cod.

297 This description of the physical appearance of the *Armouchiquois* is of course nonsense. Perhaps the more experienced Prévert was having fun with the novice Champlain, or Prévert was told these stories by the Mi'kmaq to get French support against their enemies. In any case, it is a further example of Champlain's uncritical acceptance of what Frenchmen with more experience in Canada were telling him.

298 Here Prévert is correct. The *Souricois*, who are the Mi'kmaq, were periodically at war with their southern neighbours.

Mais avec l'asseura*n*ce que ledit Sieur de Prevert leur donna, il les mena jusques à ladite mine, où les Sauvages le guidere*nt*:* C'est une fort haute montaigne advа*n*çant[a] quelque peu sur la mer, qui est fort reluisante au Soleil, où il y a quantité de verd de gris qui procede de ladite mine de cuivre.* Au pied de /[f. 34v] ladite[b] montaigne, il dict, que de basse eau y avoit en quantité de morceaux de cuivre,* comme il nous en a monstré,[c] lequel tombe du haut de la montaigne. Passant[d] trois ou quatre lieues plus outre tirant à la coste d'Arcadie, il y a une autre mine,* & une petite riviere qui va quelque peu dans les terres, tirant au Su, où il y a une montaigne qui est d'une peinture noire,* dequoy se peignent les Sauvages: puis à quelque six lieues de la seconde mine, en tirant à la mer[e] environ une lieue[f] proche de la coste d'Arcadie, il y a une isle où se trouve une maniere de metail* qui est co*m*me brun obscur, le couppant il est blanc, dont anciennement[g] ils usoient pour leurs fleches, & cousteaux, qu'ils battoie*n*t avec des pierres,[h] ce qui me faict croire, que ce n'est estain, ny plomb, estant si dur co*m*me[i] il est, & leur ayant monstré de l'argent, ils dirent que celuy de ladite[j] isle est semblable, lequel ils trouvent dedans la terre, comme à[k] un pied ou deux. Ledict Sieur Prevert a donné aux Sauvages des coins & cizeaux, & autres choses necessaires pour tirer de ladicte mine, ce qu'ils ont promis de faire, & l'annee qui vient d'en apporter, & le donner audict sieur[l] Prevert.

* RM] *Discours que m'a faict le Sieur Preuert de S. Malo sur la descouuerture de la coste d'Arcadie.*

a montaigne aduāçant] A2 montagne, aduançant

* RM] *Verd de gris en quantité.*

b ladite] A2 ladicte

* LM] *Cuiure en quantité.*

c il [le sieur Prevert] nous en [du cuivre] a monstré,] A2 il [le cuivre] nous a esté monstré,

d Passant] A1 Passans

* LM] *D'vne autre mine.*

* LM] *Peinture noire.*

e mer] A2 mer,

f lieue] A2 lieuë

* LM] *Vne isle où il y a d'vne maniere d'autre metail.*

g anciennement] A1 canciēnement

h pierres,] A2 pierres:

i estant si dur cōme] A2 estāt si dur comme

j ladite] A2 ladicte

k à] A2 a

l sieur] A2 Sieur

But with the assurances which the said *Sieur de Prevert* gave them, he brought them as far as the said mine, to which the *Sauvages* guided him. It is a very high mountain, jutting a little way into the sea, which sparkles brightly in the sun where there is a large quantity of verdigris that comes out of the said copper mine. He said that at the foot of /[f. 34^{v}] the said mountain at low water, there was an abundance of copper pieces, as he showed us, which fall from the height of the mountain. Moving three or four leagues on, drawing towards the coast of *Arcadie*, there is another mine and a small river that goes a little way inland, shooting towards the south, where there is a mountain of a black pigment, with which the *Sauvages* paint themselves. Then at some six leagues from the second mine, drawing towards the sea about a league off the coast of *Arcadie*, there is an island where a kind of metal is found that is darkish brown – cutting it, it is white – which formerly they used for their arrows and knives, which they beat out with stones; which makes me think that, being as hard as it is, it is neither tin nor lead. And having shown them some silver, they said that the metal from the said island is similar, which they find within the ground about a foot or two down. The said *Sieur Prevert* gave the *Sauvages* wedges and chisels and other things necessary for extraction from the said mine; which they have promised to do, and next year to bring some of it and give it to the said *sieur Prevert*.[299]

299 As Champlain found out in 1604 and 1605, Prévert's descriptions of the "mines" he said he had discovered or was told about by his men and Mi'kmaq informants were exaggerated. Champlain describes his visits to these "mines" in *Les Voyages, 1613*, 23–8, 39, and 100.

Ils disent aussi qu'à quelques cent ou /[f. 35] 120. lieües,[a] il y a d'autres mines,* mais ils n'osent y aller,[b] s'il n'y a des François parmy eux pour faire la guerre à leurs ennemis qui la tiennent en leur possession. Cedict lieu où est la mine qui est par les 44. degrez & quelque minutte proche de ladicte coste de l'Arcadie, comme de cinq ou six lieues,* c'est une maniere de baye; qui en son entree peut tenir quelques lieues de large, & quelque peu d'avantage de long, où il y a trois Rivieres qui viennent tomber en la grand' Baye[c] proche de l'isle de sainct Jean, qui a quelques trente ou trente cinq lieues de long, & à quelque six lieues de la terre du Su. Il y a aussi une autre petite Riviere[d] qui va tomber comme à moitié chemin de celle par où revint ledict sieur Prevert, où sont co*mm*e[e] deux manieres de lacs en ceste-dicte Riviere.[f] Plus y a aussi une autre petite Riviere[g] qui va à la painture:[h] toutes ces Rivieres tombent en ladicte Baye au Su est,[i] environ de ladicte isle que lesdits Sauvages disent y avoir de ceste mine blanche. Au costé du Nort de ladicte baye sont les mines de cuivre, où il y a bon port pour des vaisseaux,* & une petite isle[j] à l'entree du port, le fonds est vase & sable, où l'on peut eschoüer les vais-[f. 35v]seaux. De ladicte mine jusques au commencement de l'entree desdictes Rivieres, il y a quelque 60. ou 80. lieues par terre: Mais du costé de la mer, selon mon jugement, depuis la sortie de l'isle de S. Laurens[k] & terre ferme, il ne peut y avoir plus de 50. ou 60. lieues jusques à ladicte mine. Tout ce païs[l] est tres-beau,* & plat, où il y a de toutes les sortes d'arbres que nous avons veues[m] allant au pre-

a lieües,] A2 lieues,
* RM] *Autres mines.*
b aller,] A2 aller
* RM] *Description du lieu où est ladite mine.*
c grand' Baye] A1 grand Baye
d Riuiere] A2 riuiere
e cõme] A2 comme
f Riuiere.] A2 riuiere.
g Riuiere] A2 riuiere
h painture:] A2 peincture:
i Su est,] A2 Su-est,
* RM] *Bon port pour les vaisseaux.*
j isle] A2 Isle
k Laurens] A2 Laurent
l païs] A2 pays
* LM] *Bon païs.*
m veues] A2 veus

They also say that some hundred or /[f. 35] 120 leagues away there are some other mines, but they dare not go there, unless there are some Frenchmen among them to make war on their enemies, who hold the mine in their possession. This said place where the mine is, which is at 44° and some minutes,[300] about five or six leagues from the said coast of *Arcadie*, is a kind of bay, which may extend a few leagues in width at its entrance and somewhat more in length, where there are three rivers that come to fall into the *grand' Baye*, near the *isle de sainct Jean*,[301] which is some thirty or thirty-five leagues in length and about six leagues from the land to the south. There is also another small river, which descends about halfway from that river by which the said *sieur Prevert* returned, where there are two lake-like expanses in this said river. Furthermore, there is also another small river that leads to the pigment. All these rivers fall into the said bay to the southeast, around the said island where the said *Sauvages* say there is this white mine. On the north side of the said bay are the copper mines, where there is a good harbour for vessels and a small island at the harbour's entrance. The bottom is mud and sand, where one can beach vessels. /[f. 35^{v}] From the said mine up to the beginning of the mouth of the said rivers, it is some 60 or 80 leagues overland. But on the side of the sea, according to my judgment, from the exit from the *isle de S. Laurens* and the mainland, there can be no more than 50 or 60 leagues to the said mine. This entire land is very beautiful and flat, where there are all the kinds of trees, which we saw going to the first rapids of the great *Riviere de Canadas*, but

300 The copper deposits found by Champlain near Advocate Harbour, Nova Scotia, in 1604 were at about 45°15'N.

301 This was the name for Prince Edward Island during the French regime.

mier sault de la grande Riviere[a] de Canadas, fort peu de sapins & cyprez. Voylà[b] au certain ce que j'ay apprins[c] & ouy dire audit[d] sieur Prevert.

Chap. XIII.[e]

D'un monstre espouvantable que les Sauvages appellent Gougou, & de nostre bref & heureux retour en France.

Il y a encore une chose estrange digne de reciter, que plusieurs Sauvages m'o*n*t asseuré estre vray; C'est que proche de la baye de Chaleurs tirant au Su, est une isle, où fait[f] residence un monstre espouvantable,[g*] que les Sauvages appellent *Gougou*, & m'ont dit[h] qu'il avoit la forme d'une femme: mais fort effroyable, & d'une telle grandeur, qu'ils me disoient que le bout des mats de nostre vaisseau ne luy fust pas venu jusques à la ceinture, tant ils le peignent grand: & que souvent il a devoré & /[f. 36] devore, beaucoup de Sauvages,[i] lesquels il met dedans une grande poche[j] quant il les peut attraper[k] & puis les mange: & disoient ceux qui avoient esvité le peril de ceste malheureuse beste, que sa poche estoit si grande, qu'il y eust peu mettre nostre vaisseau: Ce monstre fait des bruits horribles dedans ceste isle, que les Sauvages appellent le Gougou: Et[l] quand ils en parlent ce n'est que avec une peur si estrange, qu'il ne se peut dire de plus, & m'ont asseuré plusieurs l'avoir veu: Mesme ledit sieur[m] Prevert de sainct Malo en allant à la descouverture des mines ainsi que nous avons dit au chapitre preced*e*nt,[n] m'a dit[o]

a grande Riuiere] A2 grand' Riuiere
b cyprez. Voylà] A1 cyprez: voylà
c apprins] A2 appris
d audit] A2 audict
e CHAP. XIII. *below title*] A2 CHAP. XIII. *moved up to the same level as the title*
f fait] A2 faict
g espouuantable,] A2 espouuātable,
* LM] *Monstre espouuentable.*
h dit] A2 dict
i Sauuages,] A2 Sauuages
j grande poche] A1 grand poche
k attraper] A2 attrapper
l Et] A2 &
m sieur] A2 Sieur
n precedēt,] A2 precedēt.
o m'a dit] A2 m'a dict

very few fir trees and cedars. This is unquestionably what I learned and heard from the said *sieur Prevert*.[302]

Chapter 13

Of a frightful monster which the Sauvages *call* Gougou, *and of our short and felicitous return to France*

There is yet another strange thing worthy of recounting, which many *Sauvages* have assured me is true.[303] It is that near the *baye de Chaleurs*, towards the south, is an island[304] where a frightful monster makes its abode, which the *Sauvages* call *Gougou*; and they told me it had the shape of a woman, but most terrifying and of such a size that, according to them, the top of the masts of our ship would not have come up to its waist, so big do they depict it. And they said that it has often devoured /[f. 36] and does yet devour many *Sauvages*, whom it puts into a great pocket when it can catch them, and then eats them. And those who had escaped the perils of this miserable beast said that its pocket was so big that it could have put our ship into it. This monster, which the *Sauvages* call the *Gougou*, makes horrible noises in this island, and when they speak of it, it is only with such a strange terror that no more can be said; and several assured me that they had seen it. Even the said *sieur Prevert de sainct Malo* told me that while going to explore the mines, as we said[305] in the preceding chapter, he passed so near the haunt of this

302 See footnote 295.

303 Here begins the infamous story of the *Gougou*. This story and Champlain's description of the *Armouchiquois* earned him the ridicule of his contemporary, Marc Lescarbot, and others for his naivety (Grant, *History of New France*, 2:168–76).

304 Perhaps Île Miscou or Île Lamèque at the southern entrance of Baie des Chaleurs

305 we said: *nous avons dit*. This is a rare instance of Champlain using the first person plural to refer to himself.

avoir passé si proche de la demeure de ceste effroyable beste, que luy & tous ceux de son vaisseau entendoient[a] des sifflements estranges du bruit qu'elle faisoit: & que les Sauvages qu'il avoit avec luy, luy dirent, que c'estoit la mesme beste, & avoie*nt*[b] une telle peur, qu'ils se cachoient de toutes parts, craignant qu'elle fust venüe à eux pour les emporter: & qui me faict croire ce qu'ils disent:[c] C'est que tous les Sauvages en general la craignent & en parlent si estrangement, que si je mettois tout ce qu'ils en disent, l'on le tiendroit pour[d] fables: mais je tiens que ce soit la reside*n*ce de quelque Diable qui /[f. 36v] les tourme*n*te de la façon. Voilà[e] ce que j'ay apprins[f] de ce Gougou.

Premier que partir de Tadousac, pour nous en retourner en France, un des Sagamoz[g] des Montagnez no*m*mé Bechourat, donna son fils au sieur du Pont pour l'amener[h] en Fra*n*ce,[i] & luy fut fort reco*m*mandé par le grand Sagamo Anadabijou, le priant de le bien traiter, & luy faire voir ce que les autres deux Sauvages que nous avio*n*s remenez avoie*n*t veu.[j] Nous leur demandasmes une femme des Irocois qu'ils vouloient manger laquelle ils nous do*n*nerent, & l'avons aussi amenée[k] avec ledit Sauvage. Le sieur de Prevert a aussi amené quatre Sauvages: Un ho*m*me, qui est de la coste d'Arcadie, une femme & deux enfans des Canadiens.[l]

Le 24. jour d'Aoust[m] nous partismes de Gachepay, le vaisseau dudit sieur Prevert & le nostre;[n] le 2. jour de Septembre[o] nous faisions estat d'estre aussi ava*n*t q*ue* le Cap de rase. Le 5. jour dudit mois nous entrasmes sur le banc où ce fait[p] la pesche du poisson. Le 16. dudict mois nous estions à la sonde, qui peut estre

a entendoient] A2 entēdoiēt
b auoiēt] A2 auoient
c disent:] A2 disent,
d pour] A2 pou
e Voilà] A2 Voila
f apprins] A2 aprins
g Sagamoz] A1 Sagamo
h l'amener] A1 l'emmener
i Frãce,] A2 France,
j veu.] A1 A2 veu,
k amenée] A2amenee
l Canadiens.] A2 Canadiens,
m d'Aoust] A2 d'Aoust,
n nostre;] A1 A2 nostre,
o Septembre] A2 Septembre,
p ce fait] A2 se faict

frightful beast that he and all those of his vessel heard strange hissings from the noise that it was making; and that the *Sauvages* whom he had with him told him that it was the same creature, and they were so afraid that they hid themselves on all sides, fearing that it would come to carry them away. And what makes me believe what they say is the fact that all the *Sauvages* in general fear it, and speak of it so strangely that if I were to put down everything that they say about it, it would be taken for lies; but I hold that this may be the haunt of some Devil which /[f. 36^{v}] torments them in this way. That is what I learned about this *Gougou.*[306]

Before setting out from *Tadousac* to return from there to France, one of the *Sagamoz* of the *Montagnez*, named Bechourat,[307] gave his son to the *sieur du Pont* to be taken away to France. And he was highly recommended to him by the *grand Sagamo* Anadabijou, who begged him to treat him well and to have him see what the two other *Sauvages*, whom we had brought back, had seen.[308] We asked them for a woman of the *Irocois* whom they wanted to eat, whom they gave to us, and we have also brought her back with the said *Sauvage*. The *sieur de Prevert* has also brought along four *Sauvages*: one man who comes from the coast of *Arcadie*, one woman and two children of the *Canadiens*.[309]

On the 24th day of August, we, the vessel of the said *sieur Prevert* and our own,[310] set out from *Gachepay*,[311] and on the 2nd day of September we reported being as far as the *Cap de rase*.[312] On the 5th day of the said month, we entered upon the bank where fishing is carried on. On the 16th of the said month, we were

306 It is entirely possible that Champlain was being told Mi'kmaq myths about *Kuhkw* and/or *Kokwees*. According to Rand (*Legends*, 1:39), "*Kuhkw* means Earthquake; this mighty personage can pass along under the surface of the ground, making all things shake and tremble by his power," while the *Kokwees* is a giant cannibal (ibid., 2:288–9). Champlain's fault was not so much in reporting what others had told him but by implying he believed the stories by not refuting them.

307 Begourat (Bechourat), a headman of the Montagnais, should not be confused with Bessouat (Tessouat) the headman of the *Kichesipirini* Algonquins (DCB, 1:86).

308 The two "other *Sauvages*" were the men who had reported to Anadabijou and the Montagnais, on their return to Tadoussac on 27 May, about their visit with Henri IV.

309 An alternate name for the *Gaspesiens*, the western branch of the Mi'kmaq. See footnote 252.

310 The name of Prévert's ship is not known. Champlain was on *La Bonne-Renommée* with Gravé Du Pont.

311 Gaspé

312 Cape Race on the southeastern tip of the Avalon Peninsula, Newfoundland

à quelque 50. lieues d'Ouessant. Le 20. dudit[a] mois nous arrivasmes par la grace de Dieu avec contentement d'un chacun & tousjours le vent favorable au port du Havre de Grace.

FIN

a dudit] A2 dudict

in the sound that may be some 50 leagues from *Ouessant*.[313] On the 20th of the said month, we arrived, by the grace of God, to the satisfaction of each and every one, and with the wind always favourable, at the port of *Le Havre de Grace*.[314]

END

313 *Ouessans* (Ushant) is an island off the western tip of Brittany.

314 The return trip from Gaspé to *Le Havre de Grace* (Le Havre) took 27 days, in contrast to the outward trip over the same distance, which took 70 days.

DOCUMENT H

Of the French Who Have Become Accustomed to Being in Canada. Summary of *Des Sauvages* by Pierre-Victor Cayet, 1605 / *Des François qui se sont habituez en Canada.*[1] *Sommaire de* Des Sauvages *par Pierre-Victor Cayet*, 1605

INTRODUCTION TO THE DOCUMENT

Pierre-Victor Cayet[2] (1525–1610) had a chequered career as a teacher, writer, and early promoter of Calvinist and, later Catholic, causes. His connection to Henri IV began in 1562, when he was a tutor and vice-governor to the young prince. Although Henri IV was ambivalent about him, he appointed him king's chorographer[3] (*chronographe du roi*) in the late 1590s to write a documentary history of his reign. In this capacity, Cayet had access to state papers and the best libraries. One of the worthwhile series of volumes that resulted was the *Chronologie septenaire*, in which Cayet published an abbreviated version of *Des Sauvages.*

There are at least three printings of this summary of *Des Sauvages*. Its title, "Des François qui se sont habituez en Canada," is given in the index to vol. 7 on page 433 of the first printing and page 434 of the second and third printings. This title is not repeated on page 415 or 416, where the text of the summary begins. In length, the summaries are 18 printed pages (c. 4,600 words), roughly one-third of the length of the text of the original book. Like the original book, the summaries have content-oriented marginalia, and the type was reset for each subsequent printing. In the entire piece, neither Champlain nor the title of his book *Des Sauvages* is mentioned. François Gravé Du Pont is correctly identified as the leader of the expedition, and Jean Sarcel de Prévert as reconnoitring the Acadian coast.

1 Note that Cayet considers "Canada" a feminine noun here.

2 For a brief statement on Cayet, see Bernard C. Webber and Gregory de Rocher, "Personal and Political Aspects in the Correspondence of Henri IV" (www.bama.ua.edu/~gderoche/henriiv/intro.htm). For a longer biography, see Michaud, *Biographie universelle*, 7:279–80. His name is sometimes given as Pierre-Victor Palma-Cayet.

3 A chorographer is a geographer/historian dealing in description rather than analysis; a compiler of documents and factual information.

Unfortunately, Cayet repeated Champlain's story of the monstrous *Gougou*. For this he earned the following comment from Marc Lescarbot, who was trying to discredit Champlain for his credulity:

> *Or si ledit sieur Champlein a esté credule, vn sçavant personage que i'honore beaucoup pour sa grande literature, est encore en plus grand' faute, ayant mis en sa Chronologie septenaire de l'histoire de la paix imprimée l'an mil six cens cinq, tout le discours dudit sieur Champlein, sans nommer son autheur, & ayāt baillé les fables des* Armouchiquois *& du Gougou pour bonne monnoye.*[4]
> (Now if the said Sieur Champlein was indeed credulous, a learned individual whom I honour greatly for his literary output is still more at fault for having put into his *Chronologie septenaire de l'histoire de la paix*, printed in the year 1605, the entire discourse of the said Sieur Champlein, without naming its author, and for having passed off the fables of the Armouchiquois and the *Gougou* as authentic.)

Although there is little new information in Cayet's summary, it is interesting to note what he chose to reproduce. All the new information is in his brief introduction. It concerns the visit of the two Montagnais to Henri IV's court in 1602 and their story of a passage to the west. It is this information that Champlain was ordered to investigate. It is unlikely that Champlain was present at that meeting, otherwise he would have mentioned it. It is, however, likely that Cayet was present as chorographer. His introduction appears to be the only reported account of the meeting. His repeated use of the words "the *Sieur du Pont* and his" (*Le Sieur du Pont & les siens*) in this account is an indication of Cayet's effort to exclude Champlain, or at least to convert his first-person account into an objective third-person narrative with Gravé Du Pont at the centre; Champlain is included among *les siens* (his companions). Lescarbot's criticism of Cayet's failure to name the author of *Des Sauvages* demonstrates that such blatant plagiarism, if not illegal, was frowned upon.

Another early reference we have seen to Cayet's summary of *Des Sauvages* is in a footnote to all the editions of Sully's *Mémoires*, edited by the abbé de l'Écluse.[5] Sully was not in favour of the support Henri IV was giving to overseas coloniz-

4 Lescarbot, *Histoire de la Nouvelle France*, 424–5, 415. The word *fables* could also be translated as "lies."

5 Sully, *Mémoires de Sully*, 3:398. In 1745 the abbé Pierre Mathurin de l'Écluse des Loges (1716–83) published Sully's memoirs after he had rewritten them in ordinary narrative form.

ation and stated in his *Mémoires* that de Mons's voyage was made against his advice. De l'Écluse seems to support Sully's opinion by citing Cayet's version of *Des Sauvages*. Like Cayet, he does not credit Champlain with the original text; but unlike Cayet, he passes off Champlain's ethnographic descriptions as "unfaithful" and "filled with fables":

> "Voyez dans le Septénaire la description du voyage que fit en Canada le sieur du Mont. Il s'y trouve aussi une relation des mœurs des habitants de cette partie du nouveau monde, mais peu fidelle, et remplie de fables."[6] (See in the *Septénaire*, a description of the voyage that the sieur du Mont made to Canada. There is also an account of the manners of the inhabitants of this part of the new world; but it is very unfaithful, and filled with fables.[7])

In his later writings, Champlain avoided unsubstantiated accounts from other people. Both Lescarbot's criticisms and the unsolicited abridgement of his writings in a "popular" reprint such as *Chronologie septenaire* must have made him more circumspect.

INTRODUCTION TO THE ORIGINAL FRENCH DOCUMENT

The original French document exhibits the same accidental forms as the original publications of *Des Sauvages*. This is not suprising if our hypothesis is correct that Jean Richer III actually printed both *Des Sauvages* and this *Chronologie septenaire*, in the latter case without the lately deceased Claude de Monstr'œil who held the licence.[8] The same abbreviation marks are used and some of the same errors are repeated, especially minor ones involving punctuation in lists. We have corrected the most obvious errors, using the *siglum C* for Cayet to indicate the reading in the *Chronologie septenaire*. Other abnormalities concerning punctuation (e.g., commas for periods) and agreements (subject-verb; adjective-noun) have been left as they are, because they are not felt to inhibit comprehension. All the paragraph indentations follow the original edition here; none have been added.

6 Ibid.

7 For the word *fables*, see n. 4 above.

8 See part 1, B, "Establishing the French Text."

Des François qui se sont habituez en Canada

[f. 415] Voyons maintenant le succez des François, qui allerent ceste annee en la France nouvelle dicte Canada.

Le sieur du Pont dez l'an passé avoit esté en la nouvelle France dite Canada, d'où il avoit ame-/[f. 415v]né deux des Sauvages qui y habitent, lesquels il presenta au Roy: Or il apprint d'eux que la gra*n*de riviere (que l'on pensoit autres-fois n'estre qu'un Sin ou goulphe, pource qu'elle a dixhuict lieuës à son emboucheure da*ns* la mer) avoit plus de quatre cent lieuës de long, & traversoit une infinité de beaux pays & lacs, en laquelle aussi se venoit rendre une grande quantité de belles rivieres, & qu'il pourroït y aller avec les Canots dequoy les Sauvages usent pour naviger par ceste gra*n*de riviere: Il se resoult avec quelques autres Capitaines de mer (soubs le bon plaisir du Roy) d'y retourner, & voir par le moyen des Sauvages le dedans du pays aussi bien qu'ils en avoient veu les costes le long de la mer, qui ne sont que montagnes haut eslevees, où il y a peu de terre, qua*n*tité de rochers. & sables remplis de pins, cyprez, sapins, & bouilles.

Of the French Who Have Become Accustomed to Being in Canada[1]

Cayet, Pierre-Victor. *Chronologie septenaire de l'histoire de la paix entre les roys de France et d'Espagne … divisee en sept livres.* Book 7: *Contenant ce qui s'est passé l'an 1604.* Paris: Par Jean Richer, ruë S. Iean de Latran, à l'Arbre verdoyant: Et en sa boutique au Palais, sur le perron Royal, vis a vis de la galerie des prisonniers, 1605, 415–24 (fig. 3)

EDITION USED: National Library of Canada, Ottawa, Reserve, DC122, C38
MICROFORM: CIHM, 33072

[Pierre-Victor Cayet's introduction, f. 415] Now let us consider the success of the French who went in that year[2] to New France, called Canada.

The sieur du Pont had been in New France, called Canada, since the previous year, from where he had brought back /[f. 415ᵛ] two of the *Sauvages*[3] who live there, whom he presented to the King. Now he learned from them that the big river[4] (which we once thought to be only a bay or gulf, because it was eighteen leagues across its mouth at the sea) was more than 400 leagues long and cut across an infinity of fine lands and lakes, into which a great number of beautiful rivers also flowed together; and because one could go there with the canoes which the *Sauvages* use to navigate by way of this great river, he, along with a few other sea captains (with the good wishes of the King), resolved to return there and see the interior of the land with the help of the *Sauvages* in the same way as they had seen the coast alongside the ocean,[5] which is nothing but very high mountains, where there is little land, a great number of rocks, and sands filled with pine, cedar, fir, and birch trees.

1 Note that the toponyme here is simply "Canada," still feminine, although Cayet maintains the *s* when naming the *riviere de Canadas* and the *voyage de Canadas* at the end.
2 *Chronologie septenaire* was published in 1605 but purports to deal with events of 1604 (see subtitle). "That year" seems to refer to 1604 but is incorrect because the events took place in 1603. It may be that Cayet had a copy of *Des Sauvages* printed in 1604 and simply took the date from the second edition without actually reading the title page.
3 This event took place in 1602.
4 The "big river" is the St. Lawrence.
5 This is an important addition by Cayet not mentioned in *Des Sauvages.* Cayet had in fact summarized one of the aims entrusted by Henri IV to Gravé Du Pont for the 1603 voyage, namely, the "discovery of lands in Canada and the adjacent countryside." See Document E, "Decrees and Commissions, 1603." It was Champlain who was assigned to the completion of this aim.

[*Des Sauvages*, ch. 1] Pour faire ce voyage il partit de Honfleur le 15. de Mars de ceste annee, remenant quand &[a] luy les deux Sauvages, & apres avoir eu plusieurs tempestes, il arriva à l'entree de la grande riviere de Canadas le 18 Avril, où estant entré bien cent lieuës avant, il aborda en fin le 24. à Tadousac, où il trouva quantité de Sauvages cabannez.

[*Des Sauvages*, ch. 2] Ayant mis pied a terre, il fut avec aucuns des siens à la Cabanne du grand Sagamo, appellé Anadabijou, où ils le trouvere*n*t avec quelque 80. ou 100. de ses compagnons qui faisoie*n*t Tabagie, (qui veut dire festin) lequel les receut fort bien, selon leur coustume, & les fit asseoir auprez de luy, tous le Sauvages arangez les uns aupres des autres des deux costez de la cabanne. L'un des /[f. 416] Sauvages qui'ils avoient remenez commença à faire sa harangue, sur la bonne reception que leur avoit faicte le Roy, & du bon traictement qu'ils avoient receu en France, & que sa Majesté leur vouloit du bie*n*, & desiroit peupler leur terre, & faire leur paix avec leurs ennemis (qui sont les Irocois) ou leur envoyer des forces pour les vaincre: Il leur racompta aussi les beaux Chasteaux, Palais, maiso*ns*, & peuples qu'il avoit veus & la façon de vivre des François. Il fut entendu des Sauvages avec un gra*n*d silence. Or apres qu'il eut achevé de parler, le gra*n*d Sagamo l'ayant attentivement ouy, il comme*n*ça à prendre du Petum, & en donner audit sieur du Pont Gravé de S. Malo, & aux siens, & à quelques autres Sagamos qui estoient aupres de luy: Ayant bien petunné, il commença à faire sa harangue à tous, parlant posément, s'arrestant quelques-fois un peu, & puis reprenant sa parole, il leur dist, Que veritablement ils devoient estre fort contents d'avoir un tel Roy pour grand amy: à quoy tous les autres Sauvages respondirent d'une voix, *ho, ho, ho,* qui est à dire, *ouy; ouy.*

Puis le Sagamo leur dit encor', Qu'il estoit fort aise que le Roy de France peuplast leur terre, & fist la guerre à leurs ennemis, qu'il n'y avoit nation au monde à qui ils voulussent plus de bien qu'aux Fra*n*çois: puis fit ente*n*dre à ses Sauvages le

a quand &: *with* (quand & *is found in literature from Normandy*).

[*Des Sauvages*, ch. 1] In order to make this voyage, he set off from Honfleur on 15 March of this year,[6] taking back with him the two *Sauvages*. And after having experienced several storms, he arrived at the entrance of the great river of *Canadas*[7] on 18 April[8] where, having gone upstream a good 100 leagues, he finally reached *Tadousac* on the 24th,[9] where he found a large number of *Sauvages* encamped.

[*Des Sauvages*, ch. 2] Having set foot on land, he went with some of his own people to the lodge of the *grand Sagamo*, named Anadabijou, where they found him with some 80 or 100 of his companions who were making *Tabagie* (which means a feast). He received them very well according to their custom and made them sit down beside him, with all the *Sauvages* arranged one beside the other on both sides of the lodge. One of the /[f. 416] *Sauvages* whom they had brought back began to deliver his oration about the good reception that the King had given them and the good treatment they had received in France, and that His Majesty wished them well and desired to people their land and to make peace with their enemies (who are the *Irocois*) or send forces to conquer them.[10] He also told them about the fine castles, palaces, houses, and peoples he had seen and about the French manner of living. He was heard by the *Sauvages* with a great silence. Now, after he had finished speaking, the *grand Sagamo*, having listened to him attentively, began to take some tobacco and to give some of it to the said sieur du Pont Gravé of St. Malo, and his [companions],[11] and to some other Sagamos who were near him. Having had a good smoke, he began to address the whole gathering, speaking deliberately, pausing sometimes a little. And then resuming his speech, he told them that in truth they ought to be very glad to have such a king for a great friend; to which all the other *Sauvages* replied with one voice, "*ho, ho, ho*," which is to say, "*yes, yes*."

Then the *Sagamo* went on to tell them that he was quite content that the King of France would populate their land and make war on their enemies and that there was no nation in the world for which he would wish more good than the French. Then he led his *Sauvages* to understand the advantage and profit that they

6 "This year" should be 1603.

7 The river of *Canadas* is the St. Lawrence.

8 This should be 18 May, according to Champlain.

9 24 May, according to Champlain.

10 This important statement, interpreted by the Montagnais as a treaty, is verbatim from *Des Sauvages*.

11 *les siens*: his, implying his people or company. In *Des Sauvages,* Champlain states explicitly *& à moi* (lit. and to me, f. 4) at this point. Cayet's method of excluding Champlain from the narrative by lumping him in with *les siens* continues throughout the summary.

bien & utilité qu'ils pourroie*n*t recevoir de sa Majesté: Apres qu'il eut achevé sa hara*n*gue, du Pont & les siens sortirent de la Cabanne, & eux commencerent à faire leur Tabagie ou festin, qu'ils font avec des chairs d'Orignac, qui est comme /[f. 416v] bœuf, d'Ours, de Loumarins, & castors, qui sont les viandes les plus ordinaires qu'ils ont, & du gibier en qua*n*tité. Pour manger ils s'assisent des deux costez de la Cabanne avec chacun leur escuelle d'escorce d'arbre: & la viande estant cuitte, il y en a un qui fait les partages à chacun dans leurs escuelles, où ils mangent fort salleme*n*t: car quand ils ont les mains grasses, il les frottent à leurs cheveux, ou bien au poil de leurs chiens, dont ils ont quantité pour la chasse. Devant que manger ils dansent au tour de leurs chaudieres, & apres qu'ils on mangé ils recommencent leurs danses en prenant chacun la teste de leur ennemy qu'ils ont tué en bataille, laquelle leur pend par derriere.

Ils faisoient ce festin ensembleme*n*t pour la victoire par eux obtenuë sur les Irocois, dont ils en avoient tué quelque cent.

Trois nations de Sauvages estoient là assemblez, sçavoir les Estechemins, Algoumequins, & Montagnez, au nombre de mille, tous ennemis des Irocois, ausquels ils font une cruelle guerre par surpise, par ce qu'ils sont en plus grand no*m*bre qu'eux. Le 28. jour de mois, lesdits Sauvages qui estoient lors à la pointe S. Mathieu, se vindrent cabanner au port de Tadousac, où estoie*n*t les François. A la poincte du jour leur grand Sagamo sortit de sa cabanne, allant autour de toutes les autres cabannes, en criant à haute voix, qu'ils eussent à desloger pour aller à Tadousac, où estoient leurs bons amis: Tout aussi tost un chacun d'eux deffit sa cabanne, en moins d'un rien, & le grand Sagamo le premier commença /[f. 417] à prendre son canot, & le porter à la mer, où il embarqua sa femme & ses enfans,

might receive from His Majesty.[12] After he had finished his speech, Du Pont, and his [companions] left the lodge and they began[13] to make their *Tabagie*, or feast, which they make with moose meat – which is like /[f. 416v] beef – and with bear, seal, and beaver, which are the most ordinary meats that they have, and with great quantities of wild fowl. To eat, they sit along two sides of the lodge, each with his bowl made of the bark of a tree. And the meat being cooked, one of them apportions to each man his share in their bowls, out of which they feed very filthily. For when they have greasy hands, they rub them in their hair, or else in the fur of their dogs, of which they have a great number for hunting. Before eating, they dance around their cauldrons,[14] and after they have eaten, they begin their dances again, each one taking the head of their enemy whom they have killed in battle, which hangs behind them.

They made this feast together on account of the victory that they had won over the *Irocois*, of whom they had killed about a hundred.

Three nations of *Sauvages* were assembled there together; that is to say, the *Estechemins*, *Algoumequins*, and *Montagnez*,[15] to the number of a thousand, all enemies of the *Irocois* against whom they are waging a cruel war by surprises, because they[16] are greater in number than them. On the 28th day of the month,[17] the said *Sauvages*, who were then at the point *St. Mathieu,* came together to pitch camp at *Tadousac* harbour, where the French were. At daybreak, their *grand Sagamo* came out of his lodge, going around all the other lodges, crying out with a loud voice that they would have to break camp to go to *Tadousac*, where their good friends were. Immediately, each one of them took down his lodge in less than no time, and the *grand Sagamo* was the first to begin /[f. 417] to take his canoe and carry it to the sea, where he put his wife and children on board and a

12 The *Sagamo*'s speech shows that the Montagnais accepted the king's proposal.

13 The grammatical clumsiness here which suggests that Gravé Du Pont and his company both "left the lodge" and then made "their *Tabagie*" results from Cayet's rewording of Champlain's account in order to exclude Champlain. Champlain had written: "*nous sortismes de sa Cabanne & eux commencerent*" ("we went out of his lodge and they began"; f. 4v.

14 *Chaudières*, literally "cauldrons." This word is often translated as "kettles."

15 The *Irocois* are the Iroquois. The *Estechemins* (*Etechemins, Etchemin*) are the Maliseet of the St. John River and the Passamaquoddy of the Sainte-Croix River. The *Algoumequins* are the Algonquins from the Ottawa River and the *Montagnez* are the Montagnais from the Tadoussac area.

16 The pronoun "they" refers to the Iroquois. Champlain's original version is clearer: "*lesdits Iroquois, qui sont en plus grand nombre que lesdits Montagnes, Estechemains, & Algoumequins*" (f. 5v).

17 The month is May.

& qua*n*tité de fourreures, & se mirent ainsi pres de deux cents Canots, qui vont estrangement: car encore que la chaloupe du sieur du Pont fust bien armee, si alloient-ils plus viste qu'elle. Il n'y a que deux personnes qui travaillent à la nage, l'homme & la femme: Leurs Canots ont quelque huict ou neuf pas de long, & large comme d'un pas, ou pas & demy par le millieu, & vont tousjours en amoindrissant par les deux bouts: ils sont fort subjects à tourner, si on ne les sçait bien gouverner: ils sont faicts d'escorce d'arbre apellé bouille, renforcez par le dedans de petits cercles de bois bien & proprement faicts, & sont si legers qu'un homme en porte un aisément, & chacun Canot peut porter la pesa*n*teur d'une pipe: Qua*n*d ils veulent traverser la terre pour aller à quelque riviere, où ils ont affaire, ils les porte*n*t avec eux. Leurs cabannes sont basses, faictes comme des tentes couvertes d'escorce d'arbre, & laisse*n*t tout le haut descouvert comme d'un pied, d'où le jour leur vient, & font plusieurs feux droict au milieu de leur cabanne, où il sont quelques-fois dix mesnages ensemble. Ils couche*n*t sur des peaux les uns parmy les autres, les chiens avec eux. Ils estoient lors bien mille personnes, tant hommes que femmes & enfans.

[*Des Sauvages*, ch. 3] Tous ces peuples sont tenus d'une humeur assez joyeuse, ils rient le plus souve*n*t, toutesfois ils sont quelque peu Saturniens: ils parlent fort posément, comme se voulans bien faire entendre, & s'arrestent aussi tost en songeant une /[f. 417v] grande espace de temps, puis reprennent leur parole: ils usent bien souvent de ceste façon de faire parmy leurs hara*n*gues au Conseil, où il n'y a que les plus principaux, qui sont les anciens: Les femmes & enfans n'y assistent point. Tous ces peuples patissent tant quelques fois, qu'ils sont presque contraincts de se manger les uns les autres pour les grandes froidures & neiges: car les animaux & gibier dequoy ils vivent se retirent aux pays plus chauds. Qui leur monstreroit à vivre & enseigner le labourage des terres, & autres choses ils l'apprendroie*n*t fort bien: car il s'en trouve assez parmy eux qui ont bon jugeme*n*t: & respondent à propos sur ce que l'on leur demande: Ils ont une meschanceté en eux, qui est, user de vengeance & estre gra*n*ds menteurs, gens en qui il ne faict pas trop bon s'asseurer, sinon qu'avec raison & la force à la main: promettent assez

great number of furs. And thus about two hundred canoes were launched, which move in an astonishing way,[18] for although the *chaloupe* of the sieur du Pont was well manned, yet they went more swiftly than it. There are only two people who paddle: husband and wife. Their canoes are about eight or nine paces long and about a pace wide, or a pace and a half at the middle, always tapering down to both ends. They are very liable to capsize if one does not know how to manage them properly. They are made out of the bark of a tree called a birch, reinforced within by small ribs of wood, well and carefully fashioned, and are so light that a man may carry one of them easily; and each canoe can carry the weight of a pipe.[19] When they wish to go overland to get to some river where they have business, they carry them with them. Their lodges are low, made like tents covered with tree bark and they leave the whole top part uncovered for about a foot, through which the daylight comes into them. And they make several fires right in the middle of their lodge, where there are sometimes ten households together. They sleep upon skins among each other, and the dogs with them. There were at that time a good thousand, as many men as women and children.

[*Des Sauvages*, ch. 3] All these people are upheld by a rather joyful disposition. They laugh most often, yet they are somewhat saturnine.[20] They speak very deliberately, as if wishing to make themselves well understood, and stop suddenly, reflecting for a /[f. 417^{v}] long space of time, then begin to speak again. They very often resort to this way of proceeding with their speeches in council, where there are only the foremost leaders, who are the elders; the women and children are not present. All these people suffer so much sometimes that they are almost forced to eat each other because of the great cold and snows, for the animals and fowl upon which they subsist migrate to warmer climates. If anyone were to show them how to live and teach them how to work the land and other matters, they would learn it very well, for there are enough among them who have good judgment and respond quite appropriately to whatever one asks of them.[21] They have a wickedness in them, in that they resort to revenge and are great liars, a people whom it is not too good to trust, except within reason and with force at hand. They promise

18 *Étrangement*, literally "strangely" or "oddly"

19 A pipe is a measure of capacity equivalent to half a tun. It contains about 105 imperial gallons (477 litres). One pipe of water weighs about 840 pounds (380 kg).

20 *Saturniens*, literally "melancholic"

21 By excluding the words *Je tiens que* (I hold that) and *Je vous asseure qu'* (I assure you that) from this last sentence, Cayet has transformed Champlain's subjective opinion into an objective-sounding fact. Cayet commonly overrides Champlain's caution in making his opinions or estimates; e.g., by excluding words equivalent to "about" before numbers in leagues.

& tiennent peu: Ce sont la pluspart gens qui n'ont point de loy, & qui croyent qu'apres que Dieu eut fait toutes choses, il print qua*n*tité de fleches, & les mit en terre, d'où il sortit hommes & femmes, qui ont multiplié au monde jusques à present, & sont venus de ceste façon, Qu'il y a un Dieu, un Fils, une Mere, & le Soleil, qui so*n*t quatre: Nea*n*tmoins que Dieu est par-dessus to*us*: mais que le Fils est bon & le Soleil, à cause du bien qu'ils en reçoivent: & que la Mere ne vaut rien, pour ce qu'elle les mange; aussi que le Pere n'est pas trop bon. Ils ont une infinité d'autres folles creances, & ont parmy eux certains Sauvages qu'ils appellent Pilotoua, qui parlent au diable visiblement, & [il] leur dit ce qu'il faut qu'ils face*n*t, /[f. 418] tant pour la guerre, que pour autres choses: ausquels ils obeyssent à leur premier commandement.

Aussi ils croye*n*t que tous les songes qu'ils font sont veritables, & de fait, il y en a beaucoup qui disent avoir veu & songé chose qui advie*n*dront: Mais pour en parler avec verité, ce sont visions du Diable, qui les trompe & seduit.

Tous ces peuples sont bien proportionnez de leur corps, sans aucune difformité, dispos, & leurs femmes bien formees, remplies & potelees, de couleur basanee pour la quantité de certaine peinture dont ils se frotent, qui les fait devenir olyvastres. Ils sont habillez de peaux, une partie de leur corps est couverte & l'autre partie[a] descouverte: Mais l'Hiver ils remedient à tout, car ils sont habillez de bo*n*nes fourrures, comme d'Orignac, Loutre, Castors, ours-marins, Cerfs, & Biches, qu'ils ont en quantité. L'Hyver qua*n*d il y a beaucoup de neiges, il usent d'une maniere de raquette qui est grande deux ou trois fois comme celles de France, qu'ils attachent à leurs pieds, & vont ainsi da*n*s les neiges sans enfoncer, car autrement ils ne pourroient chasser ny aller en beaucoup de lieux.

a partie] *C* parie

much and perform little. They are, for the most part, a people who have no law and who believe that[22] after God had made all things, He took a great quantity of arrows and stuck them in the ground, whence He drew men and women, who have multiplied in the world up to the present time, and they came in this way; that there is a God, a Son, a Mother, and the Sun, which makes four, although God is above all, but that the Son and the Sun are good because of the benefit they receive from them, but that the Mother is not worth anything because she eats them, and that the Father is not too good. They have an infinity of other mad beliefs[23] and have among them certain *Sauvages* whom they call *Pilotoua*,[24] who speak to the devil face to face and [he] tells them what they must do /[f. 418] as much in war as in other affairs, whom they obey at their first command.

Moreover, they believe that all the dreams they have are true; and indeed there are many of them who say they have seen and dreamed something that will happen. But to tell the truth of it, these are visions of the Devil who deceives and seduces them.

All these peoples are well proportioned in their bodies, without any deformity, agile, and their wives are well made, filled out and plump, with a tawny complexion[25] on account of the great quantity of a certain pigment with which they rub themselves and which causes them to assume an olive-like hue. They are dressed in skins; one part of their body is covered and the other part uncovered. But in winter they fix everything, for they are dressed in good furs, like those of moose, otter, beavers, bears, seals, stags and deer, which they have in abundance. In the winter, when there is a lot of snow, they use a kind of racquet that they fasten to their feet, which is two to three times bigger than those of France. And thus they advance in the snow without sinking, for otherwise they could neither hunt nor get into a lot of places.

22 The entire sentence following "who believe that" was first reported by Champlain as the opinion of the *grand Sagamo* Anadabijou in response to his own questions. The past tense of Anadabijou's account (*ils croyoient*) has also been changed to the present.

23 With this statement, purely his own, Cayet dismisses the native legends that Champlain recorded.

24 According to Father Pierre Biard, *pilotoua* (*pilotois*) was a Basque word meaning "sorcerer." This passage comes directly from *Des Sauvages*, where *pilotoua* must be intended to be plural, because of its verb *parlent*, although Champlain corrects it to *Pillotois* when he repeats this passage in his *Voyages* of 1613. See Document G, "Des Sauvages," n. 74. We have added the *il* here to clarify the change in subject, as in *Des Sauvages*.

25 *Basanee*, literally "tanned."

Ils ont aussi une forme de mariage, Qua*n*d une fille est en l'aage de 14. ou 15. ans, elle aura plusieurs serviteurs & amis, & aura co*m*pagnie avec tous ceux que bon luy semblera, puis au bout de quelque cinq ou six ans, elle pre*n*dra lequel il luy plaira pour son mary, & vivront ainsi ensemble jusques à la fin de leur vie, si ce n'est qu'apres avoir esté quelque temps ensemble ils n'ont en-/[f. 418^{v}]fans: alors l'homme se peut démarier & prendre autre fe*m*me: Or depuis qu'elles so*n*t marriees, elles sont chastes, & leurs maris so*n*t la pluspart jaloux lesquels donnent des presens au pere ou parens de la fille qu'ils auront espousee. Voylà la ceremonie & façon qu'ils usent en leurs mariages.

Pour ce qui est de leurs enterremens, quand un homme ou femme meurt, ils font une fosse, où ils mettent tout le bien qu'ils auro*n*t, comme chaudro*n*s, fourrures, haches, arcs & fleches, robbes, & autres choses, & puis ils mettent le corps dedans la fosse, & le couvrent de terre, où ils mettent quantité de grosses pieces de bois dessus, & un bois debout qu'ils[a] peignent de rouge par le haut.

Ils croyent l'immortalité des ames, & disent qu'ils vont se resjouyr en d'autre pays avec leurs parens & amis quant ils sont morts.

[Transition: Cayet] Apres avoir assez traicté des meurs & coustumes de ces Sauvages, voyons comme par leur moyen le sieur du Pont & les siens furent descouvrir[b] plusieurs pays, où par cy devant autres que les Sauvages n'avoient esté.

[*Des Sauvages*, ch. 5] Le Mercredy dixhuictiesme de Juin, il partit de Tadousac, dans sa chaloupe avec quelques Sauvages qui estoient dans leurs canots, pour chercher la source de ceste gra*n*de riviere & passerent pres de l'Isle du Lievre, à sept lieuës de Tadousac.

De l'Isle au Lievre ils arrivere*n*t à l'Isle au Coudre, qui peut tenir environ deux lieuës de large: au bout de l'Ouest de ceste Isle il y a des prairies & pointes de rochers qui advancent beaucoup /[f. 419] dans la riviere: elle est quelque peu aggreable pour les bois qui l'environnent: il y a force ardoise, & la terre est graveleuse.

a qu'ils] *C* qu'il
b descouvrir] *C* desconurir

They also have a form of marriage. When a girl is 14 or 15 years old, she will have many suitors[26] and male friends, and will keep company with all those who suit her. Then, at the end of some five or six years, she will take whichever one she pleases for her husband, and they will live together in this way until the end of their lives unless it happens that after having been together some time, they have no /[f. 418^{v}] children; then the man may dissolve the marriage and take another wife. Now, as soon as the women are married, they are chaste, and their husbands are, for the most part, jealous, who give presents to the father or relatives of the girl whom they have married. You see the ceremony and the way they practise marriage.

As for their burials, when a man or a woman dies, they dig a pit into which they place all the goods they will possess, such as cauldrons, furs, hatchets, bows and arrows, robes, and other things, and then they place the body in the pit and cover it with earth, upon which they place a great many large pieces of wood and one stake standing upright. which they paint red on the upper part.

They believe in the immortality of souls, and say that they are going to rejoice in other lands with their family and friends when they are dead.

After having said enough about the habits and customs of these *Sauvages*, let us see how, by means of them, the sieur du Pont and his [companions] went to discover several lands where formerly none other than the *Sauvages* had been.

[*Des Sauvages*, ch. 5][27] On Wednesday, the eighteenth day of June, he set out from *Tadousac* in his shallop[28] with some *Sauvages* who were in their canoes[29] to look for the source of this great river, and they passed near the *Isle du Lievre* at seven leagues from *Tadousac*.

From the *Isle au Lievre*,[30] they arrived at the *Isle au Coudre*, which may extend about two leagues in width. At the western end of this island, there were meadows and rocky points that jut out far /[f. 419] into the river. It is somewhat pleasant because of the woods that surround it. There is much slate and the land is gravely.

26 *Serviteurs*, literally "man servants," although suitors (*soupirant*) must have been meant.
27 Cayet omitted any reference to chapter 4.
28 Cayet inferred that the exploring party went in a *chaloupe* (shallop) because Champlain had mentioned earlier that canoes could travel faster than the sieur Du Pont's shallop. Champlain did not say what type of vessel they were travelling in until he got to the Rivière Richelieu, and he then called it a *barque*. In a later passage he called them "medium pinnaces of 12 to 15 tuns" (*Les Voyages, 1632*, pt. 1, 40).
29 The accompaniment of the Natives in canoes is not mentioned by Champlain at this point.
30 Île aux Lièvres is called both *Isle du Lievre* and *Isle au Lievre* by Cayet.

Le Jeudy ensuivant ils moüillerent l'ancre à une anse dangereuse du costé du Nort, où il y a quelques prairies, & une petite riviere, où les Sauvages cabannent quelque-fois.

Le Dimanche vingt deuxiesme ils arriverent à l'isle d'Orleans du costé du Su: Ceste Isle est à une lieuë de la terre du Nort fort plaisante & unie contenant de long huict lieuës: Le costé de la terre du Su est terre basse, quelques deux lieuës avant en terre; lesdites terres co*m*mence*n*t à estre basses à l'e*n*droit de ladite isle, & y faict fort dangereux aborder pour les ba*n*cs de sable, & rochers qui sont entre ladite isle & la grande[a] terre, laquelle asseiche presque toute de basse mer.

De l'Isle d'Orleans ils furent moüiller l'ancre à Quebec qui est un destroit de la riviere de Canadas, qui a quelques 300. pas de large: ce pays est uny & beau, où[b] ils veirent de bonnes terres pleines d'arbres, co*m*me chesnes, cypres, boulles, sapins, & tre*m*bles, & autres arbres fruictiers, sauvages, & vignes: Le long de la coste dudict Quebec il se trouve des Diamans dans des rochers d'ardoise, qui sont meilleurs que ceux d'Alençon.

[*Des Sauvages*, ch. 6] Le Lundy 23. ils[c] partirent de Quebec, où la riviere commence à s'eslargir quelques fois d'une lieue, puis de lieue & demie ou deux lieues au plus, le pays va de plus en plus en embellissant, estant toutes terres basses, sans rochers, que fort peu, Il y a quelques petites rivieres qui ne sont /[f. 419^{v}] point navigables, si ce n'est pour les Canos des Sauvages, à cause de la qua*n*tité des saults, qu'il y peut avoir.

De Quebec ils arriverent à saincte Croix, qui est une pointe basse qui va en haussant des deux costez: Le pays est beau & uny, toutes bo*n*nes terres, avec qua*n*-tité de bois: mais fort peu de sapins & cypres: il s'y trouve en quantité de vignes, poires, noysettes, cerizes, groyzelles, rouges & vertes, & de certaines petites racines de la grosseur d'une petite nois, ressemblant au goust co*m*me treffes, qui sont tres-bonnes rosties & bouillies: Toute ceste terre est noir, sans aucuns rochers, sinon qu'il y a grande quantité d'ardoise: elle est fort tendre, & propre a cultiver: Du costé du Nort il y a une riviere qui s'appelle Batiscan, qui va fort avant en terre, & une autre du mesme costé à trois lieues dudit saincte Croix sur

a grande] *C* grand
b où] *C* ou
c ils] *C* il

The following Thursday, they cast anchor at a dangerous cove on the north coast where there are some meadows and a little river, where the *Sauvages* sometimes set up camp.

On Sunday, the twenty-second, they arrived at the *isle d'Orleans* on the south side. This island is very pleasant and level, measuring eight leagues in length, and is situated a league from the land to the north. The southern shore is lowland for some two leagues inland. The said lands begin to be low at the area of the said island and make it very dangerous to approach on account of the sandbanks and rocks that lie between the said island and the mainland, which almost completely dries up at low tide.

From the *Isle d'Orleans* they went on to cast anchor at Quebec, which is a strait of the *riviere de Canadas* and has a width of some 300 paces. This country is level and beautiful, where they saw some good lands full of trees, such as oak, cedar, birch, fir, and aspen, and other wild fruit-bearing trees and vines. Along the shore of the said Quebec, some diamonds are found in the slate rocks which are better than those of Alençon.[31]

[*Des Sauvages*, ch. 6] On Monday the 23rd, they set out[32] from Quebec, where the river begins to broaden, sometimes by a league, then by a league and a half or two leagues at the most. The country becomes more and more beautiful, being all lowlands, without rocks except for a very few. There are some small rivers that are /[f. 419^{v}] not navigable except with the canoes of the *Sauvages* on account of the many rapids that might be there.

From Quebec they arrived at *saincte Croix*, which is a low point that rises up on both sides. The land is fine and level, all good soil with extensive woods, but very few fir trees and cedars. In these are found a great many vines, pears, hazelnuts, cherries, red and green currants, and certain small roots, the size of a small nut, tasting like truffles, which are very good roasted or boiled. All this soil is black, without any rocks, except that there is a good deal of slate. The soil is very soft and suitable for cultivation. On the north side there is a river called Batiscan that goes far inland, and another on the same side at three leagues from the said *saincte*

31 The so-called diamonds of Alençon, France, are quartz crystals.

32 This is the second occurrence of the singular pronoun *il* with a plural verb, which suggests that it was perhaps an error of transposition from *nous* to *il* (i.e., sieur Du Pont), not just the carelessness of a typesetter. The first was "*il sont quelques-fois dix mesnages ensemble*" ("there are sometimes ten households together") near the end of Cayet's chapter 2.

le chemin de Quebec qui est en celle où fut Jacques Quartier au commencement de la descouverture qu'il en fit, & ne passa point plus outre, ny autre apres luy, qu'e*n* ce voyage. Ladite riviere est plaisante, & va assez ava*n*t da*n*s les terres. Tout ce costé du Nort est fort uny & agreable.

Le Mercredy quatorziesme dudit mois, ils partirent de saincte Croix, plus ils alloient en ava*n*t, plus ils trouverent le pays beau: Ils passere*n*t pres d'une petite Isle, qui estoit remplie de vignes, & mouillerent l'Ancre, à la bande de Su, pres d'un petit coustau: & avec les canaux des sauvages ils fure*n*t en une infinité de petites rivieres, où il y a forces isles plaisantes à voir, les terres esta*n*s pleines d'arbres, qui ressemblent à des noyers, & en [f. 420] ont la mesme odeur.

Retournez à leur Chaloupe, ils passerent plus outre, & rencontrere*n*t une isle, qu'ils[a] appellere*n*t sainct Eloy, & le Vendredy ensuivant, costoya*n*s tousjours la bande du Nort tout proche terre, qui est basse, & pleine de tous bons arbres, & en quantité[b] ariverent aux trois rivieres, où il commence d'y avoir te*m*perature du temps quelque peu dissemblable à celuy de Saincte Croix. Des trois rivieres jusques à saincte Croix il y a quinze lieues: En l'une des rivieres il y a six isles, trois desquelles so*n*t fort petites, & les autres de quelque cinq à six cens pas de long, fort plaisantes & fertilles, pour le peu qu'elles contienne*n*t. Il y en a une au milieu de la riviere qui regarde le passage de celle de Canadas, & commande aux autres esloignees de la terre, tant d'un costé que d'autre de quatre à cinq ce*n*s pas : Elle est eslevee du costé du Su, & va quelque peu en baissant du costé du Nort: Ce lieu

a qu'ils] *C* qu'ls
b quantité] *C* quantitè

Croix on the way from Quebec, which is the one[33] that *Jacques Quartier*[34] was on at the beginning of the discovery that he made of it, and he did not proceed farther, nor anyone after him except on this voyage.[35] The said river is attractive and goes rather far inland. This whole north shore is very level and pleasant.

On Wednesday, the fourteenth of the said month,[36] they set out from *saincte Croix* where the farther they proceeded, the more they found the country to be beautiful. They passed close by a little island, which was covered with vines, and cast anchor on the southern stretch near a small hillside, and with the canoes of the *Sauvages* they went into countless small rivers, where there are many islands pleasing to see,[37] the land being covered with trees that resemble walnut trees and /[f. 420] have the same smell as they do.

Having returned to their shallop,[38] they moved on and encountered an island which they called *sainct Eloy*. And the following Friday, still coasting the north shore quite near the land, which is low and filled with all sorts of good trees, and in great number, they arrived at *trois rivières*, where the temperature of the air begins to be somewhat unlike that of *saincte Croix*. From *trois rivieres* up to *saincte Croix* there are fifteen leagues. In one of the rivers there are six islands, three of which are very small, and the others of some five to six hundred paces in length, very pleasant and fertile for their little size. There is one of them in the middle of the river that faces opposite the channel of the *Canadas* river and dominates the others, removed by four to five hundred paces from the land on both sides. It is high on the south shore becoming a little lower on the north side. This location

33 *rivière*: fem. n.; the wording is awkward here, since the antecedent *une autre* (meaning "another river") is far removed from *qui est celle où fut*. Possibly to clarify the fact that *celle* stands for the second river, Cayet or the compositor added *en*: *qui est en celle*. This makes it even more awkward to translate and has therefore been omitted.

34 This is the regular spelling of Cartier from sixteenth-century France; see Bideaux, *Relations*, and Lestringant, *Le Brasil d'André Thevet*.

35 This statement is of course in error. Not only Cartier but many other Frenchmen had been west of the Rivière Batiscan before 1603. It is notable that Champlain did not claim this himself, as his detractors would complain, but stopped his sentence at "he did not proceed farther."

36 A misprint for "On Wednesday the 24th" in *Des Sauvages*; also an error, 24 June being a Tuesday.

37 Champlain actually wrote that there were "some islands … very pleasing to see" (*quelques isles … fort plaisantes à voir*). The compositor mistook *fort* for *force* and misplaced the new adjective before *isles*.

38 The mention of "their shallop" (*leur Chaloupe*) is pure invention, not found in *Des Sauvages*.

fut reputé propre pour habiter & lequel on pourroit fortifier pro*m*ptement, car sa situation est forte de soy, & proche d'un grand lac qui n'en est qu'à quelque quatre lieues, lequel presque joinct la riviere du Saguenay, selon le rapport des Sauvages qui vo*n*t pres de cent lieues au Nort, & passe*n*t no*m*bre de saults puis vont par terre quelque cinq ou six lieues, & entrent dedans un lac, d'où ledict Saguenay prend la meilleure part de sa source, & par où les Sauvages viennent dudit lac à Tadousac.

[*Des Sauvages*, ch. 7] Le Samedy ensuivant le sieur du Pont & les siens partirent des trois Rivieres & vindrent mouiller l'ancre à un lac où il y a quatre lieues, /[f. 420v] Tout ce pays depuis les trois rivieres jusques à l'entree dudict lac, est toute terre bonne à fleur d'eau: les bois y sont clairs, qui fait que l'on y pourroit traverser aiseme*n*t. Le lendemain[a] 29. de Juin, ils entrere*n*t dans le lac, qui a quelque 15. lieues de long, & quelques 7. ou 8. lieues de large qu'ils traverserent le mesme jour, & vindrent mouiller l'ancre enviro*n* deux lieues da*n*s la riviere qui va au sault, à l'entree de laquelle il y a tre*n*te petites isles, les unes de deux lieues, d'autres de lieue & demie & quelques unes moindres, lesquelles so*n*t re*m*plies de quantité de Noyers, & de vignes sur le bord desdites isles: mais quand les eaues sont grandes, la plus-part d'icelles sont couvertes d'eau: Le dernier de Juin il passerent à l'entree de la riviere des Irocois, où estoie*n*t caba*n*nez & fortifiez les Sauvages qui leurs alloie*n*t faire la guerre: Leur forteresse est faicte de qua*n*tité de bastons fort pressez les uns contres les autres, laquelle vie*n*t joindre d'un costé sur le bord de la grand' riviere, & l'autre sur le bord de la riviere des Irocois, & leurs Canos arra*n*gez les uns contre les autres sur le bord, pour pouvoir pro*m*pteme*n*t fuyr, si d'adventure ils sont surprins des Irocois: Car leur forteresse est couverte d'escorce de chesnes, & ne leur sert q*ue* pour avoir le te*m*ps de s'embarquer. Ils fure*n*t dans la riviere des Irocois quelques cinq ou six lieues, & où ils ne peurent passer plus outre avec leur barque, à cause du grand cours d'eau qui y descend; Toute ceste riviere est large de quelque trois à quatre cent pas, & va comme au Sorouest. Les Sauvages disent, qu'à quelque quinze lieues, il y a un sault /[f. 421] qui vient de fort haut, où ils porte*n*t leurs Canos pour le passer enuiro*n* un quart

a Le lendemain] *C* Lel'endemain

was reputed to be suitable for habitation,[39] and one could fortify it quickly, for its situation is strong in itself and near a great lake that is only some four leagues away. It almost joins the *riviere du Saguenay*, according to the report of the *Sauvages* who travel nearly a hundred leagues to the north and pass by a number of rapids, then go by land some five or six leagues and enter a lake from which the said Saguenay takes the better part of its source, and by which the *Sauvages* come from the said lake to *Tadousac*.

[*Des Sauvages*, ch. 7] On the following Saturday, the sieur du Pont and his [companions] set out from *trois Rivieres* and came to drop anchor at a lake some four leagues distant. /[f. 420ᵛ] All this region from *trois rivieres* to the entrance of the lake is all good land at water level. The woods there are open, which would make them easy to go through. The next day, the 29th of June, they entered the lake, which is some 15 leagues long and some 7 or 8 leagues wide, which they crossed the same day. And they came to drop anchor about two leagues up the river that leads to the rapids,[40] at whose mouth there are thirty small islands – some of two leagues, others a league and a half, and some less – which abound in walnut trees and vines on the shores of the said islands; but when the waters are high, the majority of them are covered with water. On the last day of June, they passed into the mouth of the *riviere des Irocois*[41] where, encamped and fortified, were the *Sauvages* who were going to make war on them. Their stronghold is made of a great many stakes set very close together and comes to an end on the bank of the great river on one side, and on the bank of the *riviere des Irocois* on the other, and their canoes arranged one against the other on the bank so that they may quickly take flight if by chance they are surprised by the *Irocois*; for their stronghold is covered with the bark of the oak and serves only to give them time to embark. They went some five or six leagues up the *riviere des Irocois*, and where they could pass no farther with their *barque*[42] because of the great current of water that descends there. This entire river is some three to four hundred paces wide and runs about southwest. The *Sauvages* say that, at some fifteen leagues, there is a rapids /[f. 421] that descends from very high up, where they carry their

39 It was Champlain who made the suggestion in *Des Sauvages* that this could be the site for a fortified habitation.

40 Cayet wrote "*qui va au sault*," while Champlain wrote "*qui va au hault*." Of the two versions, Cayet's makes more sense in that the St. Lawrence River leads to the Lachine "rapids." It is probable that the *hault* in *Des Sauvages* is a misprint for *sault*.

41 The *riviere des Irocois* is the Rivière Richelieu.

42 Following Champlain, Cayet now calls the *chaloupe* a *barque*. The French becomes more awkward with the addition of "where" (*où*) following "and," when Champlain only used the latter conjunction.

de lieue, & entre*n*t deda*ns* un lac, où à l'e*n*tree il y a trois Isles: & esta*n*t dedans, ils en rencontrent encores quelques unes, il peut co*n*tenir quelque quara*n*te ou cinqua*n*te lieues de long, & de large quelque vingt cinq lieues, dans lequel descendent qua*n*tité de rivieres, jusques au nombre de dix, lesquelles porte*n*t Canots assez avant: puis venans à la fin dudit lac, il y a un autre saut, & rentrent dedans un autre lac, qui est de la gra*n*deur du premier, au bout duquel sont cabannez les Irocois: au pays desquels il y a une riviere qui va re*n*dre à la coste de la Floride, & que tout ce pays est quelque peu montagneux, neantmoins pays tres bon, temperé sans beaucoup d'hyver, que fort peu.

[*Des Sauvages*, ch. 8] De la riviere des Irocois, ils allerent mouiller l'ancre à trois lieues de là, à la bonde du Nort, Tout ce pays est une terre basse remplie de toutes sortes d'arbres & fruicts, co*m*me vignes, noix, noizettes, & une maniere de fruict qui semble à des chataignes, cerises, chesnes, tremble, pible, houblon, fresne, erabe, hestre, cypres, fort peu de pins & sapins: il y a aussi d'autres arbres, desquels il n'y en a point en Europe: Il s'y trouve quantité de fraizes, framboises, groizelles, rouges, vertes & bleues, avec force petits fruits qui y croissent parmy grande quantité d'herbages, Il y a aussi plusieurs bestes sauvages, co*m*me Orignacs, cerfs, biches,[a] dains, ours, porcs-espics, lapins, renards, castors, loutres, rats musquets[b] & quelques autres sortes d'animaux, lesquels sont bons à manger, & dequoy vivent les Sauvages. /[f. 421v]

En fin le Mercredy ensuivant, ils arrivererent à l'entree du sault, avec vent en poupe: mais ne pouvant passer plus outre à cause du grand courant d'eau qui s'y faict, ils entrerent dans un petit equif qu'ils avoient faict faire expres, pour passer ledit sault: Ils ne furent pas à trois cents pas, qu'il falut que les Matelots se missent à l'eau pour faire passer l'esquif: le Canot des Sauvages passoit aisément: & ainsi continua*ns* leur chemin costoyans plusieurs isles & rochers, ils arriverent à

a cerfs, biches,] *C* cerfs biches,

b rats musquets] *C* rats, musquets

canoes for about a quarter of a league in order to pass it by and enter a lake, where there are three islands at the mouth. And once in the lake, they still encounter some more of them. The lake may measure some forty or fifty leagues in length and about twenty-five leagues in width, into which a number of rivers descend – as many as ten – which carry canoes fairly far inland. Then, coming to the end of the said lake, there is another rapids and they enter another lake again, which is the same size as the first, at the end of which the *Irocois* are encamped, in whose lands there is a river that goes to the coast of Florida.[43] And this whole land is somewhat mountainous, nevertheless a very good region, temperate and without much of a winter, only very little.[44]

[*Des Sauvages*, ch. 8] From the *riviere des Irocois*, they went to drop anchor three leagues from there on the north shore. This whole region is lowland, replete with all kinds of trees and fruits; such as vines, walnuts, hazelnuts, and a kind of fruit that resembles the chestnut, cherries, oaks, aspen, poplar, hops, ash, maple, beech, cedar, and very few pines and firs. There are also other trees, of which there are none found in Europe.[45] One finds there large quantities of strawberries, raspberries, red, green, and blue currants, together with a great number of small fruits that grow in the thick grasslands there. There are also many wild animals, such as moose, elk, hinds,[46] deer, bears, porcupines, rabbits, foxes, beavers, otters, muskrats,[47] and some other kinds of animals that are good to eat and on which the *Sauvages* subsist.

/[f. 421ᵛ] At length, the following Wednesday, they arrived at the entrance of the rapids[48] with the wind in the stern. But not being able to go on, because of the great current of water that is made there, they got into a little skiff that they had had constructed precisely for the purpose of crossing the said rapids. They had not gone as far as three hundred paces when some sailors had to get into the water to get the skiff across. The *Sauvages*' canoe crossed easily. And thus continuing their way, coasting several islands and rocks, they arrived at a kind of lake, which may

43 The preceding is a description of the route from the St. Lawrence River through Lake Champlain and the Hudson Valley to the Atlantic seaboard.

44 Again, the abbreviation of Champlain's text is somewhat inept in linking this sentence to the previous one with *& que*. We have omitted the *que* in this translation.

45 The second half of this sentence is another awkward attempt to exclude Champlain; it replaces *que je ne cognois point* (which I don't know; trans. "with which I am not familiar").

46 Cayet omitted a comma present in *Des Sauvages* and wrote "*cerf biches*," rather than "*cerfs, biches*."

47 Both Champlain and Cayet wrote "*rat, musquets*," as if there were two animals. The *rat musqué* is the muskrat (*Ondatra zibethicus*). Rats were introduced to the Americas with European settlement (Banfield, *Mammals of Canada*, 221–3).

48 The Lachine Rapids

une maniere de lac, lequel peut contenir quelque cinq lieues de long, & presque autant de large, où il y a quantité de petites isles qui so*n*t rochers: mais venans à approcher du sault avec leur petit esquif & le canot des Sauvages, il leur fut impossible de passer plus avant, bien que le sault ne soit pas beaucoup haut, n'esta*n*t en d'aucuns lieux que d'une brasse ou de deux, & au pl*us* de trois: lequel descend comme de degré en degré, & en chasque lieu où il y a quelque peu de hauteur il s'y faict un esbouillonnement estrange de la force & roideur que va l'eau en le traversant qui peut contenir une lieue: il y a force rochers de large, & environ le milieu, il y a des Isles qui sont fort estroictes & fort longues: Il y a sault tant du costé desdites isles qui sont au Su, co*m*me du costé du Nort, ou il faict si dangereux qu'il est hors de la puissance d'homme d'y passer un basteau, pour petit qu'il soit.

Outre ce sault premier, les Canadois disent, qu'il y en a dix autres, la plus-part difficiles à passer, & ausquels on ne sçauroit aller qu'avec les Canots des Sauvages. Ledit sault est par /[f. 422] les 45, degrez & quelque minutes.

Le Sieur du Pont & les siens voyans qu'ils ne pouvoie*n*t faire d'avantage, retournere*n*t en leur barque, où ils interrogerent les Sauvages de la fin de la riviere, & de quelle partie procedoit sa source: Ils leur dirent que passé ce premier sault, ils faisoient quelque dix ou quinze lieues avec leurs Canots deda*n*s la riviere, où il y a une riviere qui va en la demeure des Algoumequins, qui sont à quelques soixa*n*tes lieues esloignez de la grande riviere, & puis ils venoie*n*t à passer cinq saults, lesquels peuvent co*n*tenir du premier au dernier huict lieues, desquels il y en a deux où ils porte*n*t leurs canots pour les passer: chasque sault peut tenir quelque demy quart de lieue, ou un quart au plus: Et puis ils viennent dedans un lac, qui peut tenir quelque quinze ou seize lieues de long. De là ils rentrent dedans une riviere, qui peut contenir une lieue de large, & font quelque deux lieues dedans, & puis rentrent dans un autre lac de quelque quatre ou cinq lieues de long, venant au bout duquel ils passent cinq autres saults, distant du premier au dernier quelque vingt cinq ou trente lieues, do*n*t il y en a trois où ils porte*n*t leurs canots pour les passer, & les autres deux ils ne les font que traisner dedans l'eau, d'autant que le cours n'y est si fort ni mauvais comme aux autres: De tous ces saults qu'aucun n'estoit si difficile à passer comme le premier qu'ils avoient veu: Et puis qu'ils arri-

measure some five leagues in length and almost as many in width, where there are a great number of little islands that are rocks. But approaching the rapids with their little skiff and the *Sauvages'* canoe, it was impossible for them to proceed, although the rapid is not very high, not being in any place more than a *brasse* or two, or three at the most. It descends as if stepwise, and in each place where there is somewhat of a height, a strange bubbling forth occurs there from the force and steepness with which the water flows when crossing it for possibly a league. There are many rocks in the open, and around the middle there are some islands that are very long and narrow. There are rapids as much on the side of the said islands that are to the south as on the north shore, where it is so dangerous that it is beyond the power of a man to get a boat past it, however small it may be.

Beyond this first rapids, the *Canadois*[49] say that there are ten more of them, for the most part difficult to pass, and to which one could not go except with the canoes of the *Sauvages*. The said rapids is at /[f. 422] 45 degrees and some minutes.

The sieur du Pont and his [companions], seeing that they could do no more, returned in their barque, when they questioned the *Sauvages* about the end of the river and about where its source came from.[50] They told them that, this first rapids having been crossed, they did some ten or fifteen leagues with their canoes in the river, to where there is a river that enters the home of the *Algoumequins*, who are removed from the great river by some sixty leagues. And then they used to go past five rapids, which may extend eight leagues from first to last, of which there are two of them where they carry their canoes to get past them. Each rapids may extend an eighth of a league or a quarter at the most. And then they come into a lake, which may extend some fifteen or sixteen leagues in length. From there, they enter a river again, which may measure a league in width, and accomplish some two leagues in it; and then they enter another lake again, of some four or five leagues in length, coming to the end of which, they pass another five rapids, distant from first to last some twenty-five or thirty leagues, of which there are three where they carry their canoes to cross them. And for the other two, all they do is drag them in the water, for as much as the current there is neither as strong nor as bad as in the others. Of all these rapids, they told them that none[51] was so difficult to cross as the first which they had seen; and then that they ar-

49 A term not used by Champlain except (*Canadiens*) as an early synonym for the *Gaspesiens*, a group of the Mi'kmaq. Cayet probably used the term as a synonym for Native Canadian.

50 Champlain obtained three Algonquin descriptions with maps of the water systems west of the Lachine Rapids (fig. 8, top). Cayet repeated only the first and best of these descriptions.

51 Cayet's introduction of an extra *que* here builds on Champlain's sporadic use of it in this passage. In all cases, we have translated the indirect discourse of the Natives that

voient dedans un lac qui peut tenir quelque 80. lieues de lo*n*g, où il y a qua*n*tité d'isles; & qu'au bout d'iceluy l'eau y est salubre, & l'hyver /[f. 422v] doux. Qu'à la fin dudict lac ils passent encor un sault, qui est quelque peu eslevé, où il y a peu d'eau laquelle descend: là qu'ils porte*n*t leurs canots par terre, environ un quart de lieue pour passer ce sault: De là qu'ils entrent dans un autre lac qui peut tenir quelque soixa*n*te lieues de lo*n*g, dont l'eau en est fort salubre: où estant à la fin ils viennent à un destroict qui co*n*tient deux lieues de large, lequel va assez ava*n*t da*n*s les terres: qu'ils n'avoient point passé plus outre; & n'avoient veu la fin d'un lac qui est à quelque quinze ou seize lieues d'où ils ont esté, ny veu homme qui l'eust veu: d'autant qu'il est si grand, qu'ils ne se hazarderont pas de se mettre au large, de peur que quelque tourmente ou coup de vent ne les surprint: & que l'eau de ce lac est tres-mauvaise comme celle de la mer. Voylà tout ce que le sieur du Pont apprit des Sauvages, touchant la grande riviere de Canadas.

[*Des Sauvages*, ch. 9] Ne pouvant passer plus outre, il partit dudit sault le Vendredy quatriesme jour de Juin, & revint par le mesme chemin qu'il y avoit esté: le Vendredy unziesme dudit mois il fut de retour, avec les siens à Tadousac, où il avoit laissé son vaisseau.

[*Des Sauvages*, ch. 10] A la descouverture d'un pays l'on demande tousjours s'il y a des mines, le sieur du Pont ne l'oublia pas à le demander: les Sauvages dirent, qu'il y en avoit; il s'y faict conduire, & pour cest effect se rembarquant dans son vaisseau, il arrive avec les sie*n*s à Gachepay distant de Tadousac de ce*n*t lieues, & continuant son chemin il arriva à la Baye des Moluës, laquelle peut tenir quel-/[f. 423]ques trois lieuës de long, autant de large à son entree: De là il vient à l'Isle Percee, qui est co*m*me un rocher fort haut, eslevé des deux costez. Tous cesdits lieux de Gachepay, Baye des Moluës, & Isle Percee, sont les lieux où se

rived in a lake, which may extend some 80 leagues long, where there are many islands, and that at the end of that lake the water is wholesome[52] and the winter /[f. 422ᵛ] mild. They told them that at the end of the said lake, they pass yet another rapid that is somewhat elevated and where there is a bit of water that flows down;[53] that there they carry their canoes overland for about a quarter of a league in order to pass this rapid; that from there they enter into another lake, which may extend some sixty leagues in length, of which the water is very wholesome; where, being at the end, they come to a strait that measures two leagues in width and leads rather far inland. They said that they had not gone any farther and had not seen the end of a lake that is located at some fifteen or sixteen leagues from where they were, nor had they seen anyone who had seen it, forasmuch as it is so big that they will not risk setting out into the open, for fear that some storm or gale might surprise them. And they say that the water of this lake is very bad, like that of the sea. This is all that the sieur du Pont learned[54] from the *Sauvages* touching the great *riviere de Canadas*.

[*Des Sauvages*, ch. 9] Not being able to go on, he set off from the said rapids on Friday, the fourth day of June,[55] and returned by the same way that he had gone. On Friday, the eleventh[56] of the said month, he was back with his [companions] in *Tadousac*, where he had left his vessel.

[*Des Sauvages*, ch. 10] Upon discovery of a land, one always asks if there are some mines. The sieur du Pont did not forget to ask this. The *Sauvages* said that there were some.[57] He gets himself guided there and to this end, re-embarking into his vessel, he arrives with his [companions] at *Gachepay*, removed from *Tadousac* by a hundred leagues, and continuing his way he arrived at the *Baye de Moluës*, which may extend some /[f. 423] three leagues in length and as much in width at its entrance. From there, he comes to the *Isle Percee*, which is like a very high rock, lofty from both sides. All these said places, *Gachepay*, *Baye de Moluës*, and *Isle Percee*,

begins above: *Ils leur dirent que* … (They told them that …) by repeating this phrase whenever *que* arises.

52 The word used by Champlain and Cayet to describe the water of this lake (Ontario) is *salubre* (wholesome). Unfortunately, it has commonly been translated as "brackish" or "salty."

53 An unfortunate description of Niagara Falls by Champlain, repeated here by Cayet.

54 Champlain tells us only that he himself had seen this and heard the Natives reply thus to their questions, although Gravé Du Pont shared the experience and the questioning with him.

55 The correct date is Friday, the fourth day of July. Champlain also wrote June.

56 Cayet corrected Champlain's date here from Friday the tenth to the eleventh, even though the month (June, not July) is still incorrect.

57 These three weak introductory sentences are pure invention, not found in *Des Sauvages*.

faict la pesche du poisson sec & verd. Passant l'Isle Percee, il arriva à la Baye de Chaleurs, & de là vint à une riviere qui s'appelle Souricoua, d'où le sieur de Prevert fut envoyé pour descouvrir une mine de cuivre qui est sur le bord de la mer du costé du Su [ch. 12], où il fut avec peine, pour la crainte que les Sauvages qu'il mena avec luy avoient de rencontrer leurs ennemis, qui sont les Armouchicois, lesquels sont hommes sauvages du tout monstrueux, pour la forme qu'ils ont: car leur teste est petite, & le corps court, les bras menus co*m*me d'un schelet, & les cuisses semblablement: les jambes grosses & lo*n*gues qui sont toutes d'une venuë, & quand ils sont assis sur leurs tallons, les genoux leur[a] passent plus d'un demy pied par dessus la teste qui est chose estrange, & semble*n*t estre hors de nature: Ils sont neantmoins fort dispos & determinez: & sont aux meilleures terres de toute la coste d'Arcadie: aussi les Souricois les craignent fort: Mais avec l'asseurance que le sieur de Pevert leur donna, il les mena jusques à ladite mine, où les Sauvages le guiderent: C'est une fort haute montagne, advançant quelque peu sur la Mer, qui est fort reluisa*n*te au Soleil, où il y a quantité de verd de gris, qui procede de la mine de cuivre. Au pied de ladite mo*n*tagne, y a quantité de morceaux de cuivre, lequel tombe du haut de la montagne: le lieu où /[f. 423v] est ceste mine est par les 44. degrez quelque minute. Passant trois ou quatre lieuës plus outre, tirant à la coste d'Arcadie, il y a une autre mine, & vne petite riviere qui va quelque peu da*n*s les terres tirant au Su, où il y a une montaigne, qui est d'une peinture noire, dequoy se peignent les Sauvages: puis à quelque six lieuës de la seconde mine, en tirant à la mer, environ une lieue proche de la coste d'Arcadie, il y a une Isle, où se trouve une maniere de metail qui est co*m*me bru*n* obscur, le couppant il est blanc, dont anciennement les Sauvages usoient pour leurs flesches & cousteaux,

a leur: *C* leu

are places where fishing for dry and green fish is done. Crossing the *Isle Percee*, he arrived in the *Baye de Chaleurs,* and from there he came to a river which is called *Souricoua*, from where the sieur de Prevert was sent to discover a copper mine[58] that lies beside the sea on the south [ch. 12],[59] where he went with difficulty on account of the fear that the *Sauvages* he took with him had of encountering their enemies, the *Armouchicois*,[60] who are wild men, totally monstrous because of the shape they have. For their heads are small and their bodies short, their arms slender like a skeleton, as well as their thighs in a similar way. Their legs are thick and long, and straight up and down. And when they are seated upon their heels, their knees go more than half a foot above their heads, which is a strange thing. And they seem to be unnatural. Nevertheless, they are very agile and determined, and inhabit the best lands on the entire coast of *Arcadie*.[61] Moreover, the *Souricois* fear them greatly.[62] But with the assurances which the sieur de Prevert gave them, he brought them as far as the said mine, to which the *Sauvages* guided him. It is a very high mountain, jutting a little way into the sea, which sparkles brightly in the sun, where there is a large quantity of verdigris[63] that comes out of the copper mine. At the foot of the said mountain, there is an abundance of copper pieces, which fall from the height of the mountain. The place where /[f. 423^{v}] this mine is situated is at 44 degrees and some minutes.[64] Moving three or four leagues on, drawing towards the coast of *Arcadie*, there is another mine and a small river that goes a little way inland, shooting towards the south, where there is a mountain of a black pigment, with which the *Sauvages* paint themselves. Then at some six leagues from the second mine, drawing towards the sea about a league off the coast of *Arcadie*, there is an island where a kind of metal is found that is darkish brown – cutting it, it is white – which formerly the *Sauvages* used for their arrows

58 The idea that Sarcel de Prévert was "sent" (*envoyé*) to discover mines is not in *Des Sauvages* but is correct.

59 Cayet now jumps from about the middle of chapter 10 to the middle of chapter 12, omitting all of chapter 11.

60 The *Armouchiquois*, the Algonquian-speaking groups along the New England coast from the Saco River west around Cape Cod. They lived in well-organized villages with a subsistence economy based on horticulture and fishing.

61 This description is, of course, nonsense, perhaps concocted by the Mi'kmaq in order to get help from Prévert and the French in fighting their southern enemies.

62 The *Armouchiquois* were periodically at war with the Mi'kmaq (*Souricois*).

63 Verdigris is copper acetate, produced over time when copper is exposed to seawater or air.

64 These copper deposits were visited by Champlain in 1604 near Advocate Harbour, NS, at about 45°15'N.

qu'ils battoient avec des pierres: ce qui me fait croire que ce n'est estain, ny plomb, estant si dur comme il est: Le sieur de Prevert leur monstra de l'argent, ils dirent, que celuy de ladite Isle estoit semblable, lequel ils trouvent dedans la terre, comme à un pied ou deux.

Apres que le sieur de Prevert eut donné aux Sauvages des coins & cizeaux, & autres choses necessaires pour tirer du metail de ladite mine, ce qu'ils luy promirent de faire, il s'en retourna à la Baye de Chaleurs retrouver le sieur du Pont pour s'en retourner en France: [ch. 13] mais en s'en retournant, il passa contre une Isle où fait residence un monstre espouventable, que les Sauvages appellent Gougou, & disent qu'il a la forme d'une femme, mais fort effroyable, & d'une telle grandeur, qu'ils asseurent que le bout des masts d'un vaisseau ne luy viendroit pas jusques à la ceinture, tant ils le peignent grand: & que souvent il a devoré & devore beaucoup de Sauvages, lesquels il met deda*n*s une gra*n*de poche qua*n*d /[f. 424] il les peut attraper & puis les mange: & disoient ceux qui avoient evité le peril de ceste malheureuse beste, que sa poche estoit si grande, qu'il y eust peu mettre un navire. Or ledit sieur de Prevert passa si proche de la demeure de ceste effroyable beste, que luy & tous ceux de son vaisseau entendoient des sifflements estranges du bruit qu'elle faisoit: si que les Sauvages qu'il avoit avec luy, avoie*n*t une telle peur qu'ils se cachoient de toutes parts, craigna*n*t qu'elle fust venue à eux pour les emporter: Tous les Sauvages en general craignent cela, & en parlent si estrangement, l'appellant la mauvaise Mere, que c'est chose esmerveillable de leur en ouïr parler: mais il faut croire que c'est la residence de quelque Diable qui les tourmente de la façon.

and knives, which they beat out with stones; which makes me[65] think that, being as hard as it is, it is neither tin nor lead. The sieur de Prevert showed them some silver. They said that the metal from the said island was similar, which they find within the ground about a foot or two down.

After the sieur de Prevert had given the *Sauvages* some wedges and chisels and other things necessary for extracting the metal from the said mine, which they promised him they would do, he returned from there to the *Baye de Chaleurs* to find the sieur du Pont again and return from there to France; [ch. 13] but while returning, he passed opposite an island where a frightful monster makes its abode, which the *Sauvages* call *Gougou*.[66] And they say that it has the shape of a woman, but most terrifying and of such a size that they assert that the top of the masts of a ship would not come up to its waist, so big do they depict it; and that it has often devoured and does yet devour many *Sauvages*, whom it puts into a great pocket when /[f. 424] it can catch them, and then eats them. And those who had escaped the perils of this miserable beast said that its pocket was so big that it could have put a ship into it. Now the said sieur de Prevert passed so near the haunt of this frightful beast that he and all those of his vessel heard strange hissings from the noise that it was making, such that the *Sauvages* whom he had with him were so afraid that they hid themselves on all sides, fearing that it would come to carry them away. All the *Sauvages* in general fear that thing and speak of it so strangely, calling it the bad Mother,[67] that it is a marvellous thing to hear them speak of it. But one must think that it is the haunt of some Devil which torments them in this way.[68]

65 This "me" is the unacknowledged sieur de Champlain, which Cayet somehow allowed to slip by.

66 Here begins the infamous story of the *Gougou* that earned Champlain and Cayet the ridicule of their contemporary, Marc Lescarbot, and others, for their naivety (Grant, *History of New France*, 2:168–76).

67 Champlain does not mention this name for the *Gougou*, "the bad Mother" (*la mauvaise Mere*), although her character conforms to the primordial Mother of whom Anadabijou spoke to Champlain (see *Des Sauvages*, ch. 3, f. 9, and f. 417ᵛ above).

68 It is entirely possible that Champlain was being told Mi'kmaq myths about *Kuhkw* and/or *Kokwees*. According to Rand (*Legends*, 1:39), "*Kuhkw* means Earthquake, this mighty personage can pass along under the surface of the ground, making all things shake and tremble by his power," while the *Kokwees* is a giant cannibal (ibid., 2:288–9). Champlain's fault was not so much in reporting what others had told him but implying that he believed the stories by not refuting them; but Cayet has misrepresented Champlain by excluding his disclaimer "*si je mettois tout ce qu'ils disent, l'on le tiendroit pour fables*" ("if I were to put down everything that they say about it, it would be taken for lies"). Cayet's fault, according to Lescarbot, was to pass these stories on without comment.

Le 24. jour d'Aoust, les vaisseaux dudict sieur de Prevert & du Pont partirent pour retourner en France, & y arriverent le 20. de Septembre, ayant eu tousjours le vent favorable, jusques au port du Havre de Grace. Voylà tout ce qui s'est passé au voyage de Canadas en ceste annee: Au livre suyvant, nous verrons comme le sieur du Mont y est arrivé, & y a basti un fort, & des choses les plus remarquables qui s'y sont passees en son voyage.

On the 24th day of August, the vessels of the said sieur de Prevert and du Pont set off to return to France and arrived there on the 20th of September, having always had a favourable wind, all the way up to the harbour of *Havre de Grace*. This is everything that took place on the voyage of *Canadas* in this year. In the following volume, we shall see how the sieur du Mont arrived there and built a fort there, and the most remarkable things that happened on his voyage.

DOCUMENT I

The Voyage of Samvel Champlaine of Brouage, made unto Canada in the yeere 1603

INTRODUCTION TO THE DOCUMENT

In 1625 Samuel Purchas (1577–1626) published the first English translation of any of Champlain's writings. Of the original text, the only section that was omitted was the entire description of the *Gougou* in chapter 13, probably because it stretched the credulity of the reader.

Purchas had obtained the translation of *Des Sauvages* from Richard Hakluyt (1552–1616), who may have been preparing a third edition of his *Principal Navigations* sometime before he died. It is likely that the translation was made by Hakluyt or by someone hired by him.[1] Both Narcisse Dionne, in his biography of Champlain, and Philéas Gagnon state that the Purchas text was based on the 1604 printing of *Des Sauvages*.[2] On what grounds they make this assertion is not known, because it is unlikely that Dionne and Gagnon had ever seen that edition. Moreover, the two editions are so similar that one cannot tell which edition was used by Hakluyt.

This translation was to have an indelible effect on all subsequent English translations of the book except those by Charles Pomeroy Otis (1880)[3] and William L. Grant's (1911) translation of Lescarbot (see below). Instead of re-examining Champlain's French, later translators and editors seemed to find the Hakluyt-Purchas version an irresistible model to follow. The word *salubre* (wholesome, from the Latin *salubris*), for example, was used by Champlain to describe the quality of the water in the lakes west of the Lachine Rapids that various Natives had told him about. In the Hakluyt-Purchas version, this word is consistently translated as "salty" or "brackish," probably because Hakluyt wanted his readers to think that there was a passage west to a salt sea. Otis and Grant both translated *salubre* correctly as "fresh" and "very good," respectively. Since all other

1 Quinn, *The Hakluyt Handbook*, 1:82, 92; Pennington, *The Purchas Handbook*, 1:64, 317

2 Dionne, *Champlain*, 14–15; Gagnon, "Notes bibliographiques," 61

3 See Slafter, *Voyages*, trans. Otis.

English translations followed Hakluyt-Purchas, their translations of Champlain's descriptions of the western lakes are faulty. In most respects, however, the Hakluyt-Purchas version is fairly faithful to the original.

Additions to the original document are given in square brackets. Original pagination is given thus: /[p. 1607]. The italics are in the original text. The only change that has been made to the original is to normalize *u*/*v* and *i*/*j*; e.g., Brouage instead of *Brovage* and Jersey instead of *Iersey*. All other Archaic English spelling has been maintained. We have not reproduced the sidebars in the Hakluyt-Purchas text because they simply give directions to the reader in place of the table of contents in Champlain's original text. In places where Purchas inserted explanations in his sidebars that are not in Champlain's text, we have rendered them as footnotes.

The Voyage of Samvel Champlaine of *Brouage, made unto* Canada *in the yeere* 1603

Purchas, Samuel, comp. and ed. *Hakluytus Posthumus or Purchas His Pilgrimes. Contayning a History of the World, in Sea Voyages & Lande Travells, by Englishmen & others.* Volume 4: *1605–1619*. London: Printed by William Stansby for Henrie Featherstone, 1625

EDITION USED: Library of Congress. Kraus Collection of Sir Francis Drake, G159, P98

ONLINE: http://hdl.loc.gov/loc.rbc/rbdk.d0404

/[p. 1605; *Des Sauvages*, ch. 1] WE departed from *Honfleur*, the fifteenth day of March 1603 [fig. 7a]. This day we put into the Roade of New *Haven*, because the winde was contrary. The Sunday following being the sixteenth of the said moneth, we set saile to proceed on our Voyage. The seventeenth day following, we had sight of *Jersey* and *Yarnsey*, which are Iles betweene the Coast of *Normandie* and *England*. The eighteenth of the said moneth, we discryed the Coast of *Britaine*. The nineteenth, at seven of the clocke at night, we made account that we were thwart of *Ushent*. The one and twentieth, at seven of clocke in the morning, we met with seven ships of *Hollanders*, which to our judgement came from the /[p. 1606] *Indies*. On Easter day, the thirtieth of the said moneth, wee were encountred with a great storme, which seemed rather to be thunder then winde, which lasted the space of seventeene dayes, but not so great as it was the two first dayes; and during the said time we rather lost way then gained.

The sixteenth day of Aprill the storme began to cease, and the Sea became more calme then before, to the contentment of all the Company; in such sort as continuing our said course untill the eighteenth of the said moneth, we met with a very high Mountaine of Ice. The morrow after we discried a banke of Ice, which continued above eight leagues in length, with an infinite number of other smaller peeces of Ice, which hindred our passage. And by the judgement of our Pilot, the said flakes or Ice were one hundred, or one hundred & twenty leagues from the country of *Canada*; and we were in 45. degrees and two third parts; & we found passage in 44. deg. The second of May, at eleven of clocke of the day, we came upon *The Banke* in 44. degrees one third part. The sixt of the said moneth, we came so neere the land that we heard the Sea beate against the shore, but we could not descrie the same through the thicknesse of the fogge, whereunto these coasts are subject; which was the cause that we put farther certaine leagues into the Sea, untill the next day in the morning, when we descried land, the weather being very cleere, which was the Cape of Saint *Marie*. The twelfth day following we were overtaken with a great flaw of winde, which lasted two dayes. The fifteenth of the said moneth, wee descried the Isles of Saint *Peter.* The seventeenth following we met with a banke of Ice neere Cape *de Raie*, sixe leagues in length, which caused us to strike saile all the night, to avoide the danger we might incurre. The next day we set saile, and descried Cape *de Raie*, and the Isles of Saint *Paul*, and Cape *de* Saint *Laurence*, which is on the South side. And from the said Cape of Saint *Laurence* unto Cape de *Raie*, is eighteene leagues, which is the breadth of the entrance of the great Gulfe of *Canada*.

The same day, about ten of the clocke in the morning, we met with another Iland of Ice, which was above eight leagues long. The twentieth of the said moneth, we discried an Isle, which containeth some five and twenty or thirty leagues in length, which is called the Isle of *Assumption*, which is the entrance of the River of *Canada*. The next day we descried *Gachepe*, which is a very high land, and began to enter into the said River of *Canada*, ranging the South coast unto the River of *Mantanne*, which is from the said *Gachepe* sixtie five leagues; from the said River of *Mantanne* we sailed as farre as the *Pike*, which is twenty leagues, which is on the South side also: from the said *Pike* we sailed over the River unto the port of *Tadousac*, which is fifteene leagues. All these Countries are very high, and barren, yeelding no commoditie. The foure and twentieth of the said moneth we cast anker before *Tadousac*, and the six and twentieth we entred into the said Port, which is made like to a creeke in the entrance of the River of *Saguenay*,[1] where there is a very strange currant and tide, for the swiftnesse and depth thereof, where sometimes strong windes do blow, because of the cold which they bring with them; it is thought that the said River is five and forty or fiftie leagues unto the first fall, and it commeth from the North North-west. The said Port of *Tadousac* is little, wherein there cannot ride above ten or twelve Ships: but there is water enough toward the East, toward the opening of the said River of *Sagenay* along by a little hill, which is almost cut off from the maine by the Sea: The rest of the Countrie are very high Mountaines, whereon there is little mould, but rockes and sands full of woods of Pines, Cypresses, Fir-trees, Burch, and some other sorts of trees of small price. There is a little Poole neere unto the said Port, enclosed with Mountaines covered with woods. At the entrance of the said Port there are two points, the one on the West side running a league into the

1 An explanatory sidebar by Purchas: "The River of *Sagenay* falleth into *Canada*. That of *Sagenay* is in *Lescarbots* Map expressed to enter on the North side of *Canada*, about 51. or 40. from thence to the Sea shoare of *Canada* is above sixtie miles, which entring into the Sea, hath 100. miles, and up to the fals (which Voyage followeth) continueth a marveilous breadth, so that it may be for greatnesse reputed greater then any other River in our world, or in the Northerne parts of the New: full also of Lakes and Ilands for greater magnificence."

Sea, which is called Saint *Matthewes* point; and the other on the South-east side, containing a quarter of a league, which is called the point of *all the Divels*. The South and South South-east, and South South-west windes doe strike into the said haven. But from Saint *Matthewes* Point, to the said Point of *all the Divels*, is very neere a league: Both these Points are dry at a low water.

[*Des Sauvages*, ch. 2] THe seven and twentieth day we sought the Savages at the Point of Saint *Matthew*, which is a league from *Tadousac*, with the two Savages whom Monsieur du Pont brought with him, to make report of that which they had seene in *France*, and of the good entertainement which the King had given them. As soone as we were landed we went to the Caban of their great *Sagamo*, which is called *Anadabijou*, where we found him with some eightie or a hundred of his companions, which were making *Tabagie*, that is to say, a Feast. Hee received us very well, according to the custome of the Countrey, and made us sit downe by him, and all the Savages sat along one by another on both sides of the said Cabine. One of the Savages which we had brought with us began to make his Oration, of the good entertainement which the King had given them, and of the good usage that they had received in *France*, and that they might assure /[p. 1607] themselves that his said Majestie wished them well, and desired to people their Countrey, and to make peace with their enemies (which are the *Irocois*) or to send them forces to vanquish them. He also reckoned up the faire Castels, Palaces, Houses, and people which they had seene, and our manner of living. He was heard with so great silence, as more cannot be uttered: Now when he had ended his Oration, the said grand *Sagamo Anadabijou*, having heard him attentively began to take Tobacco, and gave to the said Monsieur *du Pont Grave* of Saint *Malo*, and to mee, and to certaine other *Sagamos* which were by him: after he had taken store of Tobacco, he began to make his Oration to all, speaking distinctly, resting sometimes a little, and then speaking againe, saying, that doubtlesse they ought to be very glad to have his Majestie for their great friend: they answered all with one voyce, ho, ho, ho, which is to say, yea, yea, yea. He proceeding forward in his speech, said, That he was very well content that his said Majestie should people their Countrey, and make warre against their enemies, and that there was no Nation in the world to which they wished more good, then to the *French*. In fine, hee gave them all to understand what good and profit they might receive of his said Majestie. When hee had ended his speech, we went out of his Cabine, and they began to make their Tabagie or Feast, which they make with the flesh of Orignac, which is like an Oxe, of Beares, of Seales, and Bevers, which are the most ordinary victuals which they have, & with great store of wilde Fowle. They had eight or ten Kettels full of meate in the middest of the said Cabine, and they were set one from another some six paces, and each one upon a severall fire. The men sat on both sides the house (as I said before) with his dish made of the barke of a tree: and when the meate is sodden, there is one which devideth to every man his part in the same dishes, wherein they feede very filthily, for when their hands be fattie, they rub them on their haire, or else on the haire of their dogs, whereof they have store to hunt with. Before their meate was sodden, one of them rose up, and took a dog, & danced about the said Kettels from the one end of the Cabin to the other: when he came before the great *Sagamo*, he cast his dog perforce upon the ground, and then all of them with one voice, cried, ho, ho, ho, which being done, he went and sat him downe in his place, then immediately another rose up and did the like, and so they continued untill the meate was sodden. When they had ended their Feast, they began to dance, taking the heads of their enemies in their hands, which hanged upon the wall behinde them; and in signe of joy there is one or two which sing, moderating their voice by the measure of their hands, which they beate upon their knees, then they rest sometimes, and cry, ho, ho, ho; and begin againe to dance, & blow like a man that is out of breath. They made this triumph for a victory which they had gotten of the *Irocois*, of whom they had slaine some hundred, whose heads they cut off, which they had with them for the ceremony. They were three Nations when they went to war, the *Estechemins*, *Algoumequins*, and *Mountainers*, to the number of a thousand, when they went to war against the Irocois, whom they encountred at the mouth of the River of the said *Irocois*, and slew an hundred of them. The war which they make is altogether by surprises, for otherwise they would be out of hart; & they feare the said *Irocois* very much, which are in greater number then the said *Mountainers*, *Estechemins* and *Algoumequins*. The twenty eight day of the said moneth, they encamped themselves in the foresaid haven of *Tadousac*, where our Ship was; at the break of day their said great *Sagamo* came out of his Cabine, going round about all the other Cabins, and cried with a loud voice that they should dislodge to goe to *Tadousac*, where their good friends were. Immediately every man in a trice tooke down his cabin, and the said grand Captain, first began to take his canoe, & carried it to the Sea, where he embarked his wife and children, & store of furs; and in like manner did well neere two hundred canowes, which goe strangely; for though our Shallop was well manned, yet they went more swift then we. There are but two that row, the man and the wife: Their Canowes are some eight or nine pases long, and a pace, or a pace & a halfe broad in the middest, and grow sharper & sharper toward both the ends. They are very subject to overturning, if one know not how to guide them; for they are made of the barke of a Birch tree, strengthned within with little circles of wood well & handsomely framed, and are so light, that one man will carry one of them easily; and every Canowe is able to carry the weight of a Pipe:[2] when they would passe over any land to goe to some River where they have busines, they carry them with them. Their Cabins are low, made like Tents, covered with the said barke of a tree, and they leave in the roofe about a foot space uncovered, whereby the light commeth in; and they make many fires right in the midst of their Cabin, where they are sometimes ten housholds together. They lie upon skins one by another, and their dogs with them. They were about a thousand persons, men, women and children. The place of the point of S. *Matthew*, where they were first lodged, is very pleasant; they were at the bottome of a little hill, which was ful of Fir & Cypresse trees: upon this point there is a little level plot, which discovereth far off, & upon the top of the said hill, there is a Plain, a league long, and halfe a league broad, covered with trees; the soile is very sandy, and is good pasture; all the rest is nothing but Mountains of very bad rocks: the Sea beateth round about the said hil, which is dry for a large halfe league at a low water.

[*Des Sauvages*, ch. 3] THe ninth day of June the Savages began to make merrie together, and to make their feast, as I have said before, and to dance for the aforesaid victory which they had obtained against /[p. 1608] their enemies. After they had made good cheere, the *Algoumequins*, one of the three Nations, went out of their Cabins, and retired themselves apart into a publike place, and caused all their women and girles to sit downe in rankes one by the other, and stood themselves behinde, then singing all in one time, as I have said before. And suddenly all the women and maidens began to cast off their Mantles of skins, and stripped themselves starke naked, shewing their privities, neverthelesse adorned with *Matachia*,[3] which are paternosters and chaines enterlaced made of the haire of the Porkespicke, which they dye of divers colours. After they had made an end of their songs, they cried all with one voyce, ho, ho, ho; at the same instant all the women and maidens covered themselves with their Mantels, for they lye at their feete, and rest a short while; and then eftsoones beginning againe to sing, they let fall their Mantels as they did before. They goe not out of one place when they dance, and make certaine gestures and motions of the body, first lifting up one foote and then another, stamping upon the ground. While they were dancing of this dance, the *Sagamo* of the *Algoumequins*, whose name was *Besouat*, sat before the said women and virgins, betweene two staves, whereon the heads of their

2 A pipe is 380 kg (840 lb). See appendix 4. This is about correct for a 6 m (20 ft) family canoe.

3 Explanatory note by Purchas: "*Matachia*, or cordons of the haire of the Pork-pike." Given also as "Porkespicke." For the meaning of *Matachias*, see Document G, "*Des Sauvages*," footnote 57. The "hair of the Pork-pike" are the quills of a porcupine.

enemies did hang. Sometimes he rose and made a speech, and said to the Mountainers and *Estechemains*; ye see how we rejoyce for the victory which we have obtained of our enemies, ye must doe the like, that we may be contented: then they all together cried, ho, ho, ho. Assoone as hee was returned to his place, the great *Sagamo*, and all his companions cast off their Mantels, being starke naked save their privities, which were covered with a little skin, and tooke each of them what they thought good, as Matachias, Hatchets, Swords, Kettels, Fat, Flesh of the Orignac, Seales, in briefe, every one had a present, which they gave the *Algoumequins*. After all these ceremonies the dance ceased, and the said *Algoumequins* both men and women carried away their presents to their lodgings. They chose out also two men of each Nation of the best disposition, which they caused to run, and he which was the swiftest in running had a present.

All these people are of a very cheerefull complexion, they laugh for the most part, neverthelesse they are somewhat melancholly. They speake very distinctly, as though they would make themselves well understood, and they stay quickely bethinking themselves a great while, and then they begin their speech againe: they often use this fashion in the middest of their Orations in counsaile, where there are none but the principals, which are the ancients: the women and children are not present. All these people sometimes endure so great extremity, that they are almost constrained to eate one another, through the great colds and snowes; for the Beasts and Fowles whereof they live retire them selves into more hot climates. I thinke if any would teach them how to live, and to learne to till the ground, and other things, they would learne very well; for I assure you that many of them are of good judgement, and answere very well to the purpose to any thing that a man shall demand of them. They have one naughty qualitie in them, which is, that they are given to revenge, and great lyars, a people to whom you must not give too much credit, but with reason, and standing on your owne guard. They promise much and performe little. They are for the most part a people that have no Law, as farre as I could see and enforme my selfe of the said great *Sagamo*, who told me, that they constantly beleeve, that there is one God, which hath made all things: And then I said unto him, since they beleeve in one God onely, How is it that he sent them into this world, and from whence came they? he answered me, that after God had made all things, he tooke a number of Arrowes, and stucke them in the ground, from whence men and women grew, which have multiplied in the world untill this present, and had their originall on this fashion. I replied unto him, that this which hee said was false; but that indeede there was one God onely, which had created all things in the earth, and in the heavens: seeing all these things so perfect, without any body to governe this world beneath, he tooke of the slime of the earth, & thereof made *Adam*, our first Father: As *Adam* slept, God tooke a rib of the side of *Adam*, & thereof made *Eve*, whom he gave him for his companion; and that this was the truth that they and we had our originall after this manner, and not of Arrowes as they beleeved. He said nothing into me, save, that he beleeved rather that which I said, then that which he told me. I asked him also, whether he beleeved not that there was any other but one God onely? He told me, that their beliefe was, That there was one God, one Sonne, one Mother, and the Sunne; which were foure; yet that God was above them all: but that the Son was good, and the Sunne in the firmament, because of the good that they received of them; but that the Mother was naught, and did eate them, and that the Father was not very good. I shewed him his errour according to our faith, wherein he gave mee some small credit. I demanded of him, whether they had not seene, nor heard say of their ancestors, that God came into the world. He told me, that he had never seene him; but that in old time there were five men which went toward the Sunne setting, which met with God, who asked them, Whither goe ye? They said, we goe to seeke our living: God answered them, you shall finde it here. They went farther, without regarding what God had said unto them: which tooke a stone, and touched two of them with it, which were turned into a stone: And hee said againe unto the other three, Whither goe yee? and they answered as at the first: and God said to them againe, Goe no further, you shall finde it here. And seeing that nothing came unto them, they went /[p. 1609] farther: and God tooke two staves, and touched the two first therewith, which were turned into staves; and the fift staied and would goe no further: And God asked him againe, whither goest thou? I goe to seeke my living: stay and thou shalt finde it. He stayed without going any further, and God gave him meate, and he did eate thereof; after he had well fed, hee returned with other Savages, and told them all the former storie. He told them also, That another time there was a man which had store of Tobacco (which is a kinde of hearbe, whereof they take the smoake.) And that God came to this man, and asked him where his Tobacco pipe was? The man tooke his Tobacco pipe and gave it to God, which tooke Tobacco a great while: after hee had taken store of Tobacco, God broke the said pipe into many peeces: and the man asked him, why hast thou broken my pipe, and seest that I have no more? And God tooke one which hee had, and gave it him, and said unto him; loe here I give thee one, carry it to thy great *Sagamo*, and charge him to keepe it, and if he keepe it well he shall never want any thing, nor none of his companions. The said man tooke the Tobacco pipe, and gave it to his great *Sagamo*, which as long as he kept, the Savages wanted nothing in the world. But after that the said *Sagamo* lost this Tobacco pipe, which was the occasion of great famine, which sometimes they have among them. I asked him whither he beleeved all this? he said yea, and that it was true. This I beleeve is the cause wherefore they say that God is not very good. But I replied and told him, that God was wholly good; and that without doubt this was the Divell that appeared to these men, and that if they would beleeve in God as we doe, they should not want any thing needefull. That the Sunne which they beheld, the Moone and the Starres were created by this great God, which hath made heaven and earth, and they have no power but that which God hath given them. That we beleeve in this great God, who by his goodnesse hath sent us his deare Sonne, which being conceived by the holy Ghost, tooke humaine flesh the Virginall wombe of the Virgin *Marie*, having bin thirty three yeares on the earth, working in infinite miracles, raising up the dead, healing the sicke, casting out Divels, giving sight to the blinde, teaching men the will of God his Father, to serve, honour, and worship him, did shed his bloud, and suffred death and passion for us, and for our sinnes, and redeemed mankinde, and being buried, he rose againe, he descended into hell, and ascended into heaven, where he sitteth at the right hand of God his Father. That this the beleefe of all the *Christians*, which beleeve in the Father, the Sonne, and the holy Ghost, which neverthelesse are not three Gods, but one onely, and one onely God, and one Trinitie, in the which none is before or after the other, none greater or lesse then another. That the Virgin *Mary* the Mother of the Sonne of God, and all men and women which have lived in this world, doing the commandements of God, and suffring martyrdome for his name sake, and by the permission of God have wrought miracles, and are Saints in heaven in his Paradise, doe all pray this great divine Majestie for us, to pardon us our faults and our sinnes which we doe against his Law and his Commandements: and so by the prayers of the Saints in heaven, and by our prayers which we make to his divine Majestie, he giveth that which we have neede of, and the Divell hath no power over us, and can doe us no harme: That if they had this beliefe, they should be as we are, and that the Divell should be able to doe them no hurt, and should never want any thing necessary. Then the said *Sagamo* told me, that he approved that which I said. I asked him what ceremony they used in praying to their God? He told me, that they used none other ceremonies, but that every one praied in his heart as he thought good: This is the cause why I beleeve they have no law among them, neither doe they know how to worship or pray to God, and live for the most part like brute beasts, and I thinke in short space they would be brought to be good *Christians*, if their Countrie were planten, which they desire for the most part.

They have among them certaine Savages which they call *Pilotoua*, which speak visibly with the Divell, which telleth them what they must doe, as well for the warre as for other things; and if he should command them to put any enterprise in execution, either to kill a *French* man, or any other of their Nation, they would immediately obey his commandement. Also they beleeve that all the dreames which they dreame are true: and indeede there are many of them, which say that they have seene and dreamed things which doe happen or shall happen. But to speake truely of these things, they are visions of the Divell, which doth deceive and seduce them. Loe this is all their beliefe that I could learne of them, which is brutish and bestiall. All these people are well proportioned of their bodies, without any deformitie, they are well set, and the women are well shapen, fat and full, of a tawnie colour by abundance of a certaine painting wherewith they rubbe themselves, which maketh them to be of an Olive colour. They are apparelled with skins, one part of their bodies is covered, and the other part uncovered; but in the winter they cover all, for they are clad with good Furres, namely with the skins of *Orignac*, Otters, Bevers, Lea-boores, Stagges, and Deere, whereof they have store. In the winter when the Snowes are great, they make a kinde of racket which is twice or thrice as bigge as one of ours in France, which they fasten to their feete, and so goe on the Snow without sinking; for otherwise they could not hunt nor travaile in many places. They have also a kinde of Marriage, which is, that when a Maide is foureteene or fifteene yeares old, shee shall have many servants and friends, and she may have carnall company with all those which she liketh, then after five /[p. 1610] or six yeares, she may take which of them she will for her husband, and so they shall live together all their life time, except that after they have lived a certaine time together and have no children, the man may forsake her and take another wife, saying that his old wife is nothing worth, so that the Maides are more free then the married Women. After they be married they be chaste, and their husbands for the most part are jealous, which give presents to the Father or Parents of the Maide, which they have married: loe this is the ceremonie and fashion which they use in their marriages.

Touching their burials, when a man or woman dieth, they make a pit, wherein they put all the goods which they have, as Kettels, Furres, Hatchets, Bowes and Arrowes, Apparell, and other things, and then they put the corps into the grave, and cover it with earth, and set store of great peeces of wood over it, and one stake they set up on end, which they paint with red on the top. They beleeve the immortality of the Soule, and say that when they be dead they goe into other Countries to rejoyce with their parents and friends.

[*Des Sauvages*, ch. 4] THe eleventh day of June, I went some twelve or fifteene leagues up *Saguenay*, which is a faire River, and of incredible depth; for I beleeve, as farre as I could learne by conference whence it should come, that it is from a very high place, from whence there descendeth a fall of water with great impetuositie: but the water that proceedeth thereof is not able to make such a River as this; which neverthelesse holdeth not but from the said course of water (where the first fall is) unto the Port of *Tadousac*, which is the mouth of the said River of *Saguenay*, in which space are fortie five or fiftie leagues, and it is a good league and a halfe broad at the most, and a quarter of a league where it is narrowest, which causeth a great currant of water. All the Countrie which I saw, was nothing but Mountaines, the most part of rockes covered with woods of Fir-trees, Cypresses, and Birch-trees, the soyle very unpleasant, where I found not a league of plaine Countrey, neither on the one side nor on the other. There are certaine hils of Sand and Isles in the said River, which are very high above the water. In fine, they are very Desarts voide of Beasts and Birds; for I assure you, as I went on hunting through places which seemed most pleasant unto mee, I found nothing at all, but small Birds which are like Nightingales, and Swallowes, which come thither in the Summer; for at other times I thinke there are none, because of the excessiye cold which is there; this River commeth from the North-west. They reported unto me, that having passed the first fall, from whence the currant of water commeth, they passe eight other sauts or fals, and then they travaile one dayes journey without finding any, then they passe ten other sauts, and come into a Lake, which they passe in two dayes (every day they travaile at their ease, some twelve or fifteene leagues:) at the end of the Lake there are people lodged: then they enter into three other Rivers, three or foure dayes in each of them; at the end of which Rivers there are two or three kinde of Lakes, where the head of *Saguenay* beginneth: from the which head or spring, unto the said Port of *Tadousac*, is ten[4] dayes journey with their Canowes. On the side of the said Rivers are many lodgingings, whither other Nacions come from the North, to trucke with the said Mountainers, for skins of Bevers and Marterns, for other Merchandises, which the *French* Ships bring to the said Mountainers. The said Savages of the North say, that they see a Sea, which is salt. I hold, if this be so, that it is some gulfe of this our Sea, which disgorgeth it selfe by the North part between the lands; and in very deede it can be nothing else. This is that which I have learned of' the River of *Saguenay*.

[*Des Sauvages*, ch. 5] ON Wednesday the eighteenth day of June, we departed from *Tadousac*, to go to the Sault: we passed by an Ile, which is called the Ile *du lievre*, or the Ile of the *Hare*, which may be some two leagues from the Land on the North side, and some seven leagues from the said *Tadousac*, and five leagues from the South Coast. From the Ile of the *Hare* we ranged the North Coast about halfe a league, unto a point that runneth into the Sea, where a man must keepe farther off.

The said point is within a league of the Ile, which is called the Ile *du Coudre*, or the Ile of *Filberds*, which may be some two leagues in length: And from the said Ile to the Land on the North side is a league. The said Ile is somewhat even, and

4 Footnote by Purchas: "That is, 120. leagues." In giving the length of the Saguenay River Purchas simply multiplied the 10 days by 12 leagues, the distance Champlain wrote the Montagnais could travel in their canoes in a day.

groweth sharpe toward both the ends; on the West end there are Medowes and Points of Rockes which stretch somewhat into the River. The said Ile is somewhat pleasant, by reason of the Woods which environ the same. There is store of Slate, and the soyle is somewhat gravelly: at the end whereof there is a Rocke which stretcheth into the Sea about halfe a league. We passed to the North of the said Ile, which is distant from the Ile of the *Hare* twelve leagues.

The Thursday following we departed from thence, and anchored at a dangerous nooke on the Northside, where there be certaine Medowes, and a little River, where the Savages lodge sometimes. The said day wee still ranged the Coast on the North, unto a place where wee put backe by reasons of the winds which were contrary unto us, where there were many Rockes and places very dangerous: here we stayed three dayes wayting for faire weather. All this Coast is nothing but Mountaynes as well on the South side as on the North, the most part like the Coast /[p. 1611] of the River of *Saguenay*. On Sunday the two and twentieth of the said moneth wee departed to goe to the Ile of *Orleans*, in the way there are many Iles on the South shoare, which are low and covered with trees, shewing to be very pleasant, contayning (as I was able to judge) some two leagues, and one league, and another halfe a league. About these Iles are nothing but Rocks and Flats, very dangerous to passe, and they are distant some two leagues from the mayne Land on the South.

And from thence wee ranged the Ile of *Orleans* on the Southside: It is a league from the North shoare, very pleasant and levell, contayning eight leagues in length. The Coast on the South shoare is low land, some two leagues into the Countrey: the said lands begin to be low over against the said Ile, which beginneth two leagues from the South Coast: to passe by the North side is very dangerous for the bankes of Sand and Rockes, which are betweene the said Ile and the mayne Land, which is almost all dry at a low water. At the end of the said Ile I saw a fall of water, which fell from a great Mountaine, of the said River of *Canada*, and on the top of the said Mountaine the ground is levell and pleasant to behold, although within the said Countries a man may see high Mountaynes which may bee some twenty, or five and twenty leagues within the Lands, which are neere the first *Sault* of *Saguenay*. We anchored at *Quebec*, which is a Strait of the said River of *Canada*, which is some three hundred pases broad: there is at this Strait on the North side a very high Mountayne, which falleth downe on both sides: all the rest is a levell and goodly Countrey, where there are good grounds full of Trees, as Okes, Cypresses, Birches, Firre-trees and Aspes, and other Trees bearing fruit, and wild Vines: So that in mine opinion, if they were dressed, they would be as good as ours. There are along the Coast of the said *Quebec* Diamants in the Rockes of Slate, which are better then those of *Alonson*. From the said *Quebec* to the Ile of *Coudre*, or *Filberds*, are nine and twenty leagues.

[*Des Sauvages*, ch. 6] ON Munday the three and twentieth of the said moneth, we departed from *Quebec*, where the River beginneth to grow broad sometimes one league, then a league and halfe or two leagues at most. The Countrey groweth still fairer and fairer, and are all low grounds, without Rockes, or very few. The North Coast is full of Rockes and bankes of Sand: you must take the South side, about some halfe league from the shore. There are certaine small Rivers which are not navigable, but only for the Canowes of the *Savages*, wherein there be many fals. Wee anchored as high as Saint *Croix*, which is distant from *Quebec* fifteene leagues. This is a low point, which riseth up on both sides. The Countrey is faire and levell, and the soyles better then in any place that I have seene, with plenty of wood, but very few Firre-trees and Cypresses. There are in these parts great store of Vines, Peares, small Nuts, Cheries, Goose-beries, red and greene, and certaine small Roots of the bignesse of a little Nut, resembling Musheroms in taste, which are very good roasted and sod. All this soyle is blacke, without any Rockes, save that there is great store of Slate: The soyle is very soft, and if it were well manured it would yeeld great increase. On the Northside there is a River which is called *Batiscan*, which goeth farre into the Countrey, whereby sometimes the *Algoume-quins* come downe: and another on the same side three leagues from the said Saint *Croix*, in the way from *Quebec*, which is that where *Jacques Quartier* was in the beginning of the Discovery which he made hereof, and hee passed no farther. The said River is pleasant, and goeth farre up into the Countries. All this North Coast is very levell and delectable.

On Tuesday the foure and twentieth of the said moneth, wee departed from the said Saint *Croix*, where we stayed a tyde and an halfe, that we might passe the next day following by day light, because of the great number of Rockes which are thwart the River (a strange thing to behold) which is in a manner dry at a low water: But at halfe flood, a man may beginne to passe safely; yet you must take good heed, with the Lead alwayes in hand. The tyde floweth heere almost three fathomes and an halfe: the farther we went, the fairer was the Countrey. We went some five leagues and an halfe, and anchored on the North side. The Wednesday following wee departed from the said place, which is a flatter Countrey then that which we passed before, full of great store of Trees as that of Saint *Croix*. We passed hard by a little Ile, which was full of Vines, and came to an Anchor on the South side neere a little Hill: but beeing on the top thereof all is even ground.

There is another little Ile three leagues from Saint *Croix*, joyning neere the South shore. Wee departed from the said Hill the Thursday following, and passed by a little Ilee, which is neere the North shoare, where I saw sixe small Rivers, whereof two are able to beare Boats farre up, and another is three hundred pases broad: there are certaine Ilands in the mouth of it; it goeth farre up into the Countrey; it is the deepest of all the rest which are very pleasant to behold, the soyle being full of Trees which are like to Walnut-trees, and have the same smell: but I saw no Fruit, which maketh me doubt: the *Savages* told me that they beare Fruit like ours.

In passing further we met an Ile, which is called Saint *Eloy*, and another little Ile, which is hard by the North shoare: we passed betweene the said Ile and the North shore, where betweene the one and the other are some hundred and fiftie paces. From the said Ile we passed a league and /[p. 1612] an halfe, on the South side neere unto a River, whereon Canowes might goe. All this Coast on the North side is very good, one may passe freely there, yet with the Lead in the hand, to avoid certaine points. All this Coast which we ranged is moving Sand; but after you be entred a little into the Woods, the soile is good. The Friday following we departed from this Ile, coasting still the North side hard by the shoare, which is low and full of good Trees, and in great number as farre as the three Rivers, where it beginneth to have another temperature of the season, somewhat differing from that of Saint *Croix*: because the Trees are there more forward then in any place that hitherto I had seene. From the three Rivers to Saint *Croix* are fifteene leagues. In this River are sixe Ilands, three of which are very small, and the other some five or sixe hundred paces long, very pleasant and fertile, for the little quantitie of ground that they containe. There is one Iland in the middest of the said River, which looketh directly upon the passage of the River of *Canada*, and commandeth the other Ilands which lye further from the shoare, as well on the one side as on the other, of foure or five hundred paces: it riseth on the South side, and falleth somewhat on the North side. This in my judgement would be a very fit place to inhabit; and it might bee quickly fortified: for the situation is strong of it selfe, and neere unto a great Lake, which is above foure leagues distant, which is almost joyned to the River of *Saguenay*, by the report of the *Savages*, which travell almost an hundred leagues Northward, and passe many Saults, and then goe by Land some five or sixe leagues, and enter into a Lake, whence the said River of *Saguenay* receiveth the best part of his Spring, and the said Savages come from the said Lake to *Tadousac*.

Moreover, the planting of *The three Rivers* would be a benefit for the liberty of certaine Nations, which dare not come that way for feare of the said *Irocois* their enemies, which border upon all the said River of *Canada*. But this place being inhabited, we might make the *Jrocois* and the other *Savages* friends, or at least wise under the favour of the said Plantation, the said *Savages* might passe freely without feare or danger: because the said place of *The three Rivers* is a passage. All the soyle which I saw on the North shoare is sandy. Wee went up above a league into the said River, and could passe no further, by reason of the great current of water. We took a Boate to search up further, but we went not past a league, but we met a very Strait full of water, of some twelve paces, which caused us that we could not passe no further. All the ground which I saw on the bankes of the said River riseth more and more, and is full of Firre-trees and Cypresse Trees, and hath very few other Trees.

[*Des Sauvages*, ch. 7] ON the Saturday following, we departed from *The three Rivers*, and anchored at a Lake, which is foure leagues distant. All this Countrey from *The three Rivers* to the entrance of the said Lake is low ground, even with the water on the North side; and on the South side is somewhat higher. The said Countrey is exceeding good, and the most pleasant that hitherto we had seene: the Woods are very thinne, so that a man may travell easily through them. The next day being the nine and twentieth of June, we entred into the Lake, which is some fifteene leagues in length, and some seven or eight leagues broad: At the entrance thereof on the Southside within a league there is a River which is very great, and entreth into the Countrey some sixtie or eightie leagues, and continuing along the same Coast, there is another little River, which pierceth about two leagues into the Land, and commeth out of another small Lake, which may containe some three or foure leagues. On the North side where the Land sheweth very high, a man may see some twentie leagues off; but by little and little the Mountaynes beginne to fall toward the West, as it were into a flat Countrey.

The *Savages* say, that the greatest part of these Mountaynes are bad soyle. The said Lake hath some three fahoms water whereas we passed, which was almost in the middest: the length lieth East and West, and the breadth from North to the South. I thinke it hath good fish in it, of such kinds as we have in our owne Countrey. Wee passed it the very same day, and anchored about two leagues within the great River which goeth up to the *Sault*: In the mouth whereof are thirtie small Ilands, as farre as I could discerne; some of them are of two leagues, others a league and an halfe, & some lesse, which are full of Walnut-trees, which are not much different from ours; and I thinke their Walnuts are good when they bee ripe: I saw many of them under the Trees, which were of two sorts, the one small, and the others as long as a mans Thumbe, but they were rotten. There are also store of Vines upon the bankes of the said Ilands. But when the waters be great, the most part of them is covered with water. And this Countrey is yet better then any other which I had seene before.

The last day of June wee departed from thence, and passed by the mouth of the River of the *Irocois*; where the *Savages* which came to make warre against them, were lodged and fortified. Their Fortresse was made with a number of posts set very close one to another, which joyned on the one side on the banke of the great River *Canada*, and the other on the banke of the River of the *Irocois*: and their Boates were ranged the one by the other neere the shoare, that they might flie away with speed, if by chance they should bee surprised by the *Irocois*. For their

Fort is covered with the barke of Okes, and serveth them for /[p. 1613] nothing else, but to have time to embarke themselves. We went up the River of the *Irocois* some five or sixe leagues, and could passe no farther with our Pinnasse, by reason of the great course water which descendeth, and also because we cannot goe on Land, and draw the Pinnasse for the multitude of Trees which are upon the bankes.

Seeing we could not passe any further, we tooke our Skiffe, to see whether the current were more gentle, but going up some two leagues, it was yet stronger, and wee could goe no higher. Being able to doe no more we returned to our Pinnasse. All this River is some three hundred or foure hundred paces broad, and very wholsome. Wee saw five Ilands in it, distant one from the other a quarter or halfe a league, or a league at the most: one of which is a league long, which is the neerest to the mouth, and the others are very small. All these Countries are covered with Trees and low Lands, like those which I had seene before; but here are more Firres and Cypresses then in other places. Neverthelesse, the soile is good, although it bee somewhat sandy. This River runneth in a manner South-west. The *Savages* say, that some fifteene leagues from the place where we were up the River, there is a Sault which falleth downe from a very steepe place, where they carry their Canowes to passe the same some quarter of a league, and come into a Lake; at the mouth whereof, are three Ilands, and being within the same they meete with more Iles: This Lake may containe some fortie or fiftie leagues in length, and some five and twentie leagues in breadth, into which many Rivers fall, to the number of ten, which carrie Canowes very far up. When they are come to the end of this Lake, there is another fall, and they enter againe into another Lake, which is as great as the former, at the head whereof the *Irocois* are lodged. They say moreover, that there is a River, which runneth unto the Coast of *Florida*, whether it is from the said last Lake some hundred, or an hundred and fortie leagues. All the countrey of the Irocois is somewhat Mountaynous, yet notwithstanding exceeding good, temperate, without much Winter, which is very short there.

[*Des Sauvages*, ch. 8] AFter our departure from the River of the *Irocois*, wee anchored three leagues beyond the same, on the North side. All this Countrie is a lowe Land, replenished with all sorts of trees, which I have spoken of before. The first day of July we coasted the North side, where the wood is very thinne, and more thinne then wee had seene in any place before, and all good land for tillage. I went in a Canoa to the South shoare, where I saw a number of Iles, which have many fruitfull trees, as Vines, Wal-nuts, Hasel-nuts, and a kinde of fruit like Chest-nuts, Cheries, Oakes, Aspe, Hoppes, Ashe, Beech, Cypresses, very few Pines and Firre-trees. There are also other trees which I knew not, which are very pleasant. Wee found there store of Strawberries, Rasp-berries Goos-berries red, greene, and blue, with many small fruits, which growe there among great abundance of grasse. There are also many wilde beasts, as Orignas,[5] Stagges, Does, Buckes, Beares, Porkepickes, Conies, Foxes, Beavers, Otters, Muske-rats, and certaine other kindes of beasts which I doe not knowe, which are good to eate, and whereof the Savages live. Wee passed by an Ile, which is very pleasant, and containeth some foure leagues in length, and halfe a league in breadth. I saw toward the South two high Mountaines, which shewed some twentie leagues within the Land. The Savages told mee, that here beganne the first fall of the foresaid River of the *Irocois*. The Wednesday following wee departed from this place, and sayled some five or sixe leagues. Wee saw many Ilands: the Land is there very lowe, and these Iles are covered with trees, as those of the River of the *Irocois* were.

The day following, being the third of July, we ranne certaine leagues, and passed likewise by many other Ilands, which are excellent good and pleasant, through the great store of Medowes which are there about, as well on the shoare of the maine Land, as of the other Ilands: and all the Woods are of very small growth, in comparison of those which wee had passed. At length we came this very day to the entrance of the Sault or Fall of the great River of *Canada*, with favourable wind; and wee met with an Ile, which is almost in the middest of the said entrance, which is a quarter of a league long, and passed on the South side of the said Ile, where there was not past three, foure or five foot water, and sometimes a fathome or two, and straight on the sudden wee found againe not past three or foure foot. There are many Rockes, and small Ilands, whereon there is no wood, and they are even with the water. From the beginning of the foresaid Ile, which is in the middest of the said entrance, the water beginneth to runne with a great force. Although we had the wind very good, yet wee could not with all our might make any great way: neverthelesse wee passed the said Ile which is at the entrance of the Sault or Fall. When wee perceived that wee could goe no further, wee came to an anchor on the North shoare over against a small Iland, which aboundeth for the most part with those kinde of fruits which I have spoken of before. Without all delay wee made ready our skiffe, which wee had made of purpose to passe the said Sault: whereinto the said *Monsieur de Pont* and my selfe entred,[6] with certaine Savages, which we had brought with us to shew us the way. Departing from our Pinnace, we were scarse gone three hundred paces, but we were forced to come out, and cause certain Mariners to goe into the water to free our Skiffe. The Canoa of the Savages passed easily. Wee met with an infinite

5 Explanatory sidebar by Purchas: "*Orignas* are before said to bee like oxen perhaps Buffes. *Lescarbot*, that *Orignacs* are *Ellans*."

6 Explanatory sidebar by Purchas: "*Monsieur du Pont and Monsieur du Champlaine* search the Sault." Purchas identifies *Champlaine* as the writer.

number of small Rockes, which were even with the water, on which wee touched oftentimes.

/[p. 1614] There be two great Ilands, one on the North side, which containeth some fifteene leagues in length, and almost as much in breadth, beginning some twelve, leagues up within the River of *Canada*, going toward the River of the *Irocois*, and endeth beyond the Sault. The Iland which is on the South side is some foure leagues long, and some halfe league broad. There is also another Iland, which is neere to that on the North side, which may bee some halfe league long, and some quarter broad: and another small Iland which is betweene that on the North side, and another neerer to the South shoare, whereby wee passed the entrance of the Sault. This entrance being passed, there is a kinde of Lake, wherein all these Ilands are, some five leagues long and almost as broad, wherein are many small Ilands which are Rockes. There is a Mountaine neere the said Sault which discovereth farre into the Countrie, and a little River which falleth from the said Mountaine into the Lake. On the South side are some three or foure Mountaines, which seeme to be about fifteene or sixteene leagues within the Land. There are also two Rivers; one, which goeth to the first Lake of the River of the *Irocois*, by which sometimes the *Algoumequins* invade them: and another which is neere unto the Sault, which runneth not farre into the Countrey.

At our comming neere to the said Sault with our Skiffe and Canoa, I assure you, I never saw any streame of water to fall downe with such force as this doth; although it bee not very high, being not in some places past one or two fathoms, and at the most three: it falleth as it were steppe by steppe: and in every place where it hath some small heigth, it maketh a strong boyling with the force and strength of the running of the water. In the breadth of the said Sault, which may containe some league, there are many broad Rockes, and almost in the middest, there are very narrow and long Ilands, where there is a Fall as well on the side of the said Iles which are toward the South, as on the North side: where it is so dangerous, that it is not possible for any man to passe with any Boat, how small soever it be. We went on land through the Woods, to see the end of this Sault: where, after wee had travelled a league, wee saw no more Rockes nor Falls: but the water runneth there so swiftly as it is possible: and this current lasteth for three or foure leagues: so that it is in vaine to imagine, that a man is able to passe the said Saults with any Boats. But he that would passe them, must fit himselfe with the Canoas of the Savages, which one man may easily carrie. For to carrie Boats is a thing which cannot be done in so short time as it should bee to bee able to returne into *France*, unlesse a man would winter there. And beside this first Sault, there are ten Saults more, the most part hard to passe. So that it would be a matter of great paines and travell to bee able to see and doe that by Boat which a man might promise himselfe, without great cost and charge, and also to bee in danger to travell in vaine. But with the Canoas of the Savages a man may travell freely and readily into all Countries, as well in the small as in the great Rivers: So that directing himselfe by the meanes of the said Savages and their Canoas, a man may see all that is to be seene, good and bad, within the space of a yeere or two. That little way which wee travelled by Land on the side of the said Sault, is a very thinne Wood, through which men with their Armes may march easily, without any trouble; the aire is there more gentle and temperate, and the soyle better then in any place that I had seene, where is store of such wood and fruits, as are in all other places before mentioned: and it is in the latitude of 45. degrees and certaine minutes.

When we saw that we could doe no more, we returned to our Pinnace; where we examined the Savages which we had with us, of the end of the River, which I caused them to draw with their hand, and from what part the Head thereof came. They told us, that beyond the first Sault that we had seene, they travelled some ten or fifteene leagues with their Canoas in the River, where there is a River which runneth to the dwelling of the *Algoumequins*, which are some sixty leagues distant from the great River; and then they passed five Saults, which may containe from the first to the last eight leagues, whereof there are two where they carrie their Canoas to passe them: every Sault may containe halfe a quarter or a quarter of a league at the most. And then they come into a Lake, which may be fifteene or sixteene leagues long. From thence they enter againe into a River which may be a league broad, and travell some two leagues in the same; and then they enter into another Lake some foure or five leagues long: comming to the end thereof, they passe five other Saults, distant from the first to the last some five and twenty or thirty leagues; whereof there are three where they carrie their Canoas to passe them, and thorow the other two they doe but draw them in the water, because the current is not there so strong, nor so bad, as in the others. None of all these Saults is so hard to passe, as that which we saw. Then they come into a Lake, which may containe some eighty leagues in length, in which are many Ilands, and at the end of the same the water is brackish, and[7] the Winter gentle. At the end of the said Lake they passe a Sault which is somewhat high, where little water descendeth: there they carrie their Canoas by land about a quarter of a league to passe this Sault. From thence they enter into another Lake, which may be some sixty leagues long, and that the water thereof is very brackish: at the end thereof they come unto a Strait which is two leagues broad, and it goeth farre into the Countrie. They told us, that they themselves had passed no farther [fig. 7b]; and that they had not seene the end of a Lake, which is within fifteene or sixteene /[p. 1615] leagues of the farthest place where themselves had beene,

7 Footnote by Purchas: "It seemeth hereby to trend southward."

nor that they which told them of it had knowne any man that had seene the end thereof, because it is so great that they would not hazard themselves to sayle farre into the same, for feare lest some storme or gust of winde should surprise them. They say that in the Summer the Sunne doth set to the North of the said Lake,[8] and in the Winter it setteth as it were in the middest thereof: That the water is there exceeding salt, to wit, as salt as the Sea water. I asked them whether from the last Lake which they had seene, the water descended alwaies downe the Rivrer comming to *Gaschepay*? They told me, no: but said, that from the third Lake onely it descended to *Gaschepay*: But that from the last Sault, which is somewhat high, as I have said, the water was almost still, and that the said Lake might take his course by other Rivers,[9] which passe within the Lands, either to the South, or to the North, whereof there are many that runne there, the end whereof they see not. Now, in my judgement, if so many Rivers fall into this Lake, having so small a course at the said Sault, it must needs of necessitie fall out, that it must have his issue forth by some exceeding great River. But that which maketh me beleeve that there is no River by which this Lake doth issue forth (considering the number of so many Rivers as fall into it) is this, that the Savages have not seene any River, that runneth through the Countries, save in the place where they were. Which maketh me beleeve that this is the South Sea, being salt as they say: Neverthelesse we may not give so much credit thereunto, but that it must bee done with apparent reasons, although there be some small shew thereof. And this assuredly is all that hitherto I have seene and heard of the Savages, touching that which we demanded of them.

[*Des Sauvages*, ch. 9] WEe departed from the said Sault on Friday the fourth day of July, and returned the same day to the River of the *Irocois*. On Sunday the sixth of July wee departed from thence, and anchored in the Lake. The Monday following wee anchored at the three Rivers. This day wee sayled some foure leagues beyond the said three Rivers. The Tuesday following we came to *Quebec*; and the next day wee were at the end of this Ile of *Orleans*, where the Savages came to us, which were lodged in the maine Land on the North side. Wee examined two or three *Algoumequins*, to see whether they would agree with those that wee had examined touching the end and the beginning of the said River of *Canada*. They said, as they had drawne out the shape thereof, that having passed the

8 Explanatory sidebar by Purchas: “The southerne situation of a great Lake. The water is salt as seawater.”

9 Explanatory sidebar by Purchas: “Many Rivers running south and north. *Hudsons* River may be one of these.”

Sault, which wee had seene, some two or three leagues, there goeth a River into their dwelling, which is on the North side. So going on forward in the said great River, they passe a Sault, where they carrie their Canoas, and they come to passe five other Saults, which may containe from the first to the last some nine or ten leagues, and that the said Saults are not hard to passe, and they doe but draw their Canoas in the most part of the said Saults or Falls, saving at two, where they carrie them: from thence they enter into a River, which is as it were a kinde of Lake, which may containe some sixe or seven leagues: and then they passe five other Falls, where they draw their Canoas as in the first mentioned, saving in two, where they carrie them as in the former: and that from the first to the last there are some twenty or five and twenty leagues. Then they come into a Lake contayning some hundred and fifty leagues in length: and foure or five leagues within the entrance of that Lake there is a River which goeth to the *Algoumequins* toward the North; and another River which goeth to the *Irocois*, whereby the said *Algoumequins* and *Irocois* make warre the one against the other. Then comming to the end of the said Lake, they meete with another Fall, where they carrie their Canoas. From thence they enter into another exceeding great Lake, which may containe as much as the former: They have beene but a very little way in this last Lake, and have heard say, that at the end of the said Lake there is a Sea, the end whereof they have not seene, neither have heard that any have seene it. But that where they have beene, the water is not salt, because they have not entred farre into it; and that the course of the water commeth from the Sun-setting toward the East;[10] and they knowe not, whether beyond the Lake that they have seene, there be any other course of water that goeth Westward. That the Sunne setteth on the right hand of this Lake: which is, according to my judgement, at the North-west, little more or lesse; and that in the first great Lake the water freezeth not (which maketh mee judge that the climate is there temperate) and that all the Territories of the *Algoumequins* are lowe grounds, furnished with small store of wood: And that the coast of the *Irocois* is Mountainous, neverthelesse they are excellent good and fertile soyles, and better then they have seene any where else: That the said *Irocois* reside some fifty or sixty leagues from the said great Lake. And this assuredly is all which they have told mee that they have seene: which differeth very little from the report of the first Savages. This day wee came within some three leagues of the Ile of *Coudres* or *Filberds*.

On Thursday the tenth of the said moneth, wee came within a league and an halfe of the Ile *Du Lievre*, or Of the Hare, on the North side, where other Savages came into our Pinnace, among whom there was a young man, an *Algoume-*

10 Explanatory sidebar by Purchas: “It seemeth to lie southward.”

quin, which had travelled much in the said great Lake. Wee examined him very particularly, as wee had done the other Savages. Hee told us, that having passed the said Fall which wee had seene, within two or tree leagues there is a /[p. 1616] River, which goeth to the said *Algoumequins*, where they be lodged; and that passing up the great River of *Canada*, there are five Falls, which may containe from the first to the last some eight or nine leagues, whereof there bee three where they carrie their Canoas, and two others wherein they draw them: that each of the said Falls may be a quarter of a league long: then they come into a Lake, which may containe some fifteene leagues. Then they passe five other Falls, which may containe from the first to the last some twenty or five and twenty leagues; where there are not past two of the said Falls which they passe with their Canoas, in the other three they doe but draw them. From thence they enter into an exceeding great Lake, which may containe some three hundred leagues in length: when they are passed some hundred leagues into the said Lake, they meet with an Iland, which is very great; and beyond the said Iland the water is brackish: But when they have passed some hundred leagues farther, the water is yet salter: and comming to the end of the said Lake, the water is wholly salt. Farther he said, that there is a Fall that is a league broad, from whence an exceeding current of water descendeth into the said Lake. That after a man is passed this Fall, no more land can be seene neither on the one side nor on the other, but so great a Sea, that they never have seene the end thereof; nor have heard tell, that any other have seene the same. That the Sunne setteth on the right hand of the said Lake: and that at the entrance thereof there is a River which goeth to the *Algoumequins*, and another River to the *Irocois*, whereby they warre the one against the other. That the Countrie of the *Irocois* is somewhat mountainous, yet very fertile, where there is store of *Indian* Wheat, and other fruits, which they have not in their Countrie: That the Countrie of the *Algoumequins* is lowe and fruitfull. I enquired of them, whether they had any knowledge of any Mines? They told us, that there is a Nation which are called the good *Irocois*, which come to exchange for merchandises, which the *French* ships doe give to the *Algoumequins*, which say, that there is toward the North a Mine of fine Copper, whereof they shewed us certaine Bracelets, which they had received of the said Good *Irocois*: and that if any of us would goe thither, they would bring them to the place, which should bee appointed for that businesse. And this is all which I could learne of the one and the other, differing but very little; save that the second which were examined, said, that they had not tasted of the salt water: for they had not beene so farre within the said Lake, as the others: and they differ some small deale in the length of the way, the one sort making it more short, and the other more long. So that, according to their report, from the Sault or Fall where wee were, is the space of some foure hundred leagues unto the Salt Sea, which may be the South Sea, the Sunne setting where they say it doth. On Friday the tenth of the said moneth we returned to *Tadousac*, where our ship lay.

[*Des Sauvages*, ch. 10] ASsoone as wee were come to *Tadousac*, wee embarqued our selves againe to goe to *Gachepay*, which is distant from the said *Tadousac* about some hundred leagues. The thirteenth day of the said moneth we met with a companie of Savages, which were lodged on the South side, almost in the mid-way betweene *Tadousac* and *Gachepay*. Their *Sagamo* or Captaine which led them is called *Armouchides*, which is held to be one of the wisest and most hardy among all the Savages: Hee was going to *Tadousac* to exchange Arrowes, and the flesh of Orignars, which they have for Beavers and Marterns of the other Savages, the Mountainers, *Estechemains*, and *Algoumequins*.

The fifteenth day of the said moneth we came to *Gachepay*, which is in a Bay, about a league and a halfe on the North side. The said Bay containeth some seven or eight leagues in length, and at the mouth thereof foure leagues in breadth. There is a River which runneth some thirty leagues up into the Countrie: Then we saw another Bay, which is called the Bay *des Mollues*, or the Bay of *Cods*, which may be some three leagues long, and as much in bredth at the mouth. From thence we come to the Ile *Percee*, which is like a Rocke, very steepe, rising on both sides, wherein there is a hole, through which Shalops and Boats may passe at an high water: and at a lowe water one may goe from the maine Land to the said Ile, which is not past foure or five hundred paces off. Moreover, there is another Iland in a manner South-east from the Ile Percee about a league, which is called the Ile *de Bonne-adventure*, and it may bee some halfe a league long. All these places of *Gachepay*, the Bay of *Cods*, and the Ile *Percee*, are places where they make dry and greene Fish. When you are passed the Ile *Percee*, there is a Bay which is called the Bay of *Heate*, which runneth as it were West South-west, some foure and twenty leagues into the land, containing some fifteene leagues in breadth at the mouth thereof. The Savages of *Canada* say, that up the great River of *Canada*, about some sixtie leagues, ranging the South coast, there is a small River called *Mautanne*, which runneth some eighteene leagues up into the Countreys and being at the head thereof, they carrie their Canowes about a league by land, and they come into the said Bay of *Heate*, by which they goe sometimes to the Isle *Percee*. Also they goe from the said Bay to *Tregate* and *Misamichy*. Running along the said coast we passe by many Rivers, and come to a place where there is a River which is called *Souricoua*, where Monsieur *Prevert* was to discover a Mine of Copper. They goe with their Conowes up this River three or foure dayes, then they passe three or foure leagues by land, to the said Mine, which is /[p. 1617] hard upon the

Sea shoare on the South side. At the mouth of the said River, there is an Iland lying a league into the Sea; from the said Island unto the Isle *Perçee*, is some sixtie or seventie leagues. Still following the said coast, which trendeth toward the East, you meete with a Strait, which is two leagues broad, and five and twenty leagues long. On the East side is an Isle, which is called the Isle of Saint *Laurence*, where Cape *Breton* is; and in this place a Nation of Savages, called the S*ouricois*, doe winter.

Passing the Strait of the Iles of Saint *Lawrence*, and ranging the South-west Coast, you come to a Bay which joyneth hard upon the Myne of Copper. Passing farther there is a River, which runneth threescore or fourescore leagues into the Countrey, which reacheth neere to the Lake of the *Irocois*, whereby the said Savages of the South-west Coast make warre upon them. It would be an exceeding great benefit, if there might be found a passage on the Coast of *Florida* neere to the said great Lake, where the water is salt; as well for the Navigation of ships, which should not bee subject to so many perils as they are in *Canada*, as for the shortning of the way above three hundred leagues. And it is most certain; that there are Rivers on the Coast of *Florida*, which are not yet discovered, which pierce up into the Countries, where the soile is exceeding good and fertile, and very good Havens. The Countrey and Coast of Florida may have another temperature of the season, and may bee more fertile in abundance of fruites and other things, then that which I have seene: But it cannot have more even nor better sayles, then those which we have seene.

The Savages say, that in the foresaid great Bay of *Hete* there is a River, which runneth up some twentie leagues into the Countrey, at the head whereof there is a Lake, which may be about twentie leagues in compasse, wherein is little store of water, and the Summer it is dried up, wherein they find, about a foot or a foot and an halfe under the ground a kind of Metall like to silver, which I shewed them; and that in another place neere the said Lake there is a Myne of Copper. And this is that which I learned of the foresaid Savages.

[*Des Sauvages*, ch. 11] WE departed from the Ile *Percee* the nineteenth day of the said moneth to returne to *Tadousac*. When we were within three leagues of Cape *le Vesque*, or the Bishops Cape, we were encountred with a storme which lasted two dayes, which forced us to put roomer with a great creeke, and to stay for faire weather. The day following we departed, and were encountred with another storme: Being loth to put roome, and thinking to gaine way wee touched on the North shore the eight and twentieth day of July in a creeke which is very bad, because of the edges of Rockes which lie there. This creeke is in 51. degrees and certaine minutes. The next day we anchored neere a River, which is called Saint *Margarites* River, where at a full Sea is some three fathomes water, and a fathome and an halfe at a low water: this River goeth farre up into the Land. As farre as I could see within the Land on the East shoare, there is a fall of water which entreth into the said River, and falleth some fiftie or sixtie fathomes downe, from whence commeth the greatest part of the water which descendeth downe. At the mouth thereof there is a banke of Sand, whereon at the ebbe is but halfe a fathome water. All the Coast toward the East is moving Sand: there is a point some halfe league from the said River, which stretcheth halfe a league into the Sea: and toward the West there is a small Iland: this place is in fiftie degrees. All these Countries are exceeding bad, full of Firre-trees. The Land here is somewhat high, but not so high as that on the Southside. Some three leagues beyond we passed neere unto another River, which seemed to be very great, yet barred for the most part with Rockes: some eight leagues farther there is a Point which runneth a league and an halfe into the Sea, where there is not past a fathome and an halfe of water. When you are passed this Point, there is another about foure leagues off, where is water enough. All this Coast is low and sandie. Foure leagues beyond this there is a creeke where a River entreth: many ships may passe heere on the West side: this is a low point, which runneth about a league into the Sea; you must runne along the Easterne shoare some three hundred paces to enter into the same. This is the best Haven which is all along the North shoare; but it is very dangerous in going thither, because of the flats and sholds of sand, which lye for the most part all along the shoare, almost two leagues into the Sea. About six leagues from thence, there is a Bay where there is an Isle of sand; all this Bay is very shallow, except on the East side, where it hath about foure fathoms water: within the channell which entreth into the said Bay, some foure leagues up, there is a faire creeke where a River entreth. All this coast is low and sandie, there descendeth a fall of water which is great. About five leagues farther is a Point which stretcheth about halfe a league into the Sea, where there is a creeke, and from the one point to the other are three leagues, but all are shoald, where is little water. About two leagues off, there is a strand where there is a good haven, and a small River, wherein are three Islands, and where Ships may harbour themselves from the weather. Three leagues beyond this, is a sandie point which runneth out about a league, at the end whereof there is a small Islet. Going forward to *Lesquenim*, you meete with two little low Islands, and a little rocke neere the shoare: these said Ilands are about halfe a league from *Lesquenim*, which is a very bad Port, compassed with rockes, and dry at a low water, and you must fetch about a little /[p. 1618] point of a rocke to enter in, where one Ship onely can passe at a time. A little higher there is a River, which runneth a little way into the land. This is the place where the *Basks* kill the Whales; to say the truth, the haven is starke naught. Wee came from thence to the foresaid haven of *Tadousac*, the third day of August. All these Countries before mentioned are low toward the shoare, and within the land very high.

They are neither so pleasant nor fruitfull as those on the South, although they be lower. And this for a certaintie is all which I have seene of this Northerne coast.

[*Des Sauvages*, ch. 12] AT our comming to *Tadousac*, we found the Savages which wee met in the River of the *Irocois*, who met with three Canowes of the *Irocois* in the first Lake, which fought against tenne others of the Mountayners; and they brought the heads of the *Irocois* to *Tadousac*, and there was but one Mountayner wounded in the arme with the shot of an Arrow, who dreaming of something, all the other tenne must seeke to content him, thinking also that his wound thereby would mend; if this *Savage* die, his Parents will revenge his death, either upon their Nation or upon others, or at least wise the Captaines must give Presents to the Parents of the dead, to content them; otherwise as I have said, they would be revenged: which is a great fault among them. Before the said Mountayners set forth to the Warre, they assembled all, with their richest apparell of Furres, Beavers, and other Skinnes adorned with *Pater-nosters* and Chaines of divers colours, and assembled in a great publike place, where there was before them a *Sagamo* whose name was *Begourat*, which led them to the Warre, and they marched one behind another, with their Bowes and Arrowes, Mases and Targets, wherewith they furnish themselves to fight: and they went leaping one after another, in making many gestures of their bodies, they made many turnings like a Snaile: afterward they began to dance after their accustomed manner, as I have said before: then they made their Feast, and after they had ended it, the women stripped themselves starke naked, being decked with their fairest Cordons, and went into their Canowes thus naked and there danced, and then they went into the water, and strooke at one another with their Oares, and beate water one upon another: yet they did no hurt, for they warded the blowes which they strooke one at the other. After they had ended all these Ceremonies, they retired themselves into their Cabines, and the *Savages* went to warre against the *Irocois*.

The sixt day of August we departed from *Tadousac*, and the eighteenth of the said moneth we arrived at the Ile *Perçee*, where wee found *Monsieur Prevert* of Saint *Malo*, which came from the Myne, where he had beene with much trouble, for the feare which the *Savages* had to meet with their enemies, which are the *Armouchicois*, which are *Savages* very monstrous, for the shape that they have. For their head is little, and their body short, their armes small like a bone, and their thigh like; their legges great and long, which are all of one proportion, and when they sit upon their heeles, their knees are higher by halfe a foot then their head, which is a strange thing, and they seeme to be out of the course of Nature. Nevertheless, they be very valiant and resolute, and are planted in the best Countries of all the South Coast: And the *Souricois* do greatly feare them. But by the incouragement which the said *Monsieur de Prevert* gave them, hee brought them to the said Myne, to which the Savages guided him. It is a very high Mountaine, rising somewhat over the Sea, which glistereth very much against the Sunne, and there is great store of Verde-grease issuing out of the said Myne of Copper. He saith, that at the foot of the said Mountayne, at a low water there were many morsels of Copper, as was otherwise declared unto us, which fall downe from the top of the Mountaine. Passing three or foure leagues further toward the South, there is another Myne, and a small River which runneth a little way up into the Land, running toward the South, where there is a Mountaine, which is of a blacke painting, wherewith the *Savages* paint themselves: Some sixe leagues beyond the second Myne, toward the Sea, about a league from the South Coast, there is an Ile, wherein is found another kind of Metall, which is like a darke browne: if you cut it, it is white, which they used in old time for their Arrowes and Knives, and did beate it with stones. Which maketh me beleeve that it is not Tinne, nor Lead, being so hard as it is; and having shewed them silver, they said that the Myne of that Ile was like unto it, which they found in the earth, about a foot or two deepe. The said *Monsieur Prevert* gave the *Savages* Wedges and Cizers, and other things necessarie to draw out the said Myne; which they have promised to doe, and to bring the same the next yeere, and give it the said *Monsieur Prevert*. They say also that within some hundred or one hundred and twentie leagues there are other Mynes, but that they dare not goe thither unlesse they have *Frenchmen* with them to make warre upon their enemies, which have the said Mynes in their possession. The said place where the Myne is, standeth in 44. degrees and some few minutes, neere the South Coast within five or sixe leagues: it is a kind of Bay, which is certaine leagues broad at the mouth thereof, and somewhat more in length, where are three Rivers, which fall into the great Bay neere unto the Ile of Saint *John*, which is thirtie or five and thirtie leagues long, and is sixe leagues distant from the South shoare. There is also another little River, which falleth almost in the mid way of that whereby *Monsieur Prevert* returned, and there are as it were two kind of Lakes in the said River. Furthermore, there is yet another small River which /[p. 1619] goeth toward the Mountaine of the painting. All these Rivers fall into the said Bay on the South-east part, neere about the said Ile which the *Savages* say there is of this white Metall. On the North side of the said Bay are the Mynes of Copper, where there is a good Haven for Ships, and a small Iland at the mouth of the Haven; the ground is Oze and Sand, where a man may run his ship on shoare. From the said Myne to the beginning of the mouth of the said River is some sixtie or eightie leagues by Land. But by the Sea Coast, according to my judgement, from the passage of the Ile of Saint *Lawrence* and the *Firme Land*, it cannot be past fiftie or sixtie leagues to the said Myne. All this Countrey is exceeding faire and flat, wherein are all sorts of trees, which wee saw as wee went to the first Sault up the great River of *Canada*, very small store of Firre-trees and Cypresses. And this of a truth is as much as I learned and heard of the said *Monsieur Prevert*.

[*Des Sauvages*, ch. 13][11] BEfore we departed from *Tadousac*, to returne into *France*, one of the *Sagamoz* of the Mountayners named *Bechourat*, gave his Sonne to *Monsieur du Pont* to carrie him into *France*, and he was much recommended unto him by the Great *Sagamo Anadabijou*, praying him to use him well, and to let him see that, which the other two *Savages* had seene which we had brought backe againe. We prayed them to give us a woman of the *Irocois*, whom they would have eaten: whom they gave unto us, and we brought her home with the foresaid *Savage*. *Monsieur de Prevert* in like manner brought home foure *Savages*, one man which is of the South Coast, one woman and two children of the *Canadians*.

The foure and twentieth of August, we departed from *Gachepay*, the ship of the said *Monsieur Prevert* and ours. The second of September, we counted that wee were as farre as Cape *Rase*. The fift day of the said moneth we entred upon the Banke, whereon they use to fish. The sixteenth, we were come into the Sounding, which may be some fiftie leagues distant from the *Ushant*. The twentieth of the said moneth we arrived in *New Haven* by the grace of God to all our contentments, with a continuall favourable wind.

11 Hakluyt-Purchas reproduced only the last two paragraphs of the original text. They omitted the entire story about the *Gougou*.

APPENDIX I

Champlain's Birthdate and Appearance

Birthdate

It is unlikely that Champlain's date of birth will ever be known unless new documents come to light. Unfortunately, this meagre record cannot be expanded very far because the early parish records of Brouage no longer exist.[1] There is no doubt that Champlain was born at Brouage in the province of Saintonge,[2] France, sometime between 1567 and 1580. The earliest and most often quoted date of his birth is given as 1567, taken from Rainguet's 1851 biography of Champlain and the fraudulent portrait of him by Ducornet published in 1854.[3] This date cannot be corroborated and is no longer used in academic studies of Champlain. Instead, biographers adopted a neutral "circa 1570" until recently, when the issue was raised again by Jean Liebel, who tried to build the case that Champlain was born "circa 1580."[4] Liebel's main argument rests largely on a statement by Champlain: "*Que pour le Sieur du Pont I'estois son amy, & que son aage me le feroit respecter comme mon pere*"[5] ("As for the Sieur du Pont, I was his friend, and his years would lead me to respect him as I would my father.") François Gravé Du Pont, as Liebel proves, was christened on 27 November 1560. Liebel reasons that anyone who respects another person like his father must be about twenty years younger,

1 In 1867 the archivist of the Bibliothèque de La Rochelle, Leopold Delayant, and in 1875 the archivist Meschinet De Richemond of the Département Charente-Inférieure wrote that the *registres de la paroisse de Brouage* were deposited at Marennes, where they were destroyed in a fire in 1690. There are no longer any records relating to the population of Brouage earlier than August 1615 (Delayant, *Notice*, 2; Slafter, trans. Otis, *Voyages of Champlain*, 1:206).

2 Saintonge became the Department of Charente-Inférieure in 1790 and Charente-Maritime in 1941.

3 See Document A, "Early Biographies of Champlain," no. 6. Any archival papers that might have carried his birthdate no longer existed well before Rainguet wrote his biography in 1851. Why Rainguet picked 1567 is not known. For a discussion of Ducornet's lithograph, see this appendix, below.

4 Liebel, "On a vieilli Champlain," 236

5 Champlain, *Les Voyages de la Novvelle France* ... 1632, pt. 1, 224

hence Champlain was born about 1580. Morris Bishop did not have Gravé's baptismal certificate but thought he was born about 1557. He reasoned that "one ordinarily does not respect another as one's father unless there is a difference of ten years or so," and he thought that "on the whole 1567 seems about right" for Champlain's birth.[6] Whether one chooses a ten- or twenty-year age difference or some date in between is of course very arbitrary. We do not even know whether Champlain knew Gravé's real age, and in view of the older man's appearance and health problems, such as gout, he may have seemed older than he was.[7] One must also remember that Champlain had no previous experience in Canada before his first visit in 1603 with the veteran Gravé. During that first voyage and the summer on the St. Lawrence, Gravé was his major source of information, educating Champlain in Canadian conditions. To Champlain, therefore, Gravé was a knowledgeable older person in authority who deserved respect.

On the basis of a birthdate around 1580, Liebel observes that Champlain must have started serving in the Brittany army at the age of about 14 or 15 years. Without any evidence, he surmises that such a young age could have been common in an army after thirty years of war.[8] Although age data for soldiers during this period is rare, there are data for two companies of infantry active in 1567.[9] In the first of these, a veteran combat company of 163 men commanded by "Count Brissac,"[10] 56 percent of the men were under the age of 30, with a mean of 29 years and no one younger than 17. The second company was a garrison of 90 veterans, perhaps similar to the garrison Champlain served in when he was at Quimper in 1595–97. It was commanded by the governor of Pignerol, Jean de Monluc.[11] Of the 90 men, 40 percent were under 30 years of age, with a mean age of 34, with only four boys aged 13 to 16. Of the 253 men in the two companies for whom there were data, only 1.6 percent were under 16 and only "about 10% of their strength was composed of teenagers."[12] The age extremes, under 19 and over 49, were avoided in recruitment because of the rigors involved in that kind of life. The probability that Champlain was 14 to 15 years old in such army units is therefore highly unlikely.

6 Bishop, *Champlain* (1949), 340
7 See, for example, Sagard's description of Gravé (Sagard, *Histoire du Canada*, 947, 981).
8 Liebel, "On a vieilli Champlain"
9 Wood, *The King's Army*, 86–97
10 This Count Brissac was probably Timoléon de Cossé, comte de Brissac (1543–96).
11 Jean de Monluc was the son of the more famous Blaise de Monluc, who served under Henri II and III. The latter appointed him *maréchal* in 1574.
12 Wood, *The King's Army*, 93

The first known documents in which Champlain is named are the pay records of the army in Brittany for the years 1595 onwards,[13] which were issued and recorded by Gabriel Hus,[14] on orders from maréchal Jean d'Aumont[15] until d'Aumont's death, in August 1595, and after that by his successor, François d'Espinay de Saint-Luc.[16] Champlain's position in the army was that of a *fourrier* and *ayde* to the *maréchal des logis*, Jean Hardy.[17] In these pay records, ordered and issued by high-ranking officers of the crown, Champlain's name appears both as "Samuel de Champlain" and "sieur de Champlain," showing that the particle *de* was established by 1595, as was his title *sieur*. Although the particle *de* and the title *sieur* do not necessarily indicate nobility, they are marks of respect.[18] It would seem impossible for Champlain to have acquired such designations by the time he was 15 years old if he was born about 1580, as Liebel reasons.[19] Similarly, it seems unbelievable that a 15-year-old boy could hold such a demanding job as *fourrier* in the *maison du Roy* (king's household), be an *ayde* to the *maréchal des logis* in charge of Brittany, and be entrusted to carry important messages involving the king's service. Even if Champlain's father, "the deceased Anthoine de Champlain" (as he is listed on his son's wedding certificate), had been ennobled with *de* and his son had inherited the title, it seems improbable that Champlain would have been given the tasks he carried out in Brittany if he was such a young age. In the records of the *trésorier des Etats de Bretagne*, where Hus listed him as a *fourrier* who was doing confidential work in the king's service, Champlain received extra pay at least twice in March 1595 for delivering secret and confidential messages; this was not the standard work of a *fourrier*. What we are seeing here is a relationship of trust between Champlain and high-ranking officers – as well as the king himself – that was in place by 1595 and continued later in his life. In fact, Champlain reported to Henri IV whenever he returned to France. In 1598 he stated his intent to report to the king before he left Blavet for Spain, and he

13 See Document B, "Personnel and Pay Records," no. 1, "Payment of various sums."

14 Gabriel Hus, sieur de la Bouchetière, treasurer of the États de Bretagne (? – 1609)

15 Jean d'Aumont, baron d'Estrabonne, comte de Châteauroux, maréchal de France (1522–19 August 1595)

16 François d'Espinay, seigneur de Saint-Luc, governor of Brouage and Saintonge (? – 8 September 1597)

17 Jean Hardy, *maréchal des logis de la maison du Roy* (1549–13 July 1617). For an explanation of the tasks of a *fourrier* and a *maréchal des logis*, sees part 1, A, "Champlain and His Times," subsection "Champlain in the Army of Henri IV in Brittany, 1595."

18 La Roque de Roquebrune, "Particules," 9–15

19 Liebel, "On a vieilli Champlain," 236

did so after returning from the West Indies voyage sometime between 1601 and 1603. Champlain reported again to Henri IV in person late in 1603 when he returned from Canada; in fact, the king had stipulated that he do so before he went on the voyage. In 1607 Champlain reported to the king when he returned from Acadia, and again in the fall of 1609 after Quebec had been built and the sieur de Mons's trading commission had been revoked.[20] It seems that from 1595 until Henri's death in 1610, there was a pattern of Champlain making personal and confidential reports to the king, which in view of the particle *de* in his name and the tasks he was assigned in Brittany, probably had their beginnings before 1595. The fact that Champlain was already a *fourrier* in the king's household and aide to an important *maréchal* of the crown in 1595 suggests that he must have entered service in the king's household sometime before 1595. We do not know when he left Brouage, when he got to Paris, and how he received his assignments as *fourrier* and aide, but it must have been before 1595 in order for him to be given some training for the tasks he was to assume. Because of the nature of the *maison du roy*, one must assume that a young man did not simply appear looking for work but was recommended by someone with influence. If so, we do not know for certain who that was.

Eliane and Jimmy Vigé[21] hypothesize that Champlain may have gone from Brouage to Brittany as early as 1592 as a protégé of capitaine Louis de Milleaubourg, under whom Champlain was an ensign in 1597. In 1592, François Espinay de Saint-Luc, who was then governor of Brouage, received the title of *lieutenant général* of Brittany from Henri IV and was charged with raising troops in Brouage. De Milleaubourg, who was a captain under Saint-Luc, was probably among the troops raised at the time and may have taken Champlain with him. The problem with this hypothesis is that Champlain turned up in Brittany in 1595 with maréchal Jean Hardy and is not mentioned in connection with de Milleaubourg until 1597. A more plausible alternative may be that the rector of the academy in Brouage, where Champlain may have studied, or perhaps Charles Leber du Carlo were Champlain's sponsors. Both of them must have known François Espinay de Saint-Luc or his son Timoléon d'Espinay Saint-Luc when they were active in Brouage.

If Champlain was originally a Huguenot, it makes a birthdate in the 1570s even more likely – perhaps between 1572 and 1577, when Nicolas Folion was administering to the Huguenot community as the only pastor the community had.[22]

20 Biggar, *The Works*, 1:4 (1598), 3:315 (1601–03), 3:318 (1603), 4:35 (1609), 4:37 (1607)
21 Vigé and Vigé, *Brouage*, 2:282
22 Société de l'histoire, "Champlain," 275

For all of these reasons, we feel that Champlain was born before 1580, possibly during the early to mid 1570s. His location in Brittany as early as 1595 and his rank and tasks at the time make a birthdate of 1580 unlikely, and even more unlikely if Champlain got to Paris, let alone Brittany, as early as 1592. Boys of that age were not army material.

Appearance

There is no authentic picture of Champlain, yet few early explorers are depicted as often as Champlain. Most Canadians would recognize the slightly plumpish oval face, wavy long hair, large eyes, and especially the distinctive well cared-for moustache and goatee as a representation of Champlain. The original of the many variant Champlain pictures is a lithograph made by Louis-César-Joseph Ducornet in 1854, ostensibly after a picture of the explorer by the seventeenth-century artist Baltasar Montcornet, said to be in the Bibliothèque nationale in Paris.[23] The subscript on the lithograph reads: "Samuel de Champlain, Gouverneur Général du Canada N^{elle} France: Né à Brouage 1567: Fonde Québec en 1608 et meurt dans cette ville en 1635."

Numerous versions of the Ducornet picture have appeared in myriad publications. The best-known early copy of the lithograph is a beautiful oil painting by Théophile Hamel.[24] Sometimes Ducornet's portrait is retooled to make Champlain look like a young man, such as the picture by Robert Harris;[25] or as a slightly older vigorous man in his late thirties, as with the O'Neil version,[26] while in other pictures he has become aged and decidedly overweight, as depicted by Ronjat.[27] In the Laverdière edition of Champlain, the plumpish figure is passed off as being by Montcornet.[28] A more recent painting has him supervising

23 Biggar, "The Portrait," plate 1

24 The painting by Théophile Hamel, made c. 1862–64, now hangs in the Musée national des beaux-arts du Québec.

25 Grant, *Picturesque Canada*, 1:8. The picture is by Robert Harris (1848–1919), engraved by E. Brighton in 1882.

26 Dionne, *Samuel Champlain*, 1, frontispiece; Dix, *Champlain*, frontispiece. The picture was made by J.A. O'Neil in 1866 after the painting by Théophile Hamel for John G. Shea's translations of Charlevoix's *History and General Description*, 2:88, and of *First Establishment of the Faith*, 1:65.

27 Slafter, *Voyages*, 1, frontispiece. A picture by the French artist Étienne Antoine Eugène Ronjat, made in 1876.

28 Laverdière, *Œuvres*, 1, frontispiece. The artist is given as *Montcornet Ex. C.p.* This appears to be taken directly from the alleged Montcornet "original" but was reversed in the process of copying.

the construction of Quebec dressed in a theatrical costume with lace collar and flowing cape, and with an impeccable moustache and well-groomed hair.[29] The O'Neil version, originally commissioned by John Gilmary Shea, now adorns the boxes of Porter Champlain from La Brasserie Molson/O'Keefe in Quebec.[30] All these pictures have the distinctive moustache and goatee, and all are based on the lithograph by Ducornet.

As early as 1904, Victor Paltsis cast doubt on the authenticity of Ducornet's portrait, largely because the Bibliothèque nationale had no record of a Montcornet portrait of Champlain.[31] The matter was settled in 1920 when Henry P. Biggar found the Montcornet picture used by Ducornet in the Bibliothèque nationale while he was researching his edition of Champlain's *Works* for the Champlain Society.[32] However, the Montcornet portrait was not of Champlain. It was an engraving made in 1654 of Michel Particelli, a particularly unscrupulous *controlleur-général des finances* under Louis XIV. Ducornet had simply copied the engraving in reverse and passed it off as a picture of Champlain.[33] Paltsis's and Biggar's exposure of this fraud has in no way slowed the reproduction of Ducornet's picture or the later variants.

There is no authentic portrait of Champlain. The only picture of him, published during his lifetime, is a tiny figure on an engraving, depicting the battle against the Mohawk on the shore of Lake Champlain on 30 July 1609.[34] This engraving may have been prepared from a sketch done by Champlain, but it is not known to what extent the engraver changed it. The background of palm trees, hairless, naked Huron and Mohawk warriors, hammocks, and unlikely looking canoes suggest that the engraver had a very free hand. What it shows is a bearded man in breastplate and helmet firing a matchlock harquebus.

In his biography of Champlain, Morris Bishop put together a convincing description of the man: "a lean, ascetic type, dry and dark, probably rather under than over normal size."[35] Champlain came from coastal southwestern France, hence his swarthy complexion and his dark hair and eyes. He was used to great physical toil, hence a lean, ascetic body. When he was wounded by arrows in the

29 *Maclean's*, 28

30 Another company, La brasserie Champlain limitée, also makes use of this picture on its various brands, such as Bière Champlain and Select Ale.

31 Paltsis, "A Critical Examination," 306–11

32 Biggar, "The Portrait," 379–80

33 For a side-by-side illustration of the Montcornet picture of Particelli and the Laverdière picture of Champlain, see the pictures in Biggar, ibid.

34 Biggar, *The Works*, 2:101, plate 5

35 Bishop, *Champlain*, 12–13

leg and knee during the battle with the Iroquois in 1615, he was carried "on the back of one of our *Sauvages*" for several days during the retreat, bundled in a fetal position; thus it is probable that he was below average in stature and weight. On the other hand, Samuel Eliot Morison saw him as "a well built man, blond and bearded, a natural leader … a man of unusually rugged constitution."[36] Why Morison believed Champlain to be blond is baffling. In any case, there is no authentic picture of Champlain. Perhaps the best imaginative portrayals of the explorer are those by the great illustrator of Canadian history, Charles W. Jefferys.[37]

36 Morison, *Samuel de Champlain*, 22
37 Jefferys, *Dramatic Episodes*, 13, 15, and *The Picture Gallery*, 1:91–5

APPENDIX 2

Champlain's Signature and Titles: A Discussion[1]

Champlain always signed his name simply as "Champlain," never "Samuel de Champlain." With the exception of his 1607 manuscript map, where he gave "S^{r} de champlain" as author,[2] he never placed any of his titles in the context of his signature, much as the royalty of his times signed simply "Henri" or "Louis." His signature is exactly the same, from the first one extant, written in 1601 on his uncle's *donación*, to his last one, on his *testament* on 17 November 1635, twenty-eight days before he died (figs. 10a and b). The stroke he suffered before he signed his *testament* obviously affected his signature, but it is also obvious that he tried to write it exactly the way he had always written it. When viewing a document signed by Champlain, one is struck by the size and prominence of his signature. Most often it is at least twice the size of the writing on the page and any other signatures. Quite simply, his signature stands out; one does not have to search for it; one's eyes are drawn to it.[3] It is the signature of a confident person.

The first known documents in which Champlain is named are the pay records of the army in Brittany for the years 1595 onwards.[4] Champlain's position in the army was that of a *fourrier* and *ayde* to the *maréchal des logis*, Jean Hardy.[5] In these pay records, ordered and issued by high-ranking officers of the crown, Champlain's name appears both as "Samuel de Champlain" and "sieur de Champlain," showing that the particle *de* was attached to his name by 1595, as was the title *sieur*. However, *de* and *sieur* are not necessarily marks of nobility, but they

1 A list of Champlain's titles as they appear in documents and variants in the spelling of his name are given chronologically below, under "Documentary data."

2 Heidenreich, "The Mapping of Samuel de Champlain," 1540, 1548

3 There are photographs of two documents with his signature readily available. See Vachon, *Dreams of Empire*, 115, 217.

4 See Document B, "Personnel and Pay Records," no. 2, "Payment of various sums."

5 Jean Hardy, *maréchal des logis de la maison du Roy* (1549–13 July 1617]. For an explanation of the tasks of a *fourrier* and a *maréchal des logis*, see part 1, A, "Champlain and His Times to 1604," section on "Champlain in the Army of Henri IV in Brittany, 1595."

are marks of respect. In 1597 Champlain is listed as an *enseigne* attached to the sieur de Milleaubourg[6] in the garrison at Quimper, under the command of the governor, Julien du Pou, sieur de Kermouger.[7] As ensign to de Milleaubourg, he would have carried the standard of his military unit and served a similar role as an aide as he did for Jean Hardy. The reference to Champlain's presence at Quimper in 1597 seems to imply that all the men listed in the dispatch were "*capitaines* and *commandants* in the companies established in the garrison."[8] The only commandant was de Kermouger, and it seems very unlikely that an ensign, who carried the standard of his military unit and was well below the rank of a captain, could also hold the rank of a captain.

The first time Champlain's name appears in print is on the title page of his first book, *Des Sauvages*. Although it appears only as "Samuel Champlain" on the title page, he is listed three times in the *privilège* (licence to publish) as "Samuel de Champlain" and once as "Sieur de Champlain" in the title of the dedicatory poem by the alleged sieur de La Franchise.

From 1604 until late in 1607, Champlain worked for Pierre Du Gua de Mons, *lieutenant général pour le roy* in Acadia, doing the tasks of a geographer and surveyor along the coast of Acadia.[9] On his map of 1607, which resulted from these surveys, he listed himself as the author, "Sr de Champlain." Did Champlain carry the title of *géographe du Roy* (king's geographer) while he was doing these surveys? This title was first given to him by Marc Lescarbot[10] and was later repeated by Father Chrestien Le Clercq[11] in 1691. Lescarbot, who was present in Acadia with Champlain, also wrote that his superior, de Mons, designated Champlain "the one for geography" (*Chaplein ... l'un pour la geographie*).[12] The king's geographers of early-seventeenth-century France, such as Pierre Du Val and Nicolas

6 Louis de Milleaubourg, also Millambourg. Document B, "Personnel and Pay Records," no. 2, "Payment of Various Sums" (2 April 1597)

7 Julien du Pou was governor of Quimper from 1592 to 1610.

8 See: Document B, "Personnel and Pay Records," no. 2, "Payment of Various Sums" (2 April 1597). Champlain *enseigne* (early 1597).

9 De Mons (Monts) had been appointed lieutenant governor of Acadia by Henri IV on 8 November 1603 (Blanchet, *Collection*, 1:43–4). For a biography, see DCB, 1, "Du Gua de Monts, Pierre."

10 DCB, 1, "Champlain, Samuel de." Lescarbot's sonnet *Au Sieur Champlein* has the statement that Champlain was a *Geographe du Roy* directly under the title (Lescarbot, *Les Mvses*, 37).

11 Shea, *Le Clercq: First Establishment of the Faith*, 1:64

12 Lescarbot, *Histoire*, 536. The sentence carrying this phrase is only in the 1609 edition of Lescarbot's *Histoire*. It is not in the 1617 edition.

Sanson, were cartographers and compilers of information, not explorers who gathered their own information. Champlain was a good field geographer in the modern sense. He was observant, systematic, possessed of a good knowledge of the natural environment; he could survey and draft maps and had the ability to relate this diverse information to the human occupation of the land. If he had been a *géographe du Roy*, like Du Val and Sanson he could have placed this title on his maps under his name and on the title pages of his books, but nowhere does the title appear showing that he ever claimed to be a *géographe du Roy*.

When de Mons was reappointed on 7 January 1608, he chose Champlain as his *lieutenance pour le voyage* to Quebec.[13] This was Champlain's first major commission, a lieutenant to a lieutenant general who held a trade monopoly for New France between 40° and 46° north latitude. After this date he is listed as *capitaine ordinaire de la marine*. On his wedding certificate of December 1610, for example, he is termed a "*Noble homme Samuel de Champlain Sieur dudit Lieu cappitaine ordinaire de la Marine*." The term *noble homme* does not mean that Champlain had attained nobility. The title was commonly used by merchants and office holders on the brink of nobility. Shortly after being ennobled, however, the title *écuyer* was usually added to a name.[14] It is not entirely certain exactly what a *cappitaine ordinaire de la marine* was.[15] Based on seventeenth-century encyclopedias, it appears that Champlain held the rank of a naval captain, which he could exercise only in the absence of the other officers. This interpretation makes some sense. Unlike the lyric descriptions by later biographers of Champlain's prowess in navigating and commanding a ship, Champlain himself never claimed to have functioned in any of these capacities.

After Henri IV was assassinated on 14 May 1610, Champlain appealed to Marie de Medici – regent for her son, the new king, Louis XIII – to continue to support France's Canadian ventures. On 8 October 1612, Louis named Charles de Bourbon, comte de Soissons, lieutenant general of New France, and seven days later, Soissons appointed Champlain as his lieutenant.[16] Soissons died on 1 November 1612, and twelve days later Louis appointed the prince de Condé to suc-

13 Biggar, *The Works*, 2:4–8

14 For a good discussion of these terms, see Brunelle, *New World Merchants*, 103–4n53.

15 We have not seen any such rank in any sixteenth- or seventeenth-century army or navy rosters. Early-seventeenth-century encyclopedias state that the term *ordinaire* attached to a rank means "*qui servant toute année, mais seulement en l'absence des officiers de quartier*" (Ganeau, *Dictionnaire universel*, "Ordinaire").

16 Champlain's commission from Soissons was published by him (Biggar, *The Works*, 4:209–16).

ceed him.[17] Condé, however, was not a lieutenant general as Soissons had been but was elevated to be viceroy of New France. On 22 November, he appointed Champlain his lieutenant.[18] This appointment gave Champlain the powers of a governor, but without the title or a new commission. Early in 1613 the merchants of Saint-Malo issued a tract (factum), trying to discredit Champlain's role on the 1603 voyage to Canada and his elevation to lieutenant by Soissons and Condé. In this tract, they described his profession as a *peintre* (artist) who made the trip only in order to profit from it.[19] Le Blant and Baudry have suggested that Champlain may have had "a certain notoriety as a painter or draftsman, founded undoubtedly upon the illustrations of his book *Des Sauvages* [1603] and perhaps also upon the abundant series of 62 or 72 illustrations in the *Brief discours*."[20] The problem with this suggestion is that there are no illustrations in *Des Sauvages*, and the *Brief Discours*, with its little illustrations, was not published until 1859, by the Hakluyt Society. Of course, someone from Saint-Malo may have seen the manuscript before 1613, or Champlain may have been in the habit of exhibiting his art. But his book *Les Voyages, 1613*, with its many pictures and maps, was not published until early 1614,[21] eliminating it as the basis for the label *peintre*. There is no question that a cartographer had to have artistic ability and that Champlain had some ability.[22] The question is: How did the merchants of Saint-Malo know as early as 1613 that Champlain could draw? By saying that it was his profession, they were of course trying to discredit him as an explorer. In spite of their and other merchants' objections to Condé's monopoly and Champlain's commission, both were finally forced on them by the king.[23]

Shortly after his appointment by Condé, the title *escuyer* (*écuyer*) appears after his name for the first time (16 January and 5 February 1613). Because *écuyer* is an entrance level of nobility, it is likely that as lieutenant and de facto governor to a Bourbon prince, Champlain was ennobled. In the fall of 1616, Condé was impli-

17 Henri II de Bourbon, prince de Condé (1588–1646)

18 Biggar, *The Works*, 2:245

19 See Document D, "Factum."

20 Le Blant and Baudry, *Nouveaux documents*, 1:125n3

21 Champlain mentioned on two occasions that *Les Voyages, 1613*, was not printed until 1614. Although the *privilège* was assigned on 9 January 1613, *Les Voyages* contains the entire *Quatriesme Voyage*, which was written after Champlain returned on 26 August 1613 (Heidenreich and Dahl, "The Two States," 12n24).

22 Ganong, *Premier peintres* and "Champlain: Painter?"

23 For a summary of documents, see Biggar, *Trading Companies*, 195–6; and Champlain's discussion, Biggar, *The Works*, 2:243–7, 4:206–19.

cated in a conspiracy against Concino Concini, who helped Marie de Medici to govern the country. Condé was arrested and jailed, and his position as viceroy of New France was taken over in part by the marquis de Thémines, who had been appointed *maréchal de France* the year before.[24] The order in council of 18 July 1619 (see below) appointing Champlain commander of Quebec reflects the duality of Condé and Thémines as viceroy. On 20 October 1619, Condé was released from prison, and he yielded his rights as viceroy to the duc de Montmorency on 25 February 1620.[25] Soon after, the new viceroy confirmed Champlain as his lieutenant, and this was approved by Louis XIII on 7 May.[26]

Montmorency's term as viceroy came to an end early in 1625 when he sold his position to his nephew, the duc de Ventadour,[27] for 100,000 livres.[28] On 15 February, Ventadour confirmed Champlain in his office as the viceroy's lieutenant.[29] Champlain was now not only an *escuier* but also was elevated from *capitaine ordinaire de la marine* to *cappitaine pour le Roy en la Marine du Ponant*.[30] This promotion reflects his commission from Ventadour, which made him a virtual governor.

Ventadour was viceroy for only a few years. Beginning in 1624, Cardinal Richelieu[31] began to seize increasing power through his position as head of the royal council that governed France for the young Louis XIII. In 1626 he appointed himself "Grand Master and Superintendent General of Commerce and Navigation." Early in 1627 he discontinued the office of admiral of France held by Ventadour, thereby giving himself complete control of the navy and colonial affairs. By May 1627, he had founded the Compagnie de la Nouvelle France, composed of 100 associates, including Champlain, who is listed as "*Samuel Champlain, Escuyer, Cappitaine pour le Roy en la Marine.*"[32] On 21 March 1629, Richelieu appointed Champlain his lieutenant.[33] For Champlain, the commission must have

24 Lauzièrs Pons, marquis de Thémines, maréchal de France in 1626 (1552–1627)

25 Henri II, duc de Montmorency, admiral of France. Appointed viceroy on 25 February 1620

26 Biggar, *The Works*, 4:367–71

27 Henri de Lévis, duc de Ventadour (Vantadour) (1596–1651). A very devout person, de Ventadour opened the mission field of New France to the Jesuits.

28 Trudel, *The Beginnings*, 134

29 Biggar, *The Works*, 5:138–49

30 "Captain to the King and the Navy of the Western Ocean." The *marine du Ponant* was the Atlantic fleet, whereas the *marine de Levant* was the Mediterranean fleet.

31 Armand-Jean du Plessis, cardinal, duc de Richelieu (1585–1642).

32 Blanchet, *Collection*, 1:82

33 Biggar, *The Works*, 6:151–2

been a joy to read. It lauded him for the "experience that you have acquired in becoming acquainted with the country of New France and its inhabitants during your residence therein, together with the special knowledge we have of your good sense, competence, generosity, prudence, zeal for the glory of God, and affection and fidelity to the service of the King." And it charged him to: "command in the absence of my Lord the Cardinal de Richelieu … throughout the whole extent of the said country, to control and govern, as much the Aboriginals of the place as the French who currently reside there and [those who] will settle there subsequently."[34]

Champlain had become governor of New France in all but name. It is an amazing record, if only for the fact that he survived one lieutenant general (Du Gua de Mons), five viceroys (comte de Soissons, prince de Condé, marquis de Thémines, duc de Montmorency, duc de Ventadour), and finished his career as lieutenant to Cardinal Richelieu, the most powerful person in France after the king. He had risen from a *fourrier* in Brittany to de facto governor of New France.

There is considerable controversy over the designation of Samuel de Champlain as the first governor general of Canada. In *The Canadian Encyclopedia*, for example, the modern office is interpreted as descending, not from the British office, inaugurated at Confederation (1867), but from "the beginning of European settlement in Canada."[35] Although the Right Honourable Michaëlle Jean considers herself the twenty-seventh governor general of Canada, she would be sixty-fourth counting from Champlain. Champlain's title as "governor" can, however, be contested. Richelieu officially named him as his representative, *commandant en la Nouvelle-France* (commander in New France), on 21 March 1629, although Richelieu himself had assumed responsibility for the colony in the spring of 1627 (see below, entries for 1629 to 1633). In 1629 the English called him "Leieutenant Governer" of Quebec, which is what Champlain must have told them he was (see below, entry for 1629). Most historians consider Charles Huault de Montmagny, Champlain's successor, the first appointed governor to New France.[36]

Documentary data

In every case where Champlain signed a document, it is simply "*champlain*." Spelling is as in the original document.

34 Ibid.
35 Monet, "Governor General"; Trudel, "Champlain"; Mathieu, "Gouverneur"
36 *DCB*, 1:195; ibid., 1:372; Dubé, *The Chevalier de Montmagny*, 117, 120

1595 *Samuel de Champlain, aultre fourier en ladite armée* (Document B, 2: f. 194v).

Samuel de Champlain, ayde du sieur Hardy marechal des logis de l'armée du roy (ibid., f. 229v).

Sieur de Champlain (ibid., f. 523–5).

1597 *Champlain, enseigne du sieur de Milleaubourg* (ibid., [2 April 1597]).

1601 26 June. *Samuel Zamplen, franses, natural del bruaze en la provincia que llaman santonze* [Samuel Champlain, Frenchman, native of Brouage in the province they call Saintonge] (Document C, "Gift from Guillermo Elena").

1603 In *Des Sauvages*: title page, *Samuel Champlain*; privilège, *Samuel de Champlain* in three places, and *Sieur de Champlain* in the poem by La Franchise.

1607 *Sr de champlain*, manuscript map: *descr[i]psion des costs p[or]ts rades Illes de la nouuelle france* [Date altered by Champlain from 1606 to 1607] (Library of Congress, Washington. Vellum Chart Collection, no. 15).

1608 January. Appointed lieutenant to the sieur de Mons, who himself was appointed lieutenant general of New France on 7 January 1608 by Henri IV (Biggar, *The Works*, 2:4–8).

1610 27 December. … *feu Anthoine de Champlain vivant Cappitaine de la Marine et de dame Margueritte le Roy*, Champlain's marriage contract [both parents deceased] (Cathelineau, "La minute notariée," 144).

27 December. *Noble homme Samuel de Champlain Sieur dudit Lieu cappitaine ordinaire de la Marine demeurant en la ville de brouage pays de St onge*, Champlain's marriage contract (ibid.)

1612 Late 1612. *Sievr de Champlain Saint Tongois, Cappitaine Ordinaire Povr le Roy en la Marine*, title of the 1612 map, plate 81, *Carte Geographiqve de la Novvelle Franse faict len 1612* (Champlain, *Les Voyages, 1613*).

15 October. *Sieur Samuel de Champlain, Capitaine ordinaire pour le Roy en la marine*, commission by Charles de Bourbon, comte de Soissons, naming Champlain *nostre Lieutenant, pour representer nostre personne audit pays de la nouuelle France* (Biggar, *The Works*, 4:209–16).

22 November. Henri II de Bourbon, prince de Condé, appoints Champlain his *Lieutenance* for New France. Soissons died 1 November, and on 13 November Condé succeeded him (Biggar, *The Works*, 2:245).

[1613] *sieur Champlain* and *Champlain ... sa profession de peintre*, factum of Saint-Malo merchants, no date (Le Blant and Baudry, *Nouveaux documents*, 1:245).

1613 9 January. *Sieur de Champlain, Xainctongeois, capitaine ordinaire pour le Roy, en la marine*, title page, *Les Voyages*, 1613 [date from the *privilège*].

16 January. *Samuel Champelain escuyer, sieur dudit lieu, cappitaine pour le roy en la marine et lieutenant de Monseigneur le prince en la Nouvelle-France, demeurent à Paris, rue Troussevache, paroisse Sainct-Jacques-de-la-Boucherie.* [*Sieur de Champlain* for most of rest of document; also *Monssieur Champellain*] (Le Blant and Baudry, *Nouveaux documents*, 1:250).

5 February. *Samuel Champelain escuyer, sieur dudict lieu, cappitaine pour le roy en la marine et lieutenant de Monseigneur le prince en la Nouvelle-France*, agreement with Mathieu Georges (ibid., 256).

9 February. *Sieur Samuel Champlain*, annulment of agreement with Mathieu George (ibid., 262).

15 November. *Sieur Samuel de Champelain, capitaine ordinaire pour le Roy en la marine, pour le present, demeurant en ceste dicte ville de Paris, en la maison du Miroir* [*Sieur de Champelain* for rest of document, and in the appendix to the document; he is *sieur de Champlain* and *Champelain*], Act incorporating the Compagnie du Canada (ibid., 310).

21 November. *Samuel de Champlain cappitaine ordinaire pour le Roy en la marine*, Champlain assigns his rights in the Compagnie du Canada to de Mons (ibid., 322).

Sieur de Champlain, capitaine ordinaire pour le Roy en la marine, et Lieutenant de Monseigneur le Prince de Condé en la Nouvelle France, title page, *Quatriesme Voyage*, 1613.

Late 1613. *faictte par le Sr Champlain Cappi^{ne} po^{r} le Roy en la marine – i6i3.* [authorship of map: *Carte geographique de la Nouuelle franse en son vraymeridiein*], published with the *Quatriesme Voyage*, in Champlain, *Les Voyages, 1613*.

1614 10 January; 14 February. *Samuel de Champlain; sieur de Champlain*, document disinheriting Hélène Boullé, Champlain's wife (Le Blant and Baudry, *Nouveaux documents*, 1:330, 333).

1615 18 March. *Samuel de Champlain, escuyer, cappitaine pour le Roy en la marine et lieutenant de Monseigneur le prince en la Nouvelle-France*, Champlain's power of attorney to Jean Ralluau (ibid., 348).

1617 22 July. *Noble homme Samuel de Champlain, cappitainne ordinaire du Roy en la Marine de ponent*; also Sr *de Champlain*, Champlain hires a servant (Charavay, *Documents inédits*, 4–5; Biggar, *The Works*, 2:324).

1618 No date. *Le Sieur de Champlain*, Champlain's letter to the Paris Chamber of Commerce (ibid., 2:326–45).

9 February. *Sieur de Champlain*, letter from Chamber of Commerce to Champlain (ibid., 2:346–51).

12 March. *Sieur de Champlain*, order of the king [Louis XIII] to Champlain (ibid., 4:364–5).

29 December. *Samuel Champlin* $Capp^{ne}$*. de la marine de Ponant*, receipt for Champlain's salary (ibid., 5:330).

1619 14 January. *Samuel de Champlain*, appendix to Champlain's marriage contract (ibid., 4:372–3).

Sieur de Champlain, cappitaine ordinaire pour le Roy en la Mer du Ponant, title page of *Voyages et Descouvertures*, 1619.

15 March. *Samuel de Champlain, cappitaine pour le Roy de la marine de Ponant et lieutenant pour Monseigneur le prince en la Nouvelle-France*, lease of a house from Champlain to Hersan and Camaret (Le Blant and Baudry, *Nouveaux documents*, 1:374).

15 March. *Samuel de Champlain, cappitaine pour le Roy de la marine de Ponant*, establishing the rent for the above house (ibid., 376).

18 July. *Samuel de Champlain, lieutenant de Monseigneur le prince de Condé et du sieur marechal de Themines en la Nouvelle France*, order in council appointing Champlain commander of the Quebec habitation (ibid., 394).

1620 23 February. *Samuel de Champlain, cappitaine ordinaire pour le Roy en la marine de Ponant et lieutenant de Monseigneur le prince en la Nouvelle-France*, sale of half a house by Hersan and Camaret to Champlain (ibid., 397).

2 March. *Samuel de Champlain, cappitaine ordinaire pour le Roy en la marine de Ponant et lieutenant de Monseigneur le prince de Condé en la Nouvelle-France*, sale of same house as above, with receipt for payment (ibid., 399).

7 May. *Champlain*, letter from Louis XIII addressed to Champlain making him lieutenant to the duc de Montmorency, who is his viceroy (Champlain, *Les Voyages, 1632*, 228–9; also in Blanchet, *Collection*, 1:60; Biggar, *The Works*, 4:370).

1621 2 February. *Monsieur Champlain*, letter from Henry, duc de Montmorency, to Champlain (Blanchet, *Collection*, 1:61).

Trudel mentions (no reference given) a document in which Champlain is called "noble homme" (DCB, 1:187).

1625 15 February. ... *sieur Samuel de Champlain, Captaine pour le Roy en la marine ... Avons commis, ordonné, deputé, commettons, ordonnons, & deputons par ces presentes, nostre Lieutenant, pour representer nostre personne, audit pays de la Nouuelle France*, commission by de Ventadour appointing Champlain his lieutenant (Biggar, *The Works*, 5:142–9).

29 December. *Samuel Champelain, escuier, cappitaine pour le Roy en ma[sic] marine du Ponant et lieutenant pour monseigneur le duc de Ventadour en la Nouvelle France*, gift of Allene's estate near La Rochelle to du Carlo (Leymarie, "Inédit sur le foundateur," 83).

1626 25 March. *Samuel Champlain, escuier, cappitaine pour le Roy en la Marine du Ponant et lieutenant pour Monseigneur le duc de Ventadour en la Nouvelle France*, as above, gift to du Carlo (ibid., 85).

1628 27 April. ... *nostre cher & bien amé le sieur de Champlain, commandant en la Nouuelle France, en l'absence de nostre tres-cher & bien-amé cousin le Cardinal Richelieu*, commission by Louis XIII to Champlain to make an inventory, map, and report of Quebec (Biggar, *The Works*, 5:288–91).

1629 21 March. ... *sieur de Champlain l'un des Associez en ladite Compagnie ... en l'absence de Monsieur le Cardinal de Richelieu ... commander en toute l'estendue dudit pays, regir & gouverneur tant les Naturels des lieux que les François qui y resident de present, & s'y habitueront cy après*, commission written by the directors of the Company of New France, appointing Champlain representative of Cardinal Richelieu in New France (ibid., 6:151–2).

14 April. *Sr Champlain*, letter from Louis XIII to chevalier de Montigny mentioning Champlain (ibid., 6:358–60).

17 May. *Samuel Champlain, Escuyer, Cappitaine pour le Roy en la Marine*, list of names on roster of Compagnie de la Nouvelle France (Blanchet, *Collection*, 1:82).

... *Samuell Champlein of Browages ... late Lieutenant govournor of the forte called the St. Lewis at Kebecke ... Mr Champlaine ... Samuell Shamplin, Leieutenant Governer ... Monsr. Champlaine Governor of the Fort ... Monsr. Champlayne*, variants of Champlain's name on English documents relating to the capture of Quebec in 1629, written between 9 November 1629 and 9 April 1632 (Laverdière, *Œuvres*, 6: "Pièces justificatives," 3–29).

[1630] *Sieur de Champlain*, undated letter by Champlain to Louis XIII pointing out to him the value of restoring New France to France (Biggar, *The Works*, 6:361–74).

1630 27 September. *Samuel de Champelain cappictaine pour le Roy en la marine commandant pour Monseigneur le Cardinal* [Richelieu] *en la Nouvelle France dicte Canada*, sale of two houses by Champlain in Brouage (Delafosse, "Séjour de Champlain," 574).

14 April. *Sr Champlain*, letter of instruction for a voyage to Canada by Louis XIII in which Champlain is mentioned (Biggar, *The Works*, 6:357–60).

1632 *Sieur de Champlain, Xainctongeois, Capitaine pour le Roy en la Marine du Ponant*, title page of *Les Voyages*, 1632.

1633 23 January. *Samuel CHAMPELAIN, Capitaine pour le Roy, en la marine* (*RHAF*, "Documents inédits," 9:596).

Le sieur de Champlain Lieutenant general pour le Roy en la nouuelle France, title given to Champlain in the *Table du Contenu*, or summary, of his 1633 voyage (Jean Richer, *Le Dixneufiesme Tome Du Mercure Françoise*, Paris 1633, vol. 19, unnumbered p. 14). Full text of the voyage is on pp. 803–67.

1635 17 November. *Samuel de Champlain*, in the text of his *Testament de Champlain* (Campeau, *Monumenta*, 3:29).

1648 No date. *Défunt Samuel de Champelain, vivant capitaine de la marine du Ponant, lieutenant general pour le Roy en la Nouvelle France et Gouverneur pour S. M. audit pays*, letter by the Bishop of Meaux certifying that Helaine [sic] Boullé wishes to establish a convent at Meaux (Charavay, *Documents inédits*, 7–8).

APPENDIX 3

Chronology to 1604

DATE	EVENT
1573	
23 December	Mention of Anthoyne Chappelin (also Anthine Chappelin), *pilote à Brouage.* Father of Champlain
c. 1575	Birth of Samuel de Champlain
1595	
March–Dec.	Champlain *fourrier* and *ayde du Sieur Hardy, maréchal des logis du Roy,* in Quimper, Brittany
1597	
2 April	Champlain, *enseigne du sieur de Milleaubourg.* Stationed in Quimper, Brittany. May have met his maternal uncle, Guillaume Allene, at Quimper, where he made a will.
1598	
2 May	Henri IV signs Treaty of Vervins, ending the war between France and Spain.
August (9 Sept.)	Champlain meets Allene in Blavet and departs with him on the *San Julián*, chartered by Spain to return Spanish troops to Cádiz. One-eighth of the ship was owned by Allene.
1599	
January (3 Feb.)	General Don Francisco de Coloma charters the *San Julián* for his Spanish fleet on a voyage to the West Indies. Allene unable to go. Champlain receives permission from Coloma to represent Allene (*me commist la charge)* on the *San Julián*.
1600	
Spring	Chauvin de Tonnetuit, accompanied by Gravé Du Pont and de Mons, trades on St. Lawrence and erects a trading post at Tadoussac.
1601	
March	Champlain returns to Spain, arriving at Seville.

DATE	EVENT
26 June	Champlain in Cádiz to witness his uncle Allene's gift willed to him.
3 July	Champlain orders sale of one-eighth of the *San Julián*.
1602	
Late 1601 or 1602	Champlain visits court of Henri IV to report on his voyage to the West Indies. Champlain is granted a pension by the king.
Late 1602	Gravé Du Pont returns from Canada with two Montagnais men, whom he presents to Henri IV.
1603	
February (early)	Chauvin de Tonnetuit dies. Henri IV designates Aymar de Chaste to succeed to Chauvin's trading privilege in Canada.
February–March	Champlain visits de Chaste to offer his services. De Chaste accepts and asks Henri IV to give Champlain permission to join the expedition. Henri IV orders Champlain to join it and to report to him on his return.
Thurs., 13 March	Commission to trade and discover lands in Canada given to Colombier, Prévert, and Gravé Du Pont by Henri IV.
Sat., 15	Champlain leaves Honfleur on the *Bonne-Renommée*, captained by Gravé Du Pont; unfavourable winds; puts in at Le Havre.
Sun., 16	They depart from Le Havre.
Mon., 17	Sight Alderney and Guernsey.
Tues., 18	Sight coast of Brittany.
Wed., 19	Sight Ushant.
Fri., 21	See Flemish ships.
Sun., 30	Easter Sunday, caught in storm for 17 days. Lose ground.
Wed., 16 April	Milder weather.
Mon., 28	Encounter a "very high" iceberg.
Tues., 29	Ice floe 8 leagues long; "infinite number" of smaller ones, 100–120 leagues from Canada at lat. 45°40'N; passage at 44°N.
Fri., 2 May	Grand Banks at 44°N
Sat., 3	Aymar de Chaste dies.
Tues., 6	Hear waves break on shore in heavy fog; put out to sea.
Sun., 11	Reach Cape St. Mary's.

DATE	EVENT
Mon., 12	Gale for two days.
Thurs., 15	Reach St. Pierre Islands.
Sat., 17	Ice floe 6 leagues long near Cape Ray.
Sun., 18	See Cape Ray, St. Paul Islands, Cape St. Lawrence; ice floe 8 leagues long.
Tues., 20	Sight Anticosti Island.
Wed., 21	Sight Gaspé.
Sat., 24	Anchor before Tadoussac.
Mon., 26	Enter Tadoussac harbour.
Tues., 27	*Tabagie* with Montagnais at Pointe aux Alouettes.
Wed., 28	Montagnais move to Tadoussac.
Mon., 9 June	*Tabagie* by Montagnais to celebrate their victory over the Iroquois.
Wed., 11	Explore Saguenay River 12–15 leagues from Tadoussac.
Wed., 18	Depart from Tadoussac for Lachine Rapids; reach Île aux Coudres.
Thurs., 19	Depart from Île aux Coudres and anchor on north shore; strong winds.
Sun., 22	Depart for Quebec and anchor at Quebec.
Mon., 23	Depart from Quebec; anchor at Pointe au Platon.
Tues., 24	Depart from Pointe au Platon; anchor on north shore, 5.5 leagues away.
Wed., 25	Depart; anchor near Île St-Eloi.
Thurs., 26	In and around St-Eloi.
Fri., 27	Depart from Île St-Eloi; anchor at Trois-Rivières.
Sat., 28	Depart from Trois-Rivières; reach Lac Saint-Pierre and anchor.
Sun., 29	Cross Lac Saint-Pierre; anchor two leagues upstream from the lake.
Mon., 30	Depart and row up Richelieu River to rapids at Saint-Ours; return and anchor three leagues west of mouth of Richelieu on north shore of St. Lawrence.
Tues., 1 July	Depart and explore Îles de Verchères.
Wed., 2	Depart from Îles de Verchères; reach Lachine Rapids and Montreal Island.
Thurs., 3	Explore around Lachine Rapids and Montreal Island.
Fri., 4	Depart from Montreal Island and reach the Richelieu River.

DATE	EVENT
Sun., 6	Depart from the Richelieu; anchor on Lac Saint-Pierre.
Mon., 7	Reach an anchorage three leagues downstream from Trois-Rivières.
Tues., 8	Arrive at Quebec.
Wed., 9	Anchor near Île aux Coudres.
Thurs., 10	Anchor 1.5 leagues from Île aux Lièvres.
Fri., 11	Reach Tadoussac and depart for Gaspé.
Sun., 13	Midway between Tadoussac and Gaspé.
Tues., 15	Reach Gaspé.
Sat., 19	Leave Percé for Tadoussac; anchor at Rivière Madeleine; storm for two days.
Tues., 22	Depart for north shore in heavy winds
Mon., 28	Cast anchor in a cove on the north shore of the St. Lawrence.
Tues., 29	Depart; anchor in Rivière Sainte-Marguerite.
Mon., 3 August	Reach Tadoussac.
Sat., 16	Depart from Tadoussac.
Mon., 18	Reach Percé.
Sun., 24	Leave Gaspé for France.
Tues., 2 September	Sight Cape Race.
Fri., 5	Enter Grand Banks.
Tues., 16	Fifty leagues from Ushant.
Sat., 20	Expedition reaches Le Havre.
September ?	Arrive Honfleur, where they hear of Aymar de Chaste's death on 3 May.
October ?	Champlain meets with Henri IV to show him his map of the St. Lawrence and to give him an account of the findings of the expedition. The king promises to continue to support ventures into Canada.
Sat., 8 November	Henri IV appoints Pierre Du Gua de Mons lieutenant général of Acadia between lat. 40° and 46°N.
Sat., 15	Licence granted to publish *Des Sauvages*.
Thurs., 18 December	Henri IV grants de Mons a ten-year monopoly on the fur trade in Acadia between late 40° and 46°N, with the obligation to establish settlements.

APPENDIX 4

French Measures of Distance, Weight, and Coinage

Measures of distance used by Champlain

It is important to understand that all the distances Champlain gave in his writings were estimates. He had three ways of determining distance: through "dead reckoning," through a rough form of triangulation by means of a compass and "dead reckoning," and, at sea, by applying the principles of the "rule to raise or lay a degree of latitude," by means of a compass and a latitude-finding device.[1] Although he mentioned the use of the *pied* (foot), *pas* (pace), *toise* (linear measure of 6 *pieds*), *brasse* (fathom), and *lieue* (league), he never once mentioned which of the many leagues in vogue during this period he actually used.[2] An examination of his small-scale maps and navigational examples in his *Traitté de la marine* demonstrates that over large areas portrayed on maps or long dis-

1 "Dead reckoning" is distance estimated from the speed of a ship. Triangulation was enacted by taking compass bearings of a distant place from two locations and determining the distance between those locations by measurement, or "dead reckoning." This procedure yields two interior angles of a triangle and the distance between them. The "rule to raise or lay a degree of latitude" is based on the principles of plane sailing (nautical triangle) where the course of the ship, measured with a compass, is the hypotenuse of a right-angled triangle, and the other two sides are one degree of latitude and longitudinal distance traversed. Since lines of latitude and longitude cross each other at right angles and latitudinal distance can be measured with a cross-staff, or astrolabe, one side (latitudinal distance) and the two interior angles (course bearing and the right angle formed where latitude and longitude intersect) are known, permitting the other sides to be calculated or taken from a table (Heidenreich and Dahl, "Samuel de Champlain's Cartography," 321–31). For Champlain's navigational examples, see Heidenreich, *Explorations*, 115–26.

2 The classic comparative study of measures *anciens et modernes* is Doursther, *Dictionnaire universel.*

TABLE A1 LINEAR MEASURES USED BY CHAMPLAIN

Measure	*Metric*	*Imperial*
Pied de roy (French foot)	324.8 mm	12.8 in.
Pas ordinaire (2.5 *pieds de roy*; pace)	0.8 m	2.7 ft.
Toise (6 *pieds de roy*; linear measure)	1.95 m	6.4 ft.
Brasse (6 *pieds de roy*; fathom)	1.95 m	6.4 ft.
Lieue géographique ou marine de 17.5 au degré (Spain)	5.55 km	3.45 mi.*

* The "English mile" referred to here is the statute mile of 5,280 feet.

TABLE A2 ADDITIONAL MEASURES CHAMPLAIN MAY HAVE USED

French measure	*Metric*	*Imperial*
Petite lieue de 30 au degré	3.27 km	2.03 mi.
Lieue commune de 25 au degré	3.91 km	2.43 mi.
Lieue moyen de 24 au degré	4.07 km	2.53 mi.
Lieue d'une heure de chemin de 20 au degré	4.91 km	3.05 mi.
Lieue marin ou grande lieue de 20 au degré	4.91 km	3.05 mi.

tances at sea, he used the Spanish marine league exclusively, although he never called it that by name.[3]

In examining a sample (N) of 105 estimates of distances taken from Champlain's *Les Voyages, 1632*, and comparing them with the same distances taken from topographic maps, it is evident that Champlain's estimates varied widely in accuracy.[4] The 105 observations averaged 4.7 km (2.9 mi.) to a league and carried a standard deviation (s) of 1.5 km (0.93 mi.). In a sense, this was to be expected because of the difficulty of estimating distances and the conditions in which many of the estimates were made. It is disconcerting, however, that Champlain may have been using different leagues in different circumstances without alerting the readers of his books and maps. In the Gulf of St. Lawrence, his estimates under 10 leagues along coastal areas averaged 4.3 km to the league (s = 0.58 km; N = 10), while those over 10 leagues on the high seas averaged 5.5 km (s = 1.5 km; N =

3 Heidenreich, *Explorations*, 43. There are 17.5 Spanish marine leagues to a degree of latitude. Champlain divided the bar scales on his small-scale maps accordingly.

4 Heidenreich, *Explorations*, 42–50

24). The first of these may be the *lieue commune, moyen*, or *marin*, and the latter, made probably in part with instruments, is the Spanish marine league. Along the St. Lawrence River, between Tadoussac and the Lachine Rapids, the average league is 4.5 km (s = 1.04 km; N = 45), while inland over rivers and small lakes to the Huron country on the southern part of Georgian Bay, his average league is only 3.4 km (s = 1.40 km; N = 26). The large standard deviations indicate a horrendous inconsistency in his estimates, but the diminishing length of the league from an average of 5.5 km (3.4 mi.), to about 4.5 km (2.8 mi.) and eventually 3.4 km (2.1 mi.) may indicate the use of different leagues or estimates made in different conditions; i.e., the Spanish marine league on the high seas, one of the shorter leagues along coastal areas and on the St. Lawrence, and the *petite lieue* on journeys overland. His map scales, however, indicate that he used only the Spanish marine league of 5.5 km. In order to aid English readers, some editors and translators have confused the issue by simply converting all Champlain's estimates to miles by multiplying them by three, on the assumption that he was using a league of three miles. In fact, an average league based on all 105 observations comes to 2.9 mi. (4.7 km), but the huge standard deviation (s = 1.5 km; 0.9 mi.) makes this average meaningless and any conversions a misleading and senseless exercise.

Champlain's distances in Des Sauvages

Champlain gave a large number of distance estimates in *Des Sauvages*.[5] Those estimates extracted for analysis are only those that he, and not someone else, made and over land or sea that he actually travelled, rather than guessing the distance to some inaccessible location. Unlike his later work along the coast of Acadia, he does not mention any attempts at triangulation, which suggests that all these estimates were made by "dead reckoning" or an "educated guess."

Of 37 distance estimates Champlain made between locations that can be identified on topographic maps, his average league was 3.6 km (s = 1.3 km). If one reduces his estimates to 33 – by dropping the uniformly bad estimates made on his return from the Gaspé along the north shore of the St. Lawrence to Tadoussac – the average league becomes 3.25 km (s = 1.2 km). Superficially, it looks as if he may have used the *petite lieue* of 3.27 km to make his estimates, but the large standard deviations make it difficult to identify whatever league he may have had in mind. His distançe estimates must therefore be used with great caution. Fortunately, his identification of most landmarks is good enough to permit one

5 See Document G, "*Des Sauvages*." Champlain's distances in leagues are given in kilometres in the footnotes when it was possible to determine the distances he estimated.

to identify the route over which he travelled without having to resort to using his distance estimates.

Returning to Champlain's estimates between Tadoussac and the Lachine Rapids from *Les Voyages, 1632*, it is interesting to note that his average league has risen to 4.5 km (s = 1.04 km; N = 45), with a slightly lower standard deviation over the same route as in *Des Sauvages*. It is probable that his observations improved by repeated travelling over the same route, but it is also possible that he used a different league later in his career. It may be that in 1603 he was using the short land league (*petite lieue*), which he would have used as a *fourrier* in Brittany, and that he switched to one of the longer French marine leagues after 1604, having gained experience in coastal situations.

Native estimates of distance

In order to understand what lay ahead for an exploring party, or to draw maps from Native accounts, explorers had to become familiar with Native concepts of distance. Since no one ever recorded concepts of precise measurements among Natives of northeastern North America and since no references to such concepts can be found in Native vocabularies, it is fairly certain that they did not exist and had to be borrowed from Europeans if necessary. Indeed, it would be surprising if Algonquin or Montagnais hunting and fishing cultures had such concepts, because there is nothing in their culture or their interaction with the environment that would make it necessary for them to measure anything precisely. Yet these Native groups did, of course, have concepts of distance. Most Europeans who travelled with Natives in eastern Canada claimed that they had good notions of distance, based on travel time over familiar routes. Their concept of distance was, in effect, a concept of time in portions of a day or days of travel. What was meaningful in Native life was how long it took to get to a place, not how far away it was in precisely measured terms. Absolute distance covered in a day could vary with the speed at which one travelled, depending on variables such as the size and composition of the travelling group, the weather, and the difficulty of the terrain being traversed. Therefore time meant something, but distance did not.

The first European to record this concept of distance was Champlain, when on 11 June 1603 the Montagnais at Tadoussac told him about the route up the Saguenay to the as yet "undiscovered" Hudson Bay.[6] The entire description is in terms of days of travel time. Champlain, who had to try to convert this information into cartographic form, noted that "each day they can easily make some twelve to fifteen leagues." In 1609, while travelling with a war party over a much easier

6 Document G, "*Des Sauvages*," ch. 4

route than the northern rivers to Hudson Bay, Champlain estimated that "their usual rate" was 25 to 30 leagues per day.[7] In 1657, Father Gabriel Druillettes wrote that in converting days of travel time estimated by Natives on northern rivers, he reckoned "15 leagues a day going down stream … and seven or eight leagues going up."[8] Sometime days were given as *grandes journées* (long days) or *petites journées* (short days). It seems pointless to establish a measure of distance based on some notion of a day's travel, because each day could be unique; nevertheless, such a measure did exist, stemming no doubt from the necessity to produce maps from Native descriptions of areas not yet visited by Europeans. From the late seventeenth century on, a measure called the *journée de voyage* or *journée de sauvage* was cited on map scales.[9] There were usually three *journées* assigned to a degree of latitude, or about 37 km (23 mi.) to a *journée*. Demonstrating the difficulty of such conversions, a map of 1692 places a *journée* at 1.5 to a degree or about 74 km (46 mi.).

In giving the length of Lake Ontario on his map of 1612, based on Algonquin accounts, Champlain made it "15 *Journees des canaux des sauvages*" (fig. 7). Since the actual length of Lake Ontario is about 290 km (183 mi.) from Wolfe Island to Hamilton harbour, this estimate implies that the canoe parties were moving at a speed of 19 km (12 mi.) per day,[10] which is nowhere near 12 to 15 leagues per day, let alone the 25 to 30 leagues per day estimated by Champlain, no matter which league was meant. Demonstrating the difficulty of Native estimates even more clearly are the three Algonquin accounts Champlain received for the length of Lake Ontario at 80, 150, and 300 leagues.[11] Comparing these estimates to the map implies *journées* of 5.3, 10, or 20 leagues per day.

Tunnage

The common measure used to describe the carrying capacity of a ship was in terms of tuns (*tonneaux*) burthen, or tonnage. This was a measure of volume (capacity) not weight (as in ton). A tun was a large wooden barrel that could hold 252 old wine gallons (210 imperial gallons or 954.7 litres). The capacity of one tun was 19.5 cubic metres (404 cubic feet). One tun of water weighs approximately

7 Biggar, *The Works*, 2:105

8 *JR*, 44:239

9 Heidenreich, "Measures of Distance," 121–37

10 This speed seems very slow, especially for a group of warriors. At a good canoe-tripping summer camp in Ontario, a day's journey of 20 to 25 km over rivers, portages, and small lakes is not out of the ordinary.

11 Document G, "*Des Sauvages*," chs. 8 and 9

762 kg (1,674 lb). Since everything was stored in barrels, it was important to know how many barrels of a standard size could fit into the hold of a ship, rather than the weight of the disparate objects. A tun was equivalent to two pipes, four hogsheads, and eight barrels. Like the tun, a pipe is a measure of capacity. It contains about 105 imperial gallons (477 litres). One pipe of water weighs about 837 pounds. On seeing canoes for the first time, Champlain estimated they could hold about a pipe, or 840 pounds.

The *San Julián* was a large ship of 500 (*tonneaux*) tuns burthen. By comparison, in 1603 Champlain sailed to Canada on *La Bonne Renommée*, a ship of 120 tuns burthen. In 1608 he was on the *Don-de-Dieu*, with 150 tuns burthen. In the Arctic, of the three-well known contemporary exploration ships, Henry Hudson's *Discovery* was 55 tuns, Luke Foxe's *Charles* was 80 tuns, and Thomas James's *Henrietta Maria* was 70 tuns.

Coinage

French coinage varied somewhat between the reigns of Henri III, Henri IV, and Louis XIII.[12] During this period the following major coins were circulated: the écu (*escus*), worth 60 *sols* (sous) or 720 *deniers* (pennies or coppers). In 1602 the *écu* was abolished and replaced by the livre, worth 20 *sols*. In 1615 the *écu* returned, worth 75 *sols*. The worth of *deniers* stayed fairly steady at about twelve to one *sol* (sou).

12 Abot de Bazinghen, *Traitté des monnaies*, 1:312–18 (*denier*), 1:390–1 (*écus*), 1:635–45 (*livre*)

APPENDIX 5

Champlain's *Des Sauvages* and Edward Hayes's *Treatise*

Introduction

Hayes's propositions for northeastern North America can be summarized as follows:[1]

- England should "plant" settlements of "Christian people" in the temperate areas of the northeastern parts of America between latitude 40 and 44 degrees.
- These lands have not been claimed by any "Christian prince" but would have if the French had not been preoccupied by their religious wars. The lands are really "the rightful inheritance" of Queen Elizabeth I because they were first discovered by John Cabot and his sons during the reign of Henry VII.
- Settlements should first be made at the mouths of the large rivers that come from the interior. The local Natives ("naturall inhabitants") should be questioned about the interior geography, and it is likely that one will learn as much from them as from "our owne navigations."
- The French, especially Jacques Noël, followed such a course by travelling up the St. Lawrence and across a number of rapids to a "mighty lake," whose entrance has fresh water but is salt water at its western end. This lake is at forty-four degrees latitude and is called *Tadouac* by the Natives ("Salvages"). Similarly, the English should look for mighty rivers that flow from the interior to the coast of Norumbega.
- Among the many resources available through the settlement of the East Coast, Hayes mentioned that in the Bay of Menan[2] (Fundy) there were copper and silver deposits and "also Salt as good as that of Buruage" (Brouage) in France.

1 This summary is based on Hayes's "A Treatise," in Quinn and Quinn, *English New England Voyages*, 168–203.

2 The Bay of Menan (Menon, Menim) is the Bay of Fundy, so called by Bellenger in 1583 (Ganong, *Crucial Maps*, 458–9).

A "mighty lake," fresh at one end and salt at the other

References to a western sea began with Jacques Cartier, who in 1535 was told by the St. Lawrence Iroquoians of a western *mer doulce* (freshwater sea).[3] Similarly, in 1587 his nephew, Jacques Noël, described the existence of the westward passage and of rapids filling the St. Lawrence beyond the Lachine Rapids and entering a huge lake.[4] More recently, in 1602, the Montagnais who were interviewed at the court of Henri IV had repeated a similar geography about the St. Lawrence River beyond the Lachine Rapids.[5] None of these accounts said anything about salt water, but not so the English accounts as rendered by Edward Hayes. Hayes, claiming that he based his information on accounts by Jacques Noël, wrote: "Neither upon the discoveries of Iacques Noel, who having passed beyond the three Saults, where Iacques Cartier left to discover, finding the river of S. Laurence passable on the other side or branch; and afterwards, understood of the inhabitants, that the same river did lead into a mighty lake, which at the entrance was fresh, but beyond, was bitter or salt; the end whereof was unknowen."[6] The last part of this quotation, is startlingly similar to the two sentences in the story collected by Champlain from the young Algonquin: "… they enter a very large lake … [where] the water is drinkable. But he told us that, continuing on some hundred leagues farther [into the lake], the water is still worse.[7] Arriving at the end of the said lake, the water is totally salty; … [and beyond it] … a sea so great that they have not seen the end of it nor heard tell of anyone who may have seen it."[8]

Comparing these two accounts, one might rightfully think Champlain was aware of Hayes's *Treatise*.[9] What Noël actually wrote in the first of two of his surviving documents is: "I myself have knowledge thereof [the course of the St. Lawrence River] as farre as to the *Saults* [Lachine Rapids] where I have bene: The

3 Biggar, *Jacques Cartier*, 202

4 Ibid., 313–14. These and other French accounts were published by Richard Hakluyt in his *Principal Navigations*, 3:201–42. In fact, the English had access to Cartier's first two voyages as early as 1580 through the translation by John Florio, *A Shorte and Briefe Narration*, taken from the Italian version by Ramusio, *Terzo volume delle navigationi …* 1556.

5 Document H, "Of the French Who Have Become Accustomed to Being in Canada," Introduction

6 Quinn and Quinn, *English New England Voyages*, 177

7 Champlain wrote "still worse" without explaining what it was "still worse" than. In the preceding sentence he was writing about "drinkable" water.

8 Document G, *"Des Sauvages,"* ch. 9

9 For an exploration of the relationship between Champlain and Hayes, see Hunter, "Was New France Born in England?" Also Hunter, *God's Mercies*, 197–9. We are indebted to Douglas Hunter for first commenting on these relationships.

height [latitude] of which *Saults* is in 44. degrees."[10] In the second document, Noël commented on a map he was sent, stating that the "Great Lake" on the map is not on the maps of Jacques Cartier and that his (Noël's) knowledge about the map is "according as the Savages have advertised us, which dwell at the sayd *Saults*." He went on to say: "I have bene upon the toppe of a mountaine [Mount Royal], which is at the foot of the *Saults*, where I have seene the said River [St. Lawrence] beyond the sayd saultes, which shewed unto us to be broader [Lac St-Louis] than it was where we passed it. The people of the Countrey advertised us, that there are ten dayes iourney from the Saults unto this *Great Lake*."[11]

In neither Cartier's writings nor Noël's is there any claim to have travelled to the "Great Lake,"[12] nor is there any mention of salt water. Either Hayes fabricated the story or he had additional unrecorded information from Noël. In view of what Hakluyt did in his translation of *Des Sauvages*, the former is more likely. In order to reinforce the English thesis that the Great Lakes are salty, Richard Hakluyt, who did the translation, or Samuel Purchas, who published it, changed the word *salubre* (wholesome, from the Latin *salubris*) in Champlain's text to *salty* and *brackish*.[13] We think that this translation was not an error but was a deliberate falsification, because both Hakluyt and Purchas were fluent in French and Latin.

A river connection from Acadia to the "inland sea"

In one of his rare attempts at theorizing about the geography of New France, Champlain used some of the same arguments as Hayes – that there was probably a river connection between the coast of Florida (Acadia, Norumbega) and the interior salt sea west of the Lachine Rapids: "It would be a great asset if someone could find some passage on the coast of Florida that would come out near the above-said great lake where the water is salty [western Lake Ontario and Lake Erie], both for the navigation of the vessels that would not be subjected to so many perils as they are in *Canadas*, and for the shortening of the route by more than three hundred leagues. And it is most certain that there are rivers on the

10 Biggar, *Jacques Cartier*, 260

11 Ibid., 313–14

12 It is likely that, like Champlain, Noël walked past the Lachine Rapids and saw the widening of the river to Lac Saint-Louis, but not that he managed to get to Lake Ontario. The geopolitical situation between the warring groups to the north and south of the upper St. Lawrence, as well as unfamiliarity with canoes, would have made such a journey impossible.

13 Document I, "The Voyage of Samvel Champlaine," chs. 8 and 9

coast of Florida that have not been discovered yet, and that go inland where the land is very good and fertile, with very good harbours."[14]

Was Champlain aware of the Gosnold voyage, the Hayes treatise, and Edward Wright's map? We think he was.

Prévert's "copper mines" in Hayes's "Bay of Menan"

John Brereton's account of Gosnold's voyage contained a long passage on the coastal Natives and the "very redde and some of paler colour" copper ornaments and implements in their possession.[15] He added that the expedition did not have the time to look for their mines.[16] Hayes took these observations a step further by combining them with information most likely obtained from Étienne Bellenger's voyage of 1583: "I could give large information of the rich copper mines in the East side of the Bay of Menan within 30 or 40. leagues to the Southwest of Cape Breton, whereof I my selfe have seene above an hundred pieces of the copper, and have shewed some part thereof to divers knightes of qualitie, as also of Salt as good as that of Buruage in France, found neere that Bay, and could make proofe of the testimonie of the Salvages touching a Silver mine in another Bay within two or three leagues to the west of the aforesaid Bay of Menan."[17]

The reference to the "Bay of Menan" (Bay of Fundy) and the salt deposits were from Bellenger, as probably were the pieces of copper that Hayes claimed to have seen. Although copper deposits exist in the Bay of Fundy area, it is certain that the copper represented by the artifacts seen among the Natives encountered on the Gosnold and other voyages was not mined locally but was obtained directly, or through Mi'kmaq intermediaries, from Basque traders.[18] It is likely that just as Champlain was ordered in 1603 to discover the details about the westward passage, Prévert was sent out to question the Mi'kmaq about the copper and silver deposits mentioned by Hayes and, if possible, to discover their location. In 1604 Champlain was given the task of completing Prévert's discoveries by locating and sampling the same deposits. Maistre Jacques, the Slovenian miner he had with him at the time, was not impressed by what they found. Although Prévert claimed to have travelled overland with Mi'kmaq guides from Northumberland Strait to the Bay of Fundy, Champlain discovered that he had not gone himself to look for the "mines" but had sent some of his men overland with Mik'maq

14 Document G, "*Des Sauvages*," ch. 10

15 Quinn and Quinn, *English New England Voyages*, 155–7

16 Ibid., 157

17 Ibid., 203, and note 2. Also DCB, 1, "Bellenger, Étienne"; Quinn, "The Voyage of Étienne Bellenger"

18 Bourque and Whitehead, "Trade and Alliances," 131–8

guides, who had found little. According to Champlain, Prévert had lied and had exaggerated his reports about the quality of these deposits.[19]

Preparation for the 1604 expedition of de Mons

It is a certainty that Gravé Du Pont's and Champlain's reports after the 1603 voyage were positive regarding the possibility of settlement in New France. According to Champlain, de Mons was less certain: "[T]he little he had seen had taken away any desire of his to enter the great river St. Lawrence, having on the voyage seen only a forbidding country. He desired to go farther south in order to enjoy a softer and more agreeable climate. And not satisfied with the accounts that had been given to him, he wished to seek for a spot the situation and climate of which he knew only by imagination and reason, which believe that the farther south one goes the warmer it becomes."[20]

It is possible that de Mons was dissatisfied with the reports of Champlain and Gravé, for he was in a position to make up his own mind as a result of his earlier visits to the St. Lawrence. Spelled out in the commission that de Mons received on 31 October 1603 from Lord High Admiral Charles de Montmorency, there was a second important factor swinging the thrust of the next expedition south.[21] After the deciding factors already mentioned by Champlain (climate, goodness of soil, and "good mines"), the commission cited factors that clearly look like a French reaction to some of the sentiments expressed in the Brereton account and the Hayes *Treatise* summarized at the beginning of this appendix, to whit: "And having also received information from various quarters that certain strangers design to go to set up colonies and plantations in and about the said country of La Cadie, should it remain much longer as it has hitherto remained, deserted and abandoned."[22]

Hayes suggested that the English settle between 40 and 44 degrees, while the commission given to de Mons stated 40 to 46 degrees in order to include the southern Gulf of St. Lawrence. Besides commerce and settlement, geopolitical rivalries were becoming more important. De Mons was undoubtedly sent to stake claims and begin a settlement in order to forestall the English. The reason the English were not mentioned by name was probably because France and England were at peace at this time. The case for moving south was therefore ironclad. Acadia was to be settled, and the St. Lawrence valley was to become a source of revenue from the fur trade.

19 Biggar, *The Works*, 1:260–3, 278, 374–5
20 Ibid., 3:320
21 See Grant, *History of New France by Lescarbot*, 2:217–20
22 Ibid., 218

REFERENCES

MANUSCRIPT ARCHIVES

Canada. Library and Archives Canada, Ottawa
France. Bibliothèque nationale de France, y compris l'Arsenal, Paris
– Bibliothèque Sainte-Geneviève, Paris
– Département d'Ille-et-Vilaine, Archives régionales de Bretagne, Rennes
Spain. Archivo Histórico Provincial de Cádiz

PRINTED SOURCES

Abot de Bazinghen, François-André. *Traitté des monnaies, et de la jurisdiction de la cour des monnaies, en forme de dictionnaire …* Paris: Chez Guillyn, 1764

Adams, Arthur T. *The Explorations of Pierre Esprit Radisson: From the Original Manuscript in the Bodleian Library and the British Museum.* Minneapolis: Ross & Haines, 1961

Alonso, Martin. *Enciclopedia del idioma: diccionario histórico y moderno de la lengua española (siglos XII al XX), etimológico, tecnológico, regional e hispanoamericano.* 3 vols. Madrid: Aguilar, SA, de Ediciones Juan Bravo, 1947

Arbour, Roméo. "Raphaël du Petit Val, de Rouen, et l'édition des textes littéraires en France (1587–1613)." *Revue française de l'histoire du livre* 5, no. 9 (1975): 87–141

Articles accordés entre les deputés du roy, & ceux du roy d'Espagne, à Vervins, avec ceux du duc de Saoye, pour negociation du traitté de paix. Grenoble: Guillaume Verdier, 1599

Atkinson, Geoffroy. *La littérature géographique de la Renaissance: répertoire bibliographique.* Paris: Auguste Picard, 1927

Aubin, Nicolas. *Dictionnaire de marine contenant les termes de la navigation …* 3rd edn. Den Haag: Adrien Moetjens, 1742

Augereau, Laurence. "La vie intellectuelle à Tours pendant la Ligue (1589–1594)." Doctoral thesis, Université François-Rabelais, Tours, 2003. [Extracts kindly shared by Laurence Augereau on 18 June 2008]

Axtell, James. *The Invasion Within: The Contest of Cultures in Colonial North America.* New York: Oxford University Press, 1985

Babelon, Jean-Pierre. *Henri IV.* Paris: Librairie Arthème Fayard, 1982

Balsamo, Jean, and Michel Simonin. *Abel L'Angelier & Françoise de Louvain (1574–1620): suivi du catalogue des ouvrages publiés par Abel L'Angelier (1574–1610) et la veuve L'Angelier (1610–1620).* Geneva: Librairie Droz, 2002

Balsamo, Jean, Michel Magnien, and Catherine Magnien-Simonin. *Les Essais: Michel de Montaigne.* Paris: Éditions Gallimard, 2007

Banfield, A.W. Frank. *The Mammals of Canada*. Toronto: University of Toronto Press, 1981

Barbiche, Bernard. "Henri IV and the World Overseas: A Decisive Time in the History of New France." In Raymonde Litalien and Denis Vaugeois, eds., *Champlain: The Birth of French America*, 24–32. Montreal: McGill-Queen's University Press and Éditions du Septentrion, 2004

Barbiche, Bernard, and Monique Chatenet. *L'édition des textes anciens XVI–XVIII siècle*. Paris: Inventaire général des monuments et des richesses artistiques, 1990

Barker, William. "Pandora's Box: To Call a Spade a Spade, and the Problems of Proverb Distortion in Erasmus' *Adages*." Unpublished presentation to the Canadian Society for Renaissance Studies. Ottawa, 1998

Barthélemy, Anatole de, ed. *Choix de documents inédits sur l'histoire de la Ligue en Bretagne*. Nantes: Société des bibliophiles bretons et de l'histoire de Bretagne, 1880

Baudry, Jeanne. *La Fontenelle le ligeur et le brigandage en Basse-Bretagne pendant la Ligue (1574–1602)*. Nantes: L. Durance, 1920

Beauchesne, Th. "Brouage a l'époque de Samuel de Champlain." *Nova Francia* 1 (1925): 57–61

Beaulieu, Alain. "The Birth of the Franco-American Alliance." In Raymonde Litalien and Denis Vaugeois, eds., *Champlain: The Birth of French America*, 153–61. Montreal: McGill-Queen's University Press and Éditions du Septentrion, 2004

Beaulieu, Alain, and Réal Ouellet, eds. *Champlain: Des Sauvages*. Montreal: Éditions TYPO, 1993

Belèze, Guillaume-Louis. *Dictionnaire des noms de baptême*. Paris: L. Hachette, 1863

Benedict, Friedrich, ed. *Graesse, orbis latinus*. 1861. Reprint of 2nd edn., 1909. Berlin, Paris & Milan, 1972

Bersier, Eugène. *Coligny: The Earlier Life of the Great Huguenot*. Trans. Annie Harwood Holmden. London: Hodder and Stoughton, 1884

Berton, Pierre. *My Country: The Remarkable Past*. Toronto: McClelland & Stewart, 1976

Biblia sacra, iuxta vulgatam versionem. 2 vols. Stuttgart: Württembergische Bibelanstalt, 1969

Bideaux, Michel, ed. *Relations: Jacques Cartier*. Montreal: Presses de l'Université de Montréal, 1986

Biggar, Henry Percival. *The Early Trading Companies of New France*. Toronto: University of Toronto Library, 1901

– "The Portrait of Champlain." *Canadian Historical Review* 1 (1920): 379–80

– ed. *The Voyages of Jacques Cartier*. Ottawa: Publications of the Public Archives of Canada, no. 11, 1924

– gen. ed. *The Works of Samuel de Champlain*. 6 volumes. Toronto: Champlain Society, 1922–36

– ed. *A Collection of Documents Relating to Jacques Cartier and the Sieur de Roberval*. Ottawa: Publications of the Public Archives of Canada, no. 14, 1930

Bird, Edward A., ed. *Cyrano de Bergerac: comédie héroique en cinq actes en vers*. Toronto: Methuen, 1968

[Blanchet, Jean]. *Collection de manuscrits contenant letters, mémoires, et autres documents historiques relatifs a la Nouvelle-France.* Vol. 1. Quebec: A. Coté, 1883

Bourde de la Rogerie, Henri. "Hélène Boullé, femme de Samuel de Champlain." *Extrait des Mémoires de la Société archéologique du Département d'Ille-et-Vilaine*, 63:1–15. Rennes, 1937

Bourne, Edward Gaylord, ed. *The Voyages and Explorations of Samuel de Champlain (1604–1616) Narrated by Himself ... Together with the Voyage of 1603, reprinted from Purchas His Pilgrimes.* Trans. Annie N. Bourne. 2 vols. New York: A.S. Barnes, 1906

– *Algonquians, Hurons, and Iroquois: Champlain Explores America, 1603–1616, Being ... The Voyage of 1603, reprinted from Purchas His Pilgrimes.* Dartmounth, NS: Brook House Press, 2000

Bourque, Bruce J., and Ruth H. Whitehead. "Trade and Alliances in the Contact Period." In Emerson W. Baker et al., *American Beginnings: Exploration, Culture, and Cartography in the Land of Norumbega*, 131–47. Lincoln: University of Nebraska Press, 1994

Bowers, Fredson. *Principles of Bibliographic Description.* 5th rpt, Winchester, UK: St. Paul's Bibliographies, [1949], 1998

Bradley, James W. *Evolution of the Onondaga Iroquois: Accommodating Change, 1500–1655.* Rochester: Syracuse University Press, 1987

Brereton, John. *A Briefe and True Relation of the Discoverie of the North Part of Virginia* ... Londini: Geor. Bishop, 1602

Broekema, C. *Maps of the Canary Islands Published before 1850: A Checklist.* London: Map Collectors' Circle, 1971

Bruchési, Jean. "Champlain a-t-il menti?" *Les Cahiers des Dix* 15 (1950): 39–53

Brunelle, Gayle K. *The New World Merchants: Merchants of Rouen, 1559–1630.* Kirksville, MO: Sixteenth Century Publishers, 1991

Buisseret, David. *Henry IV.* London: George Allen & Unwin, 1984

– "Monarchs, Ministers, and Maps in France before the Accession of Louis XIV." In David Buisseret, ed., *Monarchs, Ministers, and Maps: The Emergence of Cartography as a Tool of Government in Early Modern Europe*, 99–123. Chicago: University of Chicago Press, 1992

– *Ingénieurs et fortifications avant Vauban: l'organisation d'un service royal aux XVIe–XVIIe siècles.* Paris: Ministère de la jeunesse, de l'éducation nationale et de la recherche, Comité des travaux historiques et scientifiques, 2002

– "French Cartography: The *ingénieurs du roi*, 1500–1650." In David Woodward, ed., *The History of Cartography: Cartography of the Renaissance.* Vol. 3, pt. 2: *1504–21.* Chicago: University of Chicago Press, 2007

Bulletin du Comité de la langue, de l'histoire des arts de la France. Vols. 2 and 3. Paris: Imprimerie Impériale, 1856, 1857

Bulletin et mémoires de la Société archéologie du Département d'Ille-et-Vilaine. Vol. 42, pt. 2. Rennes: Journal de Rennes, 1913

Campeau, Lucien, ed. *Monumenta novae franciae.* Vol 1: *La première mission d'Acadie (1602–1615).* Quebec: Monumenta historica Societatis Iesu and Presses de l'Université Laval, 1967

– ed. *Monumenta novae franciae.* Vol. 3: *Fondation de la mission huronne (1635–1637).* Quebec: Monumenta historica Societatis Iesu and Presses de l'Université Laval, 1987

Carné, Gaston de, ed. *Documents sur la ligue de Bretagne. Correspondance du duc de Mercoeur et des ligueurs bretons avec l'Espagne: extraite des Archives nationales, et publiée avec une preface historique et des notes par Gaston de Carné.* Vols. 11 and 12. Nantes: Société des bibliofiles bretons et de l'histoire de Bretagne, 1899

Carpin, Gervais. *Histoire d'un mot: l'ethnonyme* Canadien *de 1535–1691.* Sillery, QC: Septentrion, 1995

Cartier, Jacques. *Discours du voyage fait par le capitaine Jaques Cartier aux Terres-neufves de Canadas, Norembergue, Hochelage, Labrador, & pays adjacens, dite Nouvelle France* [*etc.*]. Rouen: Raphaël du Petit Val, 1598

Catach, Nina. *Histoire de l'orthographe française.* Paris: Honoré Champion, 2001

Cathelineau, M. Emmanuel de. "La minute notariée du contrat de marriage de Champlain." *Nova Francia* 5 (1930): 142–55

Cayet, Pierre-Victor, comp. *Chronologie septenaire de l'histoire de la paix entre les roys de France et d'Espagne ... divisee en sept livres,* 7:415–24. Paris: Jean Richer, 1605

Champlain, Samuel de. *Des Savvages, ov, Voyage de Samvel Champlain, de Brovage, fait en la France nouuelle, l'an mil six cens trois: contenant les mœrs, façon de viure, mariages, guerres, & habitations des Sauuages de Canadas.* Paris: Claude de Monstr'œil, [1603]

– *Des Savvages, ov, Voyage de Samvel Champlain, de Brovage, faict en la France nouuelle, l'an mil six cens trois: contenant les mœrs, façon de viure, mariages, guerres, & habitations des Sauuages de Canadas.* Paris: Claude de Monstr'œil, 1604

– *Les Voyages dv Sievr de Champlain Xaintongeois, capitaine ordinaire pour le Roy, en la marine ...* Paris: Jean Berjon, 1613. [Copy examined at Royal Ontario Museum, Rare Books, FC 332 C43 1613 Canadiana]

– *Voyages et Descovvertvres faites en la Novvelle France, depuis l'année 1615 iusques à la fin de l'année 1618.* Paris: Claude Collet, 1619. Reprinted with changes by Collet, 1620, 1627

– *Les Voyages de la Novvelle France Occidentale, dicte Canada ...* Paris: Louis Sevestre; Paris: Claude Collet; Paris: Pierre Le-Mur, 1632. Reprinted with some changes by Le-Mur and Collet, 1640

– *Voyages du Sieur de Champlain, ou Journal ès découvertes de la Nouvelle France ...* 2 vols. Paris: Imprimé aux frais du gouvernement pour procurer du travail aux ouvriers typographes, 1830

Charavay, Étienne. *Documents inédits sur Samuel de Champlain ...* Paris: Librairie J. Charavay Ainé, 1875

Charlevoix, Pierre-François-Xavier de. *Histoire et description generale de la Nouvelle France ...* 6 vols. Paris: Rolin fils, 1744

Chateaubriand, François-Rene. *Œuvres complètes de Chateaubriand: mémoires d'outre-tombe.* Vol. 1. Ed. Charles-Augustin Sainte-Beuve, and Edmond Biré. Paris: Garnier Frères, 1904. Reprint, Nendeln: Kraus, 1975

[CHS] Canadian Hydrographic Service. *St. Lawrence River Pilot from Cap Des Rosiers (South Shore) and Rivière St. Jean (North Shore) to Kingston.* Ottawa: Canadian Hydrographic Service, Marine Sciences Branch, 1966

Clapin, Sylva. *Dictionnaire canadien-française*. 1894. Reprint of original edition, Quebec: Presses de l'Université Laval, 1974

Codignola, Luca. "Le prétendu voyage de Samuel de Champlain aux Indies occidentales, 1599–1601." In Madelaine Frédéric and Serge Jasumain, eds., *Actes du séminaire de Bruxelles. La relation de voyage: un document historique et littéraire*, 61–80. Université libre de Bruxelles, Centre d'Études canadiennes, 1999

Cook, Ramsay, ed. *The Voyages of Jacques Cartier*. Toronto: University of Toronto Press, 1993

Crété, Liliane. *Coligny*. Paris: Librairie Arthème Fayard, 1985

Cuignet, Jean-Claude. *L'itinéraire d'Henri IV: les 20,597 jours de sa vie*. Bizanos: Éditions Héraclès, 1997

Cuoq, Jean André. *Lexique de la langue algonquine*. Montreal: J. Chapleau & fils, 1886

D'Aubigné, Théodore Agrippa. "Les avantures du baron Fæneste." *Œuvres complètes de Théodore Agrippa d'Aubigné publiées pour la première fois d'après les manuscrits originaux* … Vol. 2 of 6 vols. Ed. Eugène Réaume and F. de Caussade. Geneva: Slatkine Reprints, 1967

[DCB] *Dictionary of Canadian Biography*. Vol. 1: *1000–1700*, gen. ed. George W. Brown. Toronto: University of Toronto Press, 1966

Delafosse, Marcel. "Séjour de Champlain à Brouage en 1630." *Revue d'histoire de l'Amérique française* 9 (1956): 571–8

– "L'oncle de Champlain." *Revue d'histoire de l'Amérique française* 12 (1958–59): 208–16

Delayant, Leopold. *Notice sur Champlain: né à Brouage, 1567, mort à Québec, le 25 décembre 1635*. Niort: L. Clouzot, 1867

Deschamps, Hubert. *Les voyages de Samuel Champlain, saintongais, père du Canada*. Paris: Presses Universitaires de France, 1951

Deslandres, Dominique. "Samuel de Champlain and Religion." In Raymonde Litalien and Denis Vaugeois, eds., *Champlain: The Birth of French America*, 191–204. Montreal: McGill-Queen's University Press and Éditions du Septentrion, 2004

[DHLF] *Dictionnaire historique de la langue française*. 3 vols. Alain Rey et al. Paris: Le Robert, 2000

Dickason, Olive Patricia. *Canada's First Nations: A History of Founding Peoples from Earliest Times*. Toronto: McClelland & Stewart, 1992

Dionne, Narcisse E. *Samuel Champlain: foundateur de Québec et père de la Nouvelle-France*. Vol. 1. Quebec: A. Coté, 1891

– *Champlain*. Toronto: Morang, 1906

Dix, Edwin A. *Champlain: The Founder of New France*. New York: D. Appleton, 1903

Doursther, Horace. *Dictionnaire universel des poids et mesures anciens et modernes, contenant des tables des monnaies de tous pays*. 1840. Reprint, Amsterdam: Meridian, 1965

Doussinet, Raymond. *Grammaire saintongeaise: étude des structures d'un parler régional*. La Rochelle: Éditions Rupella, 1971

Dubé, Jean-Claude. *The Chevalier de Montmagny, 1601–1657: First Governor of New France*. Trans. Elizabeth Rapley. Ottawa: University of Ottawa Press, 2005

Du Fouilloux, Jacques. *La vénerie de Jacques Fouilloux, seigneur dudit lieu, gentilhomme du pays de Gastine en Poictou* … Paris: Vve Abel L'Angelier, 1614

Éveillé, M.A. *Glossaire saintongeais*. Paris: H. Champion, 1887

Fiquet, Nathalie. "Brouage in the Time of Champlain: A New Town Open to the World." In Raymonde Litalien and Denis Vaugeois, eds., *Champlain: The Birth of French America*, 33–41. Montreal: McGill-Queen's University Press and Éditions du Septentrion, 2004

Fitzgerald, William R., Laurier Turgeon, Ruth H. Whitehead, and James W. Bradley. "Late Sixteenth-Century Basque Banded Copper Kettles." *Historical Archaeology* 27 (1993): 44–57

Flores, Robert Morales, ed. *Cervantes, Don Quixote de la Mancha: An Old-Spelling Control Edition Based on the First Editions of Parts I and II*. Vol. 1. Vancouver: University of British Columbia Press, 1988

Florio, John, trans. *A Shorte and briefe narration of the two Navigations and Discoveries to the Northweast partes called Newe France* ... London: H. Bynneman, 1580

Furetière, Antoine, comp. *Dictionnaire Universel, contenant generalement tout les Mots François tant vieux que modernes, & les Termes des toutes les Sciences et Des Arts* ... Rotterdam: Arnaut and Reinier Leers, 1690

Gagnon, François-Marc. *Premier peintres de la Nouvelle-France*. Quebec: Ministère culturelles du Québec, 1976, 2:9–55

– "Is the *Brief Discours* by Champlain?" In Raymonde Litalien and Denis Vaugeois, eds., *Champlain: The Birth of French America*, 83–92. Montreal: McGill-Queen's University Press and Éditions du Septentrion, 2004

– "Champlain: Painter?" In Raymonde Litalien and Denis Vaugeois, eds., *Champlain: The Birth of French America*, 302–11. Montreal: McGill-Queen's University Press and Éditions du Septentrion, 2004

Gagnon, Philéas. "Notes bibliographiques sur les écrits de Champlain manuscripts et imprimés." *Bulletin de la Société de géographie de Québec* 3, no. 2 (1908)

Ganeau, Étienne. *Dictionnaire universel françois et latin, contenant* ... Paris: Julien-Michel Gandouin, 1732

Ganong, William Francis, trans. and ed. 1908. *Denys: The Description and Natural History of the Coasts of North America (Acadia)*. Toronto: Champlain Society. Original edn., Paris: Loüis Billaine, 1672

– "The Identity of the Animals and Plants Mentioned by the Early Voyagers to Eastern Canada and Newfoundland." *Transactions of the Royal Society of Canada*, session 2 (1909): 197–242

– *Crucial Maps in the Early Cartography and Place-Nomenclature of the Atlantic Coast of Canada*. Toronto: University of Toronto Press, 1964. Originally printed as a series of papers in the *Transactions of the Royal Society of Canada*, 3rd series, 1929 to 1937

Garcias Rivas, Manuel. "En el IV centenario del fallecimiento de Pedro Zubiaur, un marino vasco del siglo XVI." *Itsas memoria: revista de estudios maritimos del pais Vasco*, 5. Donostia–San Sebastiàn: Untzi Museoa–Museo Naval, 2006

Gaskell, Philip. *A New Introduction to Bibliography*. Rev. edn. New Castle, DE: Oak Knoll Press, 1995

Gautier, Michel. *Grammaire du Poitevin-Saintongeais: parlers de Vendée, Deux-Sèvres, Vienne et Charente, Charente-Maritime.* Mougon: Geste éditions, 1993

Gehring, Charles T., and William A. Starna. *A Journey into Mohawk Country, 1634–1635.* Syracuse: Syracuse University Press, 1988

Giraudet, Eugène. *Une association d'imprimeurs et de libraires réfugiés à Tours au XVIe siècle: Jamet Mettayer, Marc Orry, Claude de Monstr'œil, Jean Richer, Matthieu Guillemot, Sébastien du Molin, Georges de Robet, Abel Langellier.* Tours: Rouillé-Ladevèze, 1877

Giraudo, Laura. "The Manuscripts of the *Brief Discours.*" In Raymonde Litalien and Denis Vaugeois, eds., *Champlain: The Birth of French America*, 63–82. Montreal: McGill-Queen's University Press and Éditions du Septentrion, 2004

– "Research Report: A Mission to Spain." In Raymonde Litalien and Denis Vaugeois, eds., *Champlain: The Birth of French America*, 93–7. Montreal: McGill-Queen's University Press and Éditions du Septentrion, 2004

Gleason, Henry A. *The New Britton and Brown Illustrated Flora of the Northeastern United States and Adjacent Canada.* 2nd edn. 3 vols. Lancaster, PA: New York Botanical Gardens, 1958

Glénisson, Jean. *La France d'Amérique: voyages de Samuel Champlain, 1604–1629.* Paris: Nationale éditions, 1994

– "Champlain's Voyage Accounts." In Raymonde Litalien and Denis Vaugeois, eds., *Champlain: The Birth of French America*, 279–83. Montreal: McGill-Queen's University Press and Éditions du Septentrion, 2004

Godefroy, Frédéric. *Dictionnaire de l'ancienne langue française et de tous ses dialectes du IXe au XVe siècle.* 10 vols. Paris: Vieweg, 1881–1902

Gooding, S. James. *The Canadian Gunsmiths, 1608 to 1900.* West Hill, ON: Museum Restoration Service, 1962

Gougenheim, Georges. *Grammaire de la langue française du seizième siècle.* Lyon: Edition IAC for the CNRS, 1951

Grant, George M. *Picturesque Canada: The Country as It Was and Is.* 2 vols. New York: James Clarke, 1882

Grant, William L., trans. and ed. Intro. by Henry P. Biggar. *The History of New France by Marc Lescarbot.* 3 vols. Toronto: Champlain Society, 1907, 1911, 1914. Trans. of original 3rd edn., Paris: Adrian Perier, 1618

Great Britain. Hydrographic Department. *The St. Lawrence Pilot: Comprising Sailing Directions for the Gulf and River St. Lawrence, Including the Western Side of Cabot Strait, the Strait of Belle Isle, Chedabucto Bay, and the Gut of Canso.* Originally compiled by H.W. Bayfield. 7th edn. London: Eyre and Spottiswoode, 1906

Greetham, David C. *Textual Scholarship: An Introduction.* New York & London: Garland, 1994

Grossman, Edith, ed. *Miguel de Cervantes: Don Quixote.* New York: HarperCollins, 2003

Guyot, Joseph-Nicolas, ed. *Traité des droits, functions, franchises, exemptions, prerogatives et privileges annexes en France à chaqque dignité …* Vol. 1. Paris: Visse, 1786

Hakluyt, Richard. *The Principal Navigations, Voyages, Traffiques and Discoveries of the English Nation ...* Vol. 3: *The Third and Last Volume of the Voyages, Navigations, Traffiques ...* London: George Bishop, Ralfe Newberie, and Robert Barker, 1600
– *The Principal Navigations, Voyages, Traffiques and Discoveries of the English Nation.* 12 vols. 1589. Revised edn. New York: AMS Press, 1989
Harbage, Alfred, ed. *William Shakespeare: The Complete Works.* New York: Viking Press, 1969
Harris, R. Cole, ed., and Geoffrey J. Matthews, cart. *Historical Atlas of Canada.* Vol. 1: *From the Beginning to 1800.* Toronto: University of Toronto Press, 1987
Heidenreich, Conrad E. "Measures of Distance Employed on 17th and early 18th Century Maps of Canada." *Canadian Cartographer* 12 (1975): 121–37
– *Explorations and Mapping of Samuel de Champlain, 1603–1632.* Monograph no. 17. Cartographica. Toronto: University of Toronto Press, 1976
– "History of the St. Lawrence–Great Lakes Area to A.D. 1650." In Chris J. Ellis and Neal Ferris, eds., *The Archaeology of Southern Ontario to A.D. 1650.* Occasional Publications of the London Chapter, OAS, no. 5, 475–92. London, ON: OAS, 1990
– "An Analysis of the 17th Century Map 'Novvelle France.'" *Journal of the Canadian Institute of Surveying and Mapping. Cartography in Canada, 1987–1991.* Spring 1991, 33–63
– *Champlain and the Champlain Society.* Occasional Paper no. 3. Toronto: Champlain Society, 2006
– "The Mapping of Samuel de Champlain, 1603–1635." In David Woodward, ed., *The History of Cartography: Cartography of the Renaissance*, vol. 3, pt. 2, 1538–49. Chicago: University of Chicago Press, 2007
– "The Skirmish with the Mohawk on Lake Champlain: Was Champlain a 'Trigger-Happy Thug' or 'Just Following Orders"? *de Halve Maen* 82, no. 1 (2009): 11–20
Heidenreich, Conrad E., and Edward H. Dahl. "The Two States of Champlain's Carte Geographique." *Canadian Cartographer* 16 (1979): 1–16
– "Samuel de Champlain's Cartography, 1603–32." In Raymonde Litalien and Denis Vaugeois, eds., *Champlain: The Birth of French America*, 312–32. Montreal: McGill-Queen's University Press and Éditions du Septentrion, 2004
Helm, June, ed. *Handbook of North American Indians.* Vol. 6: *Subarctic.* Washington: Smithsonian Institution, 1981
Henri IV, Roi de France. *Recueil des letters missives de Henri IV publié par M. Berger de Xivrey [Jules].* Vol. 6. Paris: Imprimerie impériale, 1853
Hoffman, Bernard G. *Cabot to Cartier: Sources for a Historical Ethnography of Northeastern North America, 1497–1550.* Toronto: University of Toronto Press, 1961
Holtz, Grégoire. "Pierre Bergeron et l'écriture du voyage à la fin de la Renaissance (Les récits de Jean Mocquet, François Pyrard de Laval et Vincent Le Blanc)." Doctoral thesis, Université de Paris, 2006
Hooker, Brian, trans. *Cyrano de Bergerac by Edmond Rostand: An Heroic Comedy.* 1923. Revd. edn. New York: Bantam Books, 1959
Huchon, Mireille. *Le français au temps de Jacques Cartier.* Presentation by Claude de la Charité. Rimouski: Tangence, 2006

[HU] Huguet, Edmond. *Dictionnaire de la langue française du seizième siècle.* 7 vols. Paris: Librairie ancienne Honoré Champion, 1925–67

Hunter, Douglas. "Was New France Born in England?" *Beaver*, June 2006, 39–44

– *God's Mercies: Rivalry, Betrayal, and the Dream of Discovery.* Scarborough: Doubleday, 2007

Imbs, Paul, ed. *Trésor de la langue française: dictionnaire de la langue du XIXe siècle et du XXe siècle (1789–1960).* 16 vols. Paris: Gallimard, 1971–94

Jaenen, Cornelius J. *Friend and Foe: Aspects of French-Amerindian Cultural Contact in the Sixteenth and Seventeenth Centuries.* Toronto: McClelland & Stewart, 1973

Jamet, Denis. "Relation du Père Denis Jamet, Récollet de Québec, au Cardinal de Joyeuse." In Robert Le Blant and René Baudry, eds., *Nouveaux documents sur Champlain et son époque*, vol. 1: *1560–1622*, no. 15, 349–54. Ottawa: Public Archives of Canada, 1967

Jamieson, James Bruce. "The Archaeology of the St. Lawrence Iroquoians." In Chris J. Ellis and Neal Ferris, eds., *The Archaeology of Southern Ontario to A.D. 1650.* Occasional Publications of the London Chapter, OAS, no. 5, 385–404. London, ON: OAS, 1990

Jarman, Beatriz G., R. Russell, C.S. Carvajal, and J. Horwood, eds. *Oxford Spanish Dictionary.* 3rd edn. New York: Oxford University Press, 2003

Jefferys, Charles W. *Dramatic Episodes in Canada's History.* Toronto: Hunter-Rose, 1930

– *The Picture Gallery of Canadian History.* 3 vols. Toronto: Ryerson Press, 1942

Jennings, Francis. *The Invasion of America: Indians, Colonialism, and the Cant of Conquest.* New York: W.W. Norton, 1976

Jonain, P. *Dictionnaire du patois saintongeais.* Paris: Royan, 1869

Jouanna, Arlette, et al. *Histoire et dictionnaire des guerres de religion.* Paris: Robert Laffont, 1998

[*JR*] *The Jesuit Relations and Allied Documents.* Gen. ed. Reuben Gold Thwaites. 73 vols. Cleveland: Burrows Brothers, 1896–1901

Julien-Labruyere, François. *Dictionnaire biographique des Charentais.* Paris: Le Croit Vif, 2005

Keiser, Rut, ed. *Thomas Platter D.J.: Beschreibung der Reisen durch Frankreich, Spanien, England und die Niederlande, 1595–1600.* Basel: Schwabe, 1968

Kemp, Peter, ed. *The Oxford Companion to Ships and the Sea.* London: Oxford University Press, 1976

Kenyon, Ian. "Sagard's 'Rassade Rouge' of 1624." *KEWA* (Newsletter of the London Chapter, Ontario Archaeological Society) 4 (1984): 2–15

Kenyon, Walter A. *The Grimsby Site: A Historic Neutral Cemetery.* Publications in Archaeology. Toronto: Royal Ontario Museum, 1982

Klump, André, ed. *César Oudin: Grammaire et observations de la langue espagnolle recueillies & mises en françois* [Paris, 1604]. Hildesheim: Georg Olms Verlag, 2004

Knecht, Robert Jean. *Renaissance Warrior and Patron: The Reign of Francis I.* Cambridge: Cambridge University Press, 1994

– *The Rise and Fall of Renaissance France, 1483–1610.* 2nd edn. Oxford: Blackwell, 2001

Koeman, Cornelis. "The Theatrum Universae Galliae, 1631." *Imago Mundi* 17 (1963): 62–72

Kuhn, Robert D., Robert Funk, and James F. Pendergast. "The Evidence for a Saint Lawrence Iroquoian Presence on Sixteenth Century Mohawk Sites." *Man in the Northeast* 45 (1993): 77–86

Labarre, Albert. 1976. *Répertoire bibliographique des livres imprimés en France au seizième siècle.* Livraison, 23; Blois, 132; Saint-Denis, 153; Tours, Bibliotheca bibliographica aureliana, 63. Baden-Baden: Librairie Valentin Koerner, 1993

La Force, Jacques-Nompar de Caumont. *Mémoires authentiques de Jacques Nompar de Caumont, duc de la Force, maréchal de France ...* Paris: Charpentier, 1843

La Lande de Calan, Charles de. *Documents inédits relatifs aux États de Bretagne de 1491 à 1589.* Vol. 2. Rennes: Archives de Bretagne, Societé des bibliophiles bretons, [1843], 1909

La Roque de Roquebrune, R. "Particules, surnoms, titres et armoiries." *Nova Francia* 5 (1930): 9–15

Laverdière, Charles-Honoré, ed. *Œuvres de Champlain publiés sous le patronage de l'Université Laval.* 2nd edn. 6 vols. Quebec: Geo.-E. Desbarats, 1870

Lazarad, Madeleine. *Agrippa d'Aubigné.* Paris: Arthème Fayard, 1998

Leacock, Eleanor. "Seventeenth-Century Montagnais Social Relations and Values." In William C. Sturtevant and June Helm, eds., *Handbook of North American Indians*, Vol. 6: *Subarctic*, 190–5. Washington: Smithsonian Institution, 1981

Le Blant, Robert, and René Baudry, eds. *Nouveaux documents sur Champlain et son époque.* Vol. 1: *1560–1622*, no. 15. Ottawa: Public Archives of Canada, 1967

Lemoine, Georges. *Dictionnaire français-montagnais ...* Boston: W.B. Cabot and P. Cabot, 1901

Lepage, Yvan G. *Guide de l'édition de textes en ancien français.* Paris: Champion, 2001

Le Roux de Lincy. *Le livre des proverbes français*, 2nd edn. 2 vols. Paris: Adolphe Delahays, 1859

Lescarbot, Marc. *Histoire de la Novvelle France ...* Paris: Iean Milot, 1609

– *Les Mvses De La Novvelle France. A Monseignevr Le Chancellier.* Paris: J. Millot, 1612

Lestringant, Frank. "Nouvelle-France et fiction cosmographique dans l'œuvre d'André Thevet." *Étude Littéraires* 10 (1977): 145–73

– ed. *Le Brésil d'André Thevet: Les singularités de la France antarctique.* 1557. Paris: Éditions Chandeigne, 1997

Leymarie, A.-Léo. "Inédit sur le foundateur de Québec." *Nova Francia* 1 (1925): 80–5

Liebel, Jean. "On a vieilli Champlain." *Revue d'histoire de l'Amérique française* 32 (1978): 229–37

Litalien, Raymonde. "Historiography of Samuel Champlain." In Raymonde Litalien and Denis Vaugeois, eds., *Champlain: The Birth of French America*, 11–16. Montreal: McGill-Queen's University Press and Éditions du Septentrion, 2004

Litalien, Raymonde, and Denis Vaugeois, eds. *Champlain: The Birth of French America.* Trans. Käthe Roth. Montreal: McGill-Queen's University Press and Éditions du Septentrion, 2004

Littré, Émile. *Dictionnaire de la langue française.* 7 vols. Levallois: Gallimard-Hachette, 1966–67

Lorenzo, Miguel Angel Puche. *El español del siglo XVI en textos notariales.* Murcia: Universidad de Murcia, 2003

McClelland, John. *Body and Mind: Sport in Europe from the Roman Empire to the Renaissance.* London & New York: Routledge, 2007

McKenzie, Donald F. *Making Meaning: "Printers of the Mind" and Other Essays.* Studies in Print Culture and the History of the Book. Amherst: University of Massachusetts Press, 2002

Maclean's. "Samuel de Champlain," July 1998, 28

Macklem, Michael, trans. Intro. by Edward Miles. *Voyages to New France: Being a Narrative of the Many Remarkable Things ... 1599–1601, with an Account ... in the year 1603.* [Ottawa]: Oberon, 1973

Mandea, Mioara, and Pierre-Noël Mayaud. "Guillame Le Nautonier: un précurseur dans l'histoire du géomagnétisme." *Revue d'histoire des sciences* 57 (2004): 161–73

Martín, D. Manuel Ravina. *Pasaron por Cádiz: personas y cosas. Catálogo de la exposición.* Cádiz: Archivo Histórico Provinicial de Cádiz, 2005

Mathieu, Jacques. "Gouverneur." In *The Canadian Encylopedia*, gen. ed. James H. Marsh, 915. Edmonton: Hurtig, 1988

Mellottée, Paul. *Histoire économique de l'imprimerie.* Vol. 1: *L'imprimerie sous l'ancien Régime, 1439–1789.* Paris: Hachette, 1905

Michaud, Louis-Gabriel, gen. ed. *Biographie universelle ancienne et moderne.* 45 vols. New edn. Paris: A. Thoisnier Desplaces, 1844–18?

Michelant, Henri-Victor, and Alfred Ramé, eds. *Relation originale du voyage de Jacques Cartier au Canada en 1534: Documents inédits sur Jacques Cartier et Canada.* New series. Paris: Tross, 1867

Monet, Jacques, SJ. "Governor General." In *The Canadian Encyclopedia*, gen. ed., James H. Marsh, 918–19. Edmonton: Hurtig, 1988

Montaigne, Michel de. *Essais.* 2nd edn. Bordeaux 1582. Facsimile edn. with intro by Philippe Desan. Paris: Société des textes français moderne, 2005

Morison, Samuel Eliot. *Samuel de Champlain: Father of New France.* Boston: Little, Brown, 1972

Morissonneau, Christian. *Le langue géographique de Cartier et de Champlain: choronymie, vocabulaire, et perception.* Quebec: Presses de l'Université Laval, 1978

Muller, Jean. *Répertoire bibliographique des livres imprimés en France au seizième siècle.* Baden-Baden: Verlag Librairie Heitz Gmbh., 1970

Muller, John. *A Treatise Containing the Elemetary Part of Fortification, Regular and Irregular.* London: J. Nourse, 1746

Musset, Georges. *Glossaire des patois et des parlers de l'Aunis & de la Saintonge.* 5 vols. La Rochelle: Masson & Renaud, 1948

Nautonier, Guillaume Le. *Mecometrie de l'eymant, c'est a dire la maniere de mesurer les longitudes par le moyen de l'eymant.* 1st edn. privately printed by author at Vénès, 1603. 2nd edn. Toulouse and Vénès: Raimond Colomies & Antoine de Courtneful, 1604

New Catholic Encyclopedia. 2nd edn. Detroit: Catholic University of America, 2003

Nicot, Jean. *Thresor de la Langue francoyse, tant ancienne que moderne.* Paris: David Douceur, 1606

Nouveau dictionnaire biographique universel et historique des personages célèbres ... Paris: Mme Huzard, 1836

Pallier, Denis. *Recherches sur l'imprimerie à Paris pendant la ligue (1585–1594).* Histoire et civilisation du livre, 9. Geneva: Droz, 1975

Paltsis, Victor Hugo. "A Critical Examination of Champlain's Portraits." *Acadiensis* (special Champlain number), 4, nos. 3–4 (1904): 306–11

Pennington, Loren E., ed. *The Purchas Handbook.* 2 vols. London: Hakluyt Society, 1997

Pioffet, Marie-Christine, ed. *Marc Lescarbot: voyages en Acadie (1604–1607) suivis de la description des mœurs souriquoises comparées à celles d'autres peuples.* Quebec: Presses de l'Université Laval, 2007

Pivetea, Vianney. *Dictionnaire poitevin-saintongeais: parlers de Vendée, Deux-Sèvres, Vienne, Charente, Charente-Maritime, nord Gironde, et Loire-Atlantique.* La Crèche, Deux-Sèvres: Geste éditions, 1996

Pocquet, Barthélemy. *Histoire de Bretagne.* Vol. 5. Rennes: Plihon et Harvié, 1905

Poirier, Pascal. *Le glossaire acadien: édition critique établie par Pierre M. Gerin.* Moncton: Édition d'Acadie, 1994

Pope, Mildred K. *From Latin to Modern French with Especial Consideration of Anglo-Norman.* Manchester: Manchester University Press, 1934

Potier, Pierre. *Huron Manuscripts from Rev. Pierre Potier's Collection.* Facsimile edn. of manuscripts written in 1745 and 1751. Fifteenth report of the Bureau of Archives for the Province of Ontario, 1918–19. Toronto: Clarkson W. James, 1920

Purchas, Samuel, comp. and ed. *Hakluytus Posthumus or Purchas His Pilgrimes. Contayning a History of the World, in Sea Voyages & Lande Travells, by Englishmen & others.* 4 vols. London: Printed by William Stansby for Henrie Featherstone, 1625. Reprinted in 20 vols. Glasgow: MacLehose, 1905–06

Quinn, David B., ed. *The Voyages and Colonizing Enterprises of Sir Humphrey Gilbert.* 2 vols. London: Hakluyt Society, 1939

– "The Voyage of Étienne Bellenger to the Maritimes in 1583: A New Document." *Canadian Historical Review* 43 (1962): 328–43

– ed. *The Hakluyt Handbook.* 2 vols. London: Hakluyt Society, 1974

Quinn, David B., and Alison M. Quinn, eds. *The English New England Voyages, 1602–1608.* London: Hakluyt Society, 1983

Rainguet, Pierre Damien. "Champlain." *Biographie saintongeaise; ou, Dictionnaire historique de tous les personagesqui se sont illustre parleurs ecrits ou ...* Saintes: Niox, 1851

Ramé, Alfred. "Documents inédits sur le Canada." In Henri-Victor Michelant and Alfred Ramé, eds. *Voyage de Jacques Cartier au Canada en 1534.* Paris: Tross, 1865

– "Documents inédits sur le Canada." In Henri-Victor Michelant and Alfred Ramé, eds. *Relation originale du voyage de Jacques Cartier au Canada en 1534.* Paris: Tross, 1867

Ramusio, Giovanni Battista. *Terza volume, delle navigatione et viaggi nel quale si contengono le navigationi al Mundo Nuovo ...* Venis: Giunti, 1556

Rand, Silas Tertius. *Legends of the Micmac*. 2 vols. New York: Longmans, Green, 1894
Rayburn, Alan. *Naming Canada*. Toronto: University of Toronto Press, 1994
Real Academia Española. *Diccionario de autoridades*. 3 vols. Madrid: Gredos, 1561–1963
Renouard, Philippe. *Les marques typographiques parisiennes des* XVe *et* XVIe *siècles*. Revue des bibliothèques, suppl. 13. Paris: Librairie ancienne Honoré Champion, 1926
– *Répertoire des imprimeurs parisiens, libraires, fondeurs de caractères et correcteurs d'imprimerie depuis l'introduction de l'Imprimerie à Paris (1470) jusqu'à la fin du seizième siècle*. Paris: M.J. Minard, 1965
RHAF, "Documents inédits." *Revue d'histoire de l'Amérique française* 9 (1955–56): 594–7
Richer, Jean. *Le Mercure françois ov svitte de l'histoire de la paix*. Paris: Jean Richer, 1610. Toronto: University of Toronto electronic file 546160, mercurefrancois.ehess.fr/picture
Roberts, Kenneth G., and Philip Shackelton. *The Canoe: A History of the Craft from Panama to the Arctic*. Toronto: Macmillan, 1983
Rogers, Edward S., and Eleanor Leacock. "Montagnais-Naskapi." In William C. Sturtevant and June Helm, eds., *Handbook of North American Indians*. Vol. 6: *Subarctic*, 169–89. Washington: Smithsonian Institution, 1981
Roques, Mario. "L'établissement de règles pratiques pour l'édition des anciens textes français et provençaux." *Romania* 52 (1926)
Runnalls, Graham A. *Études sur les mystères: un recueil de 22 études sur les mystères français, suivi d'un répertoire du théâtre religieux du Moyen Age et d'une bibliographie*. Paris: Honoré Champion, 1998
– *Les mystères français imprimés: une étude sur les rapports entre le théâtre religieux et l'imprimerie à la fin du Moyen Age français suivie d'un répertoire complet des mystères français imprimés (ouvrages, éditions, exemplaires), 1484–1630*. Paris: Honoré Champion, 1999
Sagard Theodat, Gabriel. *Histoire du Canada et voyages que les Freres Mineurs Recollects y ont faicts pour la conversion des Infidelles*. Paris: Claude Sonnius, 1636
Scarratt, David J., ed. *Canadian Atlantic Offshore Fishery Atlas*. Special publication of Fisheries and Aquatic Sciences, no. 47 (revised). Ottawa: Fisheries and Aquatic Sciences, 1982
Schlesinger, Roger, and Arthur P. Stabler, eds. *André Thevet's North America: A Sixteenth-Century View*. Montreal: McGill-Queen's University Press, 1986
Screech, M.A. 1987. *The Essays*. Michel de Montaigne. London: Penguin Books, 2004
Sée, Henri. "Les États de Bretagne au XVIe siècle." *Annales de Bretagne* (Faculté des lettres, Rennes), 10, no. 1 (Nov. 1894)
Shea, John Gilmary, ed. and trans. *History and General Description of New France by the Rev. P.F.X. De Charlevoix, S.J.* 6 vols. New York: John Gilmary Shea, 1866
– ed. and trans. *First Establishment of the Faith in New France, by Father Christian Le Clerq*. 2 vols. New York: John G. Shea, 1881
Simoni-Arembou, Marie-Rose. "En quelle langue a écrit Samuel de Champlain." *La diversité linguistisque: langue française et langues partenaires de Champlain Senghor*. Travaux de la XXe biennale de la langue française, La Rochelle, et du Colloque International, Paris, Sorbonne, 2002

Slafter, Edmund, ed. *Voyages of Samuel De Champlain.* Trans. Charles Pomeroy Otis. 3 vols. Boston: Prince Society, 1880

Société de l'histoire du protestantisme français. "Champlain était-il huguenot?" *Bulletin.* 61 (1912): 274–7

Société des archives historiques de la Saintonge et de l'Aunis. *Bulletin.* 20 (1900)

Steckley, John. *A Huron-English / English-Huron Dictionary, Listing Both Words and Nouns and Verb Roots.* Lewiston: Edwin Mellon Press, 2007

Sully, Maximilien de Béthune, duc de. *Mémoires de Sully, principal ministre de Henri-le-grand. Nouvelle édition plus exacte et plus correcte que les précédents.* Rewritten and edited by abbé Pierre Mathurin de l'Écluse des Loges. 8 vols. Paris: Jean-François Bastien, 1788

– *Mémoires des sages et royals œconomies d'estat, domestiques, politiques et militaries de Henry le Grand ...* In Joseph-François Michaud and Jean-Joseph-François Poujoulat, eds. *Nouvelle collection des mémoires pour server à l'histoire de France.* 2 vols. Paris: L'éditeur du commentaire analytique du Code civil, 1837

Tanguay, l'abbé Cyprien. *Dictionnaire généalogique des familles canadiennes depuis la fondation de la colonie jusqu'à nos jours.* 8 vols. 1871–90. New edn., Quebec: Eusèbe-Senécal, 1975

Taylor, Eva G.R., ed. *The Original Writings and Correspondence of the Two Richard Hakluyts.* 2 vols. London: Hakluyt Society, 1935

Thevet, F. Andre. Les Singvlaritez de la France antarctique, avtrement nommee Amerique, & de plusieurs Terres & Isles decouuertes de nostre temps. Antwerp: Christophle Plantin, 1558. [Copy examined at Royal Ontario Museum, Rare Books, E141 T4 1558 Canadiana]

Thierry, Éric. "Champlain and Lescarbot: An Impossible Friendship." In Raymonde Litalien and Denis Vaugeois, eds., *Champlain: The Birth of French America*, 121–34. Montreal: McGill-Queen's University Press and Éditions du Septentrion, 2004

– *La France de Henri IV en Amérique du Nord.* Paris: Honoré de Champion, 2008

[TLF] *Trésor de la langue française.* 15 vols. Paris: Gallimard, 1971

Tooley, Ronald Vere. *Maps and Map-Makers.* 6th edn. New York: Crown Publishers, 1982

Trigger, Bruce G. *The Children of Aataentsic: A History of the Huron People to 1660.* 2 vols. Montreal: McGill-Queen's University Press, 1976

– *Handbook of North American Indians.* Gen. ed. William C. Sturtevant. Vol. 15: *Northeast.* Washington: Smithsonian Institution, 1978

Trinquier, Pierre. *Proverbes et dictons de la langue d'oc d'après le* Dictionnaire languedocien-français *de l'abbé Boissier de Sauvages (1785).* Montpellier: Presses du Languedoc, 1993

Trudel, Marcel. *The Beginnings of New France, 1524–1663.* Toronto: McClelland & Stewart, 1973

Turgeon, Laurier. "French Fishers, Fur Traders, and Amerindians during the Sixteenth Century: History and Archaeology." *William and Mary Quarterly*, 3rd series, 55, no. 4 (1998): 585–610

– "French Beads in France and Northeastern North America during the Sixteenth Century." *Historical Archaeology* 35, no. 4 (2001): 58–82

Vachon, André, Victorin Chabot, and André Desrosiers. *Dreams of Empire: Canada before 1700* (issued in French as *Rêves d'empire*). Ottawa: Public Archives of Canada, 1982

Valois, M. Noël. *Inventaires et documents publiés par l'administration des archives nationales: inventaire des arrêts du conseil d'État ("règne de Henri IV")*. Vol. 2. Paris: Imprimerie Nationale, 1893

Vigé, Eliane, and Jimmy Vigé. *Brouage: capitale du sel et patrie de Champlain*. Vol. 2. Saint-Jean d'Angély, 1990

Vigneras, Louis-André. "Le voyage de Samuel Champlain aux Indes occidentales." *Revue d'histoire de l'Amérique française* 11 (1957–58): 163–200

– "Encore le capitaine provençal." *Revue d'histoire de l'Amérique française* 13 (1959–60): 544–9

Villalba, José Miguel López. "Normas españolas para la transcripción y edición de colecciones diplomáticas." *Espacio, Tiempo y Forma*, series 3, *H.ª Medieval* 11 (1998): 285–306

Wallace, Anthony F.C. "Dreams and Wishes of the Soul: A Type of Psychoanalytic Theory among the Seventeenth Century Iroquois." *American Anthropologist* 60 (1958): 234–48

Wartburg, Walter von. *Französisches Etymologisches Wörterbuch; eine Darstellung des gallo-romanischen Sprachschatzes*. Vols. 6–14. Basel: Zbinden Druck und Verlag AG, 1961–69

Warwick, Jack. "Humanisme chrétien et bons sauvages (Gabriel Sagard, 1623–1636)." *XVIIe siècle revue* 97 (1972): 25–49

Waters, David W. *The Art of Navigation in England in Elizabethan and Early Stuart Times*. London: Hollis Carter, 1958

Wood, James B. *The King's Army: Warfare, Soldiers, and Society during the Wars of Religion in France, 1562–1576*. Cambridge: Cambridge University Press, 1966

Wrong, George M., ed. *The Long Journey to the Country of the Hurons by Father Gabriel Sagard*. Trans. Hugh H. Langton. Original edn., Paris: Denys Moreau, 1632. Toronto: Champlain Society, 1939

MAPS AND ATLASES

Titles of printed maps are in italics; titles of manuscript maps are in quotation marks.

LAC: Library and Archives Canada, Ottawa
BnF: Bibliothèque nationale de France, Paris

Aubry, Joseph, SJ. "Carte Pour les hauteurs des terres, et pour servir de limitte, suivant la Paix, entre la france et l'angleterre ..." 1715. LAC, NMC 6364

Bayfield, Henry Wolsey. *Plan of the Harbour and Basin of Quebec ... October 1827*. London: Hydrographic Office of the Admiralty, 1829

Bellin, Nicolas. *Carte Du Cours Du Fleuve De Saint Laurent Depuis Quebec jusqu'a la Mer* ... 1761. LAC, NMC 19314

Bouchette, Joseph. *To His Most Excellent Majesty King William IV: This Topographical Map of the Districts of Quebec, Three Rivers, S[t]. Francis and Gaspé, Lower Canada* ... 1831. LAC, NMC 17998 (6 sections)

Canada. Department of Energy Mines and Resources. *Canada Gazetteer Atlas.* Ottawa: Macmillan of Canada, 1980

Champlain, Samuel de. "descrpsion des costs p[or]ts rades Illes de la nouuelle france ..." [Date altered by Champlain from 1606 to 1607]. Library of Congress, Washington, 1607. Vellum Chart Collection no. 15

– *Carte Geographiqve De La Novvelle Franse Faictte Par Le Sievr De Champlain Saint Tongois Cappitaine Ordinaire Povr Le Roy En La Marine. faict len 1612.* In Samuel de Champlain, *Les Voyages dv Sievr de Champlain Xaintongeois, capitaine ordinaire pour le Roy, en la marine* ... Paris: Jean Berjon, 1613. Map bound into book between *Privilege* and page 1, or at end of book after the *Table Des Chapitres* of the *Quatriesme Voyage*

– *Carte de la novvelle france, augmentée depuis la derniere, servant a la nauigation ... 1632.* In Samuel de Champlain, 1632. *Les Voyages de la Novvelle France Occidentale, Dicte Canada* ... Paris: Louis Sevestre; Paris: Claude Collet; Paris: Pierre Le-Mur, 1632

Chaussegros de Léry, Gaspard-Joseph [père]. "Carte du Lac Ontario et du fleuve St Laurens depuis le lac Erie jusques au dessus de L'Isle de Montreal. Celle de Niagara et partied du lac erie 1726. Donné par Mr. de Lery fils 1752." LAC, NMC 6429

Chaussegros de Léry, Gaspard-Joseph [père] and J.-B.B. D'Anville. *Le Fleuve Saint-Laurent Représenté plus en detail que dans l'étendue de la Carte.* [1755]. LAC, NMC 25348

Desceliers, Pierre. "[World Map] Faicte a Arques, l'an 1550." British Library, London, 1550. Add. Ms. 24065

Deshayes, Jean. *La Grande Riviere De Canada Appelée par les Europeens De S[t]. Laurens.* Paris: N[icolas] de Fer [1715]. LAC, NMC 22665

Franquelin, Jean-Baptiste-Louis. "Carte Du Grand Fleuve St Laurens dressée & designée sur les memoires, & observations que le Sr Iolliet a tres-exactement faites en barq: & en canot en 46. voyages pendant plusieurs années ..." 1685. LAC, NMC 16686

– "Partie De L'Amerique Septentrionalle Ou Est Compris La Nouvelle France Nouvelle Angleterre, N. Albanie, Et La N. Yorc, La Pensilvanie, Virginie, Caroline Floride, Et La Louisiane, Le Golfe Mexique, Et Les Isles Quele Bordent A L'Orient ..." 1699. LAC, NMC 8475

[Franquet, Louis]. "Carte Particuliere De L'Ishme De L'Acadie Ou Sont Situés Les Forts De Beauseiou[r] Gaspareaux Et Fort Laurent Anglois." [1751]. LAC, NMC 221

Laure, Pierre-Michel, SJ. "Carte Du Domaine Du Roy En Canada Dressée par le Pere Laure Missionaire J. 1731. Augmentée de nouveau revue et corrigée avec grand soin en attendant un exemplaire complet l'automne 1732." LAC, NMC 1029. [There are four versions of this manuscript map dating between 1731 and 1733, differing somewhat in detail as Father Laure acquired more information.]

Levasseur, Guillemme. ["Carte de l'océan Atlantique ..."] A Dieppe par Guillemme Levasseur Le 12 de Juillet 1601. BnF, Ge SH Arch 5. 1601. LAC, NMC 17966

Sanson d'Abeville, Nicolas. *Atlas Du Monde: Cartes Generales De Tovtes Les Parties Dv Monde* … Paris: Pierre Mariette, 1665. Facsimile reprint, with notes by Mireille Pastoureau. Paris: Sand & Conti, 1988

Wyld, James. *A Map of the Provinces of New Brunswick, and Nova Scotia* … 1825. LAC, NMC 18092

Wytfliet, Corneille. *Descriptionis Ptolemaicae augmentum*. Louvain, 1597. 3rd edn. Douay, 1603

– *Nova Francia Et Canada* … 1597. LAC, NMC 6322

A SELECTION OF BIOGRAPHIES OF SAMUEL DE CHAMPLAIN

* *Published in English and French*

1863. Faillon, Etienne-Michel. *Histoire de la Colonie Française en Canada*. Vol. 1. Villemarie: Bibliothèque Paroissiale

1865. Parkman, Francis. *Pioneers of France in the New World*. Boston: Little, Brown

1865–67. [Margry, Pierre]. "Samuel Champlain." *Recueil des actes de la Commission des arts et monuments de la Charente-Inférieure*. Saintes: Imprimerie Hus

1867. Delayant, Leopold. *Notice sur Samuel Champlain: né à Brouage 1567, mort à Québec, le 25 décembre 1635*. Niort: L. Clouzot

1870. Laverdière, Charles-Honoré, ed. *Œuvres de Champlain publiés sous le patronage de l'Université Laval*. 2nd edn. 6 vols. Quebec: Imprimé au Séminaire par Geo.-E. Desbarats

1875. Charavay, Étienne. *Samuel de Champlain: foundateur de Québec*. Paris: Librairie J. Charavay

1880. Slafter, Edmund, ed. *Voyages of Samuel De Champlain*. 3 vols. Trans. Charles Pomeroy Otis. Boston: Prince Society

1891. Dionne, Narcisse E. *Samuel de Champlain, foundateur de Québec et père de la Nouvelle-France: histoire de sa vie et de ses voyages*. 2 vols. Quebec: Augustin Côté

1893. Audiat, Louis. *Samuel de Champlain de Brouage, foundateur de Québec, 1567–1635*. Saintes: Z. Montreuil

1898. Casgrain, L'abbé H.-R. *Champlain: sa vie et son charactère*. Quebec: Imprimerie de L.-J. Demers & Frere

1900. Gravier, Gabriel. *Vie de Samuel de Champlain: foundateur de la Nouvelle-France (1567–1635)*. Paris: Librairie orientale & americaine

1902. Sedgwick, Henry D. *Samuel de Champlain*. Boston: Houghton, Mifflin

1903. Dix, Edwin A. *Champlain: The Founder of New France*. New York: D. Appleton

1906. Dionne, Narcisse E. *Champlain*. Toronto: Morang*

1908. Saguenay, Jean du. *Le foundateur de la Nouvelle-France: Champlain*. Quebec: Action sociale

1920. Colby, Charles W. *The Founder of New France: A Chronicle of Champlain*. Toronto & Glasgow: Brook

1924. Flenley, Ralph. *Samuel de Champlain, Founder of Canada.* Toronto: Macmillan of Canada
1929. Micard, Étienne. *L'effort persérvérant de Champlain.* Paris: Éditions Pierre Roger
1931. Constantin-Weyer, Maurice. *Champlain.* Paris: Plon
1949. Bishop, Morris. *Champlain: The Life of Fortitude.* London: Macdonald
1951. Deschamps, Hubert J. *Les voyages de Samuel de Champlain, Saintongeais Père du Canada.* Paris: Presses universitaires de France
1956. Trudel, Marcel. *Champlain.* Montreal: Fides. 2nd edn., 1968
1967. Campeau, Lucien. *Monumenta Novæ Franciæ. 1: La première mission d'Acadie (1602–1616).* Rome: Monumenta Historica Societatis Iesu; and Quebec: Presses de l'Université Laval
1972. Morison, Samuel E. *Samuel de Champlain: Father of New France.* Boston: Little, Brown
1973. Trudel, Marcel. *The Beginnings of New France, 1524–1663.* Toronto: McClelland & Stewart
1987. Armstrong, Joe C.W. *Champlain.* Toronto: Macmillan of Canada. Trans. Norman Paiement with the collaboration of Christiane Lacroix. *Samuel de Champlain.* Montreal: Éditions de l'Homme, 1988
1994. Glénisson, Jean. *La France d'Amérique: voyages de Samuel Champlain, 1604–1629.* Paris: Imprimerie nationale
2004. Capella, Émile. *Champlain, le fondateur de Québec.* Paris: Magellan
2004. Legaré, Francine. *Samuel de Champlain: Father of New France.* Montreal: XYZ Pub.*
2004. Litalien, Raymonde, and Denis Vaugeois, eds., *Champlain: The Birth of New France,* Montreal: McGill-Queen's University Press and Éditions du Septentrion*
2004. Montel-Glénisson, Caroline. *Champlain: la découverte du Canada.* Paris: Nouveau Monde
2008. Fischer, David Hackett. *Champlain's Dream.* Toronto: Knopf Canada

A SELECTION OF BIOGRAPHIES OF HÉLÈNE (BOULLÉ) DE CHAMPLAIN

1888. LeMoine, James M. *Les heroines de la Nouvelle-France.* Lowell, MA: R. Renault
1938. Bourde de la Rogerie, Henri. *Hélène Boullé: femme de Samuel de Champlain.* Extrait des mémoires de la société archéologique du Département d'Ille-et-Vilaine. Vol. 63. Rennes: Imprimerie Centrale
1968. Baudry, René. *Madame de Champlain.* Montreal: Editions des Dix
2003, 2005. Fyfe-Martel, Nicole. *Hélène de Champlain.* 2 vols. Montreal: Hurtubise HMH

A SELECTION OF NOVELS AND BIOGRAPHIES FOR YOUNGER AUDIENCES

1927. Macdonald, Adrian. *Samuel de Champlain.* Toronto: Ryerson Press
1941. McDowell, Franklin D. *The Champlain Road.* Toronto: Macmillan
1944. Tharp, Louise H. *Champlain, Northwest Voyager.* Boston: Little, Brown

1952. Syme, Ronald. *Champlain of the St. Lawrence.* New York: Morrow
1959. Kent, Louise A. *He Went with Champlain.* Boston: Houghton Mifflin
1961. Ritchie, Cicero T. *The First Canadian: The Story of Champlain.* Toronto: Macmillan
1963. Wilson, Charles M. *Wilderness Explorer: The Story of Samuel de Champlain.* New York: Hawthorn Books
1981. Garrod, Stan. *Samuel de Champlain.* Don Mills, ON: Fitzhenry & Whiteside*
2004. Moore, Christopher. *Champlain.* Toronto: Tundra Books

PLAYS AND POEMS

1890. Dawson, Samuel E. *Champlain.* Montreal: "Written for the Montreal Pen and Pencil Club." Privately printed
1908. Harper, John M. *Champlain: A Drama in Three Acts.* Toronto: W. Briggs
1942. Hooke, Hilda Mary. "Hélène of New France." *One Act Plays from Canadian History.* Toronto: Longmans, Green

INDEX